CW01551289

RAIL GUIDE 2022

Pip Dunn

Crécy Publishing Ltd
www.crecy.co.uk

First published 2010
Reprinted 2010
Revised editions 2011, 2012, 2013, 2014, 2015, 2016, 2017, 2018, 2019, 2020, 2021

This Thirteenth Edition published 2022
by Crécy Publishing Ltd

ISBN 9781800351394

© Crécy Publishing Ltd 2022

All rights reserved. No part of this book may be reproduced or transmitted in any form or by any means, electronic or mechanical, including photocopying, recording, scanning or by any information storage and retrieval system, on the internet or elsewhere, without permission from the Publisher in writing.

Printed in Bulgaria by Multiprint

Crécy Publishing Ltd
1a Ringway Trading Estate
Shadowmoss Road
Manchester
M22 5LH
Tel +44 (0) 161 499 0024

www.crecy.co.uk

Front cover top: 91101 *Flying Scotsman* stands at King's Cross with the 1403 to Leeds on 10 July 2021. *Pip Dunn*

Front cover bottom: On the first known visit by Class 50s to Stranraer, 50049 *Defiance* waits to leave with the 1352 return charter to Tame Bridge Parkway on September 11 2021. *Pip Dunn*

Back cover top left: Northern Ireland Railways (or NIR) CAF built three coach railcar set No 3008 arrives at Whitehead on a service from Great Victoria Street station in Belfast. *Tom Ferris*

Back cover top right: On 25 November 2021, 180114 waits to leave with the 1457 King's Cross-Bradford Interchange. There are ten of these Adelante units in the GC fleet and they replaced HSTs as well as allowing more services. *Pip Dunn*

Back cover bottom: Early on 20 November 2021, 37116 has arrived with the 0209 King's Cross-Derby Test train and will now propel into the RTC for stabling. *Pip Dunn*

Maps are based on the official National Rail Network maps

Rail Guide information is correct to 12 March 2022

Welcome to the 2022 edition of the Crecy Rail Guide, which includes all of the material included in both the 2021 guide and its supplement, plus a few extra new sections. As always, if you have any comments, input, updates or corrections then please do get in touch via the publisher.

Once again, this has been an exceptionally challenging book to put together because of the continued Covid-19 pandemic, although at least since June it has been possible to get out and about on the railways again and the signs are we are 'coming out the other side' of it now and life is returning to some semblance of normality.

However, there is no doubt that there has been long-term damage caused to the railways, with passenger numbers down dramatically. Even as more people return to work, the demands for both commuting and leisure travel has been seriously reduced. It's hard to see patronage getting back to pre-pandemic levels any time soon, especially as many people have been able to work from home and wish to continue doing so given the huge savings it can make on their travel bills, as well, as – potentially – giving them a better work-life balance.

Since the last edition of the Rail Guide, more TOCs have returned to state running, or are due so, and the Shapps-Williams Rail Review will see more companies moving to a risk-averse cost-plus contractual arrangement. Over the next few years, passenger operations will, effectively be renationalised, although with many private companies still managing the operations. However, while the Department for Transport has already been pulling many strings for the last few years, such as in train specification and procurement, over the next couple of years the move to a unified Great British Railways will be more obvious.

That should at least bring some benefits – the potential for connectivity between operators, greater uniformity in ordering of trains and the like – some of the benefits of a nationalised railway that were lost in the 1990s break-up of the network.

But unlike returning to a fully, old BR-style, nationalised railway, other operations will remain wholly private such as freight and Open Access Train Operators and Private Train Operating Companies (such as West Coast Railways, Locomotive Services and the like). These will be able to continue as before as none take any subsidies from the taxpayer. While Grand Central's Blackpool-Euston service was pulled before it got started, purely down to the pandemic, it has retained its ECML operation, while new entrant Lumo started to run its trains from October 2021 using brand new Class 803 units between Edinburgh and King's Cross.

Crossrail is still yet to open fully, but is edging ever closer to doing so. Construction of the first phase of HS2 is ongoing, although the project has been dramatically cut back with a major government U-turn. After promising that HS2 would reach Leeds, the government reneged on this in November and announced it would not, but a 'package of improvements' to the existing infrastructure would be implemented, many of which were projects to which commitments had already been made.

The aim of the Rail Guide remains as in previous years; to provide a handy reference resource for the current British Isles railway. That includes Northern Ireland and the Republic of Ireland.

It is split into passenger Train Operating Companies, Open Access Operators, Channel Tunnel Operators, Private Train Operating Companies, Northern Ireland, Light rail – London Underground and trams, rolling stock providers, Freight Operating companies and maintenance and spot hire companies.

It also deals with off-lease vehicles still in the UK, Network Rail, those ex-UK locos now in use (or stored) abroad, main-line preservation groups and other key industry bodies. It's not intended as a 'mark them off' spotters' book.

Of course, there is some overlap – especially with loco hire where vehicles are owned by FOCs but used by TOCs. In these instances, the vehicles are listed under the owner, the FOC, but mention is made of them in the relevant TOC chapter.

Official Guidelines

Train Operators and Network Rail welcome rail enthusiasts and photographers, but in today's safety-led railway and with the continued concerns about possible transport terrorism, guidelines are very important and we encourage all to follow these published guidelines as much as possible. They are available to view and download from the National Rail and ROG websites, but are reproduced in full below to assist you with this information.

The Official Railway Enthusiast's Guidelines

- Network Rail welcomes rail enthusiasts to our stations.
- The following guidelines are designed to help you to have a safe and enjoyable experience. Please keep them with you when you are at Network Rail-managed stations.
- You may also wish to take a copy of the Railway by-laws which are available from the Office of Public Sector Information website.

Before you enter the platform

- When you arrive at a station, please let the staff at the Network Rail Reception Desk know that you are on the station. This will help keep station staff informed so that they can go about their duties without concern as to your reasons for being there.
- You may require a platform ticket to allow access to platforms.

While you are on the platform

- You need to act safely and sensibly at all times.
  - Stay clear of the platform edge and stay behind the yellow lines where they are provided.
  - Be aware of your surroundings.

Please DO NOT:

  - Trespass on to the tracks or any other part of the railway that is not available to passengers.
  - Use flash photography because it can distract train drivers and train despatch staff and so is potentially very dangerous.
  - Climb on any structure or interfere with platform equipment.
  - Obstruct any signalling equipment or signs which are vital to the safe running of the railway.
  - Wear anything which is similar in colour to safety clothing, such as high-visibility jackets, as this could cause confusion to drivers and other railway employees.
  - Gather together in groups at busy areas of the platform (e.g. customer information points, departure screens, waiting areas, seating etc.) or where this may interfere with the duties of station staff.
- If possible, please try to avoid peak hours which are Monday – Friday 6:00am (06.00) – 10:30am (10.30) and 3:30pm (15.30) – 7:30pm (19.30).

Extra eyes and ears

- If you see anything suspicious or notice any unusual behaviour or activities, please tell a member of staff immediately.
- For emergencies and serious incidents, either call: The British Transport Police on 0800 40 50 40. Or text a message to 61016. The Police on 999, or 101.
- Your presence at a station can be very helpful to us as extra 'eyes and ears' and can have a positive security benefit.

Photography

- You can take photographs at stations provided you do not sell them. However, you are not allowed to take photographs of security-related equipment, such as CCTV cameras.
- Flash photography on platforms is not allowed at any time. It can distract train drivers and train despatch staff and so is potentially very dangerous.
- Tripod legs must be kept away from platform edges and behind the yellow lines. On busy stations, you may not be allowed to use a tripod because it could be a dangerous obstruction to passengers.

Railway by-laws

For safety and ease of travel on the railway system (which includes passengers, staff, property and equipment), the by-laws must be observed by everyone. A copy of the by-laws can be obtained at stations or downloaded from the Office of Public Sector Information website.

General

Train operators must put the safety of their passengers and staff first. You may very occasionally be asked by station staff to move to another part of the station or to leave the station altogether. Station staff should be happy to explain why this is necessary. If you are travelling by train, they may ask you to remain in the normal waiting areas with other passengers. If this occurs, please follow their instructions with goodwill as staff have many things to consider, including the safety and security of all passengers, and are authorised to use judgement in this regard.

Pendolino 390042 passes Abington with the 0940 Glasgow Central-Euston on 1 June 2021. *Paul Shannon*

Contents

About This Book

This book is aimed at providing an overview of the ownership of rolling stock on the ever-changing modern UK railway system with key facts and fleet details. The railway is hugely complex and relies on hundreds of companies, and it is easy to get confused and bogged down with ownership, uses and operators, although it should, hopefully, be getting a little easier!

As regards the Train Operating Companies, and indeed most of the Open Access Operators, trains are owned by Rolling Stock Leasing companies (ROSCOs) and when a franchise changes company, the rolling stock changes with them. Franchises are coming to an end, and now management contracts will oversee the operations of the trains. This should eliminate the wasteful resources – time and money – put into bidding to win franchises.

The structure of the railway's operations is changing, as detailed on page 10, with the franchised Train Operating Companies being phased out in favour of management contracts as ownership of the railway slowly returns to public ownership.

What it means – currently – for fleets is not likely to be a great deal, other than it should in theory be easier to redeploy rolling stock, even on a short-term basis, between operations.

On the freight side, it is business as usual with locos owned or leased by the Freight Operating Companies (FOCs) – with leasing far more prevalent than when the railways were privatised in the 1990s. Since vehicles may be on long-term hire, as opposed to leased, other vehicles may simply be spot-hired to meet fluctuating and seasonal demand.

This book is split into several chapters, or sections, but the main premise is an operator-by-operator breakdown. It includes all UK operators, including those in Northern Ireland and the Republic of Ireland.

Vehicles are listed by their current number, old number (if relevant), pool code if relevant, livery, owner and place of allocation. If a loco or unit carries a name then that is listed, though for multiple units it is related to the unit rather than the individual vehicle.

There are, as is to be expected, a number of grey areas and overlaps. Those TOCs that hire in locomotives on a contract basis – as opposed to leasing them – do not have these locos listed as such in their section, although of course reference is made to them. The locos themselves are listed under their owning FOC or supplier. The exception are those vehicles that are owned by private individuals or preservation groups but are on longer-term use by a particular FOC. The same applies to the Class 66s owned by DB Cargo but now used by DRS.

Those locos, such as Class DRS Class 68s that are used by Transpennine or Chiltern Railways are listed under their FOC. This is because DRS only needs to supply a set number of locos a day to these TOCs and any locos in these pools not being used by the TOCs are therefore free to be used by DRS for its own traffic.

Vehicles listed as (Q) are under test, be it static, driver training or fault-free mileage accumulation, but could enter traffic at any time. Those vehicles listed as (Z) are being rebuilt. (U) means the vehicle is 'stored unserviceable', which implies it is unlikely to return to use without major expenditure. Those listed as (S) are 'stored serviceable' at the time of going to press and these could return to traffic much sooner. Main-line locos listed as (I) are in industrial use only and not passed to work on the national network.

Of course, the status of any vehicle can change relatively quickly, but these statuses are intended as a guide to help readers.

Information is correct to 12 March 2022. If you have any corrections, additional information or comments for future editions, or would like to submit images for the 2023 Rail Guide, please do so via pip.dunn@eastfieldmedia.com.

Changes since the 2021 Rail Guide

On the loco front, construction has started on Rail Operations Group's fleet of ten Class 93s, while GB Railfreight is poised to place an order for 20 tri-mode Co-Co Class 99s, with options for 30 further locos. The latter has also sourced a few more Class 66s, with another 11 locos likely to be added to the fleet in 2022. It has also taken two Class 67s from Beacon Rail that were formerly with Colas Rail. GBRf's first Class 69s – rebuilt Class 56 with GM engines – have also now entered traffic and work continues at Longport to build the others, with the full fleet of 16 now agreed.

DRS has withdrawn more of its older ex-BR locos, and sold several Class 20/37/57s. It now has no 20s left on its books, while its 37 fleet has also been trimmed and more withdrawals are expected. It is likely to regain the use of many of its 68s if, as has been suggested, Chiltern, and then Transpennine, end their use of these locos.

DB Cargo has started to repatriate some of the 60-plus Class 66s from France and reintegrated them into its UK fleet. Additionally, a few Class 67s have received overhauls for use with Transport for Wales but overall its Class 60, 66, 67, 90 and 92 fleets remain relatively static.

Freightliner has had all its 13 ex-Greater Anglia Class 90s repainted, while it has returned some of its stored Class 70s back to traffic to release Class 66s for Mendip work. This was especially important as the Class 59s are beginning to show their age.

There is a new FOC in business now, SLC Operations, which currently has no locos of its own but works with spot hire companies such as 20189 Ltd, Hanson & Hall and HNRC to provide it with traction to perform stock moves.

Locomotive Services Limited continues to acquire heritage locos and coaches and return them to stock, while West Coast Railways' fleet remains pretty much as it was in 2021.

As well as the structural changes to the railway caused by the Covid pandemic, there have been many operational changes since the last edition of the Rail Guide was published some 16 months ago.

New trains being delivered or entering traffic have been the Class 197 and 231 with Transport for Wales; the former cleared for use, the latter just starting testing. West Midlands Trains now has its 196s and 730s cleared for use, and although both fleets remain on test, their introduction into service will not have happened before this book hits the shelves. Construction has started on new trains for Avanti West Coast and East Midlands Railway, while the order for the initial tranche of HS2 rolling stock has been made, with 54 eight-car EMUs to be built by an Alstom/Hitachi joint venture.

Some fleets have been retried, most notably the Class 142/143/144 railbuses from Transport for Wales, Northern and Great Western Railway. The Class 153s from EMR, Northern and West Midlands Trains have also been redeployed or withdrawn, leaving the type in use with TfW and ScotRail. The plan to return Class 442s for South Western Railway has been shelved and the units scrapped, despite several being refurbished for their re-introduction! Class 180 and 360s have entered use with EMR, as have Mk 4 hauled sets using Class 67s for TfW. Class 769s are also now in use with TfW and Northern, but their introduction with GWR remains up in the air.

ScotRail has ended its Class 314 operations and also added five Class 153 single cars as converted cycle carriers. All its HSTs are also now in use, although one set was written off in a derailment and has not been replaced. However, 17 of the 25 four-car sets are being increased to five-cars.

New trains continue to be delayed into traffic, such as SWR's Class 701s. Merseyrail's Class 777s are delayed, while Crossrail's 345s have finally all been delivered and most have been set up as nine-car trains. Their use is limited until the Elizabeth line finally opens in its entirety, expected fairly soon now. Greater Anglia's Class 720s are only now finally coming on stream.

The pandemic forced Grand Central to shelve its plans for a Blackpool North-Euston Open Access Operation using DB Cargo Class 90s and ex-LNER Mk 4 coaches despite several vehicles being repainted and driver training started. The coaches have been taken on by TfW. First East Coast – under the Lumo brand – however – has started its King's Cross-Edinburgh operation with brand new Class 803 EMUs, although it suffers from a limited number of places it can actually stop the trains at! Heathrow Express has rid itself of Class 332s, all of which bar one have been sent for scrap, and replaced them with Class 387s hired from GWR.

Various unit cascades have been possible as new trains enter traffic, with some, or all, of Classes 314, 315, 317, 319, 321, 322, 332, 360, 379, 442, 455, 456, 458, 465, 466, 483, 507 and 508 being withdrawn, redeployed or stored off lease.

As regards the network, work continues on building HS2, but there has also been the opening of the Werrington dive under near Peterborough, which allows freight trains from East Anglia to head north along the GNGE joint line without having to cross over the ECML.

Stations that opened in 2020-21 have been Horden, Bow Street, Kintore and Soham, while stations due to open in 2022 should be the Paddington to Abbey Wood section of Crossrail, with interchange stations with the LUL at Bond Street, Tottenham Court Road, Farringdon, Liverpool Street while there were new Crossrail stations at Canary Wharf, Woolwich, Custom House and Abbey Wood.

Other new stations expected to open on the network in 2022 are Portway Park and Ride on the Severn Beach Branch, Dalcross (Inverness Airport), Reading Green Park, Barking Riverside, Brent Cross West, Marsh Barton, near Exeter, Reston on the ECML north of Berwick-upon-Tweed and Thanet Parkway.

Other projects of note ongoing are the plan to reopen the Levenmouth branch in Fife, while work continues on relaying the East-West line from Bicester to Bletchley.

Franchise Overview

As I write this, in mid-February 2022, after two years, the country is looking like it is coming out of the Covid-19 global pandemic.

To say the pandemic has affected the railways badly would be a gross understatement, and the truth is the privatised railway could not function. It simply had to have government support on a scale not seen since the 1990s. Timetables were – understandably – slashed, as patronage fell to less than 10 per cent of pre-pandemic levels. For months trains were running with a handful of people on them; mainly key workers travelling to their jobs. Leisure travel was non-existent.

The result has been the private companies made no revenue but incurred similar costs, which was unsustainable.

The result has been a restructuring of the railways, which will essentially lead to renationalisation. The infrastructure, privatised as Railtrack via a public flotation in 1994, was renationalised in 2002 as Network Rail. Now the train operating company franchises are either being taken directly back under governmental control or will be allowed to lapse at the end of their terms.

The management of running the trains will remain contracted out to private companies on a cost-plus deal, so they will make a level of profit. But their operational framework will come totally from the Department of Transport. It will dictate what timetables can be run, what level of services will be offered and what trains will be acquired. Rolling stock will, it seems, for the time being still be leased from private companies.

As has been said, when the pandemic hit, the Department for Transport stepped in to keep train services running for key workers and essential supplies. In September 2020 the DfT renewed the support with new arrangements, called Emergency Recovery Measures Agreements (ERMAs).

ERMAs, which were due to run for up to 18 months, were designed to bring the rail franchising system to an end. Coming into force on 19 September 2020, they contained provisions to cancel current franchises when these agreements expire.

The DfT said this was the first step in creating a new kind of railway, one that is customer-focused, easy to use, good value and where the trains run on time – interestingly similar arguments it used to rid us of 'big bad old BR' and let the private sector in to run trains in the first place all those years ago!

The DfT also said the ERMAs 'keep the best elements of the private sector, including competition and innovation, that drive growth but go further by delivering greater leadership, direction and accountability'. Open Access operators did not qualify for EMAs, ERMAs, or NRCs, and only received furlough support.

New contracts

Operators have now been placed on far more demanding management agreements, with tougher performance targets and lower management fees. Management fees will now be a maximum of 1.5 per cent of the cost base of the franchise before the pandemic began.

The new contracts allow the DfT to make an early start on key reforms, including requiring operators to co-ordinate better with each other and driving down the railways' excessive capital costs.

The DfT says the railway will 'have a renewed and much sharper focus on delivering a reliable service which passengers and freight users can trust'. It will also link to Keith Williams' root-and-branch review of the railway and will pave the way for a white paper on the wider future of the railway during the ERMA period.

Until passenger numbers return, significant taxpayer support will still be needed. But the ERMA arrangements pave the way for wider rail industry reform that the DfT hopes will 'put passenger priorities at the forefront and will enable substantial medium and longer-term savings for the taxpayer. The railway will have a new and greater focus on delivering a reliable service which passengers can trust.'

The franchising of rail operations has always been fraught with debate. Short franchises restricted the chances for TOCs to innovate and invest, while long franchises, it was argued, could lead to complacency.

Over the last 25 years, however, many franchises were terminated due to the franchisee running into financial difficulties. Too much money – and time – was wasted in preparing franchise bids, with typically three out of four bound to fail. Some franchise bids were fanciful and wholly undeliverable, and took no account of larger global issues – such as the economic crash of 2008 and – as we have seen – the global pandemic.

Fares have skyrocketed under privatisation, which was not the intention, and real competition has been minimal – Open Access Operators were restricted in where they could stop their trains for fear of taking revenue away from the incumbent franchised TOC! True competition only existed on a handful of routes, mostly into London. The real competition was with road, yet if rail fares went up, road became a better option in many cases!

And different TOCs, run by different parent companies, had no incentive to interact and connect with rival TOCs even if they were on different routes. Train types were ordered differently, affecting commonality of vehicles and staff training. It is often said the privatised railway cost three times as much to run as the nationalised BR, which, admittedly had been run on a tight budget, but equally was prone to waste resources and had been hamstrung by the unions.

Since privatisation 25 years ago the railway has, in most cases, seen huge growth, which is no bad thing. Some of the TOCs have introduced more trains, better timetables with trains running earlier and later in the day, new routes and new rolling stock, some of which was much better than BR, some that was not. Assets were also made to last longer and there are still many trains in use that were ordered and indeed built by BR. However, regular refurbishments have, to a degree, kept them fit for purpose (as I write this, I am travelling on a Class 156 DMU that was bult in 1987 and so is now 35 years old.)

The DfT has set up Great British Railways and revived the famous BR double arrow badge. It was retained at privatisation as a symbol that people were 'entering the railway', so it was prominent at stations and on signs directing people to stations. However, it is now being more widely used, and there's every chance it will start to appear on rolling stock and publicity material again.

The DfT hopes GBR will simplify ticketing for passengers, create more interaction between the different routes and 'operators', which in turn will bring cost savings. There will, inevitably, be some trimming of duplicate jobs – especially station staff – and it will be interesting to see how government can work with the unions, which if anything have gained power during the last 25 years if pay deals are anything to go by!

Key abbreviations for TOCs

EMA	Emergency Measures Agreement
ERMA	Emergency Recovery Measures Agreements
NRC	National Rail Contract
OCFA	Operating Contract Franchising Agreement
PSC	Public service contract

TOC trading name	Operator(s)	Current status	Franchise start date	Pre-ERMA franchise end date	Extension option
Avanti West Coast	First Group/Trenitalia	ERMA to October 2022 8/12/19	31/3/26	Was originally due to run until 31/3/31	
Caledonian Sleeper	Serco	ERMA to February 2022*	31/3/15	31/3/30	
c2c	Trenitalia	NRC to 25/7/23	10/2/17	10/11/29	10/5/30
Chiltern Railways	Arriva	NRC to 1/12/27	3/3/02	11/12/21	31/12/27
CrossCountry Trains	Arriva	ERMA to 15/10/23	11/11/07	1/10/23	Two-years
Elizabeth Line (Crossrail)	MTR	Concession to 31/5/23	31/5/15	27/5/23	27/5/25
East Midlands Railway	Abellio	ERMA to 31/10/22	18/8/19	31/3/22	Was originally due to run until 21/8/27
Govia Thameslink	Go-Ahead/Keolis	ERMA to March 2022	14/9/14	31/3/22	Now extended for three years
Greater Anglia	Abellio/Mitsui	NRC to 19/9/24	16/10/16	11/10/25	Two-years
Great Western Railway	First Group	EMA until 25/6/22	20/9/15	31/3/23	Originally due to run until March 2020
London North Eastern Railway	DfT OLR Holdings	Direct award franchise to 31/6/23	24/6/18	n/a	Option to 2025
London Overground	Arriva	Concession to 25/5/24	14/11/16	25/5/24	Two-years
Merseyrail	Serco/Abellio	Concession to 22/7/28	20/7/03	22/7/28	
Northern Rail	DfT OLR Holdings	Direct award franchise to 29/2/24, option to 2027	1/3/20	n/a	
ScotRail	DfT OLR Holdings	EMA to 31/3/22	1/4/15	31/3/22	Government-owned operator from 1/4/22
Southeastern	DfT OLR Holdings	Direct award franchise until further notice	17/10/21	n/a	
South Western Railway	First Group/MTR	NRC to 31/5/23	20/8/17	1/4/23	Originally to August 2024, replaced by direct award in 2020
Transpennine Express	First Group	NRC to 31/5/23	1/4/16	31/3/23	Two-years
Transport for Wales	DfT OLR Holdings (was KeolisAmey)	Government-owned operator took over from KeolisAmey Wales in February 2021	7/2/21		
West Midlands Trains	Abellio/JR East/Mitsui	NRC until 19/9/24	10/12/17	31/3/26	Two-years

Integrated Rail Plan

The government announced its Integrated Rail Plan (IRP) in late 2021 in response to its backtracking on a manifesto commitment to build HS2 to serve Leeds.

Instead it announced it was spending a £96bn investment on the IRP, which it claimed was the 'the largest ever single Government investment in the rail network'.

In reality a lot of the money had already committed on previously announced projects, so little of it was actually new.

Some £42.5bn was for the completion of HS2 Phase One (London to Birmingham) and Phase 2a (West Midlands to Crewe), a project that was actually already under way and already budgeted for.

A total of £17bn was for the HS2 Phase 2b – the Western Leg from Crewe to Manchester, via Manchester Piccadilly and Manchester Airport. This then continued to connect with the West Coast Main Line near Wigan.

A further £12.8bn was for the HS2 East Core Network, which is a new high-speed line from the West Midlands to East Midlands Parkway, running on an existing route planned for the HS2 line to Leeds, and this will serve Nottingham and Derby via a station near Toton. This should also include electrification of the Midland Main Line to Sheffield via Derby and to Nottingham.

The rest of the IRP is earmarked for upgrades on the East Coast Main Line from King's Cross to Leeds and the North-East.

In addition, £100m is earmarked to 'start work' on the West Yorkshire Mass Transit System but more money would be needed to complete construction of such a system.

A £17.2bn allocation is planned for the Northern Powerhouse Rail Core Network. Northern Powerhouse Rail has been a big carrot dangled to northern constituent MPs from the days of the coalition but will now focus on the 'NPR Core Network', between Liverpool, Manchester and Leeds.

This plan should see 40 miles of new electrified line between Warrington, Manchester and Marsden (near Huddersfield), and the rest of the trans-Pennine route to York upgraded and electrified. Leeds to York will have four tracks added for some of the route. Additionally, the Calder Valley Line between Leeds and Bradford will also be upgraded and electrified.

This is all part of the £5.4bn Transpennine Route Upgrade, which is now considered as Phase 1 of NPR and includes digital signalling throughout. There will be gauge upgrades to allow intermodal freight trains.

It is also hoped that in the fullness of time a new link to Manchester Piccadilly will be put in place, allowing for more additional trains.

Finally, £1.5bn of the IRP is earmarked for smaller rail schemes in the North and Midlands. These are scheduled to be undertaken by 2025, but the plan lacks any clarity as to what they actually are.

Great British Railways

Website: gbrtt.co.uk

In response to the pandemic, when the private train operates could not function without government support due to the major loss of patronage forced by lockdowns and travel restrictions, it was clear the franchises system for train operation was no longer viable.

As a result, the Department for Transport is taking back the franchises and running them on a management contract basis. They will be part of the newly formed nationalised system Great British Railways (GBR), the state-owned public body that will oversee rail transport from 1 April 2024.

GBR will also replace Network Rail, which has been state-owned since 2002 after the publicly floated Railtrack collapsed, to control the infrastructure, including its operation, enhancements, maintenance and upgrades. It will control the management contracts issued to the train operators, as well as the setting of fares, revenue collection and planning of timetables.

Andrew Haines and Sir Peter Hendy, the current CEO and chair of Network Rail respectively, will oversee the creation of Great British Railways, with Hendy expected to run the new organisation.

GBR was proposed under the Williams-Shapps Rail Review, which was carried out by Keith Williams and published as a white paper on 20 May 2021.

It will be made up of five regional divisions: Scotland, North West & Central, Eastern, Wales & Western and Southern – the latter including HS1.

GBR does not affect Northern Ireland, but will include the devolved rail operators of ScotRail and Transport for Wales, as well as London Overground, Crossrail and Merseyrail. GBR will own all the stations.

In a blast from the past, the familiar and popular British Rail 'Double Arrow' designed by Gerry Barney in 1965 is being revived. And although it never actually went away as it was retained for signage to advise where people were 'entering the railway', it will be more prominent once again, although it may be tweaked.

A 'competition' is being run to determine where GBR's headquarters will be, most likely outside of London and probably in one of the famous railway towns or cities such as Crewe, York, Swindon or Doncaster.

Avanti West Coast

Contact details

Website: www.avantiwestcoast.co.uk
Twitter: @AvantiWestCoast

Key personnel

Managing Director: Phil Whittingham
Executive Director Operations: Jonathan Dunster

Overview

AWC operates the inter-city trains along the West Coast Main Line from London Euston to Birmingham New Street, Shrewsbury, Liverpool Lime Street, Manchester Piccadilly, Blackpool North, Holyhead, Glasgow Central and Edinburgh.

The fleet is currently reliant on two main train types – five-car Bombardier Class 221 DEMUs – and a mix of nine and 11-car Alstom Class 390 tilting Pendolino units.

Both fleets were inherited from Virgin Trains, which ran the InterCity West Coast franchise from 1997 until 2019 until transfer to a joint venture between First Group and Trenitalia from 8 December.

The fleet has been reliveried in Avanti graphene dark green livery. During the life of Virgin, new services to Shrewsbury, Wrexham and Blackpool were introduced.

AWC had been looking to introduce new services, with Walsall and Llandudno due to start in 2021 and Gobowen from 2022, but they have yet to occur due to the pandemic and it remains to be seen if they will indeed happen. Likewise, planned additional trains to Liverpool from 2022 will probably only happen if there is the demand as the railway recovers. To meet these additional services, AWC has ordered 13 Class 805 five-car bi-mode Hitachi AT300 units, and ten Class 807 seven-car 25kV AC EMUs from the same source. The first bodyshells for the 805s have arrived in the UK from Japan and are at Newton Aycliffe for assembly.

The company has a contract with Direct Rail Services to provide four Class 57/3s to act as rescue 'Thunderbird' locos – 57304/307/308/309 – and these are strategically stabled at key points along the WCML. However, the normal practice is to rescue any failed Class 221 or 390 with a sister train.

The West Coast Partnership Development (WCPD) will also be responsible for operating the first trains on the new HS2 line from Euston to Birmingham Curzon Street from 2029-2033, and new trains for this will be necessary. A fleet of 54 eight-car units are on order from an Alstom/Hitachi joint venture.

Class 221

Bombardier-built tilting Voyager DEMUs ordered by Virgin CrossCountry. Twenty sets were transferred to Virgin West Coast in December 2007, when some CrossCountry routes were added to the West Coast Franchise and with the rest of the fleet – 221119-141/144 transferring to Arriva's CrossCountry Trains franchise.

These trains work all AWC services that operate for any part of their journey off the 25kV AC network, such as to the North Wales Coast, but are also used to provide extra capacity alongside 390s on the electrified AWC network, including some journeys from Edinburgh to Euston.

All are in the HFHQ pool.

Avanti West Coast Voyager 221103 works the 0852 Edinburgh-Euston through Wigan North Western on 9 September 2021. *Tom McAtee*

	Livery	Owner	Depot	DMS	MS	MS	MSRMB	DMF	
221101	AWC	BEA	CZ	60351	60951	60851	60751	60451	*101 Squadron*
221102	AWC	BEA	CZ	60352	60952	60852	60752	60452	
221103	AWC	BEA	CZ	60353	60953	60853	60753	60453	
221104	AWC	BEA	CZ	60354	60954	60854	60754	60454	
221105	AWC	BEA	CZ	60355	60955	60855	60755	60455	
221106	AWC	BEA	CZ	60356	60956	60856	60756	60456	
221107	AWC	BEA	CZ	60357	60957	60857	60757	60457	
221108	AWC	BEA	CZ	60358	60958	60858	60758	60458	
221109	AWC	BEA	CZ	60359	60959	60859	60759	60459	
221110	AWC	BEA	CZ	60360	60960	60860	60760	60460	
221111	AWC	BEA	CZ	60361	60961	60861	60761	60461	
221112	AWC	BEA	CZ	60362	60962	60862	60762	60462	
221113	AWC	BEA	CZ	60363	60963	60863	60763	60463	
221114	AWC	BEA	CZ	60364	60964	60864	60764	60464	*Royal Air Force Centenary 1918-2018*
221115	AWC	BEA	CZ	60365	60965	60865	60765	60465	
221116	AWC	BEA	CZ	60366	60966	60866	60766	60466	*City of Bangor/ Dinas Bangor*
221117	AWC	BEA	CZ	60367	60967	60867	60767	60467	
221118	AWC	BEA	CZ	60368	60968	60868	60768	60468	
221142	AWC	BEA	CZ	60392	60992	60892	60792	60492	
221143	AWC	BEA	CZ	60393	60993	60893	60793	60493	

Class 390

Alstom-built tilting 25kV AC EMUs. Fifty-three sets were ordered by Virgin West Coast as nine-car sets and entered traffic from 2002 – some as eight-car sets – followed by an additional order for four 11-car sets delivered in 2010.

Thirty-one of the original sets were then extended by two cars in 2010-11 to make them 11-cars and renumbered from 3900xx to 3901xx. Unit 390033 was written off in the Grayrigg derailment in February 2007.

They operate most AWC trains from Euston to Birmingham, Liverpool, Manchester and Scotland, although some trains may be worked by Class 221s.

Only 390155/156 carry full AWC livery on all vehicles, while others listed as in this livery currently only have the wrap on the driving cars with the remaining vehicles still in plain white. 390119 is in a Pride wrap, while 390121 was unveiled in a climate awareness wrap in late October 2021 in conjunction with the COP26 conference in Glasgow.

A refurbishment programme has started on the Class 390 fleet, with 390125 the first to be treated and 390123 the second.

All are in the HFHQ pool.

The diverted 0737 Glasgow Central-Euston passes Manchester Oxford Road on 2 April 2021, worked by an Avanti West Coast Pendolino. *Tom McAtee*

Class 390/0

Original nine-car sets, some delivered as eight-car sets.

	Livery	Owner	Depot	DMRF	MF	PTF	MS	TS	MS	PTSRMB	MS	DMS	
390001	AWC	ANG	MA	69101	69401	69501	69601	68801	69701	69801	69901	69201	*Bee Together*
390002	AWC	ANG	MA	69102	69402	69502	69602	68802	69702	69802	69902	69202	*Stephen Sutton*
390005	AWC	ANG	MA	69105	69405	69505	69605	68805	69705	69805	69905	69205	*City of Wolverhampton*
390006	AWC	ANG	MA	69106	69406	69506	69606	68806	69706	69806	69906	69206	*Rethink Mental Illness*
390008	AWC	ANG	MA	69108	69408	69508	69608	68808	69708	69808	69908	69208	*Charles Rennie Mackintosh*
390009	AWC	ANG	MA	69109	69409	69509	69609	68809	69709	69809	69909	69209	*Treaty of the Union*
390010	AWC	ANG	MA	69110	69410	69510	69610	68810	69710	69810	69910	69210	*Cumbrian Spirit*
390011	AWC	ANG	MA	69111	69411	69511	69611	68811	69711	69811	69911	69211	*City of Lichfield*
390013	AWC	ANG	MA	69113	69413	69513	69613	68813	69713	69813	69913	69213	*Blackpool Belle*
390016	AWC	ANG	MA	69116	69416	69516	69616	68816	69716	69816	69916	69216	
390020	AWC	ANG	MA	69120	69420	69520	69620	68820	69720	69820	69920	69220	
390039	AWC	ANG	MA	69139	69439	69539	69639	68839	69739	69839	69939	69239	*Lady Godiva*
390040	AWC	ANG	MA	69140	69440	69540	69640	68840	69740	69840	69940	69240	
390042	AWC	ANG	MA	69142	69442	69542	69642	68842	69742	69842	69942	69242	
390043	AWC	ANG	MA	69143	69443	69543	69643	68843	69743	69843	69943	69243	
390044	AWC	ANG	MA	69144	69444	69544	69644	68844	69744	69844	69944	69244	*Royal Scot*
390045	AWC	ANG	MA	69145	69445	69545	69645	68845	69745	69845	69945	69245	
390046	AWC	ANG	MA	69146	69446	69546	69646	68846	69746	69846	69946	69246	
390047	AWC	ANG	MA	69147	69447	69547	69647	68847	69747	69847	69947	69247	*CLIC Sargent*
390049	AWC	ANG	MA	69149	69449	69549	69649	68849	69749	69849	69949	69249	
390050	AWC	ANG	MA	69150	69450	69550	69650	68850	69750	69850	69950	69250	

Class 390/1

Original nine-car sets increased to 11-car sets; 390154-157 were additional sets delivered in 2010-12 as 11-cars.

	Old No.	Livery	Owner	Depot	DMRF	MF	PTF	MS	TS	MS	TS	MS	PTSRMB	MS	DMS	
390103	390003	AWC	ANG	MA	69103	69403	69503	69603	65303	68903	68803	69703	69803	69903	69203	
390104	390004	AWC	ANG	MA	69104	69404	69504	69604	65304	68904	68804	69704	69804	69904	69204	*Alstom Pendolino*
390107	390007	AWC	ANG	MA	69107	69407	69507	69607	65307	68907	68807	69707	69807	69907	69207	
390112	390012	AWC	ANG	MA	69112	69412	69512	69612	65312	68912	68812	69712	69812	69912	69212	
390114	390014	AWC	ANG	MA	69114	69414	69514	69614	65314	68914	68814	69714	69814	69914	69214	*City of Manchester*
390115	390015	AWC	ANG	MA	69115	69415	69515	69615	65315	68915	68815	69715	69815	69915	69215	*Crewe – All Change*
390117	390017	AWC	ANG	MA	69117	69417	69517	69617	65317	68917	68817	69717	69817	69917	69217	*Blue Peter*
390118	390018	AWC	ANG	MA	69118	69418	69518	69618	65318	68918	68818	69718	69818	69918	69218	
390119	390019	PRI	ANG	MA	69119	69419	69519	69619	65319	68919	68819	69719	69819	69919	69219	*Progress*
390121	390021	CLI	ANG	MA	69121	69421	69521	69621	65321	68921	68821	69721	69821	69921	69221	
390122	390022	AWC	ANG	MA	69122	69422	69522	69622	65322	68922	68822	69722	69822	69922	69222	*Penny the Pendolino*
390123	390023	AWC	ANG	MA	69123	69423	69523	69623	65323	68923	68823	69723	69823	69923	69223	
390124	390024	AWC	ANG	MA	69124	69424	69524	69624	65324	68924	68824	69724	69824	69924	69224	
390125	390025	AWC	ANG	MA	69125	69425	69525	69625	65325	68925	68825	69725	69825	69925	69225	*Virgin Stagecoach*
390126	390026	AWC	ANG	MA	69126	69426	69526	69626	65326	68926	68826	69726	69826	69926	69226	
390127	390027	AWC	ANG	MA	69127	69427	69527	69627	65327	68927	68827	69727	69827	69927	69227	
390128	390028	AWC	ANG	MA	69128	69428	69528	69628	65328	68928	68828	69728	69828	69928	69228	*City of Preston*
390129	390029	AWC	ANG	MA	69129	69429	69529	69629	65329	68929	68829	69729	69829	69929	69229	*City of Stoke-on-Trent*
390130	390030	AWC	ANG	MA	69130	69430	69530	69630	65330	68930	68830	69730	69830	69930	69230	*City of Edinburgh*
390131	390031	AWC	ANG	MA	69131	69431	69531	69631	65331	68931	68831	69731	69831	69931	69231	*City of Liverpool*
390132	390032	AWC	ANG	MA	69132	69432	69532	69632	65332	68932	68832	69732	69832	69932	69232	*City of Birmingham*
390134	390034	AWC	ANG	MA	69134	69434	69534	69634	65334	68934	68834	69734	69834	69934	69234	*City of Carlisle*
390135	390035	AWC	ANG	MA	69135	69435	69535	69635	65335	68935	68835	69735	69835	69935	69235	*City of Lancaster*
390136	390036	AWC	ANG	MA	69136	69436	69536	69636	65336	68936	68836	69736	69836	69936	69236	*City of Coventry*
390137	390037	AWC	ANG	MA	69137	69437	69537	69637	65337	68937	68837	69737	69837	69937	69237	
390138	390038	AWC	ANG	MA	69138	69438	69538	69638	65338	68938	68838	69738	69838	69938	69238	*City of London*
390141	390041	AWC	ANG	MA	69141	69441	69541	69641	65341	68941	68841	69741	69841	69941	69241	
390148	390048	AWC	ANG	MA	69148	69448	69548	69648	65348	68948	68848	69748	69848	69948	69248	*Flying Scouseman*
390151	390051	AWC	ANG	MA	69151	69451	69551	69651	65351	68951	68851	69751	69851	69951	69251	*Unknown Soldier*
390152	390052	AWC	ANG	MA	69152	69452	69552	69652	65352	68952	68852	69752	69852	69952	69252	
390153	390053	AWC	ANG	MA	69153	69453	69553	69653	65353	68953	68853	69753	69853	69953	69253	
390154		AWC	ANG	MA	69154	69454	69554	69654	65354	68954	68854	69754	69854	69954	69254	*Matthew Flinders*
390155		AWC	ANG	MA	69155	69455	69555	69655	65355	68955	68855	69755	69855	69955	69255	*Railway Benefit Fund*
390156		AWC	ANG	MA	69156	69456	69556	69656	65356	68956	68856	69756	69856	69956	69256	*Pride and Prosperity*
390157		AWC	ANG	MA	69157	69457	69557	69657	65357	68957	68857	69757	69857	69957	69257	*Chad Varah*

Class 805

Avanti West Coast has placed orders with Hitachi for 13 five-car Class 805 BMMUs and ten seven-car Class 807 EMUs. They will be financed through Rock Rail and Standard Life Aberdeen.

The first bodyshells were delivered to Newton Aycliffe in early 2022 and the first completed units should be ready before the end of the year. The Class 805s will replace Class 221 DEMUs.

Both new fleets will be maintained by Alstom and Hitachi staff at Oxley Depot.

	Livery	Owner	Depot	PTDS	MS	MS	MS	PDTF
805001		ROC		861001	862001	863001	864001	865001
805002		ROC		861002	862002	863002	864002	865002
805003		ROC		861003	862003	863003	864003	865003
805004		ROC		861004	862004	863004	864004	865004
805005		ROC		861005	862005	863005	864005	865005
805006		ROC		861006	862006	863006	864006	865006
805007		ROC		861007	862007	863007	864007	865007
805008		ROC		861008	862008	863008	864008	865008
805009		ROC		861009	862009	863009	864009	865009
805010		ROC		861010	862010	863010	864010	865010
805011		ROC		861011	862011	863011	864011	865011
805012		ROC		861012	862012	863012	864012	865012
805013		ROC		861013	862013	863013	864013	865013

Class 807

Seven-car IEP-type high-speed trains, these are electric units for working on electrified lines.

	Livery	Owner	Depot	PTDS	MS	MS	TS	MS	MC	PDTF
807001		ROC		871001	872001	873001	874001	875001	876001	877001
807002		ROC		871002	872002	873002	874002	875002	876002	877002
807003		ROC		871003	872003	873003	874003	875003	876003	877003
807004		ROC		871004	872004	873004	874004	875004	876004	877004
807005		ROC		871005	872005	873005	874005	875005	876005	877005
807006		ROC		871006	872006	873006	874006	875006	876006	877006
807007		ROC		871007	872007	873007	874007	875007	876007	877007
807008		ROC		871008	872008	873008	874008	875008	876008	877008
807009		ROC		871009	872009	873009	874009	875009	876009	877009
807010		ROC		871010	872010	873010	874010	875010	876010	877010

Caledonian Sleeper

Contact details

Website: www.sleeper.scot
Twitter: @CalSleeper

Key personnel

Managing Director: Kathryn Darbandi
Operations Director: Magnus Conn

Overview

Caledonian Sleeper is part of the Serco group and leases its Mk 5 coaches from Lombard Finance. However, it does not own or lease its own locomotives and instead it hires traction and drivers from GB Railfreight to work its trains (see page 282) from Euston to Glasgow Central/Edinburgh and to Inverness/Aberdeen/Fort William. The trains use Class 92s south of Edinburgh or Glasgow Central and north of Edinburgh they are worked by Class 73/9s, usually in multiple with GBRf 66s or Class 67s, which must lead the train as they do not have Dellner couplers. Pairs of 73/9s can also be used if required. 73966-971 and 92006/010/014/018/023/033/038 are all in CS livery.

The new Spanish-built Mk 5s were introduced in 2019. They have Dellner couplers, so can only be hauled by locos fitted with similar couplers, which means just Class 73/9 and 92s.

There are four sets of coaches in use every night from Sundays to Fridays, two trains leaving from Euston northbound, one leaving Glasgow Central and another leaving Inverness. The former has a portion from Edinburgh added to it at Carstairs, while the latter has portions from Aberdeen and Fort William added to it at Edinburgh. The northbound trains have corresponding portions detached at Carstairs and Edinburgh.

Caledonian Sleeper hires traction and traincrew from GBRf. Class 92s work its trains south of Edinburgh, and 92020 heads the diverted 2026 Inverness-Euston past Walton, near Peterborough on 31 May 2021. *Stuart West*

Each train from London works as 16 coaches, with each portion comprising a half brake/half-seated accommodation, a lounge car, a fully accessible sleeping car and varying numbers of sleeping cars. The fleet is maintained at Polmadie by Alstom.

No changes are expected to the operation in the next year and most of the company's attention will be focused on rebuilding the business in the wake of the pandemic.

Mk 5 brake/seated coaches

15001	CAL	LOM	PO
15002	CAL	LOM	PO
15003	CAL	LOM	PO
15004	CAL	LOM	PO
15005	CAL	LOM	PO
15006	CAL	LOM	PO
15007	CAL	LOM	PO
15008	CAL	LOM	PO
15009	CAL	LOM	PO
15010	CAL	LOM	PO
15011	CAL	LOM	PO

Mk 5 lounge cars

15101	CAL	LOM	PO
15102	CAL	LOM	PO
15103	CAL	LOM	PO
15104	CAL	LOM	PO
15105	CAL	LOM	PO
15106	CAL	LOM	PO
15107	CAL	LOM	PO
15108	CAL	LOM	PO
15109	CAL	LOM	PO
15110	CAL	LOM	PO

Mk 5 accessible sleeping cars

15201	CAL	LOM	PO
15202	CAL	LOM	PO
15203	CAL	LOM	PO
15204	CAL	LOM	PO
15205	CAL	LOM	PO
15206	CAL	LOM	PO
15207	CAL	LOM	PO
15208	CAL	LOM	PO
15209	CAL	LOM	PO
15210	CAL	LOM	PO
15211	CAL	LOM	PO
15212	CAL	LOM	PO
15213	CAL	LOM	PO
15214	CAL	LOM	PO

Mk 5 sleeping cars

15301	CAL	LOM	PO
15302	CAL	LOM	PO
15303	CAL	LOM	PO
15304	CAL	LOM	PO
15305	CAL	LOM	PO
15306	CAL	LOM	PO
15307	CAL	LOM	PO
15308	CAL	LOM	PO
15309	CAL	LOM	PO
15310	CAL	LOM	PO
15311	CAL	LOM	PO
15312	CAL	LOM	PO
15313	CAL	LOM	PO
15314	CAL	LOM	PO
15315	CAL	LOM	PO
15316	CAL	LOM	PO
15317	CAL	LOM	PO
15318	CAL	LOM	PO
15319	CAL	LOM	PO
15320	CAL	LOM	PO
15321	CAL	LOM	PO
15322	CAL	LOM	PO
15323	CAL	LOM	PO
15324	CAL	LOM	PO
15325	CAL	LOM	PO
15326	CAL	LOM	PO
15327	CAL	LOM	PO
15328	CAL	LOM	PO
15329	CAL	LOM	PO
15330	CAL	LOM	PO
15331	CAL	LOM	PO
15332	CAL	LOM	PO
15333	CAL	LOM	PO
15334	CAL	LOM	PO
15335	CAL	LOM	PO
15336	CAL	LOM	PO
15337	CAL	LOM	PO
15338	CAL	LOM	PO
15339	CAL	LOM	PO
15340	CAL	LOM	PO

Brake force runner vehicles

6392	81588, 92183	CAL	Mk 1
6397	81600, 92190	CAL	Mk 1
96604	86337, 96156	CAL	GUV
96606	86324, 96213	CAL	GUV
96608	86385, 96216	CAL	GUV
96609	86327, 96217	CAL	GUV

Fully accessible Mk 5 sleeping car 15208 passes Auchengray on the Edinburgh to Carstairs line on 8 August as part of the 0741 Edinburgh-Polmadie ECS move. *Robin Ralston*

Sleeping Car 15335 is one of 40 such vehicles in the Caledonian Sleeper fleet. It passes Auchengray on 11 March 2021 in the 0920 Fort William-Polmadie ECS move. *Robin Ralston*

c2c

Contact details

Website: www.c2c-online.co.uk
Twitter: @c2c_Rail

Key personnel

Managing Director: Rob Mullen
Engineering Director: Jeff Baker

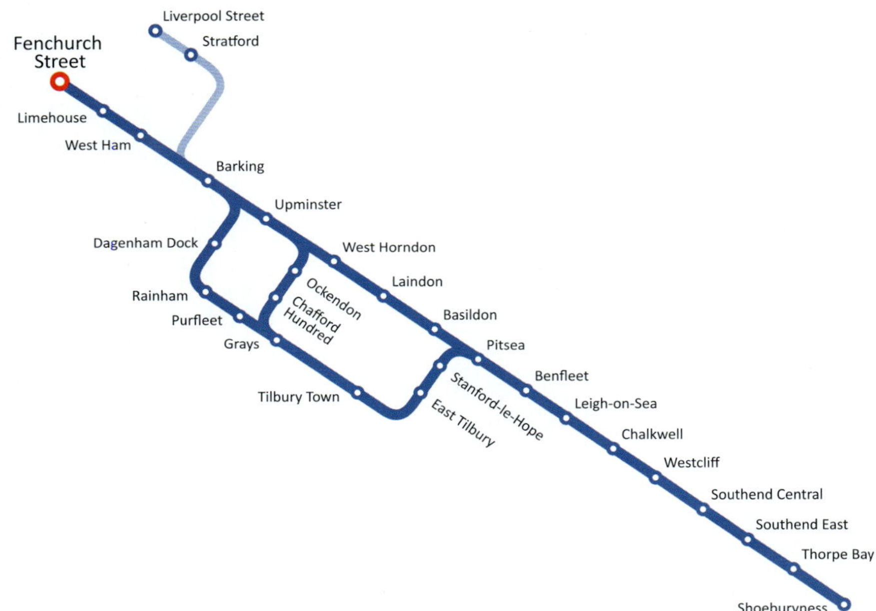

Overview

Until recently, c2c fleet was formed solely of four-car Class 357 Bombardier Electrostars delivered in 1999-2002 to operate from Fenchurch St-Southend / Shoeburyness, via Tilbury. They were ordered by the original franchise holder in the late 1990s to replace the London Tilbury Southend line's collection of old slam-door Class 302 and 310 / 312 units and sliding door Class 317s. The sub-classes differentiate the owning ROSCO, while the renumbered 357 / 3s are those units with fewer seats for high-density commuter operations. The original order of 357 / 0s was for 44 units but two additional units were added to the production run as 'compensation' for late delivery.

In 2016, six four-car Class 387 / 3s joined the fleet to allow extra seats but the company has 12 six-car Class 720 Bombardier Aventra units on order, which were expected to be delivered in 2021 but are delayed and not expected to be with the company until late 2022 at the earliest.

The 387s were only a short-term stopgap to meet increased demand for c2c and will be returned to Porterbrook when the 720 / 6s are delivered. Three have recently been on sub-lease to GWR and in the fullness of time are expected to move to GWR on a permanent basis.

The franchise is pretty self-contained, being a mass people mover for commuters in the week yet also offering leisure travel to the likes of Southend and Leigh-on-Sea.

Class 357

All are in the HTHQ pool.

Class 357/0

Porterbrook units.

	Livery	Owner	Depot	DMS	MS	PTS	DMS	Name
357001	C2C	POR	EM	67651	74151	74051	67751	*Barry Flaxman*
357002	C2C	POR	EM	67652	74152	74052	67752	*Arthur Lewis Stride 1841-1922*
357003	C2C	POR	EM	67653	74153	74053	67753	*Southend city.on.sea*
357004	C2C	POR	EM	67654	74154	74054	67754	*Tony Amos*
357005	C2C	POR	EM	67655	74155	74055	67755	*Southend: 2017 Alternative City of Culture*
357006	C2C	POR	EM	67656	74156	74056	67756	*Diamond Jubilee 1952-2012*
357007	C2C	POR	EM	67657	74157	74057	67757	*Sir Andrew Foster*
357008	C2C	POR	EM	67658	74158	74058	67758	
357009	C2C	POR	EM	67659	74159	74059	67759	
357010	C2C	POR	EM	67660	74160	74060	67760	
357011	C2C	POR	EM	67661	74161	74061	67761	*John Lowing*
357012	C2C	POR	EM	67662	74162	74062	67762	
357013	C2C	POR	EM	67663	74163	74063	67763	
357014	C2C	POR	EM	67664	74164	74064	67764	
357015	C2C	POR	EM	67665	74165	74065	67765	
357016	C2C	POR	EM	67666	74166	74066	67766	
357017	C2C	POR	EM	67667	74167	74067	67767	
357018	C2C	POR	EM	67668	74168	74068	67768	*Remembering our Fallen 88 1914-1918*
357019	C2C	POR	EM	67669	74169	74069	67769	
357020	C2C	POR	EM	67670	74170	74070	67770	
357021	C2C	POR	EM	67671	74171	74071	67771	
357022	C2C	POR	EM	67672	74172	74072	67772	
357023	C2C	POR	EM	67673	74173	74073	67773	
357024	C2C	POR	EM	67674	74174	74074	67774	
357025	C2C	POR	EM	67675	74175	74075	67775	
357026	C2C	POR	EM	67676	74176	74076	67776	
357027	C2C	POR	EM	67677	74177	74077	67777	
357028	C2C	POR	EM	67678	74178	74078	67778	*London Tilbury & Southend Railway 1854-2004*
357029	C2C	POR	EM	67679	74179	74079	67779	*Thomas Whitelegg 1840-1922*
357030	C2C	POR	EM	67680	74180	74080	67780	*Robert Harben Whitelegg 1871-1957*
357031	C2C	POR	EM	67681	74181	74081	67781	
357032	C2C	POR	EM	67682	74182	74082	67782	
357033	C2C	POR	EM	67683	74183	74083	67783	
357034	C2C	POR	EM	67684	74184	74084	67784	
357035	C2C	POR	EM	67685	74185	74085	67785	
357036	C2C	POR	EM	67686	74186	74086	67786	
357037	C2C	POR	EM	67687	74187	74087	67787	
357038	C2C	POR	EM	67688	74188	74088	67788	
357039	C2C	POR	EM	67689	74189	74089	67789	
357040	C2C	POR	EM	67690	74190	74090	67790	
357041	C2C	POR	EM	67691	74191	74091	67791	
357042	C2C	POR	EM	67692	74192	74092	67792	
357043	C2C	POR	EM	67693	74193	74093	67793	
357044	C2C	POR	EM	67694	74194	74094	67794	
357045	C2C	POR	EM	67695	74195	74095	67795	
357046	C2C	POR	EM	67696	74196	74096	67796	

A pair of c2c Class 357s, led by 357204, arrive at Barking with a Fenchurch Street to Shoeburyness train on 28 January 2014. *Tony Miles*

Class 357/2

The same as 357/0s, just financed by Angel Trains. Those renumbered in the 357/3 series have high-density interiors with fewer seats.

	Old no.	Livery	Owner	Depot	DMS	MS	PS	DMS	Name
357201		C2C	ANG	EM	68601	74701	74601	68701	*Ken Bird*
357202		C2C	ANG	EM	68602	74702	74602	68702	*Kenny Mitchell*
357203		C2C	ANG	EM	68603	74703	74603	68703	*Henry Pumfrett*
357204		C2C	ANG	EM	68604	74704	74604	68704	*Derek Fowers*
357205		C2C	ANG	EM	68605	74705	74605	68705	*John D'silva*
357206		C2C	ANG	EM	68606	74706	74606	68706	*Martin Aungier*
357207		C2C	ANG	EM	68607	74707	74607	68707	*John Page*
357208		C2C	ANG	EM	68608	74708	74608	68708	*Dave Davis*
357209		C2C	ANG	EM	68609	74709	74609	68709	*James Snelling*
357210		C2C	ANG	EM	68610	74710	74610	68710	
357211		C2C	ANG	EM	68611	74711	74611	68711	
357312	357212	C2C	ANG	EM	68612	74712	74612	68712	
357313	357213	C2C	ANG	EM	68613	74713	74613	68713	*Upminster IECC*
357314	357214	C2C	ANG	EM	68614	74714	74614	68714	
357315	357215	C2C	ANG	EM	68615	74715	74615	68715	
357316	357216	C2C	ANG	EM	68616	74716	74616	68716	
357317	357217	C2C	ANG	EM	68617	74717	74617	68717	*Allan Burnell*
357318	357218	C2C	ANG	EM	68618	74718	74618	68718	
357319	357219	C2C	ANG	EM	68619	74719	74619	68719	
357320	357220	C2C	ANG	EM	68620	74720	74620	68720	
357321	357221	C2C	ANG	EM	68621	74721	74621	68721	
357322	357222	C2C	ANG	EM	68622	74722	74622	68722	
357323	357223	C2C	ANG	EM	68623	74723	74623	68723	
357324	357224	C2C	ANG	EM	68624	74724	74624	68724	
357325	357225	C2C	ANG	EM	68625	74725	74625	68725	
357326	357226	C2C	ANG	EM	68626	74726	74626	68726	
357327	357227	C2C	ANG	EM	68627	74727	74627	68727	*Southend United*
357328	357228	C2C	ANG	EM	68628	74728	74628	68728	

Class 387/3

Similar to the Class 357s from the same Electrostar platform, these units differ by having corridor connections allowing them to run in eight- and 12-car formations with through connections between units for staff and passengers alike. Three units, 387301/302/306, have recently been sub-leased to GWR but have yet to return to c2c operation.

All are in the HTHQ pool.

While on hire to GWR, 387302 calls at Reading West with the 1212 Reading-Newbury on 25 May 2021. *Mark Pike*

	Livery	Owner	Depot	DMS	MS	PTS	DMS
387301	C2C	POR	EM	421301	422301	423301	424301
387302	C2C	POR	EM	421302	422302	423302	424302
387303	C2C	POR	EM	421303	422303	423303	424303
387304	C2C	POR	EM	421304	422304	423304	424304
387305	C2C	POR	EM	421305	422305	423305	424305
387306	C2C	POR	EM	421306	422306	423306	424306

Class 720/6

The original order of six ten-car units has been changed to 12 six-car units and are another variant of the Bombardier Aventra platform in use with Greater Anglia and others. They are numbered 720601-612 and are high-density commuter trains funded through Porterbrook. They will replace the Class 387s as well as allowing extra seats.

All will be in the HTHQ pool.

	Livery	owner	depot	DMS	PMS	MS	MS	DMS
720601	C2C	POR		450601	451601	452601	453601	454601
720602	C2C	POR		450602	451602	452602	453602	454602
720603	C2C	POR		450603	451603	452603	453603	454603
720604	C2C	POR		450604	451604	452604	453604	454604
720605	C2C	POR		450605	451605	452605	453605	454605
720606	C2C	POR		450606	451606	452606	453606	454606
720607	C2C	POR		450607	451607	452607	453607	454607
720608	C2C	POR		450608	451608	452608	453608	454608
720609	C2C	POR		450609	451609	452609	453609	454609
720610	C2C	POR		450610	451610	452610	453610	454610
720611	C2C	POR		450611	451611	452611	453611	454611
720612	C2C	POR		450612	451612	452612	453612	454612

Chiltern Railways

Contact details

Website: www.chilternrailways.co.uk
Twitter: @chilternrailway

Key personnel

Managing Director: Richard Allan
Engineering and Safety Director: Ian Hyde

Overview

Owned by Arriva, Chiltern Railways is one of the few success of the franchised rail operators and the only one to be granted a long, 20-year franchise back in 2002. Its franchise was due to expire on 31 December 2021, but on 3 December a new six-year extension was announced. It runs from London Marylebone to Birmingham Moor Street and Kidderminster, Marylebone to Aylesbury/Oxford and Aylesbury to Princes Risborough.

It has been a pretty static fleet since it was privatised, inheriting 39 ex-BR Class 165s of two- and three-car types before ordering Bombardier Class 168s.

Chiltern has frequently ordered small batches of new trains, and grown to a fleet of 32 new trains of two-, three- and four-car types. The four most recent acquisitions are four two-car Class 172/1s, which differed in being diesel mechanical as opposed to diesel hydraulics units.

The firm also operates loco-hauled trains between Marylebone and Birmingham Moor St/Kidderminster. These comprise six modified Mk 3 coaches – with sliding doors – plus a driving van trailer, sets powered by a pool of eight dedicated Stadler Class 68s hired from Direct Rail Services; 68008-015, of which 68010-015 are in Chiltern livery (see page 269). The locos are on the country end of the trains. They are maintained by DRS at Crewe with routine running exams undertaken at Wembley. Three sets stable at Stourbridge Junction each evening. Class 68s replaced Class 67s from DB Cargo in 2015. The coaches were previously used by open access operator Wrexham, Shropshire and Marylebone Railway with Class 67s.

As CR looks to reduce its carbon emissions, there is a suggestion that the Class 68-hauled sets could be removed from the operation, but as yet this has not been confirmed.

The DMU fleet is all based at Aylesbury. Chiltern also has a Ruston 0-6-0 shunter based at Aylesbury but this is not passed to run on Network Rail infrastructure and so is excluded from this book.

Class 165

BREL York Networker Turbo units built in 1991-92 and using two- or three-car sets. These units work from Marylebone to Aylesbury and Banbury and also can work through to Birmingham and Kidderminster.

All are in the HOHQ pool.

	Livery	Owner	Depot	DMSL	MS	DMS
165001	CRW	ANG	AL	58801		58834
165002	CRW	ANG	AL	58802		58835
165003	CRW	ANG	AL	58803		58836
165004	CRW	ANG	AL	58804		58837
165005	CRW	ANG	AL	58805		58838
165006	CRW	ANG	AL	58806		58839
165007	CRW	ANG	AL	58807		58840
165008	CRW	ANG	AL	58808		58841
165009	CRW	ANG	AL	58809		58842
165010	CRW	ANG	AL	58810		58843
165011	CRW	ANG	AL	58811		58844
165012	CRW	ANG	AL	58812		58845
165013	CRW	ANG	AL	58813		58846
165014	CRW	ANG	AL	58814		58847
165015	CRW	ANG	AL	58815		58848
165016	CRW	ANG	AL	58816		58849
165017	CRW	ANG	AL	58817		58850
165018	CRW	ANG	AL	58818		58851
165019	CRW	ANG	AL	58819		58852
165020	CRW	ANG	AL	58820		58853
165021	CRW	ANG	AL	58821		58854
165022	CRW	ANG	AL	58822		58855
165023	CRW	ANG	AL	58873		58867
165024	CRW	ANG	AL	58874		58868
165025	CRW	ANG	AL	58875		58869
165026	CRW	ANG	AL	58876		58870
165027	CRW	ANG	AL	58877		58871
165028	CRW	ANG	AL	58878		58872
165029	CRW	ANG	AL	58823	55404	58856
165030	CRW	ANG	AL	58824	55405	58857
165031	CRW	ANG	AL	58825	55406	58858
165032	CRW	ANG	AL	58826	55407	58859
165033	CRW	ANG	AL	58827	55408	58860
165034	CRW	ANG	AL	58828	55409	58861
165035	CRW	ANG	AL	58829	55410	58862
165036	CRW	ANG	AL	58830	55411	58863
165037	CRW	ANG	AL	58831	55412	58864
165038	CRW	ANG	AL	58832	55413	58865
165039	CRW	ANG	AL	58833	55414	58866

Note: 165004 is being converted to a low-emission electric hybrid unit by Magtec and Loramin Rotherham. After testing it will return to the CR fleet.

Class 168

Branded as Clubman units, despite being from the same platform as the Turbostar units, the different sub-classes are mainly dependent on which batch of units they were ordered in rather than their configuration or owner. The first five units differed in cab design, but on units delivered from 168106 the cab has been the same as the standard Turbostar family. Class 168s of all sub-classes work from Marylebone through to Birmingham and Kidderminster as well as trains to Aylesbury.

All are in the HOHQ pool.

On 12 September 2021, 168002 waits to set off with the 0847 Birmingham Moor St-Leamington Spa. This was one of the original Cass 168 Clubman units ordered by Chiltern Railways in 1996. *Pip Dunn*

Class 168/0

Original four-car Clubman units with old-style cabs.

	Livery	Owner	Depot	DMSL	MS/MSL	MSL/MS	DMSL
168001	CRS	POR	AL	58151	58651	58451	58251
168002	CRS	POR	AL	58152	58652	58452	58252
168003	CRS	POR	AL	58153	58453	58653	58253
168004	CRS	POR	AL	58154	58654	58454	58254
168005	CRS	POR	AL	58155	58655	58455	58255

Class 168/1

Later-build units featuring Class 170 Turbostar cabs.

	Livery	Owner	Depot	DMSL	MSL	MS	DMSL
168106	CRS	POR	AL	58156	58756	58456	58256
168107	CRS	POR	AL	58157	58757	58457	58257
168108	CRS	POR	AL	58158		58458	58258
168109	CRS	POR	AL	58159		58459	58259
168110	CRS	POR	AL	58160		58460	58260
168111	CRS	EVS	AL	58161		58461	58261
168112	CRS	EVS	AL	58162		58462	58262
168113	CRS	EVS	AL	58163		58463	58263

Class 168/2

Later-build units featuring Class 170 Turbostar cabs.

	Livery	Owner	Depot	DMSL	MSL	MS	DMSL
168214	CRS	POR	AL	58164		58464	58264
168215	CRS	POR	AL	58165	58365	58465	58265
168216	CRS	POR	AL	58166	58366	58466	58266
168217	CRS	POR	AL	58167	58367	58467	58267
168218	CRS	POR	AL	58168		58468	58268
168219	CRS	POR	AL	58169		58469	58269

Two-car 168324 waits its next duty at Stratford-upon-Avon on 8 August 2021. This unit started life as 170304 for South West Trains, then moved to Transpennine Express before moving to Chiltern. *Pip Dunn*

Class 168s are exclusive to Chiltern Railways but are essentially the same as Class 170 Turbostar units. 168321 heads across Saunderton Viaduct on 3 July 2018. *Rob France*

Class 168/3

Former South West Trains and then Transpennine Express Class 170/3 two-car units. 168329 has been modified as a HybridFLEX unit to run on batteries and diesel power and carries a promotional livery.

	Old No(s).	Livery	Owner	Depot	DMSL	DMSL
168321	170301	CRS	POR	AL	50301	79301
168322	170302	CRS	POR	AL	50302	79302
168323	170303	CRS	POR	AL	50303	79303
168324	170304	CRS	POR	AL	50304	79304
168325	170305	CRS	POR	AL	50305	79305
168326	170306	CRS	POR	AL	50306	79306
168327	170307	CRS	POR	AL	50307	79307
168328	170308	CRS	POR	AL	50308	79308
168329	170309, 170399	ADV	POR	AL	50399	79399

168329 is wrapped in blue/green with HybridFLEX branding.

Loco-hauled coaches

Chiltern has a fleet of six driving van trailers and Mk 3 coaches for Marylebone to Birmingham services. Five sets are in use daily, and three trains a day start or end at Kidderminster and are stabled overnight at Stourbridge Junction.

68010-015 the preferred locos, as they are painted in Chiltern livery. However, 68008/009 are also modified to work with the coaches yet remain in DRS livery.

Driving van trailers

82301	CRS	ARV	AL
82302	CRS	ARV	AL
82303	CRS	ARV	AL
82304	CRS	ARV	AL
82305	CRS	ARV	AL
82309	CRS	ARV	AL

Driving Van Trailer 82304 stands at Birmingham Moor Street prior to leading the 1255 to Marylebone, powered by 68011 on the rear, on 18 September 2018. *Pip Dunn*

Mk 3 coaches

KBF kitchen buffet first Mk 3.

10271	CRS	ARV	AL
10272	CRS	ARV	AL
10273	CRS	ARV	AL
10274	CRS	ARV	AL

SO – standard open

12602	CRS	ARV	AL
12603	CRS	ARV	AL
12604	CRS	ARV	AL
12605	CRS	ARV	AL
12606	CRS	ARV	AL
12607	CRS	ARV	AL
12608	CRS	ARV	AL
12609	CRS	ARV	AL
12610	CRS	ARV	AL
12613	CRS	ARV	AL
12614	CRS	ARV	AL
12615	CRS	ARV	AL
12616	CRS	ARV	AL
12617	CRS	ARV	AL
12618	CRS	ARV	AL
12619	CRS	ARV	AL
12620	CRS	ARV	AL
12621	CRS	ARV	AL
12623	CRS	ARV	AL
12625	CRS	ARV	AL
12627	CRS	ARV	AL

Mk 3 set formations

	DVT	KBF	TSO	TSO	TSO	TSO	TSO
AL01	82305	10272	12609	12606	12610	12627	12620
AL02	82303	10271	12605	12608	12607	12615	12616
AL03	82304	10274	12602	12617	12604	12621	12623
AL04	82302	10273	12603	12614	12618	12625	12619

68012 calls at Leamington Spa on 4 April 2017 with the 1555 Birmingham Moor St-Marylebone. The Class 68s are owned and maintained by DRS and six are painted in CR colours, with two others available for use if required. *Pip Dunn*

Chiltern Railways is – currently – one of the few TOCs to use loco haulage, with Class 68 hired from Direct Rail Services. 68011 stands at Birmingham Snow Hill on 7 November 2016 with the 1615 Marylebone-Kidderminster. *Pip Dunn*

CrossCountry Trains

Contact details

Website: www.crosscountrytrains.co.uk
Twitter: @CrossCountryUK

Key personnel

Managing Director: Tom Joyner
Service Delivery Director: Jon Fenn

Overview

CrossCountry Trains is a nationwide operator of inter-city trains from Penzance to Aberdeen, although its operation is much changed from the days when the franchise was first let to Virgin in 1996 with it no longer serves the WCML north of Manchester, nor to Brighton. However, it does also now include the Birmingham to Stansted and Nottingham to Cardiff routes that were previously part of Central Trains.

The routes it currently operates are Penzance to Aberdeen via Birmingham New Street and Newcastle, Birmingham-Bournemouth, Birmingham-Manchester, Birmingham-Stansted and Cardiff-Nottingham.

The XCT fleet relies on three main traction types; Class 220/221 Bombardier Voyager DEMUs, which work most of the long-distance routes, supplemented by five HST sets (with a pool of 12 Class 43s) on the Edinburgh-Plymouth route. Class 170 Turbostars work the Cardiff-Nottingham and Birmingham-Stansted corridors.

The six centre cars from ex-West Midlands Trains Class 170/6s have been inserted in XCT's six Class 170/5s to make all but seven of its 29 Class 170s three-car sets, with just 170111-117 remaining as two-car sets.

CrossCountry Trains has 12 HST powers cars and five sets of Mk 3 trailers. 43366 approaches Westbury leading a diverted 0925 Penzance-Edinburgh, passing 43378/357 working the 0606 Edinburgh-Plymouth on 11 August 2021. *Glen Batten*

Class 43/2

MTU-powered Class 43 HST sets. The Class 43s are now allocated to Laira with their maintenance undertaken under contract by Great Western Railway.

All are in the EHPC pool.

	Old No.	Livery	Owner	Depot
43207	43007	XCT	ANG	LA
43208	43008	XCT	ANG	LA
43239	43039	XCT	ANG	LA
43285	43085	XCT	POR	LA
43301	43101	XCT	POR	LA
43303	43103	XCT	POR	LA
43304	43104	XCT	ANG	LA
43321	43121	XCT	POR	LA
43357	43157	XCT	POR	LA
43366	43166	XCT	ANG	LA
43378	43178	XCT	ANG	LA
43384	43184	XCT	ANG	LA

HST trailers

XCT's HST sets run as five seven-vehicle sets with a TF, TCK, four TSs and a TGS and there are six spare trailer vehicles.

TF – trailer first

41026	XCT	ANG	LA
41035	XCT	ANG	LA
41193	XCT	POR	LA
41194	XCT	POR	LA
41195	XCT	POR	LA

TS – trailer standard

42036	XCT	ANG	LA
42037	XCT	ANG	LA
42038	XCT	ANG	LA
42051	XCT	ANG	LA
42052	XCT	ANG	LA
42053	XCT	ANG	LA
42097	XCT	ANG	LA
42234	XCT	POR	LA
42290	XCT	POR	LA
42342	XCT	ANG	LA
42366	XCT	POR	LA
42367	XCT	POR	LA
42368	XCT	POR	LA
42369	XCT	POR	LA
42370	XCT	POR	LA
42371	XCT	POR	LA
42372	XCT	POR	LA
42373	XCT	POR	LA
42374	XCT	POR	LA
42375	XCT	POR	LA
42376	XCT	POR	LA
42377	XCT	POR	LA
42378	XCT	POR	LA
42379	XCT	ANG	LA
42380	XCT	ANG	LA

TGS – trailer guard's standard

44012	XCT	ANG	LA
44017	XCT	ANG	LA
44021	XCT	POR	LA
44052	XCT	POR	LA
44072	XCT	POR	LA

TCK – trailer composite kitchen

45001	XCT	POR	LA
45002	XCT	POR	LA
45003	XCT	POR	LA
45004	XCT	POR	LA
45005	XCT	POR	LA

CrossCountry trains HST formations

XC01	41193	45001	42342	42097	42377	42380	44021
XC02	41194	45002	42053	42037	42234	42371	44072
XC03	41195	45003	42375	42378	42036	42376	44052
XC04	41026	45004	42290	42369	42038	42366	44012
XC05	41035	45005	42373	42368	42370	42379	44017
Spare		42052	42370	42367	42372	42378	

Class 170/1

Former Midland Mainline two- or three-car sets, delivered in 1998 as two-car sets, with ten units strengthened to three-cars.

All are in the EHXC pool.

	Livery	Owner	Depot	DMSL	MS	DMCL
170101	XCT	POR	TS	50101	55101	79101
170102	XCT	POR	TS	50102	55102	79102
170103	XCT	POR	TS	50103	55103	79103
170104	XCT	POR	TS	50104	55104	79104
170105	XCT	POR	TS	50105	55105	79105
170106	XCT	POR	TS	50106	55106	79106
170107	XCT	POR	TS	50107	55107	79107
170108	XCT	POR	TS	50108	55108	79108
170109	XCT	POR	TS	50109	55109	79109
170110	XCT	POR	TS	50110	55110	79110
170111	XCT	POR	TS	50111		79111
170112	XCT	POR	TS	50112		79112
170113	XCT	POR	TS	50113		79113
170114	XCT	POR	TS	50114		79114
170115	XCT	POR	TS	50115		79115
170116	XCT	POR	TS	50116		79116
170117	XCT	POR	TS	50117		79117

The Class 170/1 Turbostars were ordered by Midland Mainline and delivered in 1998 but have all since moved to CrossCountry Trains. 170101 calls at Peterborough on 12 February 2022 with the 1627 Stansted Airport-Coleshill Parkway. *Pip Dunn*

Class 170/3

Former Central Trains three-car sets.

All are in the EHXC pool.

	Livery	Owner	Depot	DMSL	MS	DMCL
170397	XCT	POR	TS	50397	55397	79397
170398	XCT	POR	TS	50398	55398	79398

Class 170/6

Former Central Trains three-car sets. 170618-623 were previously two-car sets that have recently been increased to three-car sets by adding centre cars from ex-WMR Class 170/5s.

All are in the EHXC pool.

		Livery	Owner	Depot	DMSL	MS	DMCL	
170618	170518	XCT	POR	TS	50518	56630	79518	
170619	170519	XCT	POR	TS	50519	56631	79519	
170620	170520	XCT	POR	TS	50520	56632	79520	
170621	170521	XCT	POR	TS	50521	56633	79521	
170622	170522	XCT	POR	TS	50522	56634	79522	*Pride of Leicester*
170623	170523	XCT	POR	TS	50523	56635	79523	
170636		XCT	POR	TS	50636	56636	79636	
170637		XCT	POR	TS	50637	56637	79637	
170638		XCT	POR	TS	50638	56638	79638	
170639		XCT	POR	TS	50639	56639	79639	

Class 220

Former Virgin CrossCountry four-car Voyager DEMUs, maintained by Bombardier at Central Rivers depot.

All are in the EHXC pool.

	Livery	Owner	Depot	DMS	MS	MS	DMF	
220001	XCT	BEA	CZ	60301	60701	60201	60401	
220002	XCT	BEA	CZ	60302	60702	60202	60402	
220003	XCT	BEA	CZ	60303	60703	60203	60403	
220004	XCT	BEA	CZ	60304	60704	60204	60404	
220005	XCT	BEA	CZ	60305	60705	60205	60405	
220006	XCT	BEA	CZ	60306	60706	60206	60406	
220007	XCT	BEA	CZ	60307	60707	60207	60407	
220008	XCT	BEA	CZ	60308	60708	60208	60408	
220009	XCT	BEA	CZ	60309	60709	60209	60409	*Hixon January 6th 1968*
220010	XCT	BEA	CZ	60310	60710	60210	60410	
220011	XCT	BEA	CZ	60311	60711	60211	60411	
220012	XCT	BEA	CZ	60312	60712	60212	60412	
220013	XCT	BEA	CZ	60313	60713	60213	60413	

On a bright 25 October 2021, 220016 and a sister unit head north past Teignmouth on a Plymouth to Edinburgh train. *Pip Dunn*

220014	XCT	BEA	CZ	60314	60714	60214	60414	
220015	XCT	BEA	CZ	60315	60715	60215	60415	
220016	XCT	BEA	CZ	60316	60716	60216	60416	*Voyager 20*
220017	XCT	BEA	CZ	60317	60717	60217	60417	
220018	XCT	BEA	CZ	60318	60718	60218	60418	
220019	XCT	BEA	CZ	60319	60719	60219	60419	
220020	XCT	BEA	CZ	60320	60720	60220	60420	
220021	XCT	BEA	CZ	60321	60721	60221	60421	
220022	XCT	BEA	CZ	60322	60722	60222	60422	
220023	XCT	BEA	CZ	60323	60723	60223	60423	
220024	XCT	BEA	CZ	60324	60724	60224	60424	
220025	XCT	BEA	CZ	60325	60725	60225	60425	
220026	XCT	BEA	CZ	60326	60726	60226	60426	
220027	XCT	BEA	CZ	60327	60727	60227	60427	
220028	XCT	BEA	CZ	60328	60728	60228	60428	
220029	XCT	BEA	CZ	60329	60729	60229	60429	
220030	XCT	BEA	CZ	60330	60730	60230	60430	
220031	XCT	BEA	CZ	60331	60731	60231	60431	
220032	XCT	BEA	CZ	60332	60732	60232	60432	
220033	XCT	BEA	CZ	60333	60733	60233	60433	
220034	XCT	BEA	CZ	60334	60734	60234	60434	

Class 221

Former Virgin CrossCountry four- or five-car tilting Super Voyager DEMUs. The tilting function is now isolated.

All are in the EHXC pool.

	Livery	Owner	Depot	DMS	MS	MS	MS	DMF
221119	XCT	BEA	CZ	60369	60769	60969	60869	60469
221120	XCT	BEA	CZ	60370	60770	60970	60870	60470
221121	XCT	BEA	CZ	60371	60771	60971	60871	60471
221122	XCT	BEA	CZ	60372	60772	60972	60872	60472
221123	XCT	BEA	CZ	60373	60773	60973	60873	60473
221124	XCT	BEA	CZ	60374	60774	60974	60874	60474
221125	XCT	BEA	CZ	60375	60775	60975	60875	60475
221126	XCT	BEA	CZ	60376	60776	60976	60876	60476
221127	XCT	BEA	CZ	60377	60777	60977	60877	60477
221128	XCT	BEA	CZ	60378	60778	60978	60878	60478
221129	XCT	BEA	CZ	60379	60779	60979	60879	60479
221130	XCT	BEA	CZ	60380	60780	60980	60880	60480
221131	XCT	BEA	CZ	60381	60781	60981	60881	60481
221132	XCT	BEA	CZ	60382	60782	60982	60882	60482
221133	XCT	BEA	CZ	60383	60783	60983	60883	60483
221134	XCT	BEA	CZ	60384	60784	60984	60884	60484
221135	XCT	BEA	CZ	60385	60785	60985	60885	60485
221136	XCT	BEA	CZ	60386	60786		60886	60486
221137	XCT	BEA	CZ	60387	60787	60987	60887	60487
221138	XCT	BEA	CZ	60388	60788	60988	60888	60488
221139	XCT	BEA	CZ	60389	60789	60989	60889	60489
221140	XCT	BEA	CZ	60390	60790		60890	60490
221141	XCT	BEA	CZ	60391	60791	60991		60491
221144	XCT	BEA	CZ	60394	60794	60990		60494

Crossrail (Elizabeth Line)

Contact details

Website: crossrail.co.uk
Twitter: @Crossrail

Key personnel

Chief Executive: Nigel Holness
Engineering Director: Kevin Jones

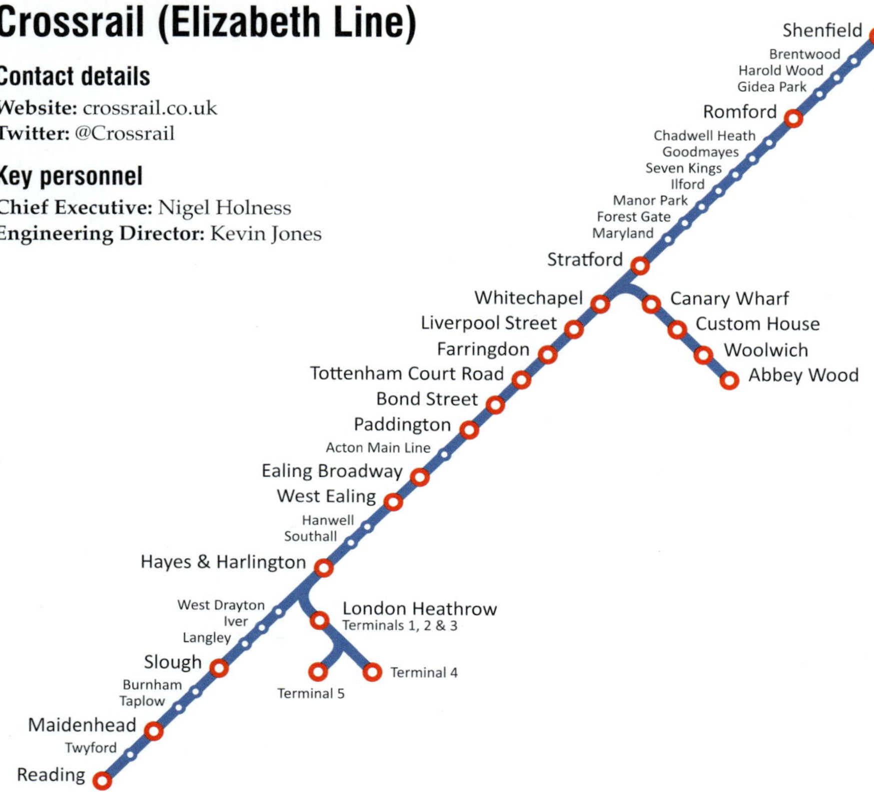

Overview

At the time of going to press, TfL Rail just operates between Hayes & Harlington-Paddington (since May 2018) and Liverpool St-Shenfield (since June 2017) using some of its brand new fleet of 70 nine-car Class 345 EMUs. However the main section of Crossrail – which will be marketed as the Elizabeth line and passed to Transport for London operation, under the city of London – is expected to open in the spring of 2022, several years later than initially planned.

All the Bombardier Aventra units have now been delivered but with the main part of the route still not open, many were in store at Old Oak Common. At the time of going to press, a handful of elderly Class 315s were still being used by TfL Rail for local services but these will be replaced by 345s in 2022.

When fully open, Crossrail will also serve a branch in the east via Canary Wharf and under the Thames to north-east London at Abbey Wood that uses much of the old National Rail line trackbed of the old North Woolwich branch.

Opposite: On 12 January 2019, 345012 waits to head west at Paddington. These Class 345s were delivered as seven-car sets but soon increased to their planned nine cars. *Rob France*

Class 315

Ex-BR four-car 25kV AC units dating from 1980-81, these units are being withdrawn as more Class 345s enter traffic. By early 2022 the remaining eight sets were operated in four pairs with just a single turn and this was expected to end imminently with the last units to be withdrawn and sent for scrap.

All are in the EXHQ pool.

	Livery	Owner	Depot	DMS	TS	PTS	DMS
315837	TFL	EVS	IL	64533	71317	71425	64534
315838	TFL	EVS	IL	64535	71318	71426	64536
315839	TFL	EVS	IL	64537	71319	71427	64538
315847	TFL	EVS	IL	64553	71327	71435	64554
315848	TFL	EVS	IL	64540	71328	71436	64556
315853	TFL	EVS	IL	64565	71333	71441	64566
315856	TFL	EVS	IL	64571	71336	71444	64572
315857	TFL	EVS	IL	64573	71337	71445	64574

Class 345

Brand new 25kV AC Bombardier Aventra units, some of which were delivered as seven-car sets. These nine-car sets will be the only trains used on the Elizabeth line when it is fully opened. There is an option for 14 extra units if required. They are owned by Rail for London.

All are in the EXHQ pool.

	Livery	Owner	Depot	DMS	PMS	MS	MS	TS	MS	MS	PMS	DMS
345001	XRL	RFL	OC	340101	340201	340301	340401	340501	340601	340701	340801	340901
345002	XRL	RFL	OC	340102	340202	340302	340402	340502	340602	340702	340802	340902
345003	XRL	RFL	OC	340103	340203	340303	340403	340503	340603	340703	340803	340903
345004	XRL	RFL	OC	340104	340204	340304	340404	340504	340604	340704	340804	340904
345005	XRL	RFL	OC	340105	340205	340305	340405	340505	340605	340705	340805	340905
345006	XRL	RFL	OC	340106	340206	340306	340406	340506	340606	340706	340806	340906
345007	XRL	RFL	OC	340107	340207	340307	340407	340507	340607	340707	340807	340907
345008	XRL	RFL	OC	340108	340208	340308	340408	340508	340608	340708	340808	340908
345009	XRL	RFL	OC	340109	340209	340309	340409	340509	340609	340709	340809	340909
345010	XRL	RFL	OC	340110	340210	340310	340410	340510	340610	340710	340810	340910
345011	XRL	RFL	OC	340111	340211	340311	340411	340511	340611	340711	340811	340911
345012	XRL	RFL	OC	340112	340212	340312	340412	340512	340612	340712	340812	340912

345013	XRL	RFL	OC	340113	340213	340313	340413	340513	340613	340713	340813	340913
345014	XRL	RFL	OC	340114	340214	340314	340414	340514	340614	340714	340814	340914
345015	XRL	RFL	OC	340115	340215	340315	340415	340515	340615	340715	340815	340915
345016	XRL	RFL	OC	340116	340216	340316	340416	340516	340616	340716	340816	340916
345017	XRL	RFL	OC	340117	340217	340317	340417	340517	340617	340717	340817	340917
345018	XRL	RFL	OC	340118	340218	340318	340418	340518	340618	340718	340818	340918
345019	XRL	RFL	OC	340119	340219	340319	340419	340519	340619	340719	340819	340919
345020	XRL	RFL	OC	340120	340220	340320	340420	340520	340620	340720	340820	340920
345021	XRL	RFL	OC	340121	340221	340321	340421	340521	340621	340721	340821	340921
345022	XRL	RFL	OC	340122	340222	340322	340422	340522	340622	340722	340822	340922
345023	XRL	RFL	OC	340123	340223	340323	340423	340523	340623	340723	340823	340923
345024	XRL	RFL	OC	340124	340224	340324	340424	340524	340624	340724	340824	340924
345025	XRL	RFL	OC	340125	340225	340325	340425	340525	340625	340725	340825	340925
345026	XRL	RFL	OC	340126	340226	340326	340426	340526	340626	340726	340826	340926
345027	XRL	RFL	OC	340127	340227	340327	340427	340527	340627	340727	340827	340927
345028	XRL	RFL	OC	340128	340228	340328	340428	340528	340628	340728	340828	340928
345029	XRL	RFL	OC	340129	340229	340329	340429	340529	340629	340729	340829	340929
345030	XRL	RFL	OC	340130	340230	340330	340430	340530	340630	340730	340830	340930
345031	XRL	RFL	OC	340131	340231	340331	340431	340531	340631	340731	340831	340931
345032	XRL	RFL	OC	340132	340232	340332	340432	340532	340632	340732	340832	340932
345033	XRL	RFL	OC	340133	340233	340333	340433	340533	340633	340733	340833	340933
345034	XRL	RFL	OC	340134	340234	340334	340434	340534	340634	340734	340834	340934
345035	XRL	RFL	OC	340135	340235	340335	340435	340535	340635	340735	340835	340935
345036	XRL	RFL	OC	340136	340236	340336	340436	340536	340636	340736	340836	340936
345037	XRL	RFL	OC	340137	340237	340337	340437	340537	340637	340737	340837	340937
345038	XRL	RFL	OC	340138	340238	340338	340438	340538	340638	340738	340838	340938
345039	XRL	RFL	OC	340139	340239	340339	340439	340539	340639	340739	340839	340939
345040	XRL	RFL	OC	340140	340240	340340	340440	340540	340640	340740	340840	340940
345041	XRL	RFL	OC	340141	340241	340341	340441	340541	340641	340741	340841	340941
345042	XRL	RFL	OC	340142	340242	340342	340442	340542	340642	340742	340842	340942
345043	XRL	RFL	OC	340143	340243	340343	340443	340543	340643	340743	340843	340943
345044	XRL	RFL	OC	340144	340244	340344	340444	340544	340644	340744	340844	340944
345045	XRL	RFL	OC	340145	340245	340345	340445	340545	340645	340745	340845	340945
345046	XRL	RFL	OC	340146	340246	340346	340446	340546	340646	340746	340846	340946
345047	XRL	RFL	OC	340147	340247	340347	340447	340547	340647	340747	340847	340947
345048	XRL	RFL	OC	340148	340248	340348	340448	340548	340648	340748	340848	340948
345049	XRL	RFL	OC	340149	340249	340349	340449	340549	340649	340749	340849	340949
345050	XRL	RFL	OC	340150	340250	340350	340450	340550	340650	340750	340850	340950
345051	XRL	RFL	OC	340151	340251	340351	340451	340551	340651	340751	340851	340951
345052	XRL	RFL	OC	340152	340252	340352	340452	340552	340652	340752	340852	340952
345053	XRL	RFL	OC	340153	340253	340353	340453	340553	340653	340753	340853	340953
345054	XRL	RFL	OC	340154	340254	340354	340454	340554	340654	340754	340854	340954
345055	XRL	RFL	OC	340155	340255	340355	340455	340555	340655	340755	340855	340955
345056	XRL	RFL	OC	340156	340256	340356	340456	340556	340656	340756	340856	340956
345057	XRL	RFL	OC	340157	340257	340357	340457	340557	340657	340757	340857	340957
345058	XRL	RFL	OC	340158	340258	340358	340458	340558	340658	340758	340858	340958
345059	XRL	RFL	OC	340159	340259	340359	340459	340559	340659	340759	340859	340959
345060	XRL	RFL	OC	340160	340260	340360	340460	340560	340660	340760	340860	340960
345061	XRL	RFL	OC	340161	340261	340361	340461	340561	340661	340761	340861	340961
345062	XRL	RFL	OC	340162	340262	340362	340462	340562	340662	340762	340862	340962
345063	XRL	RFL	OC	340163	340263	340363	340463	340563	340663	340763	340863	340963
345064	XRL	RFL	OC	340164	340264	340364	340464	340564	340664	340764	340864	340964
345065	XRL	RFL	OC	340165	340265	340365	340465	340565	340665	340765	340865	340965
345066	XRL	RFL	OC	340166	340266	340366	340466	340566	340666	340766	340866	340966
345067	XRL	RFL	OC	340167	340267	340367	340467	340567	340667	340767	340867	340967
345068	XRL	RFL	OC	340168	340268	340368	340468	340568	340668	340768	340868	340968
345069	XRL	RFL	OC	340169	340269	340369	340469	340569	340669	340769	340869	340969
345070	XRL	RFL	OC	340170	340270	340370	340470	340570	340670	340770	340870	340970

East Midlands Railway

Contact details

Website: eastmidlandsrailways.co.uk
Twitter: @EastMidRailway

Key personnel

Managing Director: Will Rogers
Fleet Director: Neil Bamford

Still in the old Stagecoach-inspired East Midland trains livery, 222003 works the 1229 Sheffield-St Pancras past Sawley on 17 December 2020. These units are all now reliveried into EMR's aubergine colours. *Tom McAtee*

Overview

The routes operated by EMR are the inter-city services from St Pancras to Nottingham, Derby and Sheffield; semi-fast trains from St Pancras to Corby, long-distance regional trains from Liverpool to Norwich and local regional trains from Peterborough to Lincoln and Doncaster, Derby to Skegness, Derby or Matlock, Derby to Crewe; Nottingham to Worksop and Nottingham to Grimsby.

For so long a diesel-only fleet, East Midlands Railway has now started using cascaded Class 360 Desiro EMUs on the St Pancras-Corby service following the extension of the electrification of the Midland Main Line north of Bedford.

Since the last Rail Guide, all Class 43 HST Power cars and Mk 3 trailer vehicles have been withdrawn from the route and replaced by multiple units.

On the DMU fleet, the single-car Class 153 units have all now been retired and some of its Class 156s are to move to Northern following the arrival of Class 170s from ScotRail, TfW and West Midlands Trains. Some of the latter are still in use with WMT, albeit painted in EMR livery prior to their transfer. Now Class 196s have been approved for use by WMR, their transfer is expected soon to allow the last 156s to head north.

Four ex-Hull Trains Class 180s were added to the fleet, three running as five-car sets, one temporarily reduced to a four-car set. They have bolstered Midland Main Line operations and allowed the HSTs to be withdrawn.

The fleet is being reliveried into a new aubergine livery, although some 222s are just having these new colours on their cab ends. White and purple is the new livery for 156/158s.

Looking ahead, 33 five-car Class 810 Hitachi AT300 bi-mode Aurora multiple units are on order to replace Class 180s and 222s.

Class 08

Five ex-BR Class 08 shunters, four at Neville Hill and one at Derby Etches Park, were inherited by EMR, although the Leeds-based locos do not see any use presently and have been replaced by hired-in locos.

All are in the EMSL pool.

08525	EMT	EMB	NL (S)	*Duncan Bedford*
08690	EMT	EMB	NL (S)	*David Thirkill*
08899	EMT	MID	DY	*Midland Counties Railway 175 Years 1839-2014*
08908	EMT	EMB	NL (S)	*Ivan Stephenson*
08950	EMT	EMB	NL (S)	*David Lightfoot*

On 4 August 2021, 156912 leaves the rural station at Ancaster with the 1114 Skegness-Nottingham. *Pip Dunn*

Class 156

The two-car Class 156 fleet has been strengthened with the addition of nine units cascaded from Greater Anglia. These are renumbered as 156/9s due to modified toilets. All 156/9 units are to revert to the previous 156/4 numbers and are due to move to Northern in 2022 along with 156403/404/413/414.

All are in the EMHQ pool.

		Livery	Owner	Depot	DMSL	DMS
156403		EMT	POR	DY	52403	57403
156404		EMT	POR	DY	52404	57404
156405		EMR	POR	DY	52405	57405
156406		EMR	POR	DY	52406	57406
156408		EMT	POR	DY	52408	57408
156410		EMT	POR	DY	52410	57410
156411		EMT	POR	DY	52411	57411
156413		EMB	POR	DY	52413	57413
156414		EMB	POR	DY	52414	57414
156422	156922	NOR	POR	DY	52422	57422
156470		EMT	POR	DY	52470	57470
156473		EMT	POR	DY	52473	57473
156497		EMT	POR	DY	52497	57497
156498		EMT	POR	DY	52498	57498
156907	156407	EMR	POR	DY	52407	57407
156909	156409	EMR	POR	DY	52409	57409
156912	156412	EMR	POR	DY	52412	57412
156916	156416	EMR	POR	DY	52416	57416
156917	156417	EMR	POR	DY	52417	57417
156918	156418	EMR	POR	DY	52418	57418
156919	156419	EMR	POR	DY	52419	57419

Class 158

Currently used for Liverpool to Norwich services, plus other EMR secondary workings as required such as Doncaster to Peterborough.

All are in the EMHQ pool.

	Livery	Owner	Depot	DMSL	DMSL
158770	EMW	POR	NM	52770	57770
158773	EMR	POR	NM	52773	57773
158774	EMR	POR	NM	52774	57774
158777	EMW	POR	NM	52777	57777
158780	EMW	ANG	NM	52783	57783
158783	EMW	ANG	NM	52783	57783
158785	EMW	ANG	NM	52785	57785
158788	EMW	ANG	NM	52788	57788

The EMR Class 158 fleet is also being reliveried into a white with purple ends. 158774 leads a sister unit still in the old colours on the 0852 Liverpool Lime Street-Norwich at Manchester Oxford Road on 24 March 2021. *Tom McAtee*

158799	EMW	POR	NM	52799	57799	
158806	EMW	POR	NM	52806	57806	
158810	EMW	POR	NM	52810	57810	
158812	EMW	POR	NM	52812	57812	
158813	EMW	POR	NM	52813	57813	
158846	EMW	ANG	NM	52846	57846	
158847	EMW	ANG	NM	52847	57847	*Lincoln Castle Explorer*
158852	EMW	ANG	NM	52852	57852	
158854	EMW	ANG	NM	52854	57854	*The Station Volunteer*
158856	EMW	ANG	NM	52856	57856	
158857	EMW	ANG	NM	52857	57857	
158858	EMW	ANG	NM	52858	57858	
158862	EMW	ANG	NM	52862	57862	
158863	EMW	ANG	NM	52863	57863	
158864	EMW	ANG	NM	52864	57864	
158865	EMW	ANG	NM	52865	57865	
158866	EMW	ANG	NM	52866	57866	
158889	EMW	POR	NM	52808	57808	

Class 170

Several Turbostar units have moved or are due to move from other TOCs of late and several WMR Class 170s are still to join the fleet when made available, having been displaced by new CAF Class 196 DMUs. The 170/6s were three-car units but the centre cars have been removed and redeployed with CrossCountry Trains.

All are in the EDHQ pool.

Class 170/2

Former Greater Anglia two-car unit.

Unit	Livery	Owner	Depot	DMCL	DMCL
170273	EMP	POR	DY	50273	79273

Several Class 170s have joined the EMR fleet, mostly from ScotRail and West Midlands Trains. However, two-car 170273 has come from Greater Anglia, via a spell in use with Transport for Wales. It waits to leave with the 0724 Derby-Matlock on 20 November 2021. *Pip Dunn*

Class 170/4

Former ScotRail three-car units.

Unit	Livery	Owner	Depot	DMCL	MS	DMCL	
170416	EMP	EVS	DY	50416	56416	79416	
170417	EMP	EVS	DY	50417	56417	79417	*The Key Worker*
170418	EMP	EVS	DY	50418	56418	79418	
170419	EMP	EVS	DY	50419	56419	79419	
170420	EMP	EVS	DY	50420	56420	79420	

Class 170/5

These two-car ex-Central Trains units are now 20 years old. They will be joined by more Class 170/5s from WMR.

	Old No.	Livery	Owner	Depot	DMSL	DMSL
170503		EMP	POR	DY	50503	79503
170511		EMP	POR	DY	50511	79511
170515		EMP	POR	DY	50515	79515
170517		EMP	POR	DY	50517	79517
170530	170630	EMP	POR	DY	50630	79630
170531	170631	EMP	POR	DY	50631	79631
170532	170632	EMP	POR	DY	50632	79632
170534	170634	EMP	POR	DY	50634	79634

Class 180

These four Adelante units were added to the EMR fleet in 2020 and work on Nottingham to St Pancras semi-fast duties. Currently 180110 is running as a four-car set after its buffet car 56910 was withdrawn due to corrosion.

All are in the EDHQ pool.

	Livery	Owner	Depot	DMSL	MFL	MSL	MSLRB	DMSL
180109	EMR	ANG	DY	50909	54909	55909	56909	59909
180110	EMR	ANG	DY	50910	54910	55910		59910
180111	EMR	ANG	DY	50911	54911	55911	56911	59911
180113	EMR	ANG	DY	50913	54913	55913	56913	59913
Spare							56910	

Class 222

Based on the Bombardier Class 220 Voyager units, the Meridians were ordered by Midland Mainline and delivered in 2003-05. Initially there were seven nine-cars and 16 four-car sets but all have been re-formed now so the fleet is a mix of five- and seven-cars. EMR's predecessor East Midlands Trains also acquired the four four-car Class 222/1s from Hull Trains.

All are in the EMHQ pool.

Class 222/0

	Livery	Owner	Depot	DMF	MF/MC	MF	MSRMB	MS	MS	DMS	
222001	EMR	EVS	DY	60241	60445	60341	60621	60561	60551	60161	*The Entrepreneur Express*
222002	EMR	EVS	DY	60242	60346	60342	60622	60562	60544	60162	*The Cutlers' Company*
222003	EMR	EVS	DY	60243	60446	60343	60623	60563	60553	60163	
222004	EMR	EVS	DY	60244	60345	60344	60624	60564	60554	60164	*Children's Hospital Sheffield*
222005	EMR	EVS	DY	60245	60347	60443	60625	60555	60565	60165	
222006	EMR	EVS	DY	60246	60447	60441	60626	60566	60556	60166	*The Carbon Cutter*
222007	EMR	EVS	DY	60247	60442		60627		60567	60167	

222008	EMR	EVS	DY	60248	60918	60628	60545	60168	*Derby Etches Park*
222009	EMR	EVS	DY	60249	60919	60629	60557	60169	
222010	EMR	EVS	DY	60250	60920	60630	60546	60170	
222011	EMR	EVS	DY	60251	60921	60631	60531	60171	*Sheffield City Battalion 1914-1918*
222012	EMR	EVS	DY	60252	60922	60632	60532	60172	
222013	EMR	EVS	DY	60253	60923	60633	60533	60173	
222014	EMR	EVS	DY	60254	60924	60634	60534	60174	
222015	EMR	EVS	DY	60255	60925	60635	60535	60175	*175 years of Derby's Railways 1839-2014*
222016	EMR	EVS	DY	60256	60926	60636	60536	60176	
222017	EMR	EVS	DY	60257	60927	60637	60537	60177	
222018	EMR	EVS	DY	60258	60928	60638	60444	60178	
222019	EMR	EVS	DY	60259	60929	60639	60547	60179	
222020	EMR	EVS	DY	60260	60930	60640	60543	60180	
222021	EMR	EVS	DY	60261	60931	60641	60552	60181	
222022	EMR	EVS	DY	60262	60932	60642	60542	60182	*Invest in Nottingham*
222023	EMR	EVS	DY	60263	60933	60643	60541	60183	

Class 222/1

Former Hull Trains four-car sets.

	Livery	Owner	Depot	DMF	MC	MSRMB	DMS
222101	EMR	EVS	DY	60271	60571	60681	60191
222102	EMR	EVS	DY	60272	60572	60682	60192
222103	EMR	EVS	DY	60273	60573	60683	60193
222104	EMP	EVS	DY	60274	60574	60684	60194

Class 360

This fleet of 21 four-car Siemens Desiro units was ordered by First Great Eastern and delivered in 2002-03. They moved to East Midlands Railway in 2020-21 to operate its new St Pancras-Corby service after modifications to allow them to run at 110mph and are being refurbished and reliveried.

All are in the EBHQ pool.

	Livery	Owner	Depot	DMC	PTS	TS	DMC
360101	EMP	ANG	BF	65551	72551	74551	68551
360102	EMP	ANG	BF	65552	72552	74552	68552
360103	EMP	ANG	BF	65553	72553	74553	68553
360104	EMP	ANG	BF	65554	72554	74554	68554
360105	FIR	ANG	BF	65555	72555	74555	68555
360106	FIR	ANG	BF	65556	72556	74556	68556
360107	EMP	ANG	BF	65557	72557	74557	68557
360108	FIR	ANG	BF	65558	72558	74558	68558
360109	EMP	ANG	BF	65559	72559	74559	68559
360110	FIR	ANG	BF	65560	72560	74560	68560
360111	FIR	ANG	BF	65561	72561	74561	68561
360112	EMP	ANG	BF	65562	72562	74562	68562

Following extension of the MML electrification from Bedford to Corby, EMR has taken the 21 four-car Siemens Desiro EMUs from Greater Anglia to work the new service. On 28 October 2021, 360121 is towed through Basingstoke by 47749 on an Eastleigh Works to Cricklewood move. *Mark Pike*

360113	FIR	ANG	BF	65563	72563	74563	68563
360114	EMP	ANG	BF	65564	72564	74564	68564
360115	FIR	ANG	BF	65565	72565	74565	68565
360116	FIR	ANG	BF	65566	72566	74566	68566
360117	FIR	ANG	BF	65567	72567	74567	68567
360118	FIR	ANG	BF	65568	72568	74568	68568
360119	FIR	ANG	BF	65569	72569	74569	68569
360120	FIR	ANG	BF	65570	72570	74570	68570
360121	EMP	ANG	BF	65571	72571	74571	68571

Class 810

New bi-mode Hitachi units on order to replace Class 180 and 222 units.

	Livery	Owner	Depot	PDTRBF	MS	TS	MS	DPTS
810001	EMP	ROC		851001	852001	853001	854001	855001
810002	EMP	ROC		851002	852002	853002	854002	855002
810003	EMP	ROC		851003	852003	853003	854003	855003
810004	EMP	ROC		851004	852004	853004	854004	855004
810005	EMP	ROC		851005	852005	853005	854005	855005
810006	EMP	ROC		851006	852006	853006	854006	855006
810007	EMP	ROC		851007	852007	853007	854007	855007
810008	EMP	ROC		851008	852008	853008	854008	855008
810009	EMP	ROC		851009	852009	853009	854009	855009
810010	EMP	ROC		851010	852010	853010	854010	855010
810011	EMP	ROC		851011	852011	853011	854011	855011
810012	EMP	ROC		851012	852012	853012	854012	855012
810013	EMP	ROC		851013	852013	853013	854013	855013
810014	EMP	ROC		851014	852014	853014	854014	855014
810015	EMP	ROC		851015	852015	853015	854015	855015
810016	EMP	ROC		851016	852016	853016	854016	855016
810017	EMP	ROC		851017	852017	853017	854017	855017
810018	EMP	ROC		851018	852018	853018	854018	855018
810019	EMP	ROC		851019	852019	853019	854019	855019
810020	EMP	ROC		851020	852020	853020	854020	855020
810021	EMP	ROC		851021	852021	853021	854021	855021
810022	EMP	ROC		851022	852022	853022	854022	855022
810023	EMP	ROC		851023	852023	853023	854023	855023
810024	EMP	ROC		851024	852024	853024	854024	855024
810025	EMP	ROC		851025	852025	853025	854025	855025
810026	EMP	ROC		851026	852026	853026	854026	855026
810027	EMP	ROC		851027	852027	853027	854027	855027
810028	EMP	ROC		851028	852028	853028	854028	855028
810029	EMP	ROC		851029	852029	853029	854029	855029
810030	EMP	ROC		851030	852030	853030	854030	855030
810031	EMP	ROC		851031	852031	853031	854031	855031
810032	EMP	ROC		851032	852032	853032	854032	855032
810033	EMP	ROC		851033	852033	853033	854033	855033

East Midlands Railway has 33 five-car Class 810 Aurora bi-mode Hitachi AT300 trains on order. They have slightly shorter bodies than the other 80x series trains. *EMR*

Govia Thameslink Railway

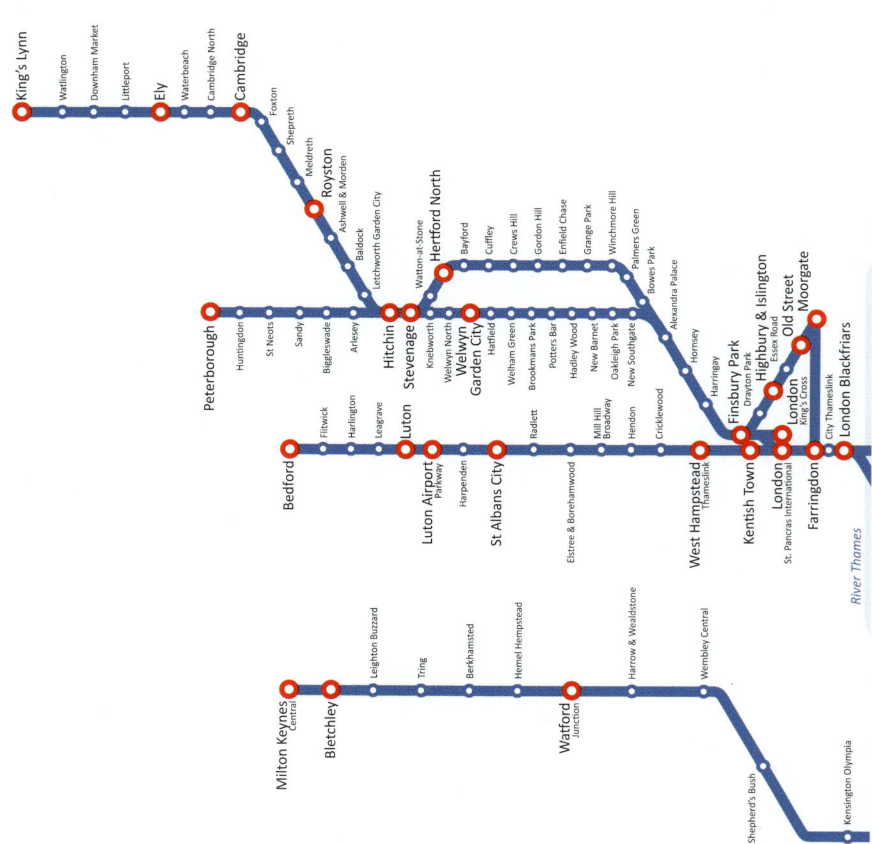

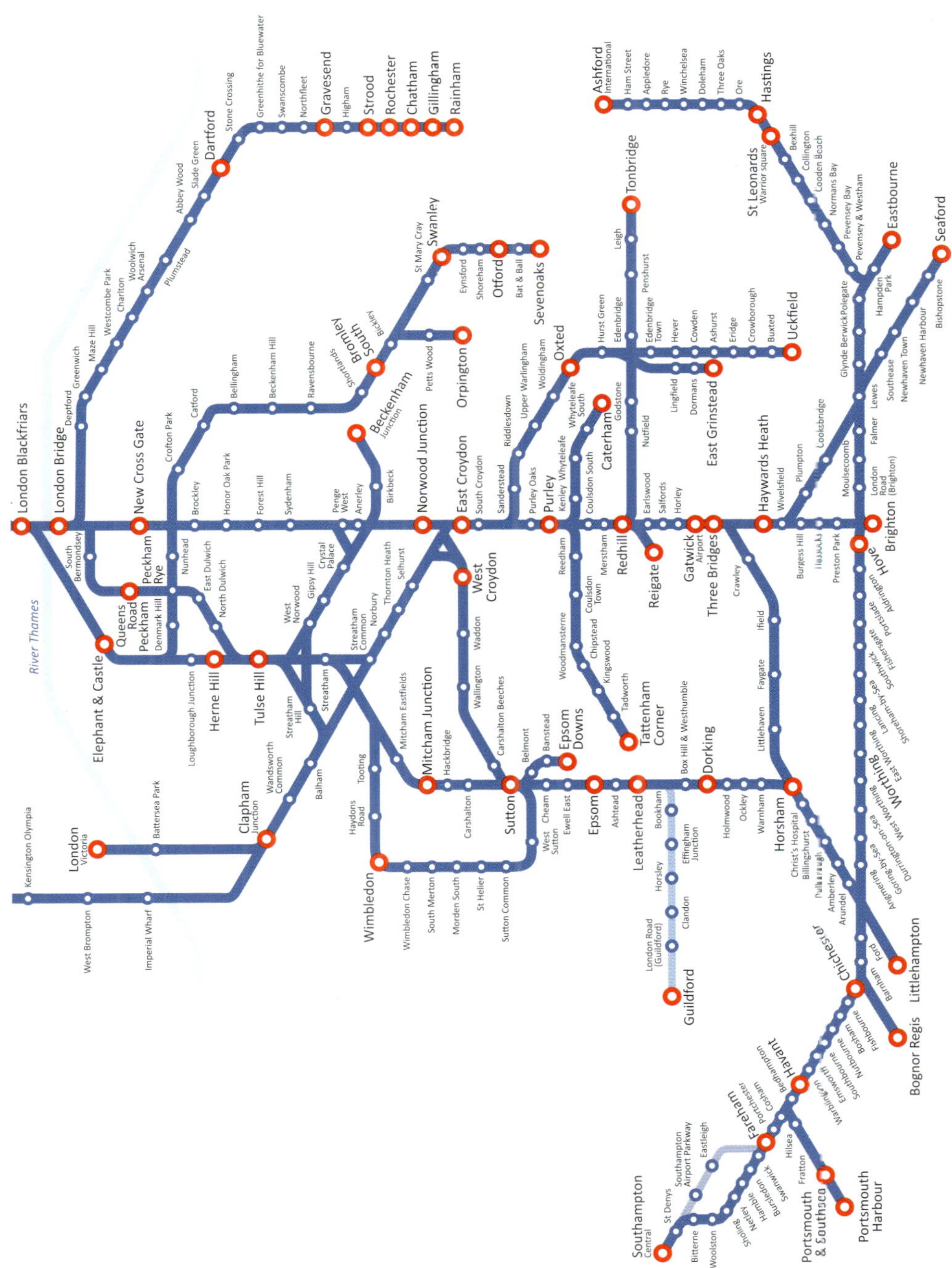
London Blackfriars
London Bridge
New Cross Gate
Deptford
Greenwich
Maze Hill
Westcombe Park
Charlton
Woolwich Arsenal
Plumstead
Abbey Wood
Slade Green
Dartford
Stone Crossing
Greenhithe for Bluewater
Swanscombe
Northfleet
Gravesend
Higham
Strood
Rochester
Chatham
Gillingham
Rainham
Crofton Park
Catford
Bellingham
Beckenham Hill
Ravensbourne
Shortlands
Bromley South
Bickley
St Mary Cray
Swanley
Eynsford
Shoreham
Otford
Bat & Ball
Sevenoaks
Petts Wood
Orpington
Beckenham Junction
Birkbeck
Norwood Junction
East Croydon
South Croydon
Sanderstead
Riddlesdown
Upper Warlingham
Woldingham
Oxted
Purley Oaks
Purley
Kenley
Whyteleafe
Whyteleafe South
Caterham
Coulsdon South
Godstone
Hurst Green
Edenbridge
Leigh
Tonbridge
Penshurst
Edenbridge Town
Hever
Cowden
Ashurst
Eridge
Crowborough
Buxted
Uckfield
Lingfield
Dormans
East Grinstead
Nutfield
Redhill
Earlswood
Salfords
Horley
Reigate
Gatwick Airport
Three Bridges
Crawley
Ifield
Faygate
Littlehaven
Horsham
Haywards Heath
Wivelsfield
Burgess Hill
Hassocks
Preston Park
Plumpton
Cooksbridge
Moulsecoomb
London Road (Brighton)
Brighton
Hove
Falmer
Lewes
Southease
Newhaven Town
Newhaven Harbour
Bishopstone
Seaford
Glynde
Berwick
Polegate
Hampden Park
Eastbourne
Pevensey & Westham
Pevensey Bay
Normans Bay
Cooden Beach
Collington
Bexhill
St Leonards Warrior square
Hastings
Ore
Three Oaks
Doleham
Winchelsea
Rye
Appledore
Ham Street
Ashford International
Brockley
Honor Oak Park
Forest Hill
Sydenham
Penge West
Anerley
South Bermondsey
Queens Road Peckham
Peckham Rye
Nunhead
East Dulwich
North Dulwich
Denmark Hill
Elephant & Castle
Loughborough Junction
Herne Hill
Tulse Hill
West Norwood
Gipsy Hill
Crystal Palace
Streatham Hill
Streatham
Streatham Common
Norbury
Thornton Heath
Selhurst
West Croydon
Waddon
Wallington
Carshalton Beeches
Mitcham Eastfields
Mitcham Junction
Hackbridge
Carshalton
Sutton
Belmont
Banstead
Epsom Downs
Cheam
Ewell East
West Sutton
Epsom
Ashtead
Leatherhead
Bookham
Effingham Junction
Horsley
Clandon
London Road (Guildford)
Guildford
Box Hill & Westhumble
Dorking
Holmwood
Ockley
Warnham
Reedham
Coulsdon Town
Woodmansterne
Chipstead
Kingswood
Tadworth
Tattenham Corner
Merstham
River Thames
Kensington Olympia
West Brompton
Imperial Wharf
London Victoria
Battersea Park
Clapham Junction
Wandsworth Common
Balham
Tooting
Haydons Road
Wimbledon
Wimbledon Chase
South Merton
Morden South
St Helier
Sutton Common
Christ's Hospital
Billingshurst
Pulborough
Amberley
Arundel
Ford
Littlehampton
Barnham
Bognor Regis
Chichester
Fishbourne
Bosham
Nutbourne
Southbourne
Emsworth
Warblington
Havant
Bedhampton
Cosham
Portchester
Fareham
Hilsea
Fratton
Portsmouth & Southsea
Portsmouth Harbour
Swanwick
Bursledon
Hamble
Netley
Sholing
Woolston
Bitterne
St Denys
Southampton Central
Southampton Airport Parkway
Eastleigh
Aldrington
Portslade
Fishersgate
Southwick
Shoreham-by-Sea
Lancing
East Worthing
Worthing
West Worthing
Durrington-on-Sea
Goring-by-Sea
Angmering

Contact details

Websites: www.thameslinkrailway.com; www.southernrailway.com; www.greatnorthernrail.com; www.gatwickexpress.com
Twitter: @GNRailUK, @SouthernRailUK, @TLRailUK, @GatwickExpress

Key personnel

Chief Executive Officer: Patrick Verwer
Chief Operating Officer: Angie Doll

Overview

This 'super franchise' was the amalgamation of the Southern, Great Northern, Gatwick Express and Thameslink operations, which has led to huge TOC although the 'separate' operations do keep their former identities to a degree.

On the Great Northern lines it runs from Peterborough and King's Lynn to King's Cross, and also local commuter trains out of Moorgate to Hitchin. The former Southern operations are Victoria or London Bridge to Portsmouth Harbour and Southampton Central, Littlehampton, Horsham via Epsom or Gatwick Airport, Epsom Downs, Tattenham Corner, Reigate, Brighton, Seaford, Eastbourne, East Grinstead, Uckfield and Caterham, plus Redhill-Tonbridge, Ashford International-Eastbourne, and Milton Keynes-Clapham Junction, the latter route requiring dual-voltage EMUs

The Thameslink network links Bedford, Cambridge and Peterborough in the north to Sutton, Rainham (Kent), Brighton, Orpington, Sevenoaks, East Grinstead, Horsham and Littlehampton.

Finally, the franchise incorporates the former InterCity Gatwick Express operation, but instead of just being a Victoria to Gatwick shuttle, this is now a Victoria to Brighton service and uses Class 387/2s rather than dedicated trains.

It has a fleet that mostly comprises of EMUs – with Class 700s on the Thameslink and Great Northern trains and Class 377s on the former Southern duties, while diesel units – Class 171 Turbostars – are used on the non-electrified lines from east Croydon to Uckfield and Ashford to Hastings.

Other fleets are Class 313/2s, which work local services out of Brighton, while Class 387/1s are used on the King's Cross-Peterborough/King's Lynn services. Class 717s – a metro version of the Class 700s – work out of Moorgate. Some 455s are also retained for local trains in London but will be withdrawn from May.

Class 73

Southern has one ex-Gatwick Express Class 73/2 loco, which sees occasional use as a Thunderbird or route-learning loco. It is in the MBED pool.

73202	SOU	POR	SL	*Graham Stenning*

Class 171

These Bombardier Turbostars, which differ from Class 170s by having a different coupling system (Dellner). The fleet has been bolstered by the recent acquisition of four units from ScotRail. These units work from London to Uckfield and also from Ashford to Hastings.

All are in the ETHQ pool.

Class 171/2

Ex-ScotRail units re-formed from three- to two-car sets.

	Old No	Livery	Owner	Depot	DMCL	DMSL
171201	170421	SOU	EVS	SU	50421	79421
171202	170423	SOU	EVS	SU	50423	79423

Govia uses Class 171 Turbostar DMUs for its Hastings to Ashford and Uckfield-East Croydon trains. On 15 July 2021, 171401 works the 0940 Eastbourne-Ashford and calls at Ham Street. *David Staines*

Class 171/4

Ex-ScotRail units re-formed from three- to four-car sets.

	Old No	Livery	Owner	Depot	DMCL	MS	MS	DMCL
171401	170422	SOU	EVS	SU	50422	56421	56422	79422
171402	170424	SOU	EVS	SU	50424	56423	56424	79424

Class 171/7

Two-car sets. Six sets were delivered as 170/7s and renumbered to 171/7s when fitted with Dellner couplers, while 171730 was formerly a SWT unit.

	Old No	Livery	Owner	Depot	DMCL	DMSL
171721	170721	SOU	POR	SU	50721	79721
171722	170722	SOU	POR	SU	50722	79722
171723	170723	SOU	POR	SU	50723	79723
171724	170724	SOU	POR	SU	50724	79724
171725	170725	SOU	POR	SU	50725	79725
171726	170726	SOU	POR	SU	50726	79726
171727		SOU	POR	SU	50727	79727
171728		SOU	POR	SU	50728	79728
171729		SOU	POR	SU	50729	79729
171730	170392	SOU	POR	SU	50392	79392

Class 171/8

Four-car sets.

	Livery	Owner	Depot	DMCL	MS	MS	DMCL
171801	SOU	POR	SU	50801	54801	56801	79801
171802	SOU	POR	SU	50802	54802	56802	79802
171803	SOU	POR	SU	50803	54803	56803	79803
171804	SOU	POR	SU	50804	54804	56804	79804
171805	SOU	POR	SU	50805	54805	56805	79805
171806	SOU	POR	SU	50806	54806	56806	79806

Southern still uses a fleet of 19 three-car Class 313 EMUs, which date from the mid-1970s, for some local turns in the Brighton area and Coastway line. On 24 March 2018, retro BR-repainted 313201 stands at Bognor Regis. *Pip Dunn*

Class 313

These ex-BR three-car 750V DC units are used mostly on local trains emanating from Brighton along the Coastway route.

All are in the ETHQ pool.

	Old no.	Livery	Owner	Depot	DMS	TS	BDMS
313201	313101	BRG	BEA	BI	62529	71213	62593
313202	313102	SOU	BEA	BI	62530	71214	62594
313203	313103	SOU	BEA	BI	62531	71215	62595
313204	313104	SOU	BEA	BI	62532	71216	62596
313205	313105	SOU	BEA	BI	62533	71217	62597
313206	313106	SOU	BEA	BI	62534	71218	62598
313207	313107	SOU	BEA	BI	62535	71219	62599
313208	313108	SOU	BEA	BI	62536	71220	62600
313209	313109	SOU	BEA	BI	62537	71221	62601
313210	313110	SOU	BEA	BI	62538	71222	62602
313211	313111	SOU	BEA	BI	62539	71223	62603
313212	313112	SOU	BEA	BI	62540	71224	62604
313213	313113	SOU	BEA	BI	62541	71225	62605
313214	313114	SOU	BEA	BI	62542	71226	62606
313215	313115	SOU	BEA	BI	62543	71227	62607
313216	313116	SOU	BEA	BI	62544	71228	62608
313217	313117	SOU	BEA	BI	62545	71229	62609
313219	313119	SOU	BEA	BI	62547	71231	62611
313220	313120	SOU	BEA	BI	62548	71232	62612

Class 377

Bombardier Electrostar units work on the former Southern routes on commuter trains across the south of England. They are a mix of three-, four- and five-car units and the Class 377/2s and 377/7s have dual-voltage capability for working to Milton Keynes. All are in the ETHQ pool.

Class 377/1

750V DC units.

	Livery	Owner	Depot	DMOS	MOS	PTOSL	DMOS
377101	SOU	POR	SU	78501	77101	78901	78701
377102	SOU	POR	SU	78502	77102	78902	78702
377103	SOU	POR	SU	78503	77103	78903	78703
377104	SOU	POR	SU	78504	77104	78904	78704

377105	SOU	POR	SU	78505	77105	78905	78705
377106	SOU	POR	SU	78506	77106	78906	78706
377107	SOU	POR	SU	78507	77107	78907	78707
377108	SOU	POR	SU	78508	77108	78908	78708
377109	SOU	POR	SU	78509	77109	78909	78709
377110	SOU	POR	SU	78510	77110	78910	78710
377111	SOU	POR	SU	78511	77111	78911	78711
377112	SOU	POR	SU	78512	77112	78912	78712
377113	SOU	POR	SU	78513	77113	78913	78713
377114	SOU	POR	SU	78514	77114	78914	78714
377115	SOU	POR	SU	78515	77115	78915	78715
377116	SOU	POR	SU	78516	77116	78916	78716
377117	SOU	POR	SU	78517	77117	78917	78717
377118	SOU	POR	SU	78518	77118	78918	78718
377119	SOU	POR	SU	78519	77119	78919	78719
377120	SOU	POR	SU	78520	77120	78920	78720
377121	SOU	POR	SU	78521	77121	78921	78721
377122	SOU	POR	SU	78522	77122	78922	78722
377123	SOU	POR	SU	78523	77123	78923	78723
377124	SOU	POR	SU	78524	77124	78924	78724
377125	SOU	POR	SU	78525	77125	78925	78725
377126	SOU	POR	SU	78526	77126	78926	78726
377127	SOU	POR	SU	78527	77127	78927	78727
377128	SOU	POR	SU	78528	77128	78928	78728
377129	SOU	POR	SU	78529	77129	78929	78729
377130	SOU	POR	SU	78530	77130	78930	78730
377131	SOU	POR	SU	78531	77131	78931	78731
377132	SOU	POR	SU	78532	77132	78932	78732
377133	SOU	POR	SU	78533	77133	78933	78733
377134	SOU	POR	SU	78534	77134	78934	78734
377135	SOU	POR	SU	78535	77135	78935	78735
377136	SOU	POR	SU	78536	77136	78936	78736
377137	SOU	POR	SU	78537	77137	78937	78737
377138	SOU	POR	SU	78538	77138	78938	78738
377139	SOU	POR	SU	78539	77139	78939	78739
377140	SOU	POR	SU	78540	77140	78940	78740
377141	SOU	POR	SU	78541	77141	78941	78741
377142	SOU	POR	SU	78542	77142	78942	78742
377143	SOU	POR	SU	78543	77143	78943	78743
377144	SOU	POR	SU	78544	77144	78944	78744
377145	SOU	POR	SU	78545	77145	78945	78745
377146	SOU	POR	SU	78546	77146	78946	78746
377147	SOU	POR	SU	78547	77147	78947	78747
377148	SOU	POR	SU	78548	77148	78948	78748
377149	SOU	POR	SU	78549	77149	78949	78749
377150	SOU	POR	SU	78550	77150	78950	78750
377151	SOU	POR	SU	78551	77151	78951	78751
377152	SOU	POR	SU	78552	77152	78952	78752
377153	SOU	POR	SU	78553	77153	78953	78753
377154	SOU	POR	SU	78554	77154	78954	78754
377155	SOU	POR	SU	78555	77155	78955	78755
377156	SOU	POR	SU	78556	77156	78956	78756
377157	SOU	POR	SU	78557	77157	78957	78757
377158	SOU	POR	SU	78558	77158	78958	78758
377159	SOU	POR	SU	78559	77159	78959	78759
377160	SOU	POR	SU	78560	77160	78960	78760
377161	SOU	POR	SU	78561	77161	78961	78761
377162	SOU	POR	SU	78562	77162	78962	78762

Class 377/2

Dual-voltage units.

	Livery	Owner	Depot	DMOS	MOSL	PTOSL	DMOS
377201	SOU	POR	SU	78571	77171	78971	78771
377202	SOU	POR	SU	78572	77172	78972	78772
377203	SOU	POR	SU	78573	77173	78973	78773
377204	SOU	POR	SU	78574	77174	78974	78774
377205	SOU	POR	SU	78575	77175	78975	78775
377206	SOU	POR	SU	78576	77176	78976	78776
377207	SOU	POR	SU	78577	77177	78977	78777
377208	SOU	POR	SU	78578	77178	78978	78778
377209	SOU	POR	SU	78579	77179	78979	78779
377210	SOU	POR	SU	78580	77180	78980	78780
377211	SOU	POR	SU	78581	77181	78981	78781
377212	SOU	POR	SU	78582	77182	78982	78782
377213	SOU	POR	SU	78583	77183	78983	78783
377214	SOU	POR	SU	78584	77184	78984	78784
377215	SOU	POR	SU	78585	77185	78985	78785

Class 377/3

750V DC units.

	Old No.	Livery	Owner	Depot	DMOS	PTOSL	DMOC
377301	375311	SOU	POR	SU	68201	74801	68401
377302	375312	SOU	POR	SU	68202	74802	68402
377303	375313	SOU	POR	SU	68203	74803	68403
377304	375314	SOU	POR	SU	68204	74804	68404
377305	375315	SOU	POR	SU	68205	74805	68405
377306	375316	SOU	POR	SU	68206	74806	68406
377307	375317	SOU	POR	SU	68207	74807	68407
377308	375318	SOU	POR	SU	68208	74808	68408
377309	375319	SOU	POR	SU	68209	74809	68409
377310	375320	SOU	POR	SU	68210	74810	68410
377311	375321	SOU	POR	SU	68211	74811	68411
377312	375322	SOU	POR	SU	68212	74812	68412
377313	375323	SOU	POR	SU	68213	74813	68413
377314	375324	SOU	POR	SU	68214	74814	68414
377315	375325	SOU	POR	SU	68215	74815	68415
377316	375326	SOU	POR	SU	68216	74816	68416
377317	375327	SOU	POR	SU	68217	74817	68417
377318	375328	SOU	POR	SU	68218	74818	68418
377319	375329	SOU	POR	SU	68219	74819	68419
377320	375330	SOU	POR	SU	68220	74820	68420
377321	375331	SOU	POR	SU	68221	74821	68421
377322	375332	SOU	POR	SU	68222	74822	68422
377323	375333	SOU	POR	SU	68223	74823	68423
377324	375334	SOU	POR	SU	68224	74824	68424
377325	375335	SOU	POR	SU	68225	74825	68425
377326	375336	SOU	POR	SU	68226	74826	68426
377327	375337	SOU	POR	SU	68227	74827	68427
377328	375338	SOU	POR	SU	68228	74828	68428

Class 377/4

750V DC units. Vehicle 78842 was removed from 377442 following fire damage. The unit was re-formed as three-car 377342 while repairs were undertaken but the unit is now back in traffic in its correct formation and unit number.

	Livery	Owner	Depot	DMOS	MSOL	PTOSL	DMS
377401	SOU	POR	SU	73401	78801	78601	73801
377402	SOU	POR	SU	73402	78802	78602	73802
377403	SOU	POR	SU	73403	78803	78603	73803
377404	SOU	POR	SU	73404	78804	78604	73804
377405	SOU	POR	SU	73405	78805	78605	73805
377406	SOU	POR	SU	73406	78806	78606	73806
377407	SOU	POR	SU	73407	78807	78607	73807
377408	SOU	POR	SU	73408	78808	78608	73808
377409	SOU	POR	SU	73409	78809	78609	73809
377410	SOU	POR	SU	73410	78810	78610	73810
377411	SOU	POR	SU	73411	78811	78611	73811
377412	SOU	POR	SU	73412	78812	78612	73812
377413	SOU	POR	SU	73413	78813	78613	73813
377414	SOU	POR	SU	73414	78814	78614	73814
377415	SOU	POR	SU	73415	78815	78615	73815
377416	SOU	POR	SU	73416	78816	78616	73816
377417	SOU	POR	SU	73417	78817	78617	73817
377418	SOU	POR	SU	73418	78818	78618	73818
377419	SOU	POR	SU	73419	78819	78619	73819
377420	SOU	POR	SU	73420	78820	78620	73820
377421	SOU	POR	SU	73421	78821	78621	73821
377422	SOU	POR	SU	73422	78822	78622	73822
377423	SOU	POR	SU	73423	78823	78623	73823
377424	SOU	POR	SU	73424	78824	78624	73824
377425	SOU	POR	SU	73425	78825	78625	73825
377426	SOU	POR	SU	73426	78826	78626	73826
377427	SOU	POR	SU	73427	78827	78627	73827
377428	SOU	POR	SU	73428	78828	78628	73828
377429	SOU	POR	SU	73429	78829	78629	73829
377430	SOU	POR	SU	73430	78830	78630	73830
377431	SOU	POR	SU	73431	78831	78631	73831
377432	SOU	POR	SU	73432	78832	78632	73832
377433	SOU	POR	SU	73433	78833	78633	73833
377434	SOU	POR	SU	73434	78834	78634	73834
377435	SOU	POR	SU	73435	78835	78635	73835
377436	SOU	POR	SU	73436	78836	78636	73836
377437	SOU	POR	SU	73437	78837	78637	73837
377438	SOU	POR	SU	73438	78838	78638	73838
377439	SOU	POR	SU	73439	78839	78639	73839
377440	SOU	POR	SU	73440	78840	78640	73840
377441	SOU	POR	SU	73441	78841	78641	73841
377442	SOU	POR	SU	73442	78842	78642	73842
377443	SOU	POR	SU	73443	78843	78643	73843
377444	SOU	POR	SU	73444	78844	78644	73844
377445	SOU	POR	SU	73445	78845	78645	73845
377446	SOU	POR	SU	73446	78846	78646	73846
377447	SOU	POR	SU	73447	78847	78647	73847
377448	SOU	POR	SU	73448	78848	78648	73848
377449	SOU	POR	SU	73449	78849	78649	73849
377450	SOU	POR	SU	73450	78850	78650	73850
377451	SOU	POR	SU	73451	78851	78651	73851
377452	SOU	POR	SU	73452	78852	78652	73852
377453	SOU	POR	SU	73453	78853	78653	73853
377454	SOU	POR	SU	73454	78854	78654	73854
377455	SOU	POR	SU	73455	78855	78655	73855
377456	SOU	POR	SU	73456	78856	78656	73856
377457	SOU	POR	SU	73457	78857	78657	73857

Class 377s were the main EMU fleet for the Southern operation. 377451 stands at Bognor Regis on 24 March 2018. *Pip Dunn*

377458	SOU	POR	SU	73458	78858	78658	73858
377459	SOU	POR	SU	73459	78859	78659	73859
377460	SOU	POR	SU	73460	78860	78660	73860
377461	SOU	POR	SU	73461	78861	78661	73861
377462	SOU	POR	SU	73462	78862	78662	73862
377463	SOU	POR	SU	73463	78863	78663	73863
377464	SOU	POR	SU	73464	78864	78664	73864
377465	SOU	POR	SU	73465	78865	78665	73865
377466	SOU	POR	SU	73466	78866	78666	73866
377467	SOU	POR	SU	73467	78867	78667	73867
377468	SOU	POR	SU	73468	78868	78668	73868
377469	SOU	POR	SU	73469	78869	78669	73869
377470	SOU	POR	SU	73470	78870	78670	73870
377471	SOU	POR	SU	73471	78871	78671	73871
377472	SOU	POR	SU	73472	78872	78672	73872
377473	SOU	POR	SU	73473	78873	78673	73873
377474	SOU	POR	SU	73474	78874	78674	73874
377475	SOU	POR	SU	73475	78875	78675	73875

Class 377/6

750V DC units.

	Livery	Owner	Depot	DMOS	MOSL	PTOSL	MOS	DMOS
377601	SOU	POR	SU	70101	70201	70301	70401	70501
377602	SOU	POR	SU	70102	70202	70302	70402	70502
377603	SOU	POR	SU	70103	70203	70303	70403	70503
377604	SOU	POR	SU	70104	70204	70304	70404	70504
377605	SOU	POR	SU	70105	70205	70305	70405	70505
377606	SOU	POR	SU	70106	70206	70306	70406	70506
377607	SOU	POR	SU	70107	70207	70307	70407	70507
377608	SOU	POR	SU	70108	70208	70308	70408	70508
377609	SOU	POR	SU	70109	70209	70309	70409	70509
377610	SOU	POR	SU	70110	70210	70310	70410	70510
377611	SOU	POR	SU	70111	70211	70311	70411	70511
377612	SOU	POR	SU	70112	70212	70312	70412	70512
377613	SOU	POR	SU	70113	70213	70313	70413	70513
377614	SOU	POR	SU	70114	70214	70314	70414	70514
377615	SOU	POR	SU	70115	70215	70315	70415	70515
377616	SOU	POR	SU	70116	70216	70316	70416	70516
377617	SOU	POR	SU	70117	70217	70317	70417	70517
377618	SOU	POR	SU	70118	70218	70318	70418	70518
377619	SOU	POR	SU	70119	70219	70319	70419	70519
377620	SOU	POR	SU	70120	70220	70320	70420	70520
377621	SOU	POR	SU	70121	70221	70321	70421	70521
377622	SOU	POR	SU	70122	70222	70322	70422	70522
377623	SOU	POR	SU	70123	70223	70323	70423	70523
377624	SOU	POR	SU	70124	70224	70324	70424	70524
377625	SOU	POR	SU	70125	70225	70325	70425	70525
377626	SOU	POR	SU	70126	70226	70326	70426	70526

Class 377/7

Dual-voltage units.

				DMOS	MOSL	PTOSL	MOS	DMOS
377701	SOU	POR	SU	65201	70601	65601	70701	65401
377702	SOU	POR	SU	65202	70602	65602	70702	65402
377703	SOU	POR	SU	65203	70603	65603	70703	65403
377704	SOU	POR	SU	65204	70604	65604	70704	65404

377705	SOU	POR	SU	65205	70605	65605	70705	65405
377706	SOU	POR	SU	65206	70606	65606	70706	65406
377707	SOU	POR	SU	65207	70607	65607	70707	65407
377708	SOU	POR	SU	65208	70608	65608	70708	65408

Class 387

Later version of the Bombardier Electrostar platform, but capable of 110mph. All are in the ETHQ pool.

Class 387/1

Dual-voltage four-car units used on the King's Cross to King's Lynn and Peterborough routes.

	Livery	Owner	Depot	DMOS	MOSL	PTOSL	MOS	
387101	TLK	POR	HE	421101	422101	423101	424101	
387102	TLK	POR	HE	421102	422102	423102	424102	
387103	TLK	POR	HE	421103	422103	423103	424103	
387104	TLK	POR	HE	421104	422104	423104	424104	
387105	TLK	POR	HE	421105	422105	423105	424105	
387106	TLK	POR	HE	421106	422106	423106	424106	
387107	TLK	POR	HE	421107	422107	423107	424107	
387108	TLK	POR	HE	421108	422108	423108	424108	
387109	TLK	POR	HE	421109	422109	423109	424109	
387110	TLK	POR	HE	421110	422110	423110	424110	
387111	TLK	POR	HE	421111	422111	423111	424111	
387112	TLK	POR	HE	421112	422112	423112	424112	
387113	TLK	POR	HE	421113	422113	423113	424113	
387114	TLK	POR	HE	421114	422114	423114	424114	
387115	TLK	POR	HE	421115	422115	423115	424115	
387116	TLK	POR	HE	421116	422116	423116	424116	
387117	TLK	POR	HE	421117	422117	423117	424117	
387118	TLK	POR	HE	421118	422118	423118	424118	
387119	TLK	POR	HE	421119	422119	423119	424119	
387120	TLK	POR	HE	421120	422120	423120	424120	
387121	TLK	POR	HE	421121	422121	423121	424121	
387122	TLK	POR	HE	421122	422122	423122	424122	
387123	TLK	POR	HE	421123	422123	423123	424123	
387124	TLK	POR	HE	421124	422124	423124	424124	*Paul McCann*
387125	TLK	POR	HE	421125	422125	423125	424125	
387126	TLK	POR	HE	421126	422126	423126	424126	
387127	TLK	POR	HE	421127	422127	423127	424127	
387128	TLK	POR	HE	421128	422128	423128	424128	
387129	TLK	POR	HE	421129	422129	423129	424129	

A fleet of 29 four-car Class 387/1s are in the Great Northern fleet and work many King's Cross-Peterborough trains. On 18 December 2021, 387113/128 wait to leave with the 0913 Peterborough-King's Cross. *Pip Dunn*

A fleet of 27 four-car Class 387/2s are used for the Gatwick Express service, but some of the units were temporarily hired to Great Western Railway or Great Northern. This explains why 387204 was leading the 1527 Paddington-Didcot Parkway on 23 September 2021, seen passing Southall. *Stuart West*

Class 387/2

Ordered to be used on the Victoria-Gatwick-Brighton route, 387201-209 are currently used on Great Northern lines. Units 387201-206 were briefly on sub-lease to GWR but now back on the GN route.

	Livery	Owner	Depot	DMOS	MOSL	PTOSL	MOS
387201	GEX	POR	HE	421201	422201	423201	424201
387202	GEX	POR	HE	421202	422202	423202	424202
387203	GEX	POR	HE	421203	422203	423203	424203
387204	GEX	POR	HE	421204	422204	423204	424204
387205	GEX	POR	HE	421205	422205	423205	424205
387206	GEX	POR	HE	421206	422206	423206	424206
387207	GEX	POR	HE	421207	422207	423207	424207
387208	GEX	POR	HE	421208	422208	423208	424208
387209	GEX	POR	HE	421209	422209	423209	424209
387210	GEX	POR	SL	421210	422210	423210	424210
387211	GEX	POR	SL	421211	422211	423211	424211
387212	GEX	POR	SL	421212	422212	423212	424212
387213	GEX	POR	SL	421213	422213	423213	424213
387214	GEX	POR	SL	421214	422214	423214	424214
387215	GEX	POR	SL	421215	422215	423215	424215
387216	GEX	POR	SL	421216	422216	423216	424216
387217	GEX	POR	SL	421217	422217	423217	424217
387218	GEX	POR	SL	421218	422218	423218	424218
387219	GEX	POR	SL	421219	422219	423219	424219
387220	GEX	POR	SL	421220	422220	423220	424220
387221	GEX	POR	SL	421221	422221	423221	424221
387222	GEX	POR	SL	421222	422222	423222	424222
387223	GEX	POR	SL	421223	422223	423223	424223
387224	GEX	POR	SL	421224	422224	423224	424224
387225	GEX	POR	SL	421225	422225	423225	424225
387226	GEX	POR	SL	421226	422226	423226	424226
387227	GEX	POR	SL	421227	422227	423227	424227

Class 455/8

Ex-BR four-car inner suburban commuter EMUs dating from 1982-84, they usually operate as eight-car sets in the peak. They are expected to be retired from May 2022.

All are in the ETHQ pool.

	Livery	Owner	Depot	DTS	MS	TS	DTS
455801	SOU	EVS	SL	77627	62709	71657	77580
455802	SOU	EVS	SL	77581	62710	71664	77582
455803	SOU	EVS	SL	77583	62711	71639	77584
455804	SOU	EVS	SL	77585	62712	71640	77586
455805	SOU	EVS	SL	77587	62713	71641	77588
455806	SOU	EVS	SL	77589	62714	71642	77590
455807	SOU	EVS	SL	77591	62715	71643	77592
455808	SOU	EVS	SL	77637	62716	71644	77594
455809	SOU	EVS	SL	77623	62717	71648	77602
455810	SOU	EVS	SL	77597	62718	71646	77598
455811	SOU	EVS	SL	77599	62719	71647	77600
455812	SOU	EVS	SL	77595	62720	71645	77626
455813	SOU	EVS	SL	77603	62721	71649	77604
455814	SOU	EVS	SL	77605	62722	71650	77606
455815	SOU	EVS	SL	77607	62723	71651	77608
455816	SOU	EVS	SL	77609	62724	71652	77633
455817	SOU	EVS	SL	77611	62725	71653	77612
455818	SOU	EVS	SL	77613	62726	71654	77632
455819	SOU	EVS	SL	77615	62727	71637	77616
455820	SOU	EVS	SL	77617	62728	71656	77618
455821	SOU	EVS	SL	77619	62729	71655	77620
455822	SOU	EVS	SL	77621	62730	71658	77622
455823	SOU	EVS	SL	77601	62731	71659	77596
455824	SOU	EVS	SL	77593	62732	71660	77624
455825	SOU	EVS	SL	77579	62733	71661	77628
455826	SOU	EVS	SL	77630	62734	71662	77629
455827	SOU	EVS	SL	77610	62735	71663	77614
455828	SOU	EVS	SL	77631	62736	71638	77634
455829	SOU	EVS	SL	77635	62737	71665	77636
455830	SOU	EVS	SL	77625	62738	71666	77638
455831	SOU	EVS	SL	77639	62739	71667	77640
455832	SOU	EVS	SL	77641	62740	71668	77642
455833	SOU	EVS	SL	77643	62741	71669	77644
455834	SOU	EVS	SL	77645	62742	71670	77646
455835	SOU	EVS	SL	77647	62743	71671	77648
455836	SOU	EVS	SL	77649	62744	71672	77650
455837	SOU	EVS	SL	77651	62745	71673	77652
455838	SOU	EVS	SL	77653	62746	71674	77654
455839	SOU	EVS	SL	77655	62747	71675	77656
455840	SOU	EVS	SL	77657	62748	71676	77658
455841	SOU	EVS	SL	77659	62749	71677	77660
455842	SOU	EVS	SL	77661	62750	71678	77662
455843	SOU	EVS	SL	77663	62751	71679	77664
455844	SOU	EVS	SL	77665	62752	71680	77666
455845	SOU	EVS	SL	77667	62753	71681	77668
455846	SOU	EVS	SL	77669	62754	71682	77670

Class 700

The Class 700 is dual-voltage Siemens Desiro City train aimed at moving large numbers of commuters into the capital, although the type does work long-distance trains such as Peterborough to Horsham and so does not have a metro-style seat arrangement. They are a mix of eight- (RLU – reduced length unit) and 12-car (FLU – Full Length unit) sets.

All are in the ETHQ pool.

Govia Thameslink is an amalgamation of Great Northern, Thameslink, Southern and Gatwick Express. At Honor Oak Park on 2 September 2021, Thameslink 700120 on the 1524 Cambridge-Brighton passes Southern 455823 leading the 1640 London Bridge-East Croydon. *Robin Ralston*

Class 700s, in eight- and 12-car units work all the Thameslink trains from the north through the heart of London and on to southern destinations. 700037 calls at Blackfriars on 10 July 2021 with a train for Sevenoaks. *Pip Dunn*

Class 700/0

Eight-car sets.

	Livery	Owner	Depot	DMC	PTS	MS	TS	TS	MS	PTS	DMC
700001	TLK	CLT	TB	401001	402001	403001	406001	407001	410001	411001	412001
700002	TLK	CLT	TB	401002	402002	403002	406002	407002	410002	411002	412002

700003	TLK	CLT	TB	401003	402003	403003	406003	407003	410003	411003	412003
700004	TLK	CLT	TB	401004	402004	403004	406004	407004	410004	411004	412004
700005	TLK	CLT	TB	401005	402005	403005	406005	407005	410005	411005	412005
700006	TLK	CLT	TB	401006	402006	403006	406006	407006	410006	411006	412006
700007	TLK	CLT	TB	401007	402007	403007	406007	407007	410007	411007	412007
700008	TLK	CLT	TB	401008	402008	403008	406008	407008	410008	411008	412008
700009	TLK	CLT	TB	401009	402009	403009	406009	407009	410009	411009	412009
700010	TLK	CLT	TB	401010	402010	403010	406010	407010	410010	411010	412010
700011	TLK	CLT	TB	401011	402011	403011	406011	407011	410011	411011	412011
700012	TLK	CLT	TB	401012	402012	403012	406012	407012	410012	411012	412012
700013	TLK	CLT	TB	401013	402013	403013	406013	407013	410013	411013	412013
700014	TLK	CLT	TB	401014	402014	403014	406014	407014	410014	411014	412014
700015	TLK	CLT	TB	401015	402015	403015	406015	407015	410015	411015	412015
700016	TLK	CLT	TB	401016	402016	403016	406016	407016	410016	411016	412016
700017	TLK	CLT	TB	401017	402017	403017	406017	407017	410017	411017	412017
700018	TLK	CLT	TB	401018	402018	403018	406018	407018	410018	411018	412018
700019	TLK	CLT	TB	401019	402019	403019	406019	407019	410019	411019	412019
700020	TLK	CLT	TB	401020	402020	403020	406020	407020	410020	411020	412020
700021	TLK	CLT	TB	401021	402021	403021	406021	407021	410021	411021	412021
700022	TLK	CLT	TB	401022	402022	403022	406022	407022	410022	411022	412022
700023	TLK	CLT	TB	401023	402023	403023	406023	407023	410023	411023	412023
700024	TLK	CLT	TB	401024	402024	403024	406024	407024	410024	411024	412024
700025	TLK	CLT	TB	401025	402025	403025	406025	407025	410025	411025	412025
700026	TLK	CLT	TB	401026	402026	403026	406026	407026	410026	411026	412026
700027	TLK	CLT	TB	401027	402027	403027	406027	407027	410027	411027	412027
700028	TLK	CLT	TB	401028	402028	403028	406028	407028	410028	411028	412028
700029	TLK	CLT	TB	401029	402029	403029	406029	407029	410029	411029	412029
700030	TLK	CLT	TB	401030	402030	403030	406030	407030	410030	411030	412030
700031	TLK	CLT	TB	401031	402031	403031	406031	407031	410031	411031	412031
700032	TLK	CLT	TB	401032	402032	403032	406032	407032	410032	411032	412032
700033	TLK	CLT	TB	401033	402033	403033	406033	407033	410033	411033	412033
700034	TLK	CLT	TB	401034	402034	403034	406034	407034	410034	411034	412034
700035	TLK	CLT	TB	401035	402035	403035	406035	407035	410035	411035	412035
700036	TLK	CLT	TB	401036	402036	403036	406036	407036	410036	411036	412036
700037	TLK	CLT	TB	401037	402037	403037	406037	407037	410037	411037	412037
700038	TLK	CLT	TB	401038	402038	403038	406038	407038	410038	411038	412038
700039	TLK	CLT	TB	401039	402039	403039	406039	407039	410039	411039	412039
700040	TLK	CLT	TB	401040	402040	403040	406040	407040	410040	411040	412040
700041	TLK	CLT	TB	401041	402041	403041	406041	407041	410041	411041	412041
700042	TLK	CLT	TB	401042	402042	403042	406042	407042	410042	411042	412042
700043	TLK	CLT	TB	401043	402043	403043	406043	407043	410043	411043	412043
700044	TLK	CLT	TB	401044	402044	403044	406044	407044	410044	411044	412044
700045	TLK	CLT	TB	401045	402045	403045	406045	407045	410045	411045	412045
700046	TLK	CLT	TB	401046	402046	403046	406046	407046	410046	411046	412046
700047	TLK	CLT	TB	401047	402047	403047	406047	407047	410047	411047	412047
700048	TLK	CLT	TB	401048	402048	403048	406048	407048	410048	411048	412048
700049	TLK	CLT	TB	401049	402049	403049	406049	407049	410049	411049	412049
700050	TLK	CLT	TB	401050	402050	403050	406050	407050	410050	411050	412050
700051	TLK	CLT	TB	401051	402051	403051	406051	407051	410051	411051	412051
700052	TLK	CLT	TB	401052	402052	403052	406052	407052	410052	411052	412052
700053	TLK	CLT	TB	401053	402053	403053	406053	407053	410053	411053	412053
700054	TLK	CLT	TB	401054	402054	403054	406054	407054	410054	411054	412054
700055	TLK	CLT	TB	401055	402055	403055	406055	407055	410055	411055	412055
700056	TLK	CLT	TB	401056	402056	403056	406056	407056	410056	411056	412056
700057	TLK	CLT	TB	401057	402057	403057	406057	407057	410057	411057	412057
700058	TLK	CLT	TB	401058	402058	403058	406058	407058	410058	411058	412058
700059	TLK	CLT	TB	401059	402059	403059	406059	407059	410059	411059	412059
700060	TLK	CLT	TB	401060	402060	403060	406060	407060	410060	411060	412060

Class 700/1

Twelve-car sets.

	Livery	Owner	Depot	DMC	PTS	MS	MS	TS	TS	TS	TS	MS	MS	PTS	DMC
700101	TLK	CLT	TB	401101	402101	403101	404101	405101	406101	407101	408101	409101	410101	411101	412101
700102	TLK	CLT	TB	401102	402102	403102	404102	405102	406102	407102	408102	409102	410102	411102	412102
700103	TLK	CLT	TB	401103	402103	403103	404103	405103	406103	407103	408103	409103	410103	411103	412103
700104	TLK	CLT	TB	401104	402104	403104	404104	405104	406104	407104	408104	409104	410104	411104	412104
700105	TLK	CLT	TB	401105	402105	403105	404105	405105	406105	407105	408105	409105	410105	411105	412105
700106	TLK	CLT	TB	401106	402106	403106	404106	405106	406106	407106	408106	409106	410106	411106	412106
700107	TLK	CLT	TB	401107	402107	403107	404107	405107	406107	407107	408107	409107	410107	411107	412107
700108	TLK	CLT	TB	401108	402108	403108	404108	405108	406108	407108	408108	409108	410108	411108	412108
700109	TLK	CLT	TB	401109	402109	403109	404109	405109	406109	407109	408109	409109	410109	411109	412109
700110	TLK	CLT	TB	401110	402110	403110	404110	405110	406110	407110	408110	409110	410110	411110	412110
700111	TLK	CLT	TB	401111	402111	403111	404111	405111	406111	407111	408111	409111	410111	411111	412111
700112	TLK	CLT	TB	401112	402112	403112	404112	405112	406112	407112	408112	409112	410112	411112	412112
700113	TLK	CLT	TB	401113	402113	403113	404113	405113	406113	407113	408113	409113	410113	411113	412113
700114	TLK	CLT	TB	401114	402114	403114	404114	405114	406114	407114	408114	409114	410114	411114	412114
700115	TLK	CLT	TB	401115	402115	403115	404115	405115	406115	407115	408115	409115	410115	411115	412115
700116	TLK	CLT	TB	401116	402116	403116	404116	405116	406116	407116	408116	409116	410116	411116	412116
700117	TLK	CLT	TB	401117	402117	403117	404117	405117	406117	407117	408117	409117	410117	411117	412117
700118	TLK	CLT	TB	401118	402118	403118	404118	405118	406118	407118	408118	409118	410118	411118	412118
700119	TLK	CLT	TB	401119	402119	403119	404119	405119	406119	407119	408119	409119	410119	411119	412119
700120	TLK	CLT	TB	401120	402120	403120	404120	405120	406120	407120	408120	409120	410120	411120	412120
700121	TLK	CLT	TB	401121	402121	403121	404121	405121	406121	407121	408121	409121	410121	411121	412121
700122	TLK	CLT	TB	401122	402122	403122	404122	405122	406122	407122	408122	409122	410122	411122	412122
700123	TLK	CLT	TB	401123	402123	403123	404123	405123	406123	407123	408123	409123	410123	411123	412123
700124	TLK	CLT	TB	401124	402124	403124	404124	405124	406124	407124	408124	409124	410124	411124	412124
700125	TLK	CLT	TB	401125	402125	403125	404125	405125	406125	407125	408125	409125	410125	411125	412125
700126	TLK	CLT	TB	401126	402126	403126	404126	405126	406126	407126	408126	409126	410126	411126	412126
700127	TLK	CLT	TB	401127	402127	403127	404127	405127	406127	407127	408127	409127	410127	411127	412127
700128	TLK	CLT	TB	401128	402128	403128	404128	405128	406128	407128	408128	409128	410128	411128	412128
700129	TLK	CLT	TB	401129	402129	403129	404129	405129	406129	407129	408129	409129	410129	411129	412129
700130	TLK	CLT	TB	401130	402130	403130	404130	405130	406130	407130	408130	409130	410130	411130	412130
700131	TLK	CLT	TB	401131	402131	403131	404131	405131	406131	407131	408131	409131	410131	411131	412131
700132	TLK	CLT	TB	401132	402132	403132	404132	405132	406132	407132	408132	409132	410132	411132	412132
700133	TLK	CLT	TB	401133	402133	403133	404133	405133	406133	407133	408133	409133	410133	411133	412133
700134	TLK	CLT	TB	401134	402134	403134	404134	405134	406134	407134	408134	409134	410134	411134	412134
700135	TLK	CLT	TB	401135	402135	403135	404135	405135	406135	407135	408135	409135	410135	411135	412135

700136	TLK	CLT	TB	401136	402136	403136	404136	405136	406136	407136	408136	409136	410136	411136	412136
700137	TLK	CLT	TB	401137	402137	403137	404137	405137	406137	407137	408137	409137	410137	411137	412137
700138	TLK	CLT	TB	401138	402138	403138	404138	405138	406138	407138	408138	409138	410138	411138	412138
700139	TLK	CLT	TB	401139	402139	403139	404139	405139	406139	407139	408139	409139	410139	411139	412139
700140	TLK	CLT	TB	401140	402140	403140	404140	405140	406140	407140	408140	409140	410140	411140	412140
700141	TLK	CLT	TB	401141	402141	403141	404141	405141	406141	407141	408141	409141	410141	411141	412141
700142	TLK	CLT	TB	401142	402142	403142	404142	405142	406142	407142	408142	409142	410142	411142	412142
700143	TLK	CLT	TB	401143	402143	403143	404143	405143	406143	407143	408143	409143	410143	411143	412143
700144	TLK	CLT	TB	401144	402144	403144	404144	405144	406144	407144	408144	409144	410144	411144	412144
700145	TLK	CLT	TB	401145	402145	403145	404145	405145	406145	407145	408145	409145	410145	411145	412145
700146	TLK	CLT	TB	401146	402146	403146	404146	405146	406146	407146	408146	409146	410146	411146	412146
700147	TLK	CLT	TB	401147	402147	403147	404147	405147	406147	407147	408147	409147	410147	411147	412147
700148	TLK	CLT	TB	401148	402148	403148	404148	405148	406148	407148	408148	409148	410148	411148	412148
700149	TLK	CLT	TB	401149	402149	403149	404149	405149	406149	407149	408149	409149	410149	411149	412149
700150	TLK	CLT	TB	401150	402150	403150	404150	405150	406150	407150	408150	409150	410150	411150	412150
700151	TLK	CLT	TB	401151	402151	403151	404151	405151	406151	407151	408151	409151	410151	411151	412151
700152	TLK	CLT	TB	401152	402152	403152	404152	405152	406152	407152	408152	409152	410152	411152	412152
700153	TLK	CLT	TB	401153	402153	403153	404153	405153	406153	407153	408153	409153	410153	411153	412153
700154	TLK	CLT	TB	401154	402154	403154	404154	405154	406154	407154	408154	409154	410154	411154	412154
700155	TLK	CLT	TB	401155	402155	403155	404155	405155	406155	407155	408155	409155	410155	411155	412155

Note: the driving cars of 700155 have Pride rainbow graphics.

Class 717

Similar to the Class 700, these dual-voltage EMUs are a metro-style version of the Siemens Desiro City. They are used on commuter trains from Moorgate and unlike the Class 700s, do not have toilets.

All are in the ETHQ pool.

The Class 717 is Siemens Desiro City design for inner-suburban use, and accordingly they have no toilets. 717001 calls at Finsbury Park with the 1020 Moorgate-Welwyn Garden City on 6 December 2019. *Pip Dunn*

	Livery	Owner	Depot	DMS	TS	TS	MS	PTS	DMS
717001	TLK	ROC	HE	451101	452101	453101	454101	455001	456001
717002	TLK	ROC	HE	451102	452102	453102	454102	455002	456002
717003	TLK	ROC	HE	451103	452103	453103	454103	455003	456003
717004	TLK	ROC	HE	451104	452104	453104	454104	455004	456004
717005	TLK	ROC	HE	451105	452105	453105	454105	455005	456005
717006	TLK	ROC	HE	451106	452106	453106	454106	455006	456006
717007	TLK	ROC	HE	451107	452107	453107	454107	455007	456007
717008	TLK	ROC	HE	451108	452108	453108	454108	455008	456008
717009	TLK	ROC	HE	451109	452109	453109	454109	455009	456009
717010	TLK	ROC	HE	451110	452110	453110	454110	455010	456010
717011	TLK	ROC	HE	451111	452111	453111	454111	455011	456011
717012	TLK	ROC	HE	451112	452112	453112	454112	455012	456012
717013	TLK	ROC	HE	451113	452113	453113	454113	455013	456013
717014	TLK	ROC	HE	451114	452114	453114	454114	455014	456014
717015	TLK	ROC	HE	451115	452115	453115	454115	455015	456015
717016	TLK	ROC	HE	451116	452116	453116	454116	455016	456016
717017	TLK	ROC	HE	451117	452117	453117	454117	455017	456017
717018	TLK	ROC	HE	451118	452118	453118	454118	455018	456018
717019	TLK	ROC	HE	451119	452119	453119	454119	455019	456019
717020	TLK	ROC	HE	451120	452120	453120	454120	455020	456020
717021	TLK	ROC	HE	451121	452121	453121	454121	455021	456021
717022	TLK	ROC	HE	451122	452122	453122	454122	455022	456022
717023	TLK	ROC	HE	451123	452123	453123	454123	455023	456023
717024	TLK	ROC	HE	451124	452124	453124	454124	455024	456024
717025	TLK	ROC	HE	451125	452125	453125	454125	455025	456025

Greater Anglia

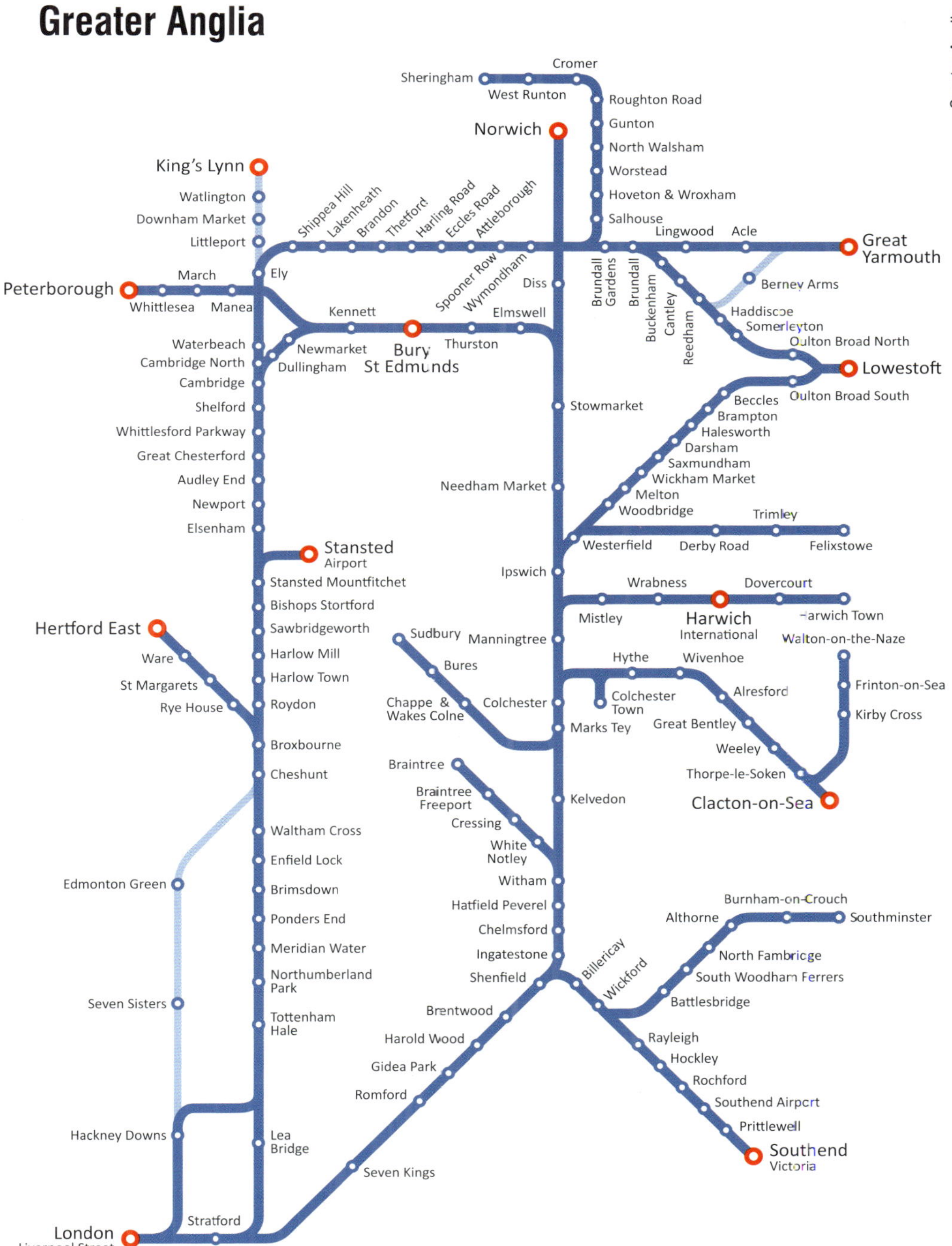

Contact details

Website: greateranglia.co.uk
Twitter: @greateranglia

Key personnel

Managing Director: Jamie Burles
Engineering Director: Martin Beable

Overview

Greater Anglia operates the inter-city route from Liverpool St to Norwich but also commuter trains from Liverpool St to Hertford East, Stansted, Southend Victoria, Southminster, Clacton, Walton-on-Naze and Harwich Town. It also operates local trains to Braintree, Ipswich-Felixstowe, Marks Tey-Sudbury and in Suffolk and Norfolk from Norwich-Cambridge, Norwich-Sheringham, Great Yarmouth and Lowestoft, Ipswich-Peterborough and Ipswich-Lowestoft.

Greater Anglia has undergone a wholesale fleet renewal in recent years with fleets of brand new trains from Stadler and Bombardier now being delivered and introduced. All rural routes are operated by a mix of 14 three-car and 24 four-car Class 755 bi-mode multiple units.

The existing Norwich to Cambridge service has been extended to Stansted Airport using Class 745s (from December 2019), while the Ipswich to Peterborough service is due to be doubled in frequency to hourly using 755s.

Ten 12-car Class 745/0s are now in use on the Liverpool Street-Norwich route, with ten 12-car Class 745/1s now in use on the Stansted Express route as well as on some GEML duties.

Class 720s are now in use and have been cleared for King's Lynn and other GA lines, and by late February 2022, 51 of the 133 sets on order had been accepted into traffic. Once all are in service, they will replace the last of the Class 317 and 321 units. The Class 379s remain on lease but have been taken out of traffic despite an initial plan for them to replace 317 and 321s before being stood down.

Class 317

The four-car ex-BR units, which date from 1981-86, are based at Ilford and work out of Liverpool Street on commuter trains in West Anglia to Hertford East, Bishop's Stortford, as well as Meridian Water to Stratford. They are steadily being returned to their ROSCO as the Class 720 fleet enters traffic, with most retired units going straight for scrap.

All are in the EBHQ pool.

Class 317/3

Outer suburban units. 317337-344/347/348 have been modified to meet current PPM standards.

	Livery	Owner	Depot	DTC	MS	TC	DTS
317337	GAW	ANG	IL	77036	62671	71613	77084
317338	GAW	ANG	IL	77037	62698	71614	77085
317339	GAW	ANG	IL	77038	62699	71615	77086
317340	GAW	ANG	IL	77039	62700	71616	77087
317342	GAW	ANG	IL	77041	62702	71618	77089
317343	GAW	ANG	IL	77042	62703	71619	77090
317347	GAW	ANG	IL	77046	62707	71623	77094

Class 317/5

Outer suburban units. Units 317501/502/504/506-508/510-514/515 have been modified to meet current PPM standards.

	Livery	Owner	Depot	DTC	MS	TC	DTS	
317501	GAW	ANG	IL	77024	62661	71577	77048	
317502	GAW	ANG	IL	77001	62662	71578	77049	
317504	GAW	ANG	IL	77003	62664	71580	77051	
317506	GAW	ANG	IL	77005	62666	71582	77053	
317507	GAW	ANG	IL	77006	62667	71583	77054	*University of Cambridge 800 Years 1209-2009*
317508	GAW	ANG	IL	77010	62697	71587	77058	
317510	GAW	ANG	IL	77012	62673	71589	77060	
317511	GAW	ANG	IL	77014	62675	71591	77062	
317515	GAW	ANG	IL	77019	62680	71596	77067	

Class 317/8

Outer suburban units. These have been modified to meet current PPM standards.

	Livery	Owner	Depot	DTC	MS	TC	DTS
317881	GAW	ANG	IL	77020	62681	71597	77068
317884	GAW	ANG	IL	77025	62686	71602	77073
317886	GAW	ANG	IL	77027	62688	71604	77075

Class 321/322

A slightly newer ex-BR EMU design, nevertheless they are still over 30 years old, having been built by BREL York in 1988-90. The remaining active units in the fleet are also based at Ilford for working commuter and local trains from Liverpool Street to Southend Victoria, the majority of trains to Chelmsford, Braintree, Clacton, Colchester, Harwich and Walton-on-Naze and some trains to Ipswich.

The fleet is steadily being stood down and returned to their ROSCOs as the Class 720 fleet is accepted into traffic. All are in the EBHQ pool.

Class 321/3

Outer suburban units.

The Class 321 Renatus fleet is a refurbishment of the existing Class 321 units being delivered by Wabtec Rail for Eversholt Rail, which features new air-conditioning and heating systems, larger vestibules, wi-fi, more space for buggies, bicycles and luggage, power sockets, energy-efficient lighting, a PRM-compliant toilet and a second controlled-emission toilet on each unit. An initial 30 units, 321301-330, have been upgraded for Greater Anglia.

	Livery	Owner	Depot	DTS	MS	TS	DTC
321301	GAW	EVS	IL	78049	62975	71880	77853
321302	GAW	EVS	IL	78050	62976	71881	77854
321303	GAW	EVS	IL	78051	62977	71882	77855
321304	GAW	EVS	IL	78052	62978	71883	77856
321305	GAW	EVS	IL	78053	62979	71884	77857
321306	GAW	EVS	IL	78054	62980	71885	77858
321307	GAW	EVS	IL	78055	62981	71886	77859
321308	GAW	EVS	IL	78056	62982	71887	77860
321309	GAW	EVS	IL	78057	62983	71888	77861
321310	GAW	EVS	IL	78058	62984	71889	77862
321311	GAW	EVS	IL	78059	62985	71890	77863
321312	GAW	EVS	IL	78060	62986	71891	77864
321313	GAW	EVS	IL	78061	62987	71892	77865

The Class 321s are rapidly being phased out by Greater Anglia, but 30 units have been refurbished and will remain in traffic for the foreseeable future. On 6 February 2022, 321306 works the 1132 Liverpool Street-Clacton-on-Sea past Great Bentley. *Paul Shannon*

321314	GAW	EVS	IL	78062	62988	71893	77866
321315	GAW	EVS	IL	78063	62989	71894	77867
321316	GAW	EVS	IL	78064	62990	71895	77868
321317	GAW	EVS	IL	78065	62991	71896	77869
321318	GAW	EVS	IL	78066	62992	71897	77870
321319	GAW	EVS	IL	78067	62993	71898	77871
321320	GAW	EVS	IL	78068	62994	71899	77872
321321	GAW	EVS	IL	78069	62995	71900	77873
321322	GAW	EVS	IL	78070	62996	71901	77874
321323	GAW	EVS	IL	78071	62997	71902	77875
321324	GAW	EVS	IL	78072	62998	71903	77876
321325	GAW	EVS	IL	78073	62999	71904	77877
321326	GAW	EVS	IL	78074	63000	71905	77878
321327	GAW	EVS	IL	78075	63001	71906	77879
321328	GAW	EVS	IL	78076	63002	71907	77880
321329	GAW	EVS	IL	78077	63003	71908	77881
321330	GAW	EVS	IL	78078	63004	71909	77882

Class 321/9

Redundant Northern units replaced by Class 331s, these units have been moved to Greater Anglia to provide additional due to the late arrival of the Class 720s but are now being stood down as new units enter traffic.

	Livery	Owner	Depot	DTS	MS	TS	DTC
321901	NBL	EVS	IL	77990	63153	72128	77993
321902	NBL	EVS	IL	77991	63154	72129	77994
321903	NBL	EVS	IL	77992	63155	72130	77995

Class 322

Like the 321/9s, the units were spare after use with Northern, were transferred to AGA use and are also being stood down as new trains come on stream.

	Livery	Owner	Depot	DTS	MS	TS	DTC
322481	NBL	EVS	IL	78163	63137	72023	77985
322482	NBL	EVS	IL	78164	63138	72024	77986
322483	NBL	EVS	IL	78165	63139	72025	77987
322484	NBL	EVS	IL	78166	63140	72026	77988
322485	NBL	EVS	IL	78167	63141	72027	77989

Class 720

The new Alstom EMUs are now being delivered and are being used for commuter trains from Liverpool St to allow 317/321s to be withdrawn and the 360 and 379s to be redeployed with other franchises, although so far just the 360s have found a new user.

The order was originally for 89 five-car 720/5 sets and 22 ten-car 720/1s but was changed to 133 five-car Class 720/5s. The vehicles are slightly longer, so each five-car set is the equivalent of a six-car formation. Some infrastructure work will be needed across the GA network, with some stations needing their platforms extended and some signals repositioned to improve sighting. These enhancements should be completed by 2022, when the full fleet is expected to have been delivered and commissioned.

All are in the EBHQ pool.

Class 720/1

These 44 five-car Class 720/1s were originally going to be 22 ten-car sets.

	Livery	Owner	Depot	DMS	PMS	MS	MS	DTS
720101	GAG	ANG		450101	451101	452101	453101	454101
720102	GAG	ANG		450102	451102	452102	453102	454102
720103	GAG	ANG		450103	451103	452103	453103	454103
720104	GAG	ANG		450104	451104	452104	453104	454104
720105	GAG	ANG		450105	451105	452105	453105	454105
720106	GAG	ANG		450106	451106	452106	453106	454106
720107	GAG	ANG		450107	451107	452107	453107	454107
720108	GAG	ANG		450108	451108	452108	453108	454108
720109	GAG	ANG		450109	451109	452109	453109	454109
720110	GAG	ANG		450110	451110	452110	453110	454110
720111	GAG	ANG		450111	451111	452111	453111	454111
720112	GAG	ANG		450112	451112	452112	453112	454112
720113	GAG	ANG		450113	451113	452113	453113	454113
720114	GAG	ANG		450114	451114	452114	453114	454114
720115	GAG	ANG		450115	451115	452115	453115	454115
720116	GAG	ANG		450116	451116	452116	453116	454116
720117	GAG	ANG		450117	451117	452117	453117	454117
720118	GAG	ANG		450118	451118	452118	453118	454118
720119	GAG	ANG		450119	451119	452119	453119	454119
720120	GAG	ANG		450120	451120	452120	453120	454120
720121	GAG	ANG		450121	451121	452121	453121	454121
720122	GAG	ANG		450122	451122	452122	453122	454122
720123	GAG	ANG		450123	451123	452123	453123	454123
720124	GAG	ANG		450124	451124	452124	453124	454124
720125	GAG	ANG		450125	451125	452125	453125	454125
720126	GAG	ANG		450126	451126	452126	453126	454126
720127	GAG	ANG		450127	451127	452127	453127	454127
720128	GAG	ANG		450128	451128	452128	453128	454128
720129	GAG	ANG		450129	451129	452129	453129	454129
720130	GAG	ANG		450130	451130	452130	453130	454130
720131	GAG	ANG		450131	451131	452131	453131	454131
720132	GAG	ANG		450132	451132	452132	453132	454132
720133	GAG	ANG		450133	451133	452133	453133	454133
720134	GAG	ANG		450134	451134	452134	453134	454134
720135	GAG	ANG		450135	451135	452135	453135	454135
720136	GAG	ANG		450136	451136	452136	453136	454136
720137	GAG	ANG		450137	451137	452137	453137	454137
720138	GAG	ANG		450138	451138	452138	453138	454138
720139	GAG	ANG		450139	451139	452139	453139	454139
720140	GAG	ANG		450140	451140	452140	453140	454140
720141	GAG	ANG		450141	451141	452141	453141	454141
720142	GAG	ANG		450142	451142	452142	453142	454142
720143	GAG	ANG		450143	451143	452143	453143	454143
720144	GAG	ANG		450144	451144	452144	453144	454144

Class 720/5

These units are now being delivered and undergoing testing. Many units are in store at Old Dalby or Worksop.

	Livery	Owner	Depot	DMS	PMS	MS	MS	DTS
720501	GAG	ANG	WK (S)	450501	451501	452501	453501	459501
720502	GAG	ANG	WK (S)	450502	451502	452502	453502	459502
720503	GAG	ANG	WK (S)	450503	451503	452503	453503	459503
720504	GAG	ANG	WK (S)	450504	451504	452504	453504	459504

720505	GAG	ANG	WK (S)	450505	451505	452505	453505	459505
720506	GAG	ANG	WK (S)	450506	451506	452506	453506	459506
720507	GAG	ANG	WK (S)	450507	451507	452507	453507	459507
720508	GAG	ANG	WK (S)	450508	451508	452508	453508	459508
720509	GAG	ANG	WK (S)	450509	451509	452509	453509	459509
720510	GAG	ANG	WK (S)	450510	451510	452510	453510	459510
720511	GAG	ANG	IL	450511	451511	452511	453511	459511
720512	GAG	ANG	WK (S)	450512	451512	452512	453512	459512
720513	GAG	ANG	WK (S)	450513	451513	452513	453513	459513
720514	GAG	ANG	WK (S)	450514	451514	452514	453514	459514
720515	GAG	ANG	IL	450515	451515	452515	453515	459515
720516	GAG	ANG	WB (S)	450516	451516	452516	453516	459516
720517	GAG	ANG	IL	450517	451517	452517	453517	459517
720518	GAG	ANG	OD (S)	450518	451518	452518	453518	459518
720519	GAG	ANG	OD (S)	450519	451519	452519	453519	459519
720520	GAG	ANG	ZD (S)	450520	451520	452520	453520	459520
720521	GAG	ANG	IL	450521	451521	452521	453521	459521
720522	GAG	ANG	IL	450522	451522	452522	453522	459522
720523	GAG	ANG	ZD (S)	450523	451523	452523	453523	459523
720524	GAG	ANG	ZD (S)	450524	451524	452524	453524	459524
720525	GAG	ANG	IL	450525	451525	452525	453525	459525
720526	GAG	ANG	ZD (S)	450526	451526	452526	453526	459526
720527	GAG	ANG	WB (S)	450527	451527	452527	453527	459527
720528	GAG	ANG	ZD (S)	450528	451528	452528	453528	459528
720529	GAG	ANG	ZD (S)	450529	451529	452529	453529	459529
720530	GAG	ANG	IL	450530	451530	452530	453530	459530
720531	GAG	ANG	IL	450531	451531	452531	453531	459531
720532	GAG	ANG	IL	450532	451532	452532	453532	459532
720533	GAG	ANG	WK (S)	450533	451533	452533	453533	459533
720534	GAG	ANG	ZD (S)	450534	451534	452534	453534	459534
720535	GAG	ANG	IL	450535	451535	452535	453535	459535
720536	GAG	ANG	IL	450536	451536	452536	453536	459536
720537	GAG	ANG	IL	450537	451537	452537	453537	459537
720538	GAG	ANG	IL	450538	451538	452538	453538	459538
720539	GAG	ANG	IL	450539	451539	452539	453539	459539
720540	GAG	ANG	IL	450540	451540	452540	453540	459540
720541	GAG	ANG	OD (S)	450541	451541	452541	453541	459541
720542	GAG	ANG	IL	450542	451542	452542	453542	459542
720543	GAG	ANG	IL	450543	451543	452543	453543	459543
720544	GAG	ANG	WK (S)	450544	451544	452544	453544	459544
720545	GAG	ANG	IL	450545	451545	452545	453545	459545
720546	GAG	ANG	IL	450546	451546	452546	453546	459546
720547	GAG	ANG	IL	450547	451547	452547	453547	459547
720548	GAG	ANG	IL	450548	451548	452548	453548	459548
720549	GAG	ANG	IL	450549	451549	452549	453549	459549
720550	GAG	ANG	IL	450550	451550	452550	453550	459550
720551	GAG	ANG	IL	450551	451551	452551	453551	459551
720552	GAG	ANG	IL	450552	451552	452552	453552	459552
720553	GAG	ANG	IL	450553	451553	452553	453553	459553
720554	GAG	ANG	IL	450554	451554	452554	453554	459554
720555	GAG	ANG	WK (S)	450555	451555	452555	453555	459555
720556	GAG	ANG	IL	450556	451556	452556	453556	459556
720557	GAG	ANG	IL	450557	451557	452557	453557	459557
720558	GAG	ANG	IL	450558	451558	452558	453558	459558
720559	GAG	ANG	IL	450559	451559	452559	453559	459559
720560	GAG	ANG	IL	450560	451560	452560	453560	459560
720561	GAG	ANG	IL	450561	451561	452561	453561	459561
720562	GAG	ANG	IL	450562	451562	452562	453562	459562

New Bombardier Class 720s are now entering traffic with Greater Anglia across its network. On 7 February 2022, 720561 works the 0802 Liverpool Street-Ipswich past Cattawade, near Manningtree. *Paul Shannon*

720563	GAG	ANG	IL	450563	451563	452563	453563	459563
720564	GAG	ANG	IL	450564	451564	452564	453564	459564
720565	GAG	ANG	IL	450565	451565	452565	453565	459565
720566	GAG	ANG	IL	450566	451566	452566	453566	459566
720567	GAG	ANG	IL	450567	451567	452567	453567	459567
720568	GAG	ANG	IL	450568	451568	452568	453568	459568
720569	GAG	ANG	IL	450569	451569	452569	453569	459569
720570	GAG	ANG	IL	450570	451570	452570	453570	459570
720571	GAG	ANG	IL	450571	451571	452571	453571	459571
720572	GAG	ANG	OY (S)	450572	451572	452572	453572	459572
720573	GAG	ANG	IL	450573	451573	452573	453573	459573
720574	GAG	ANG	IL	450574	451574	452574	453574	459574
720575	GAG	ANG	IL	450575	451575	452575	453575	459575
720576	GAG	ANG	IL	450576	451576	452576	453576	459576
720577	GAG	ANG	IL	450577	451577	452577	453577	459577
720578	GAG	ANG	IL	450578	451578	452578	453578	459578
720579	GAG	ANG	WK (S)	450579	451579	452579	453579	459579
720580	GAG	ANG	IL	450580	451580	452580	453580	459580
720581	GAG	ANG	IL	450581	451581	452581	453581	459581
720582	GAG	ANG	WK (S)	450582	451582	452582	453582	459582
720583	GAG	ANG	WK (S)	450583	451583	452583	453583	459583
720584	GAG	ANG	IL	450584	451584	452584	453584	459584
720585	GAG	ANG		450585	451585	452585	453585	459585
720586	GAG	ANG		450586	451586	452586	453586	459586
720587	GAG	ANG		450587	451587	452587	453587	459587
720588	GAG	ANG		450588	451588	452588	453588	459588

Note: the locations of stored units changes regularly.

Class 745

There are two sub-classes of these 25kV AC EMUs Stadler units, the 12-car sets 745/0s are used for Liverpool Street to Norwich inter-city services and have replaced Class 90-hauled Mk 3 sets, while the 745/1s are used on the Liverpool Street-Stansted Airport shuttles.

Class 745/0

Twelve-car sets for Liverpool St-Norwich services with first- and standard-class seats.
All are in the EBHQ pool.

	Livery	Owner	Depot	DMF	PTF	TS	TS	TS	MS	MS	TS	TS	TS	PTS	DMS
745001	GAG	ROC	NC	413001	426001	332001	343001	341001	301001	302001	342001	344001	346001	322001	312001
745002	GAG	ROC	NC	413002	426002	332002	343002	341002	301002	302002	342002	344002	346002	322002	312002
745003	GAG	ROC	NC	413003	426003	332003	343003	341003	301003	302003	342003	344003	346003	322003	312003
745004	GAG	ROC	NC	413004	426004	332004	343004	341004	301004	302004	342004	344004	346004	322004	312004
745005	GAG	ROC	NC	413005	426005	332005	343005	341005	301005	302005	342005	344005	346005	322005	312005
745006	GAG	ROC	NC	413006	426006	332006	343006	341006	301006	302006	342006	344006	346006	322006	312006
745007	GAG	ROC	NC	413007	426007	332007	343007	341007	301007	302007	342007	344007	346007	322007	312007
745008	GAG	ROC	NC	413008	426008	332008	343008	341008	301008	302008	342008	344008	346008	322008	312008
745009	GAG	ROC	NC	413009	426009	332009	343009	341009	301009	302009	342009	344009	346009	322009	312009
745010	GAG	ROC	NC	413010	426010	332010	343010	341010	301010	302010	342010	344010	346010	322010	312010

Class 745/1

Twelve-car sets for Liverpool St-Stansted Airport services with standard-class seats only.
All are in the ENHQ pool.

	Livery	Owner	Depot	DMF	PTF	TS	TS	TS	MS	MS	TS	TS	TS	PTS	DMS
745101	GAG	ROC	NC	313101	326101	332101	343101	341101	301101	302101	342101	344101	346101	322101	312101
745102	GAG	ROC	NC	313102	326102	332102	343102	341102	301102	302102	342102	344102	346102	322102	312102
745103	GAG	ROC	NC	313103	326103	332103	343103	341103	301103	302103	342103	344103	346103	322103	312103
745104	GAG	ROC	NC	313104	326104	332104	343104	341104	301104	302104	342104	344104	346104	322104	312104
745105	GAG	ROC	NC	313105	326105	332105	343105	341105	301105	302105	342105	344105	346105	322105	312105
745106	GAG	ROC	NC	313106	326106	332106	343106	341106	301106	302106	342106	344106	346106	322106	312106
745107	GAG	ROC	NC	313107	326107	332107	343107	341107	301107	302107	342107	344107	346107	322107	312107
745108	GAG	ROC	NC	313108	326108	332108	343108	341108	301108	302108	342108	344108	346108	322108	312108
745109	GAG	ROC	NC	313109	326109	332109	343109	341109	301109	302109	342109	344109	346109	322109	312109
745110	GAG	ROC	NC	313110	326110	332110	343110	341110	301110	302110	342110	344110	346110	322110	312110

The 12-car Class 745s are EMUs that replaced the Class 90-hauled Mk 3 sets on the Great Eastern Main Line. 745004 arrives at Ipswich on 22 September 2020 with the 1030 Norwich-Liverpool St. *Stuart West*

Greater Anglia

Class 755

These bi-mode Stadler multiple units are all now in traffic and used on local services on the AGA operation. They have replaced Class 153/156/170s in Norfolk and Suffolk on local lines from Norwich and Ipswich, as well as working the Sudbury to Marks Tey branch.

Although numbered as four- and five-car, the PP vehicle is non-passenger carrying and contains each unit's Deutz engines, of which there are four in the five-car sets and two in the four-car sets.

All are in the EBHQ pool.

Class 755/3

	Livery	Owner	Depot	DMS	PP	PTS	DMS
755325	GAG	ROC	NC	911325	971325	981325	912325
755326	GAG	ROC	NC	911326	971326	981326	912326
755327	GAG	ROC	NC	911327	971327	981327	912327
755328	GAG	ROC	NC	911328	971328	981328	912328
755329	GAG	ROC	NC	911329	971329	981329	912329
755330	GAG	ROC	NC	911330	971330	981330	912330
755331	GAG	ROC	NC	911331	971331	981331	912331
755332	GAG	ROC	NC	911332	971332	981332	912332
755333	GAG	ROC	NC	911333	971333	981333	912333
755334	GAG	ROC	NC	911334	971334	981334	912334
755335	GAG	ROC	NC	911335	971335	981335	912335
755336	GAG	ROC	NC	911336	971336	981336	912336
755337	GAG	ROC	NC	911337	971337	981337	912337
755338	GAG	ROC	NC	911338	971338	981338	912338

Class 755s are Stadler FLIRT bi-mode multiple units used by Greater Anglia. 755401 waits to set off with the 1750 Peterborough-Ipswich on 12 February 2022. *Pip Dunn*

Class 755/4

	Livery	Owner	Depot	DMS	PTS	PP	PTS	DMS
755401	GAG	ROC	NC	911401	961401	971401	981401	912401
755402	GAG	ROC	NC	911402	961402	971402	981402	912402
755403	GAG	ROC	NC	911403	961403	971403	981403	912403
755404	GAG	ROC	NC	911404	961404	971404	981404	912404
755405	GAG	ROC	NC	911405	961405	971405	981405	912405
755406	GAG	ROC	NC	911406	961406	971406	981406	912406
755407	GAG	ROC	NC	911407	961407	971407	981407	912407
755408	GAG	ROC	NC	911408	961408	971408	981408	912408
755409	GAG	ROC	NC	911409	961409	971409	981409	912409
755410	GAG	ROC	NC	911410	961410	971410	981410	912410
755411	GAG	ROC	NC	911411	961411	971411	981411	912411
755412	GAG	ROC	NC	911412	961412	971412	981412	912412
755413	GAG	ROC	NC	911413	961413	971413	981413	912413
755414	GAG	ROC	NC	911414	961414	971414	981414	912414
755415	GAG	ROC	NC	911415	961415	971415	981415	912415
755416	GAG	ROC	NC	911416	961416	971416	981416	912416
755417	GAG	ROC	NC	911417	961417	971417	981417	912417
755418	GAG	ROC	NC	911418	961418	971418	981418	912418
755419	GAG	ROC	NC	911419	961419	971419	981419	912419
755420	GAG	ROC	NC	911420	961420	971420	981420	912420
755421	GAG	ROC	NC	911421	961421	971421	981421	912421
755422	GAG	ROC	NC	911422	961422	971422	981422	912422
755423	GAG	ROC	NC	911423	961423	971423	981423	912423
755424	GAG	ROC	NC	911424	961424	971424	981424	912424

Great Western Railway

Great Malvern
Malvern Link
Worcester
Foregate street
Worcester
Shrub Hill
Worcester Parkway
Ashchurch
Pershore
Evesham
Honeybourne
Moreton-in-Marsh
Cheltenham Spa
Kingham
Shipton
Ascott-under-Wychwood
Charlbury
Finstock
Combe
Hanborough
Banbury
King's Sutton
Heyford
Tackley
Gloucester
Cam & Dursley
Bristol Parkway
Stonehouse
Stroud
Kemble
Yate
Oxford
Radley
Culham
Appleford
Marlow
Bourne End
Henley-on-Thames
Shiplake
Wargrave
Cookham
Furze Platt
Cholsey
Goring & Streatley
Tilehurst
Reading
Twyford
Maidenhead
Slough
Langley
Iver
West Drayton
Greenford
South Greenford
Castle Bar Park
Drayton Green
Paddington
Bath Spa
Freshford
Avoncliff
Bradford-on-Avon
Chippenham
Melksham
Trowbridge
Swindon
Didcot
Parkway
Reading West
Theale
Aldermaston
Midgham
Thatcham
Newbury Racecourse
Newbury
Kintbury
Hungerford
Bedwyn
Pewsey
Winnersh
Mortimer
Bramley
Basingstoke
Taplow
Burnham
Windsor & Eton Central
Heathrow
Airport
Hayes & Harlington
Southall
Hanwell
West Ealing
Ealing Broadway
Acton Main Line
Wokingham
Crowthorne
Sandhurst
Blackwater
Farnborough North
North Camp
Ash
Wanborough
Guildford
Shalford
Chilworth
Gomshall
Dorking West
Dorking Deepdene
Betchworth
Reigate
Redhill
Gatwick
Airport
Bruton
Frome
Westbury
Dilton Marsh
Warminster
Salisbury
Dean
Dunbridge
Romsey
Southampton
Central
St Denys
Woolston
Netley
Swanwick
Fareham
Portchester
Cosham
Fratton
Portsmouth & Southsea
Portsmouth
Harbour
Havant
Chichester
Barnham
Worthing
Shoreham-by-Sea
Hove
Brighton

Contact details

Website: www.gwr.com
Twitter: @GWRHelp

Key personnel

Managing Director: Mark Hopwood
Engineering Director: Simon Green

Overview

GWR runs all the inter-city trains from Paddington to Hereford, Cheltenham, Bristol TM, Cardiff, Swansea and Carmarthen and to Exeter, Paignton, Plymouth, Truro and Penzance.

It also runs local trains on the Exmouth to Barnstaple route, the Cornish branches, Reading to Gatwick Airport, Bristol TM to Weymouth, Portsmouth and Brighton, Bristol to Severn Beach, Cardiff and Worcester.

The GWR fleet is incredibly mixed. New Hitachi bi-mode Class 800 IETs now work all inter-city operations out of Paddington, while a fleet of short-formed HSTs has been retained and upgraded to work local trains in Devon and Cornwall and also in the Avon and Cardiff area.

A fleet of ex-BR DMUs of Classes 150/158/165/166s work local services across the entire GWR network, while new Class 387 EMUs are now used on Paddington-Reading/Swindon local commuter trains following extensive electrification of the region.

Class 57s and Mk 3 coaches work the nightly Penzance-Paddington 'Night Riviera' 'sleeper' service. GWR has four 57/6s and hires 57306 from DRS if required.

As more parts of the GWML from Paddington to Bristol TM and Cardiff have been energised, the use of electric trains has increased.

GWR is looking to assess the feasibility of a battery Class 230 VivaRail unit for the West Ealing to Greenford service. A unit will be undergoing fast-charging kit trials to evaluative its viability for this service.

Class 08

GWR retains five Class 08s, based at Laira, Penzance, Reading and Bristol St Philips Marsh.
All are in the EFSH pool.

08641	GWR	BRW	LA	*Pride of Laira*
08644	GWR	BRW	LA	*Laira Diesel Depot 50 Years 1962-2012*
08645	GWR	KER	PZ	*St Piran*
08822	GWR	ICS	PM	*Dave Mills*
08836	GWR	GWR	LA	

Class 43

Once the mainstay of the GWR inter-city operation, the popular HST sets have been replaced in top link duties by Class 800 Hitachi units.

However, several power cars and Mk 3 trailers – the latter modified to have sliding doors – are retained as 'Castle class' 2+4 sets. Initially 11 sets and 24 power cars were planned, many of the vehicles owned by First Group, others leased from Angel. However, additional HST Power Cars 43009/010/022/027/029/171/172 have been added to the fleet and been repainted into GWR green.

A handful of Porterbrook and First Group power cars are still at Laira as a source of spare parts.

All are in the EFPC pool.

43004	GWR	ANG	LA	*Caerphilly Castle*
43005	GWR	ANG	LA	*St Michael's Mount*
43009	GWR	GWR	LA	
43010	GWR	GWR	LA	
43016	GWR	ANG	LA	*Powderham Castle*
43022	GWR	GWR	LA	
43027	GWR	GWR	LA	
43029	GWR	GWR	LA	
43040	GWR	ANG	LA	*Berry Pomeroy Castle * Name allocated, not yet applied*
43041	GWR	ANG	LA	*St Catherine's Castle*
43042	GWR	ANG	LA	*Tregenna Castle*
43063	FIR	GWR	LA (S)	
43088	GWR	FIR	LA	
43091	GWR	GWR	LA (S)	
43092	GWR	FIR	LA	*Cromwell's Castle*
43093	GWA	FIR	LA	*Old Oak Common HST Depot 1976-2018*
43094	GWR	FIR	LA	*St Mawes Castle*
43097	GWR	FIR	LA	*Castle Drogo*
43098	GWR	FIR	LA	*Walton Castle*
43122	GWR	FIR	LA	*Dunster Castle*
43153	GWR	FIR	LA	*Chûn Castle*
43154	GWR	FIR	LA	*Compton Castle*
43155	GWR	FIR	LA	*Rougemont Castle*
43156	GWR	FIR	LA	
43158	GWR	FIR	LA	*Kingswear Castle*
43160	GWR	FIR	LA	
43161	GWR	GWR	LA (S)	
43162	GWR	FIR	LA	
43170	GWR	ANG	LA	*Chepstow Castle*
43171	GWR	ANG	LA	
43172	GWR	ANG	LA	
43186	GWR	ANG	LA	*Taunton Castle*
43187	GWR	ANG	LA	*Cardiff Castle * Name allocated, not yet applied*
43188	GWR	ANG	LA	*Newport Castle*
43189	GWR	ANG	LA	*Launceston Castle*
43192	GWR	ANG	LA	*Trematon Castle*
43194	GWR	FIR	LA	*Okehampton Castle*
43195	FIR	FIR	LA (U)	
43198	GWR	FIR	LA	*Driver Stan Martin/Driver Brian Cooper*

While HSTs have been replaced on the Penzance-Paddington route by Class 800s, several sets have been retained as 2+4 'Castle' sets for workings in Severnside, Devon and Cornwall. 43153 *Chûn Castle* passes Dawlish with a train for Newton Abbot on 3 June 2021. *Jack Boskett*

HST trailers

The 'Castle' HST sets rely on fixed formation four-car sets comprising three Trailer Standards and one TGS. There are two spare of each type. Authorisation for more 'Castle' HSTs will require more Mk 3 coaches to be recovered from store and overhauls and modified.

Trailer standard open

48101	GWR	FIR	LA
48102	GWR	FIR	LA
48103	GWR	FIR	LA
48104	GWR	FIR	LA
48105	GWR	FIR	LA
48106	GWR	FIR	LA
48107	GWR	FIR	LA
48108	GWR	FIR	LA
48109	GWR	FIR	LA
48110	GWR	FIR	LA
48111	GWR	FIR	LA
48112	GWR	FIR	LA
48113	GWR	FIR	LA
48114	GWR	FIR	LA
48115	GWR	ANG	LA
48116	GWR	ANG	LA
48118	GWR	ANG	LA
48117	GWR	ANG	LA
48119	GWR	ANG	LA
48120	GWR	ANG	LA
48121	GWR	ANG	LA
48122	GWR	ANG	LA
48123	GWR	ANG	LA
48124	GWR	ANG	LA
48125	GWR	ANG	LA
48126	GWR	ANG	LA
48127	GWR	ANG	LA
48128	GWR	ANG	LA
48129	GWR	ANG	LA
48130	GWR	ANG	LA
48131	GWR	FIR	LA
48132	GWR	ANG	LA
48133	GWR	ANG	LA
48134	GWR	ANG	LA
48135	GWR	ANG	LA
48136	GWR	FIR	LA
48137	GWR	FIR	LA
48140	GWR	GWR	LA
48141	GWR	GWR	LA
48142	GWR	GWR	LA
48143	GWR	GWR	LA
48144	GWR	GWR	LA
48145	GWR	GWR	LA
48146	GWR	GWR	LA
48147	GWR	GWR	LA
48148	GWR	GWR	LA
48149	GWR	GWR	LA
48150	GWR	GWR	LA

Trailer guard's standard

49101	GWR	FIR	LA
49102	GWR	FIR	LA
49103	GWR	FIR	LA
49104	GWR	FIR	LA
49105	GWR	FIR	LA
49106	GWR	ANG	LA
49107	GWR	ANG	LA
49108	GWR	ANG	LA
49109	GWR	ANG	LA
49110	GWR	ANG	LA
49111	GWR	ANG	LA
49112	GWR	FIR	LA
49113	GWR	ANG	LA
49114	GWR	GWR	LA
49115	GWR	GWR	LA
49116	GWR	GWR	LA
49117	GWR	GWR	LA

GW Castle Class HST formations

As a rule the GWR HST vehicle trailers stay in fixed formations, but the four spare vehicles may replace vehicles as required for heavy maintenance or because of faults or damage. Power cars will be swapped around as required and are not assigned to any particular set.

	TGS	TS	TS	TS
GW01	49101	48101	48102	48103
GW02	49102	48104	48105	48106
GW03	49103	48107	48108	48109
GW04	49104	48110	48111	48112
GW05	49105	48113	48114	48115
GW06	49106	48116	48117	48118
GW07	49107	48119	48120	48121
GW08	49108	48122	48123	48124
GW09	49109	48125	48126	48127
GW10	49110	48128	48129	48130
GW11	49111	48131	48132	48133
GW12	49112	48134	43135	48136
GW13	49113	48137	48149	48140
GW14	49114	48140	48141	48142
GW15	49115	48143	48144	48145
GW16	49116	48146	48147	48148
spare	49117			

Class 57s

Four Class 57/6s, Class 47s rebuilt with General Motors engines, work the Paddington-Penzance 'Night Riviera' 'sleeper' service. The fleet is sometimes supplemented by a DRS Class 57/3, usually 57306. These locos could be replaced following an invitation to tender by First Group for between 15 and 35 bi-mode locos for use on Transpennine express routes and the 'Night Riviera', but any switch would not be anytime soon.

All are in the EFOO pool.

	Donor	Livery	Owner	Depot	Name
57602	47337	GWR	POR	PZ	*Restormel Castle*
57603	47349	GWR	POR	PZ	*Tintagel Castle*

GWR still operates the 'Night Riviera' sleeper train from Paddington to Penzance using Class 57s and Mk 3 coaches. 57604 *Pendennis Castle* stands at Paddington on 19 February 2019, having arrived with the empty stock from Reading for the down train. *Stuart West*

57604	47209	GWE	POR	PZ	*Pendennis Castle*
57605	47206	GWR	POR	PZ	*Totnes Castle*

Loco-hauled coaches

A small fleet of loco-hauled Mk 3 coaches are retained for the 'Night Riviera'. They are three kitchen buffet firsts, 11 SLEPs, three TSO and three BSOs. All are based at Penzance Long Rock.

Kitchen buffet first

10217	GWR	POR	PZ
10219	GWR	POR	PZ
10225	GWR	POR	PZ

Sleeper car with pantry

10532	GWR	POR	PZ
10534	GWR	POR	PZ
10563	GWR	POR	PZ
10584	GWR	POR	PZ
10589	GWR	POR	PZ
10590	GWR	POR	PZ
10594	GWR	POR	PZ
10596	GWR	POR	PZ
10601	GWR	POR	PZ
10612	GWR	POR	PZ
10616	GWR	POR	PZ

Open standard

12100	GWR	POR	PZ
12142	GWR	POR	PZ
12161	GWR	POR	PZ

Open brake unclassified

17173	GWR	POR	PZ
17174	GWR	POR	PZ
17175	GWR	POR	PZ

Mk 3 sleeper fixed formations

There are two sets of sleeper stock formed, but vehicles are interchanged more frequently than with other train types.

	OBK	OS	KBF	SLEP	SLEP	SLEP	SLEP
PZ88	17174	12100	10217	10612	10534	10532	10584
PZ89	17173	12142	10219	10601	10563	10596	10589
spare	17175	12161	10225		10590	10594	10616

Class 150

The two-car Sprinter DMUs built by BREL in 1986-87 are based at Exeter for use on Devon and Cornish branch line workings such as Exeter to Exmouth and Barnstaple, Plymouth-Gunnislake, Liskeard-Looe, Par-Newquay, Truro-Falmouth and St Erth-St Ives.

All are in the EFHQ pool.

Class 150/2

Second production series built of the Class 150 Sprinters, these two-car units have corridor connections.

	Livery	Owner	Depot	DMSL	DMS
150202	GWR	ANG	EX	52202	57202
150207	GWR	ANG	EX	52207	57207
150216	GWR	ANG	EX	52216	57216
150219	FIR	ANG	EX	52219	57219
150221	GWR	ANG	EX	52221	57221
150232	GWR	ANG	EX	52232	57232
150233	GWR	ANG	EX	52233	57233
150234	GWR	ANG	EX	52234	57234
150238	FIR	ANG	EX	52238	57238
150239	GWR	ANG	EX	52239	57239
150243	GWR	ANG	EX	52243	57243
150244	GWR	ANG	EX	52244	57244
150246	GWR	ANG	EX	52246	57246
150247	GWR	ANG	EX	52247	57247
150248	GWR	ANG	EX	52248	57248
150249	GWR	ANG	EX	52249	57249
150261	GWR	ANG	EX	52261	57261
150263	GWR	ANG	EX	52263	57263
150265	GWR	ANG	EX	52265	57265
150266	GWR	ANG	EX	52266	57266

Note: 52233 from 150233 has Okehampton line graphics.

On 25 October 2021, two-car 150221 heads towards Teignmouth on its way to Paignton with a train from Exmouth. *Pip Dunn*

Class 158

The two- and three-car sets, built in 1990 by BREL, are used between Exeter and Penzance on those trains not worked by HSTs and from Exeter to Bristol TM. The three-car sets tend to be used on the Bristol to Weymouth, Portsmouth and Brighton routes. Although numbered in the 15895x series, these are just re-formed three-car Class 158/0s.

158763 was damaged in an accident near Salisbury in 2021 and is expected to be withdrawn.

All are in the EFHQ pool.

	Livery	Owner	Depot	DMSL	DMSL	DMSL
158745	GWR	POR	PM	52745		57745
158747	GWR	POR	PM	52747		57747
158749	GWR	POR	PM	52749		57749
158750	GWR	POR	PM	52750		57750
158760	GWR	POR	PM	52760		57760
158762	GWR	POR	PM	52762		57762
158763	GWR	POR	LM (U)	52763		57763
158765	GWR	POR	PM	52765		57765
158766	GWR	POR	PM	52766		57766
158767	GWR	POR	PM	52767		57767
158769	GWR	POR	PM	52769		57769
158798	GWR	POR	EX	52798	58715	57798
158950	GWR	POR	EX	57751	52761	57761
158951	GWR	POR	EX	52751	52764	57764
158956	GWR	POR	EX	52748	52768	57768
158957	GWR	POR	EX	57748	52771	57771
158958	GWR	POR	EX	57746	52776	57776
158959	GWR	POR	EX	52746	52778	57778

Note: 58715 is an MSL.

Class 158s are used by GWR for many local and inter-regional duties; 158762 pauses at Exeter St David's with the 1435 Barnstaple-Exeter Central on 5 August 2021. *Robin Ralston*

Class 165

Once the mainstay of local commuter duties from Paddington, many of the Network Turbo units built by BREL at York in 1991-93 have been displaced by EMUs. Nevertheless they still work the branch lines in the Thames Valley, such as to Windsor & Eton Riverside, Marlow and Henley-on-Thames, and also some Paddington-Oxford trains.

Spare units displaced from the Thames Valley have been reallocated to Bristol for working to Severn Beach, Portsmouth, Taunton and other local services in the Avon, Somerset and Devon areas.

All are in the EFHQ pool.

Class 165/1

	Livery	Owner	Depot	DMSL	MS	DMS
165101	GWR	ANG	RG	58953	55415	58916
165102	GWR	ANG	RG	58954	55416	58917
165103	GWR	ANG	RG	58955	55417	58918
165104	GWR	ANG	RG	58956	55418	58919
165105	GWR	ANG	RG	58957	55419	58920
165106	GWR	ANG	RG	58958	55420	58921
165107	GWR	ANG	RG	58959	55421	58922
165108	GWR	ANG	RG	58960	55422	58923
165109	GWR	ANG	RG	58961	55423	58924
165110	GWR	ANG	RG	58962	55424	58925
165111	GWR	ANG	RG	58963	55425	58926
165112	GWR	ANG	RG	58964	55426	58927
165113	GWR	ANG	RG	58965	55427	58928
165114	GWR	ANG	RG	58966	55428	58929
165116	GWR	ANG	PM	58968	55430	58931
165117	GWR	ANG	RG	58969	55431	58932
165118	GWR	ANG	PM	58879		58933
165119	GWR	ANG	PM	58880		58934
165120	GWR	ANG	PM	58881		58935
165121	GWR	ANG	RG	58882		58936
165122	GWR	ANG	PM	58883		58937
165123	GWR	ANG	RG	58884		58938
165124	GWR	ANG	RG	58885		58939
165125	GWR	ANG	RG	58886		58940
165126	GWR	ANG	RG	58887		58941
165127	GWR	ANG	PM	58888		58942
165128	GWR	ANG	RG	58889		58943
165129	GWR	ANG	PM	58890		58944
165130	GWR	ANG	RG	58891		58945
165131	GWR	ANG	PM	58892		58946
165132	GWR	ANG	PM	58893		58947
165133	GWR	ANG	RG	58894		58948
165134	GWR	ANG	PM	58895		58949
165135	GWR	ANG	RG	58896		58950
165136	GWR	ANG	RG	58897		58951
165137	GWR	ANG	PM	58898		58952

Former Thames Trains Class 165/166s passed to GWR many years ago. Now displaced by electric units off many of their Thames duties, they have been redeployed in Somerset, Avon and Devon. 166207 passes Rangeworthy working the 1345 Worcester Foregate St-Bristol TM on 27 September 2020. *Glen Batten*

Class 166

These ex-Thames Trains BREL Networker units are all now based at Bristol St Philips Marsh for local Avon, Somerset and Devon duties, including to Exmouth, Barnstaple and Taunton. They also work through to Brighton.

All are in the EFHQ pool.

	Livery	Owner	Depot	DMCL	MS	DMSL	
166201	FIR	ANG	PM	58101	58601	58122	
166202	FIR	ANG	PM	58102	58602	58123	
166203	FIR	ANG	PM	58103	58603	58124	
166204	GWR	ANG	PM	58104	58604	58125	*Norman Topsham MBE*
166205	GWR	ANG	PM	58105	58605	58126	
166206	GWR	ANG	PM	58106	58606	58127	
166207	FIR	ANG	PM	58107	58607	58128	

166208	GWR	ANG	PM	58108	58608	58129	
166209	FIR	ANG	PM	58109	58609	58130	
166210	GWR	ANG	PM	58110	58610	58131	
166211	FIR	ANG	PM	58111	58611	58132	
166212	GWR	ANG	PM	58112	58612	58133	
166213	GWR	ANG	PM	58113	58613	58134	
166214	GWR	ANG	PM	58114	58614	58135	
166215	FIR	ANG	PM	58115	58615	58136	
166216	GWR	ANG	PM	58116	58616	58137	
166217	GWR	ANG	PM	58117	58617	58138	
166218	GWR	ANG	PM	58118	58618	58139	
166219	GWR	ANG	PM	58119	58619	58140	
166220	GWR	ANG	PM	58120	58620	58141	*Roger Watkins – the GWR Master Train Planner*
166221	FIR	ANG	PM	58121	58621	58142	*Reading Train Care Depot*

Class 387

These Bombardier Electrostar 25kV AC EMUs work out of Paddington along the Thames corridor to Reading, Didcot and Swindon and now to Cardiff. Units 387130-141 work Paddington-Heathrow Airport shuttles and are listed on page 175 (Heathrow Express).

All are in the EFHQ pool.

	Livery	Owner	Depot	DMC	MS	PTS	DMS
387142	GWR	POR	RG	421142	422142	423142	424142
387143	GWR	POR	RG	421143	422143	423143	424143
387144	GWR	POR	RG	421144	422144	423144	424144
387145	GWR	POR	RG	421145	422145	423145	424145
387146	GWR	POR	RG	421146	422146	423146	424146
387147	GWR	POR	RG	421147	422147	423147	424147
387148	GWR	POR	RG	421148	422148	423148	424148
387149	GWR	POR	RG	421149	422149	423149	424149
387150	GWR	POR	RG	421150	422150	423150	424150
387151	GWR	POR	RG	421151	422151	423151	424151
387152	GWR	POR	RG	421152	422152	423152	424152
387153	GWR	POR	RG	421153	422153	423153	424153
387154	GWR	POR	RG	421154	422154	423154	424154
387155	GWR	POR	RG	421155	422155	423155	424155
387156	GWR	POR	RG	421156	422156	423156	424156
387157	GWR	POR	RG	421157	422157	423157	424157
387158	GWR	POR	RG	421158	422158	423158	424158
387159	GWR	POR	RG	421159	422159	423159	424159
387160	GWR	POR	RG	421160	422160	423160	424160
387161	GWR	POR	RG	421161	422161	423161	424161
387162	GWR	POR	RG	421162	422162	423162	424162
387163	GWR	POR	RG	421163	422163	423163	424163
387164	GWR	POR	RG	421164	422164	423164	424164
387165	GWR	POR	RG	421165	422165	423165	424165
387166	GWR	POR	RG	421166	422166	423166	424166
387167	GWR	POR	RG	421167	422167	423167	424167
387168	GWR	POR	RG	421168	422168	423168	424168
387169	GWR	POR	RG	421169	422169	423169	424169
387170	GWR	POR	RG	421170	422170	423170	424170
387171	GWR	POR	RG	421171	422171	423171	424171
387172	GWR	POR	RG	421172	422172	423172	424172
387173	GWR	POR	RG	421173	422173	423173	424173
387174	GWR	POR	RG	421174	422174	423174	424174

New-build Class 387 Electrostar EMUs have can now work from Paddington to Cardiff on semi-fast trains. On 10 May 2021, 387146 and 387143 make their first appearance at Swindon with a driver and train manager training run. *Jack Boskett*

Class 769

A fleet 19 Class 769 FLEX units have been converted from redundant Thameslink Class 319s by Brush Traction for the GWR franchise. They retain their dual-voltage 25kV AC and 750V DC capabilities but also have diesel engines to work 'off the wires', making them tri-mode.

They were intended for use on the Oxford-Gatwick Airport route to allow three trains an hour, however reliability issues on the Class 769s in general (they are also used by Northern and TfW), along with the inability to implement driver training due to the pandemic and a question over the business case for that level of service on the route means the units may now not see any use with GWR.

All are in the EFHQ pool.

	Old No.	Livery	Owner	Depot	DMC	MS	TS	DMC
769922	319422	GWR	POR	RG (Q)	77333	62912	71793	77332
769923	319423	GWR	POR	RG (Q)	77335	62913	71794	77334
769925	319425	GWR	POR	RG (Q)	77339	62915	71796	77338
769927	319427	GWR	POR	RG (Q)	77343	62917	71798	77342
769928	319428	GWR	POR	RG (Q)	77345	62918	71799	77344
769930	319430	GWR	POR	RG (Q)	77349	62920	71801	77348
769932	319432	GWR	POR	RG (Q)	77353	62922	71803	77352
769935	319435	GWR	POR	RG (Q)	77359	62925	71806	77358
769936	319436	GWR	POR	BU (S)	77361	62926	71807	77360
769937	319437	GWR	POR	RG (Q)	77363	62927	71808	77362
769938	319438	GWR	POR	RG (Q)	77365	62928	71809	77364
769939	319439	GWR	POR	RG (Q)	77367	62929	71810	77366
769940	319440	GWR	POR	RG (Q)	77369	62930	71811	77368
769943	319443	GWR	POR	RG (Q)	77375	62933	71814	77374
769944	319444	GWR	POR	RG (Q)	77377	62934	71815	77376
769946	319446	GWR	POR	RG (Q)	77381	62936	71817	77380
769947	319447	GWR	POR	RG (Q)	77431	62961	71866	77430
769949	319449	GWR	POR	RG (Q)	77435	62963	71868	77434
769959	319459	GWR	POR	RG (Q)	77455	62973	71878	77454

Note: Storage locations may change without notice

GWR has 19 four-car tri-mode Class 769s on order rebuilt from ex-Thameslink Class 319 EMUs. 769943 approaches the works entrance at Eastleigh, having arrived from Reading Depot on 5 July 2021. There is still a question mark as to whether these units will enter traffic with GWR. *Mark Pike*

Class 800

GWR was the first TOC to put the new Hitachi InterCity Express Trains (IET) units into traffic and the full fleet is now in use which allowed the elimination of HST workings to Paddington in 2019.

All are in the EFHQ pool.

Class 800/0

Bi-mode 25kV AC and diesel five-car units.

	Livery	Owner	Depot	EDTS	MS	MS	MS	PDTRBF	
800001	GWR	AGI	NP	811001	812001	813001	814001	815001	
800002	GWR	AGI	NP	811002	812002	813002	814002	815002	
800003	GWR	AGI	NP	811003	812003	813003	814003	815003	*Queen Elizabeth II/Queen Victoria*
800004	GWR	AGI	NP	811004	812004	813004	814004	815004	*Isambard Kingdom Brunel/ Sir Daniel Gooch*
800005	GWR	AGI	NP	811005	812005	813005	814005	815005	
800006	GWR	AGI	NP	811006	812006	813006	814006	815006	*Tulbahadhur Pun VC*
800007	GWR	AGI	NP	811007	812007	813007	814007	815007	
800008	GWR	AGI	NP	811008	812008	813008	814008	815008	
800009	GWR	AGI	NP	811009	812009	813009	814009	815009	*Sir Gareth Edwards/John Charles*
800010	GWR	AGI	NP	811010	812010	813010	814010	815010	*Michael Bond/Paddington Bear*
800011	GWR	AGI	NP	811011	812011	813011	814011	815011	
800012	GWR	AGI	NP	811012	812012	813012	814012	815012	
800013	GWR	AGI	NP	811013	812013	813013	814013	815013	
800014	GWR	AGI	NP	811014	812014	813014	814014	815014	*Megan Lloyd George CH/Edith New*
800015	GWR	AGI	NP	811015	812015	813015	814015	815015	
800016	GWR	AGI	NP	811016	812016	813016	814016	815016	
800017	GWR	AGI	NP	811017	812017	813017	814017	815017	
800018	GWR	AGI	NP	811018	812018	813018	814018	815018	
800019	GWR	AGI	NP	811019	812019	813019	814019	815019	*Joy Lofthouse/ Johnny Johnson MBE DFM*
800020	GWR	AGI	NP	811020	812020	813020	814020	815020	*Bob Woodward/Elizabeth Ralph*
800021	GWR	AGI	NP	811021	812021	813021	814021	815021	
800022	GWR	AGI	NP	811022	812022	813022	814022	815022	*Tulbahadhur Pun VC*
800023	GWR	AGI	NP	811023	812023	813023	814023	815023	*Firefighter Fleur Lombard QGM/ Kathryn Osmond*
800024	GWR	AGI	NP	811024	812024	813024	814024	815024	
800025	GWR	AGI	NP	811025	812025	813025	814025	815025	*Captain Sir Tom Moore*
800026	GWR	AGI	NP	811026	812026	813026	814026	815026	*Don Cameron*
800027	GWR	AGI	NP	811027	812027	813027	814027	815027	
800028	GWR	AGI	NP	811028	812028	813028	814028	815028	
800029	GWR	AGI	NP	811029	812029	813029	814029	815029	*Evette Wakely / Christopher Dando*
800030	GWR	AGI	NP	811030	812030	813030	814030	815030	*Henry Cleary/Lincoln Callaghan*
800031	GWR	AGI	NP	811031	812031	813031	814031	815031	*Mazen Salmou/Charlotte Marsland*
800032	GWR	AGI	NP	811032	812032	813032	814032	815032	*Iain Bugler/ Sarah Williams-Martin BEM*
800033	GWR	AGI	NP	811033	812033	813033	814033	815033	
800034	GWR	AGI	NP	811034	812034	813034	814034	815034	
800035	GWR	AGI	NP	811035	812035	813035	814035	815035	
800036	GWR	AGI	NP	811036	812036	813036	814036	815036	*Dr Paul Stephenson OBE*

Class 800/3

Bi-mode 25kV AC and diesel nine-car units.

All are in the EFHQ pool.

	Livery	Owner	Depot	PDTS	MS	MS	TS	MS	TS	MS	MF	PDTRBF	
800301	GWR	AGI	NP	821001	822001	823001	824001	825001	826001	827001	828001	829001	
800302	GWR	AGI	NP	821002	822002	823002	824002	825002	826002	827002	828002	829002	
800303	GWR	AGI	NP	821003	822003	823003	824003	825003	826003	827003	828003	829003	
800304	GWR	AGI	NP	821004	822004	823004	824004	825004	826004	827004	828004	829004	
800305	GWR	AGI	NP	821005	822005	823005	824005	825005	826005	827005	828005	829005	
800306	GWR	AGI	NP	821006	822006	823006	824006	825006	826006	827006	828006	829006	*Allan Leonard Lewis VC/Harold Day DSC*
800307	GWR	AGI	NP	821007	822007	823007	824007	825007	826007	827007	828007	829007	
800308	GWR	AGI	NP	821008	822008	823008	824008	825008	826008	827008	828008	829008	
800309	GWR	AGI	NP	821009	822009	823009	824009	825009	826009	827009	828009	829009	
800310	GWR	AGI	NP	821010	822010	823010	824010	825010	826010	827010	828010	829010	*Wing Commander Ken Rees*
800311	GWR	AGI	NP	821011	822011	823011	824011	825011	826011	827011	828210	829011	
800312	GWR	AGI	NP	821012	822012	823012	824012	825012	826012	827012	828012	829012	
800313	GWR	AGI	NP	821013	822013	823013	824013	825013	826013	827013	828013	829013	
800314	GWR	AGI	NP	821014	822014	823014	824014	825014	826014	827014	828014	829014	*Odette Hallowes GC MBE*
800315	GWR	AGI	NP	821015	822015	823015	824015	825015	826015	827015	828015	829015	
800316	GWR	AGI	NP	821016	822016	823016	824016	825016	826016	827016	828016	829016	
800317	GWR	AGI	NP	821017	822017	823017	824017	825017	826017	827017	828017	829017	*Freya Bevan*
800318	GWR	AGI	NP	821018	822018	823018	824018	825018	826018	827018	828018	829018	
800319	GWR	AGI	NP	821019	822019	823019	824019	825019	826019	827019	828019	829019	
800320	GWR	AGI	NP	821020	822020	823020	824020	825020	826020	827020	828020	829020	
800321	GWR	AGI	NP	821021	822021	823021	824021	825021	826021	827021	828021	829021	

800313 sports various slogans promoting green travel.

The Class 800/3s are nine-car bi-mode units used on many of GWR's inter-city routes. 800320 approaches Westbury with the 1054 Paignton-Paddington on 11 August 2021. *Glen Batten*

Class 802

The same as Class 800s but with modifications to work along the Dawlish sea wall.

Class 802/0

Bi-mode 25kV AC and diesel five-car units.

All are in the EFHQ pool.

	Livery	Owner	Depot	FDTS	MS	MS	MS	PDTRBF	
802001	GWR	EVS	NP	831001	832001	833001	834001	835001	
802002	GWR	EVS	NP	831002	832002	833002	834002	835002	*Steve Whiteway*
802003	GWR	EVS	NP	831003	832003	833003	834003	835003	
802004	GWR	EVS	NP	831004	832004	833004	834004	835004	
802005	GWR	EVS	NP	831005	832005	833005	834005	835005	
802006	GWR	EVS	NP	831006	832006	833006	834006	835006	*Harry Billinge MBE/ Donovan & Jenifer Gardner*
802007	GWR	EVS	NP	831007	832007	833007	834007	835007	
802008	GWR	EVS	NP	831008	832008	833008	834008	835008	*Rick Rescorla/ RNLB Solomon Browne*
802009	GWR	EVS	NP	831009	832009	833009	834009	835009	
802010	GWR	EVS	NP	831010	832010	833010	834010	835010	*Corporal George Sheard/ Kieron Griffin*
802011	GWR	EVS	NP	831011	832011	833011	834011	835011	*Sir Joshua Reynolds PRA/ Capt. Robert Falcon Scott RN CVO*
802012	GWR	EVS	NP	831012	832012	833012	834012	835012	
802013	GWR	EVS	NP	831013	832013	833013	834013	835013	*Michael Eavis CBE*
802014	GWR	EVS	NP	831014	832014	833014	834014	835014	
802015	GWR	EVS	NP	831015	832015	833015	834015	835015	
802016	GWR	EVS	NP	831016	832016	833016	834016	835016	
802017	GWR	EVS	NP	831017	832017	833017	834017	835017	
802018	GWR	EVS	NP	831018	832018	833018	834018	835018	*Preston de Mendonça/Jeremy Doyle*
802019	GWR	EVS	NP	831019	832019	833019	834019	835019	
802020	GWA	EVS	NP	831020	832020	833020	834020	835020	
802021	GWR	EVS	NP	831021	832021	833021	834021	835021	
802022	GWR	EVS	NP	831022	832022	833022	834022	835022	

Class 802/1

Bi-mode 25kV AC and diesel nine-car units.

All are in the WFHQ pool.

	Livery	Owner	Depot	PDTS	MS	MS	TS	MS	TS	MS	MF	PDTRBF	
802101	GWR	EVS	NP	831101	832101	833101	834101	835101	836101	837101	838101	839101	*Nancy Astor CH*
802102	GWR	EVS	NP	831102	832102	833102	834102	835102	836102	837102	838102	839102	
802103	GWR	EVS	NP	831103	832103	833103	834103	835103	836103	837103	838103	839103	
802104	GWR	EVS	NP	831104	832104	833104	834104	835104	836104	837104	838104	839104	
802105	GWR	EVS	NP	831105	832105	833105	834105	835105	836105	837105	838105	839105	
802106	GWR	EVS	NP	831106	832106	833106	834106	835106	836106	837106	838106	839106	
802107	GWR	EVS	NP	831107	832107	833107	834107	835107	836107	837107	838107	839107	
802108	GWR	EVS	NP	831108	832108	833108	834108	835108	836108	837108	838108	839108	
802109	GWR	EVS	NP	831109	832109	833109	834109	835109	836109	837109	838109	839109	
802110	GWR	EVS	NP	831110	832110	833110	834110	835110	836110	837110	838110	839110	
802111	GWR	EVS	NP	831111	832111	833111	834111	835111	836111	837111	838311	839111	
802112	GWR	EVS	NP	831112	832112	833112	834112	835112	836112	837112	838112	839112	
802113	GWR	EVS	NP	831113	832113	833113	834113	835113	836113	837113	838113	839113	
802114	GWR	EVS	NP	831114	832114	833114	834114	835114	836114	837114	838114	839114	

London North Eastern Railway

Contact details

Website: lner.co.uk
Twitter: @LNER

Key personnel

Managing Director: David Horne
Engineering Director: John Doughty

Overview

The main East Coast Main Line operator, which runs from King's Cross to Leeds and Edinburgh, but also operates some trains to Lincoln, Harrogate and Aberdeen and a limited number of services to Inverness, Hull, Glasgow Central and Middlesbrough. It had been hoped to introduce a daily service to Huddersfield, but this has been put on hold due to the pandemic.

The fleet of Class 800/801 Hitachi AT300 high-speed multiple units are all technically bi-mode, but the reality is the Class 801s are effectively straight electric units and only have one engine for emergency use to clear the running line if there is an issue with the electric power. The Class 800s are bi-mode for those duties that run to destinations away from the core electrified route. The new bi-mode trains have allowed LNER to not only replace HSTs on operations to Inverness, Harrogate, Hull and Aberdeen.

However, a small fleet of Class 91s has had to be retained longer than expected and should run into 2023 at least. They work mostly from King's Cross to Leeds and some trains to York. The locos have been reallocated from Bounds Green to Neville Hill.

Class 91

The fleet of 31 locos has been cut by to 12 locos. In February 2021 these locos, and their Mk 4 coaches were put into temporary warm store due to a reduced need brought about by the Covid pandemic but slowly returned to traffic later in the summer.

All are in the IECA pool.

	Old No.	Livery	Owner	Depot	
91101	91001	VFS	EVS	NL	*Flying Scotsman*
91105	91005	VEC	EVS	NL	
91106	91006	VEC	EVS	NL	
91107	91007	VEC	EVS	NL	*Skyfall*
91109	91009	VEC	EVS	NL	*Sir Bobby Robson*
91110	91010	BBM	EVS	NL	*Battle of Britain Memorial Flight*
91111	91011	FTF	EVS	NL	*For the Fallen*
91114	91014	VEC	EVS	NL	*Durham Cathedral*
91119	91019	ICS	EVS	NL	*Bounds Green InterCity Depot 1977-2017*
91124	91024	VEC	EVS	NL	
91127	91027	VEC	EVS	NL	
91130	91030	VEC	EVS	NL	*Lord Mayor of Newcastle*

LNER retains a handful of Class 91-hauled Mk 4 sets. On 25 November 2021, 91127 failed prior to working the 1003 King's Cross-Leeds and waits to run empty to Bounds Green Depot. *Pip Dunn*

Class 800

These bi-mode Azuma units of nine- and five-car sets work most of LNER's services. The latter can run as ten-car sets if required, including for some services where the trains splits.
All are in the HBHQ pool.

Class 800/1

Bi-mode 25kV AC and diesel nine-car units.

	Livery	Owner	Depot	PTDS	MS	MS	TSRB	MS	TS	MC	MF	PDTRBF
800101	LNE	AGI	DN	811101	812101	813101	814101	815101	816101	817101	818101	819101
800102	LNE	AGI	DN	811102	812102	813102	814102	815102	816102	817102	818102	819102
800103	LNE	AGI	DN	811103	812103	813103	814103	815103	816103	817103	818103	819103
800104	LNE	AGI	DN	811104	812104	813104	814104	815104	816104	817104	818104	819104
800105	LNE	AGI	DN	811105	812105	813105	814105	815105	816105	817105	818105	819105
800106	LNE	AGI	DN	811106	812106	813106	814106	815106	816106	817106	818106	819106
800107	LNE	AGI	DN	811107	812107	813107	814107	815107	816107	817107	818107	819107
800108	LNE	AGI	DN	811108	812108	813108	814108	815108	816108	817108	818108	819108
800109	LNE	AGI	DN	811109	812109	813109	814109	815109	816109	817109	818109	819109
800110	LNE	AGI	DN	811110	812110	813110	814110	815110	816110	817110	818110	819110
800111	LNE	AGI	DN	811111	812111	813111	814111	815111	816111	817111	818111	819111
800112	LNE	AGI	DN	811112	812112	813112	814112	815112	816112	817112	818112	819112
800113	LNE	AGI	DN	811113	812113	813113	814113	815113	816113	817113	818113	819113

Class 800/2

Bi-mode 25kV AC and diesel five-car units.

	Livery	Owner	Depot	PTDS	MRSB	MS	MC	PDTRBF
800201	LNE	AGI	DN	811201	812201	813201	814201	815201
800202	LNE	AGI	DN	811202	812202	813202	814202	815202
800203	LNE	AGI	DN	811203	812203	813203	814203	815203
800204	LNE	AGI	DN	811204	812204	813204	814204	815204
800205	LNE	AGI	DN	811205	812205	813205	814205	815205
800206	LNE	AGI	DN	811206	812206	813206	814206	815206
800207	LNE	AGI	DN	811207	812207	813207	814207	815207
800208	LNE	AGI	DN	811208	812208	813208	814208	815208
800209	LNE	AGI	DN	811209	812209	813209	814209	815209
800210	LNE	AGI	DN	811210	812210	813210	814210	815210

Most LNER trains are worked by Class 800 or 801 Hitachi AT300 units. 800201 has just arrived at Leeds on 11 December 2021 with the 1535 King's Cross-Harrogate and will reverse in order to complete its journey. *Pip Dunn*

Class 801

These units are essentially electric only (they do have a single diesel engine for emergency use only) and so are restricted to the King's Cross-Leeds/Edinburgh/Stirling routes only.

All are in the HBHQ pool.

Class 801/1

Bi-mode 25kV AC and single-engine diesel five-car units.

	Livery	Owner	Depot	PTDS	MRSB	MS	MC	PDTRBF
801101	LNE	AGI	DN	821101	822101	823101	824101	825101
801102	LNE	AGI	DN	821102	822102	823102	824102	825102
801103	LNE	AGI	DN	821103	822103	823103	824103	825103
801104	LNE	AGI	DN	821104	822104	823104	824104	825104
801105	LNE	AGI	DN	821105	822105	823105	824105	825105
801106	LNE	AGI	DN	821106	822106	823106	824106	825106
801107	LNE	AGI	DN	821107	822107	823107	824107	825107
801108	LNE	AGI	DN	821108	822108	823108	824108	825108
801109	LNE	AGI	DN	821109	822109	823109	824109	825109
801110	LNE	AGI	DN	821110	822110	823110	824110	825110
801111	LNE	AGI	DN	821111	822111	823111	824111	825111
801112	LNE	AGI	DN	821112	822112	823112	824112	825112

Class 801/2

Bi-mode 25kV AC and single-engine diesel nine-car units.

	Livery	Owner	Depot	PDTS	MS	MS	TRSB	MS	TS	MC	MF	PDTRBF
801201	LNE	AGI	BN	821201	822201	823201	824201	825201	826201	827201	828201	829201
801202	LNE	AGI	BN	821202	822202	823202	824202	825202	826202	827202	828202	829202
801203	LNE	AGI	BN	821203	822203	823203	824203	825203	826203	827203	828203	829203
801204	LNE	AGI	BN	821204	822204	823204	824204	825204	826204	827204	828204	829204
801205	LNE	AGI	BN	821205	822205	823205	824205	825205	826205	827205	828205	829205
801206	LNE	AGI	BN	821206	822206	823206	824206	825206	826206	827206	828206	829206
801207	LNE	AGI	BN	821207	822207	823207	824207	825207	826207	827207	828207	829207
801208	LNE	AGI	BN	821208	822208	823208	824208	825208	826208	827208	828208	829208
801209	LNE	AGI	BN	821209	822209	823209	824209	825209	826209	827209	828209	829209
801210	LNE	AGI	BN	821210	822210	823210	824210	825210	826210	827210	828210	829210
801211	LNE	AGI	BN	821211	822211	823211	824211	825211	826211	827211	828211	829211
801212	LNE	AGI	BN	821212	822212	823212	824212	825212	826212	827212	828212	829212
801213	LNE	AGI	BN	821213	822213	823213	824213	825213	826213	827213	828213	829213
801214	LNE	AGI	BN	821214	822214	823214	824214	825214	826214	827214	828214	829214
801215	LNE	AGI	BN	821215	822215	823215	824215	825215	826215	827215	828215	829215
801216	LNE	AGI	BN	821216	822216	823216	824216	825216	826216	827216	828216	829216
801217	LNE	AGI	BN	821217	822217	823217	824217	825217	826217	827217	828217	829217
801218	LNE	AGI	BN	821218	822218	823218	824218	825218	826218	827218	828218	829218
801219	LNE	AGI	BN	821219	822219	823219	824219	825219	826219	827219	828219	829219
801220	LNE	AGI	BN	821220	822220	823220	824220	825220	826220	827220	828220	829220
801221	LNE	AGI	BN	821221	822221	823221	824221	825221	826221	827221	828221	829221
801222	LNE	AGI	BN	821222	822222	823222	824222	825222	826222	827222	828222	829222
801223	LNE	AGI	BN	821223	822223	823223	824223	825223	826223	827223	828223	829223
801224	LNE	AGI	BN	821224	822224	823224	824224	825224	826224	827224	828224	829224
801225	LNE	AGI	BN	821225	822225	823225	824225	825225	826225	827225	828225	829225
801226	LNE	AGI	BN	821226	822226	823226	824226	825226	826226	827226	828226	829226
801227	LNE	AGI	BN	821227	822227	823227	824227	825227	826227	827227	828227	829227
801228	LNE	AGI	BN	821228	822228	823228	824228	825228	826228	827228	828228	829228
801229	LNE	AGI	BN	821229	822229	823229	824229	825229	826229	827229	828229	829229
801230	LNE	AGI	BN	821230	822230	823230	824230	825230	826230	827230	828230	829230

Mk 4 trailers

These former BR coaches are being retained to work with Class 91s until 2023. They run in fixed formations of nine coaches plus a DVT.

Kitchen buffet standard

10300	VEC	EVS	NL
10306	VEC	EVS	NL
10309	VEC	EVS	NL
10311	VEC	EVS	NL
10313	VEC	EVS	NL
10324	VEC	EVS	NL
10333	VEC	EVS	NL

Open first

11229	VEC	EVS	NL
11279	VEC	EVS	NL
11284	VEC	EVS	NL
11285	VEC	EVS	NL
11286	VEC	EVS	NL
11295	VEC	EVS	NL

Open first (disabled)

11306	VEC	EVS	NL
11308	VEC	EVS	NL
11312	VEC	EVS	NL
11313	VEC	EVS	NL
11315	VEC	EVS	NL
11317	VEC	EVS	NL
11318	VEC	EVS	NL
11326	VEC	EVS	NL

Open first

11406	VEC	EVS	NL
11408	VEC	EVS	NL
11412	VEC	EVS	NL
11413	VEC	EVS	NL
11415	VEC	EVS	NL
11417	VEC	EVS	NL
11418	VEC	EVS	NL
11426	VEC	EVS	NL

Open standard (end)

12205	VEC	EVS	NL
12208	VEC	EVS	NL
12212	VEC	EVS	NL
12214	VEC	EVS	NL
12220	VEC	EVS	NL
12223	VEC	EVS	NL
12226	VEC	EVS	NL
12228	VEC	EVS	NL

Open standard (disabled)

12303	VEC	EVS	NL
12309	VEC	EVS	NL
12311	VEC	EVS	NL
12313	VEC	EVS	NL
12325	VEC	EVS	NL
12328	VEC	EVS	NL
12330	VEC	EVS	NL

Open standard

12404	VEC	EVS	NL
12406	VEC	EVS	NL
12407	VEC	EVS	NL
12409	VEC	EVS	NL
12420	VEC	EVS	NL
12422	VEC	EVS	NL
12424	VEC	EVS	NL
12426	VEC	EVS	NL
12427	VEC	EVS	NL
12429	VEC	EVS	NL
12430	VEC	EVS	NL
12431	VEC	EVS	NL
12432	VEC	EVS	NL
12442	VEC	EVS	NL
12444	VEC	EVS	NL
12465	VEC	EVS	NL
12469	VEC	EVS	NL
12474	VEC	EVS	NL
12481	VEC	EVS	NL
12485	VEC	EVS	NL
12515	VEC	EVS	NL

Mk 4 DVTs

82205	VFS	EVS	NL
82208	VEC	EVS	NL
82211	VEC	EVS	NL
82212	VEC	EVS	NL
82213	VEC	EVS	NL
82214	VEC	EVS	NL
82215	VEC	EVS	NL
82222	VEC	EVS	NL
82223	VEC	EVS	NL
82225	VEC	EVS	NL

LNER Mk 4 formations

NL06	12208	12406	12420	12422	12313	10309	11279	11306	11406	82208
NL08	12205	12481	12485	12407	12328	10300	11229	11308	11408	82211
NL12	12212	12431	12404	12426	12330	10333	11284	11312	11412	82212
NL13	12228	12469	12430	12424	12311	10313	11285	11313	11413	82213
NL15	12226	12442	12409	12515	12309	10306	11286	11315	11415	82214
NL16	12213	12428	12433	12467	12312	10315	11418	11318	11416	82222
NL17	12223	12444	12427	12432	12303	10324	11288	11317	11417	82225
NL26	12220	12474	12465	12429	12325	10311	11295	11326	11426	82223
Spare	12202									82215
Spare	12214									82205

London Overground

Contact details

Website: www.arrivaraillondon.co.uk
Twitter: LDNOverground

Key personnel

Managing Director: Paul Hutchings
Engineering Director: Kate Marjoribanks

Overview

Arriva Rail London is responsible for running the London Overground network under a Concession Agreement with Transport for London (TfL). The seven-and-a-half-year deal started on 13 November 2016 and expires on 24 May 2024.

London Overground operates local commuter trains within the Greater London area. The fleet is a mix of Bombardier Class 378 EMUs and newer Class 710 EMUs – the latter have replaced the TOC's only remaining diesel trains following recent electrification of the Gospel Oak to Barking line.

Class 09

The only diesel traction left on the LOROL fleet is an ex-BR Class 09 shunter, based at Willesden depot.

09007	GRW	LOR	WN

Class 378

These new Bombardier Electrostars are metro-style trains similar in internal layout to a tube train. There were delivered as three-car sets but strengthened firstly to four-cars and now five-car sets. They work out of Euston to Watford and also on the North, East, and West London lines.

All are in the EKHQ pool.

Class 378/1

750V DC five-car units, originally delivered as four-car sets but lengthened in 2014-15.

	Livery	Owner	Depot	DMS	MS	TS	MS	DMS
378135	LOR	QWL	NG	38035	38235	38335	38435	38135
378136	LOR	QWL	NG	38036	38236	38336	38436	38136
378137	LOR	QWL	NG	38037	38237	38337	38437	38137
378138	LOR	QWL	NG	38038	38238	38338	38438	38138
378139	LOR	QWL	NG	38039	38239	38339	38439	38139
378140	LOR	QWL	NG	38040	38240	38340	38440	38140
378141	LOR	QWL	NG	38041	38241	38341	38441	38141
378142	LOR	QWL	NG	38042	38242	38342	38442	38142
378143	LOR	QWL	NG	38043	38243	38343	38443	38143
378144	LOR	QWL	NG	38044	38244	38344	38444	38144
378145	LOR	QWL	NG	38045	38245	38345	38445	38145
378146	LOR	QWL	NG	38046	38246	38346	38446	38146
378147	LOR	QWL	NG	38047	38247	38347	38447	38147
378148	LOR	QWL	NG	38048	38248	38348	38448	38148
378149	LOR	QWL	NG	38049	38249	38349	38449	38149
378150	LOR	QWL	NG	38050	38250	38350	38450	38150
378151	LOR	QWL	NG	38051	38251	38351	38451	38151
378152	LOR	QWL	NG	38052	38252	38352	38452	38152
378153	LOR	QWL	NG	38053	38253	38353	38453	38153
378154	LOR	QWL	NG	38054	38254	38354	38454	38154

Class 378/2

25kV AC and 750V DC dual-voltage five-car units. The first 24 units were originally delivered in 2008 as three-car Class 378/0s, extended to four-cars in 2010 and then the entire fleet became five-cars in 2014-15.

	Old No.	Livery	Owner	Depot	DMS	MS	PTS	MS	DMS	
378201	378001	LOR	QWL	NG	38001	38201	38301	38401	38101	
378202	378002	LOR	QWL	NG	38002	38202	38302	38402	38102	
378203	378003	LOR	QWL	NG	38003	38203	38303	38403	38103	
378204	378004	LOR	QWL	NG	38004	38204	38304	38404	38104	*Professor Sir Peter Hall*
378205	378005	LOR	QWL	NG	38005	38205	38305	38405	38105	
378206	378006	LOR	QWL	NG	38006	38206	38306	38406	38106	
378207	378007	LOR	QWL	NG	38007	38207	38307	38407	38107	

378208	378008	LOR	QWL	NG	38008	38208	38308	38408	38108	
378209	378009	LOR	QWL	NG	38009	38209	38309	38409	38109	
378210	378010	LOR	QWL	NG	38010	38210	38310	38410	38110	
378211	378011	LOR	QWL	NG	38011	38211	38311	38411	38111	
378212	378012	LOR	QWL	NG	38012	38212	38312	38412	38112	
378213	378013	LOR	QWL	NG	38013	38213	38313	38413	38113	
378214	378014	LOR	QWL	NG	38014	38214	38314	38414	38114	
378215	378015	LOR	QWL	NG	38015	38215	38315	38415	38115	
378216	378016	LOR	QWL	NG	38016	38216	38316	38416	38116	
378217	378017	LOR	QWL	NG	38017	38217	38317	38417	38117	
378218	378018	LOR	QWL	NG	38018	38218	38318	38418	38118	
378219	378019	LOR	QWL	NG	38019	38219	38319	38419	38119	
378220	378020	LOR	QWL	NG	38020	38220	38320	38420	38120	
378221	378021	LOR	QWL	NG	38021	38221	38321	38421	38121	
378222	378022	LOR	QWL	NG	38022	38222	38322	38422	38122	
378223	378023	LOR	QWL	NG	38023	38223	38323	38423	38123	
378224	378024	LOR	QWL	NG	38024	38224	38324	38424	38124	
378225		LOR	QWL	NG	38025	38225	38325	38425	38125	
378226		LOR	QWL	NG	38026	38226	38326	38426	38126	
378227		LOR	QWL	NG	38027	38227	38327	38427	38127	
378228		LOR	QWL	NG	38028	38228	38328	38428	38128	
378229		LOR	QWL	NG	38029	38229	38329	38429	38129	
378230		LOR	QWL	NG	38030	38230	38330	38430	38130	
378231		LOR	QWL	NG	38031	38231	38331	38431	38131	
378232		LOR	QWL	NG	38032	38232	38332	38432	38132	*Jeff Langston*
378233		LOR	QWL	NG	38033	38233	38333	38433	38133	*Ian Brown CBE*
378234		LOR	QWL	NG	38034	38234	38334	38434	38134	
378255		LOR	QWL	NG	38055	38255	38355	38455	38155	
378256		LOR	QWL	NG	38056	38256	38356	38456	38156	
378257		LOR	QWL	NG	38057	38257	38357	38457	38157	

The majority of London Overground's trains are worked by five-car Class 378 EMUs. Passing Kensington Olympia on 15 October 2019 is 378222 working the 1308 Stratford-Clapham Junction. *Paul Shannon*

Class 710

New Bombardier Aventra units for Gospel Oak-Barking, Liverpool Street to Enfield and Watford Junction to Euston routes. The Class 710/1 units are 25kV AC and the 710/2s and 710/3s are dual voltage.

All are in the EKHQ pool.

Class 710/1

25kV AC four-car sets.

	Livery	Owner	Depot	DMS	MS	PMS	DMS
710101	LOR	RFL	WN	431101	431201	431301	431501
710102	LOR	RFL	WN	431102	431202	431302	431502
710103	LOR	RFL	WN	431103	431203	431303	431503
710104	LOR	RFL	WN	431104	431204	431304	431504
710105	LOR	RFL	WN	431105	431205	431305	431505
710106	LOR	RFL	WN	431106	431206	431306	431506
710107	LOR	RFL	WN	431107	431207	431307	431507
710108	LOR	RFL	WN	431108	431208	431308	431508
710109	LOR	RFL	WN	431109	431209	431309	431509
710110	LOR	RFL	WN	431110	431210	431310	431510
710111	LOR	RFL	WN	431111	431211	431311	431511
710112	LOR	RFL	WN	431112	431212	431312	431512
710113	LOR	RFL	WN	431113	431213	431313	431513
710114	LOR	RFL	WN	431114	431214	431314	431514
710115	LOR	RFL	WN	431115	431215	431315	431515

Bombardier Class 710 Aventras 710114/117 work the 1118 Liverpool St-Chingford at Bethnal Green on 7 September 2021. These are 25kV AC only, but 710/2s and 710/3s are dual-voltage units. *Stuart West*

710116	LOR	RFL	WN	431116	431216	431316	431516
710117	LOR	RFL	WN	431117	431217	431317	431517
710118	LOR	RFL	WN	431118	431218	431318	431518
710119	LOR	RFL	WN	431119	431219	431319	431519
710120	LOR	RFL	WN	431120	431220	431320	431520
710121	LOR	RFL	WN	431121	431221	431321	431521
710122	LOR	RFL	WN	431122	431222	431322	431522
710123	LOR	RFL	WN	431123	431223	431323	431523
710124	LOR	RFL	WN	431124	431224	431324	431524
710125	LOR	RFL	WN	431125	431225	431325	431525
710126	LOR	RFL	WN	431126	431226	431326	431526
710127	LOR	RFL	WN	431127	431227	431327	431527
710128	LOR	RFL	WN	431128	431228	431328	431528
710129	LOR	RFL	WN	431129	431229	431329	431529
710130	LOR	RFL	WN	431130	431230	431330	431530

Class 710/2

Dual-voltage 750V DC and 25kV AC four-car sets.

	Livery	Owner	Depot	DMS	MS	PMS	DMS
710256	LOR	RFL	WN	432156	432256	432356	432556
710257	LOR	RFL	WN	432157	432257	432357	432557
710258	LOR	RFL	WN	432158	432258	432358	432558
710259	LOR	RFL	WN	432159	432259	432359	432559
710260	LOR	RFL	WN	432160	432260	432360	432560
710261	LOR	RFL	WN	432161	432261	432361	432561
710262	LOR	RFL	WN	432162	432262	432362	432562
710263	LOR	RFL	WN	432163	432263	432363	432563
710264	LOR	RFL	WN	432164	432264	432364	432564
710265	LOR	RFL	WN	432165	432265	432365	432565
710266	LOR	RFL	WN	432166	432266	432366	432566
710267	LOR	RFL	WN	432167	432267	432367	432567
710268	LOR	RFL	WN	432168	432268	432368	432568
710269	LOR	RFL	WN	432169	432269	432369	432569
710270	LOR	RFL	WN	432170	432270	432370	432570
710271	LOR	RFL	WN	432171	432271	432371	432571
710272	LOR	RFL	WN	432172	432272	432372	432572
710273	LOR	RFL	WN	432173	432273	432373	432573

Class 710/2

Dual-voltage 750V DC and 25kV AC five-car sets. The old numbers were not carried when the units were accepted into traffic.

	Old No.	Livery	Owner	Depot	DMS	MS	PMS	MS	DMS
710374	710274	LOR	RFL	OD (Q)	432174	432274	432374	432474	432574
710375	710275	LOR	RFL	WN	432175	432275	432375	432475	432575
710376	710276	LOR	RFL	WN	432176	432276	432376	432476	432576
710377	710277	LOR	RFL	WN	432177	432277	432377	432477	432577
710378	710278	LOR	RFL	OD (Q)	432178	432278	432378	432478	432578
710379	710279	LOR	RFL	OD (Q)	432179	432279	432379	432479	432579

Merseyrail

Contact details

Website: www.merseyrail.org
Twitter: @merseyrail

Key personnel

Managing Director: Andy Heath
Engineering Director: Mike Roe

Overview

Merseyrail is a successful operation serving Liverpool and the Wirral and up to Southport. Its elderly fleet of Class 507/508 ex-BR EMUs are now on borrowed time following the delivery of the first Class 777 EMUs from Stadler.

It covers routes from Liverpool to New Brighton, West Kirby, Chester, Ellesmere Port, Hunt's Cross, Ormskirk, Kirkby and Southport.

The old order on Merseyrail are elderly Class 507/508 units, which are probably in their final year of operation as Class 777s replace them. On 26 September 2021, 507033 works the 1253 New Brighton-New Brighton circle service and arrives at Liverpool James Street. *Tom McAtee*

Class 507

Ex-BR units dating from 1978-80, they are due for replacement by Class 777s, which should enter traffic imminently. The first examples have been retired and sent for scrap.

All are in the HEHQ pool.

	Livery	Owner	Depot	BDMS	TS	DMS	
507001	MER	ANG	BD	64367	71342	64405	
507002	ADV	ANG	BD	64368	71343	64406	
507003	MER	ANG	BD	64369	71344	64407	
507004	MER	ANG	BD	64370	71345	64408	*Bob Paisley*
507005	MER	ANG	BD	64371	71346	64409	
507007	MER	ANG	BD	64373	71348	64411	
507008	MER	ANG	BD	64374	71349	64412	*Harold Wilson*
507009	MER	ANG	BD	64375	71350	64413	*Dixie Dean*
507010	MER	ANG	BD	64376	71351	64414	
507011	MER	ANG	BD	64377	71352	64415	
507012	MER	ANG	BD	64378	71353	64416	
507013	MER	ANG	BD	64379	71354	64417	
507014	MER	ANG	BD	64380	71355	64418	
507015	MER	ANG	BD	64381	71356	64419	
507016	MER	ANG	BD	64382	71357	64420	*Merseyrail*
507017	MER	ANG	BD	64383	71358	64421	
507018	MER	ANG	BD	64384	71359	64422	
507019	MER	ANG	BD	64385	71360	64423	
507020	MER	ANG	BD	64386	71361	64424	*John Peel*
507021	MER	ANG	BD	64387	71362	64425	*Red Rum*
507023	MER	ANG	BD	64389	71364	64427	*Operations Inspector Stuart Mason*
507024	MER	ANG	BD	64390	71365	64428	
507025	MER	ANG	BD	64391	71366	64429	
507026	MER	ANG	BD	64392	71367	64430	*Councillor George Howard*
507027	MER	ANG	BD	64393	71368	64431	
507028	MER	ANG	BD	64394	71369	64432	
507029	MER	ANG	BD	64395	71370	64433	
507030	MER	ANG	BD	64396	71371	64434	
507031	MER	ANG	BD	64397	71372	64435	
507032	MER	ANG	BD	64398	71373	64436	
507033	MER	ANG	BD	64399	71374	64437	*Councillor Jack Spriggs*

507002 is in Liverpool Hope University white livery.

On 30 August 2020, 508130 leads 508125 past Formby. The rear unit is in an advertising livery. *Rob France*

Class 508

Ex-BR units dating from 1978-80, originally used on the Southern Region, they are also due for replacement by Class 777s. Some units have now been moved for scrap.

All are in the HEHQ pool.

	Livery	Owner	Depot	DMS	TS	BDMS	
508103	MER	ANG	BD	64651	71485	64694	
508104	MER	ANG	BD	64652	71486	64695	
508108	MER	ANG	BD	64656	71490	64699	
508111	ADV	ANG	BD	64659	71493	64702	*The Beatles*
508112	MER	ANG	BD	64660	71494	64703	
508114	MER	ANG	BD	64662	71496	64705	
508115	MER	ANG	BD	64663	71497	64706	
508117	MER	ANG	BD	64665	71499	64708	
508120	MER	ANG	BD	64668	71502	64711	
508122	ADV	ANG	BD	64670	71504	64713	
508123	MER	ANG	BD	64671	71505	64714	*William Roscoe*
508124	MER	ANG	BD	64672	71506	64715	
508125	MER	ANG	BD	64673	71507	64716	
508126	MER	ANG	BD	64674	71508	64717	
508127	MER	ANG	BD	64675	71509	64718	
508128	MER	ANG	BD	64676	71510	64719	
508130	MER	ANG	BD	64678	71512	64721	
508131	MER	ANG	BD	64679	71513	64722	
508136	MER	ANG	BD	64684	71518	64727	*Wilfred Owen MC*
508137	MER	ANG	BD	64685	71519	64728	
508138	MER	ANG	BD	64686	71520	64729	
508139	MER	ANG	BD	64687	71521	64730	
508140	MER	ANG	BD	64688	71522	64731	
508141	MER	ANG	BD	64689	71523	64732	
508143	MER	ANG	BD	64691	71525	64734	

508111 is in Beatles Story Blue livery.
508122 is in a Merseyrail promotional livery.

Class 777

New Stadler built four-car EMUs, being assembled in Poland and Switzerland, to replace the Class 507/508 fleets. They are now being delivered and the first units are undergoing testing in Merseyside. They are owned by Merseytravel and leased to Merseyrail. Seven will be battery-fitted. There is an option for a further 59 units.

All are expected to be in the HEHQ pool.

	Livery	Owner	Depot	DMS	MS	MS	DMS
777001	MER	MET	CY (Q)	427001	428001	429001	430001
777002	MER	MET	KK (Q)	427002	428002	429002	430002
777003	MER	MET	BD (Q)	427003	428003	429003	430003
777004	MER	MET	BD (Q)	427004	428004	429004	430004
777005	MER	MET	BD (Q)	427005	428005	429005	430005
777006	MER	MET	BD (Q)	427006	428006	429006	430006
777007	MER	MET	BD (Q)	427007	428007	429007	430007
777008	MER	MET	BD (Q)	427008	428008	429008	430008
777009	MER	MET	BD (Q)	427009	428009	429009	430009
777010	MER	MET	KK (Q)	427010	428010	429010	430010
777011	MER	MET		427011	428011	429011	430011
777012	MER	MET	KK (Q)	427012	428012	429012	430012
777013	MER	MET		427013	428013	429013	430013
777014	MER	MET		427014	428014	429014	430014
777015	MER	MET		427015	428015	429015	430015
777016	MER	MET	BD (Q)	427016	428016	429016	430016
777017	MER	MET		427017	428017	429017	430017
777018	MER	MET	BD (Q)	427018	428018	429018	430018
777019	MER	MET		427019	428019	429019	430019
777020	MER	MET		427020	428020	429020	430020
777021	MER	MET		427021	428021	429021	430021
777022	MER	MET		427022	428022	429022	430022
777023	MER	MET		427023	428023	429023	430023
777024	MER	MET		427024	428024	429024	430024
777025	MER	MET		427025	428025	429025	430025
777026	MER	MET		427026	428026	429026	430026
777027	MER	MET		427027	428027	429027	430027
777028	MER	MET		427028	428028	429028	430028
777029	MER	MET		427029	428029	429029	430029
777030	MER	MET		427030	428030	429030	430030
777031	MER	MET		427031	428031	429031	430031
777032	MER	MET		427032	428032	429032	430032
777033	MER	MET		427033	428033	429033	430033
777034	MER	MET		427034	428034	429034	430034
777035	MER	MET		427035	428035	429035	430035
777036	MER	MET		427036	428036	429036	430036
777037	MER	MET		427037	428037	429037	430037
777038	MER	MET		427038	428038	429038	430038
777039	MER	MET		427039	428039	429039	430039
777040	MER	MET		427040	428040	429040	430040
777041	MER	MET		427041	428041	429041	430041
777042	MER	MET		427042	428042	429042	430042
777043	MER	MET		427043	428043	429043	430043
777044	MER	MET		427044	428044	429044	430044
777045	MER	MET		427045	428045	429045	430045
777046	MER	MET		427046	428046	429046	430046
777047	MER	MET		427047	428047	429047	430047
777048	MER	MET		427048	428048	429048	430048
777049	MER	MET		427049	428049	429049	430049
777050	MER	MET		427050	428050	429050	430050
777051	MER	MET		427051	428051	429051	430051
777052	MER	MET		427052	428052	429052	430052
777053	MER	MET		427053	428053	429053	430053

The new Stadler Class 777s are being delivered and tested and are close to entering traffic. 777002 was on display at the Rail Live event at Long Marston on 16 June 2021. *Jack Boskett*

Northern Rail

Contact details

Website: www.northernrail.org
Twitter: @northernassist

Key personnel

Managing Director: Nick Donovan
Engineering Director: Jack Commandeur

Overview

Northern has been operated by the DfT via its Operator of Last Resort since 1 March 2020 and runs all the local services across large parts of Yorkshire, Merseyside, Lancashire, Cheshire and Cumbria and also into Nottinghamshire and Lincolnshire.

Northern was operated by Arriva until the spring of 2020, when due to poor performance it was stripped of its franchise. This happened just before the Covid pandemic.

The operation is currently undergoing a massive overhaul of its trains fleet. New CAF-built Class 195 DMUs – in two- and three-car formations – are now in traffic, as is a fleet of 43 Class 331 EMUs – 31 three-cars and 12 four-cars – also built by CAF and sharing the same bodyshells as the 195s. There is a tender for financing 20 more CAG units that would most likely be battery powered.

The new DMUs have allowed the last Class 142/144 Railbuses and the Class 153 single-car DMUs to be withdrawn. The rest of the DMU fleet now comprises Class 150, 155, 156, 158 and a handful of Class 170 Turbostars.

The EMU fleet, as well as the new CAF trains, comprises 16 four-car Siemens/CAF Class 333 EMUs, a few ex-Thameslink Class 319s and 17 Class 323 three-car sets. Eight bi-mode Class 769s, modified Class 319s, have also entered traffic.

Class 150

These DMUs date from 1984-87. The Class 150/0 and 150/1s have no corridor connections, a feature added to the later batch of 150/2s at construction.

All are based at Newton Heath in Manchester or Neville Hill in Leeds and are used mostly for local trains emanating from the Manchester area, although they do work into West Yorkshire.

All are in the EDHQ pool.

Class 150/0

Three-car trial units formerly with GWR. 150001/002 were original-build three-car units, while 150003-006 are four recently refurbished 'new' three-car Class 150/0s formed of vehicles from Class 150/1 and 150/2 DMUs. 150002 was numbered 154001 from March 1987, then 154002 in March 1988 before becoming 150002 in February 1992.

	Livery	Owner	Depot	DMSL	MS	DMS
150001	NOR	ANG	NH	55200	55400	55300
150002	NOR	ANG	NH	55201	55401	55301
150003	NOR	ANG	NH	52116	57209	57116
150004	NOR	ANG	NH	52112	57212	57112
150005	NOR	ANG	NH	52117	52223	57117
150006	NOR	ANG	NH	52147	57223	52147

Class 150/1

	Livery	Owner	Depot	DMSL	DMS
150101	NOR	ANG	NH	52101	57101
150102	NOR	ANG	NH	52102	57102
150103	NOR	ANG	NH	52103	57103
150104	NOR	ANG	NH	52104	57104
150105	NOR	ANG	NH	52105	57105
150106	NOR	ANG	NH	52106	57106
150107	NOR	ANG	NH	52107	57107
150108	NOR	ANG	NH	52108	57108
150109	NOR	ANG	NH	52109	57109
150110	NOR	ANG	NH	52110	57110
150111	NOR	ANG	NH	52111	57111
150113	NOR	ANG	NH	52113	57113

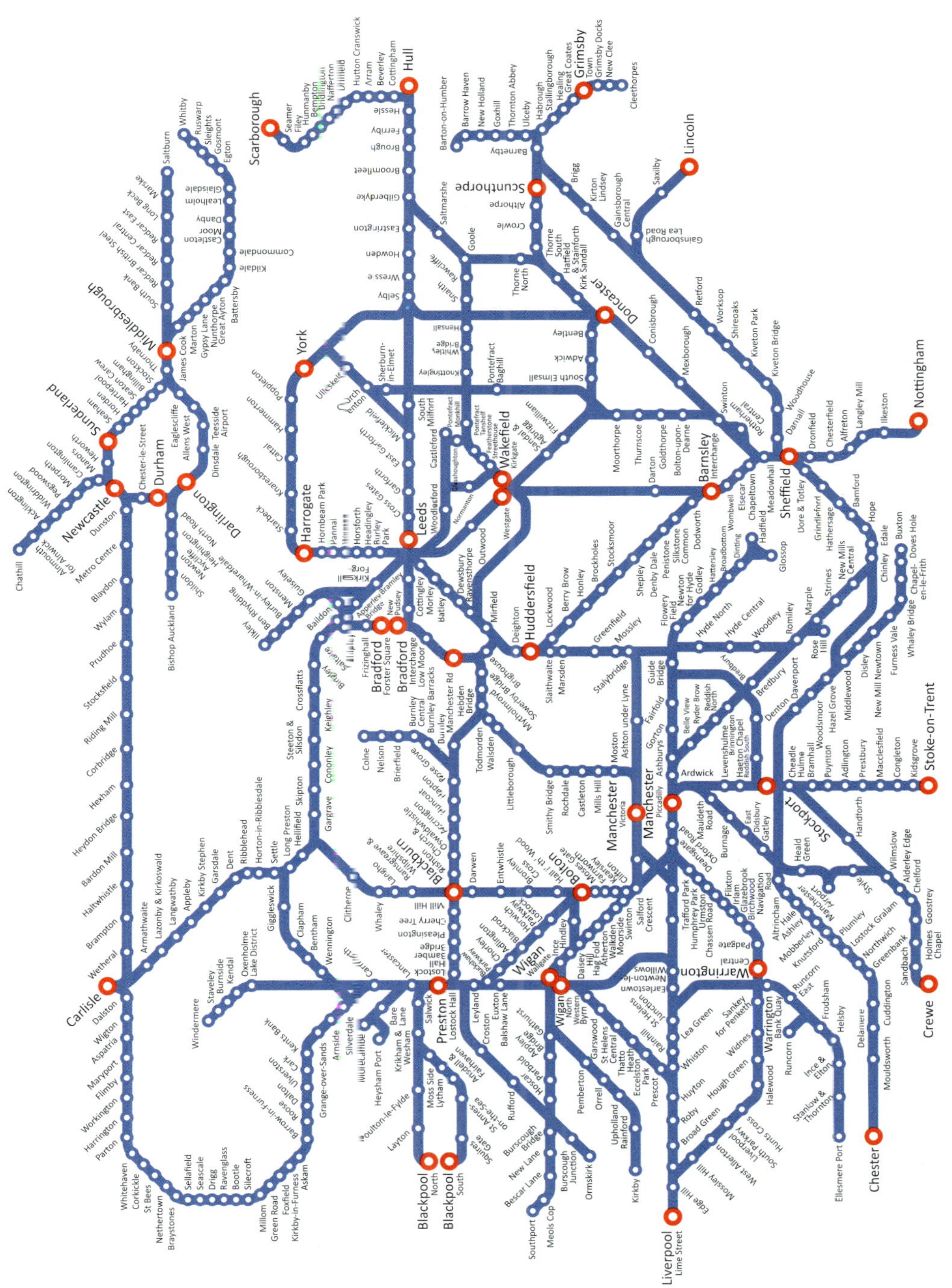
Scarborough
Hull
Grimsby Town
Lincoln
Scunthorpe
Middlesbrough
Sunderland
York
Doncaster
Newcastle
Durham
Darlington
Harrogate
Leeds
Wakefield Kirkgate
Barnsley Interchange
Sheffield
Nottingham
Bradford Forster Square
Bradford Interchange
Huddersfield
Carlisle
Manchester Victoria
Manchester Piccadilly
Stockport
Stoke-on-Trent
Blackburn
Bolton
Preston
Wigan Wallgate
Wigan North Western
Warrington Central
Warrington Bank Quay
Crewe
Blackpool North
Blackpool South
Liverpool Lime Street
Chester

150114	NOR	ANG	NH	52114	57114
150115	NOR	ANG	NH	52115	57115
150118	NOR	ANG	NH	52118	57118
150119	NOR	ANG	NH	52119	57119
150120	NOR	ANG	NH	52120	57120
150121	NOR	ANG	NH	52121	57121
150122	NOR	ANG	NH	52122	57122
150123	NOR	ANG	NH	52123	57123
150124	NOR	ANG	NH	52124	57124
150125	NOR	ANG	NH	52125	57125
150126	NOR	ANG	NH	52126	57126
150127	NOR	ANG	NH	52127	57127
150128	NOR	ANG	NH	52128	57128
150129	NOR	ANG	NH	52129	57129
150130	NOR	ANG	NH	52130	57130
150131	NOR	ANG	NH	52131	57131
150132	NOR	ANG	NH	52132	57132
150133	NOR	ANG	NH	52133	57133
150134	NOR	ANG	NH	52134	57134
150135	NOR	ANG	NH	52135	57135
150136	NOR	ANG	NH	52136	57136
150137	NOR	ANG	NH	52137	57137
150138	NOR	ANG	NH	52138	57138
150139	NOR	ANG	NH	52139	57139
150140	NOR	ANG	NH	52140	57140
150141	NOR	ANG	NH	52141	57141
150142	NOR	ANG	NH	52142	57142
150143	NOR	ANG	NH	52143	57143
150144	NOR	ANG	NH	52144	57144
150145	NOR	ANG	NH	52145	57145
150146	NOR	ANG	NH	52146	57146
150148	NOR	ANG	NH	52148	57148
150149	NOR	ANG	NH	52149	57149
150150	NOR	ANG	NH	52150	57150

Class 150/2

150201	NOR	ANG	NL	52201	57201	
150203	NOR	ANG	NL	52203	57203	
150204	NOR	ANG	NL	52204	57204	
150205	NOR	ANG	NL	52205	57205	
150206	NOR	ANG	NL	52206	57206	
150210	NOR	ANG	NL	52210	57210	
150211	NOR	ANG	NL	52211	57211	
150214	NOR	ANG	NL	52214	57214	*The Bentham Line A Dementia-Friendly Railway*
150215	NOR	ANG	NL	52215	57215	
150218	NOR	ANG	NL	52218	57218	
150220	NOR	ANG	NL	52220	57220	
150222	NOR	ANG	NL	52222	57222	
150224	NOR	ANG	NL	52224	57224	
150225	NOR	ANG	NL	52225	57225	
150226	NOR	ANG	NL	52226	57226	
150228	NOR	POR	NL	52228	57228	
150268	NOR	POR	NL	52268	57268	
150269	NOR	POR	NL	52269	57269	
150270	NOR	POR	NL	52270	57270	
150271	NOR	POR	NL	52271	57271	
150272	NOR	POR	NL	52272	57272	
150273	NOR	POR	NL	52273	57273	
150274	NOR	POR	NL	52274	57274	
150275	NOR	POR	NL	52275	57275	*The Yorkshire Regiment Yorkshire Warrior*
150276	NOR	POR	NL	52276	57276	
150277	NOR	POR	NL	52277	57277	

On 17 July 2021, 150274 calls at Penistone with a morning Sheffield to Huddersfield train. This line used to be operated by Class 142/144 railbuses but thankfully they have all been retired now. *Pip Dunn*

Class 155

A fleet of 42 Class 155s were delivered in 1987-88 and all but seven then converted into pairs of single-car Class 153s in 1991-92. The remaining seven sets – which had actually been funded by West Yorkshire Passenger Transport executive – were kept in Yorkshire as two-car units and are all based at Neville Hill and work in the West and North Yorkshire area.

All are in the EDHQ pool.

	Livery	Owner	Depot	DMSL	DMS
155341	NOR	POR	NL	52341	57341
155342	NOR	POR	NL	52342	57342
155343	NOR	POR	NL	52343	57343
155344	NOR	POR	NL	52344	57344
155345	NOR	POR	NL	52345	57345
155346	NOR	POR	NL	52346	57346
155347	NOR	POR	NL	52347	57347

Only seven Class 155s remain as the rest of the fleet was rebuilt into single-car Class 153s in the early 1990s. Those that remain as 155s have always worked in Yorkshire; 155347 waits to work the 1343 York-Hull on 5 March 2022. *Pip Dunn*

Class 156

Metro Cammell units that are based at Heaton and Newton Heath and work across the majority of the Northern network, but especially in the North-East and North-West. 156403/404/413/414 and 156907/909/912/916-919/922 are also to transfer from East Midlands Railways during 2022, the 156/9s reverting to their original 156/4 numbers.

All are in the EDHQ pool.

	Old no.	Livery	Owner	Depot	DMSL	DMS
156401		NOR	POR	NH	52401	57401
156402	156902	NOR	POR	NH	52402	57402
156415		NOR	POR	NH	52415	57415
156420		NOR	POR	NH	52420	57420
156421		NOR	POR	HT	52421	57421
156423		NOR	POR	NH	52423	57423

Class 156s are based at Heaton in Newcastle and Newton Heath in Manchester for Northern services; 156423 works the 1216 Southport-Alderley Edge through Farnworth on 7 September 2021. *Tom McAtee*

156424	NOR	POR	NH	52424	57424	
156425	NOR	POR	NH	52425	57425	
156426	NOR	POR	NH	52426	57426	
156427	NOR	POR	NH	52427	57427	
156428	NOR	POR	NH	52428	57428	
156429	NOR	POR	NH	52429	57429	
156438	NOR	ANG	HT	52438	57438	
156440	NOR	POR	HT	52440	57440	
156441	NOR	POR	HT	52441	57441	
156443	NOR	ANG	HT	52443	57443	
156444	NOR	ANG	HT	52444	57444	
156447	NOR	ANG	HT	52447	57447	
156448	NOR	ANG	HT	52448	57448	
156449	NOR	ANG	HT	52449	57449	
156451	NOR	ANG	HT	52451	57451	
156452	NOR	POR	NH	52452	57452	
156454	NOR	ANG	HT	52454	57454	
156455	NOR	POR	NH	52455	57455	
156459	NOR	POR	NH	52459	57459	
156460	NOR	POR	NH	52460	57460	
156461	NOR	POR	NH	52461	57461	
156463	NOR	ANG	HT	52463	57463	
156464	NOR	POR	NH	52464	57464	*Lancashire Dalesrail*
156465	NOR	ANG	HT	52465	57465	
156466	NOR	POR	NH	52466	57466	
156468	NOR	ANG	HT	52468	57468	
156469	NOR	ANG	HT	52469	57469	*The Royal Northumberland Fusiliers (The Fighting Fifth)*
156471	NOR	ANG	HT	52471	57471	
156472	NOR	ANG	HT	52472	57472	
156475	NOR	ANG	HT	52475	57475	
156479	NOR	ANG	HT	52479	57479	
156480	ADV	ANG	HT	52480	57480	*Spirit of the Royal Air Force*
156481	NOR	ANG	HT	52481	57481	
156482	NOR	ANG	HT	52482	57482	
156483	NOR	ANG	HT	52483	57483	*William George 'Billy' Hardy*
156484	NOR	ANG	HT	52484	57484	
156485	NOR	ANG	HT	52485	57485	
156486	NOR	ANG	HT	52486	57486	
156487	NOR	ANG	HT	52487	57487	
156488	NOR	ANG	HT	52488	57488	
156489	NOR	ANG	HT	52489	57489	
156490	NOR	ANG	HT	52490	57490	
156491	NOR	ANG	HT	52491	57491	
156496	NOR	ANG	HT	52496	57496	

156480 is in an RAF 100 years light blue livery.

Class 158

These units work across the Northern network but especially in the North-East from Newcastle, from Carlisle and Leeds.

All are in the EDHQ pool.

	Livery	Owner	Depot	DMSL	MSL	DMSL	
158752	NOR	POR	NL	52752	58716	57752	
158753	NOR	POR	NL	52753	58710	57753	
158754	NOR	POR	NL	52754	58708	57754	
158755	NOR	POR	NL	52755	58702	57755	
158756	NOR	POR	NL	52756	58712	57756	
158757	NOR	POR	NL	52757	58706	57757	

158758	NOR	POR	NL	52758	58714	57758	
158759	NOR	POR	NL	52759	58713	57759	
158782	NOR	ANG	NL	52782		57782	
158784	NOR	ANG	NL	52784		57784	*Barbara Castle*
158786	NOR	ANG	NL	52786		57786	
158787	NOR	ANG	NL	52787		57787	
158789	NOR	ANG	NL	52789		57789	
158790	NOR	ANG	NL	52790		57790	
158791	NOR	ANG	NL	52791		57791	*County of Nottinghamshire*
158792	NOR	ANG	HT	52792		57792	
158793	NOR	ANG	NL	52793		57793	
158794	NOR	ANG	NL	52794		57794	
158795	NOR	ANG	NL	52795		57795	
158796	NOR	ANG	NL	52796		57796	
158797	NOR	ANG	NL	52797		57797	*Jane Tomlinson*
158815	NOR	ANG	HT	52815		57815	

Northern has more than 50 two-car Class 158s in its fleet, which are seen right across its network. 158815 has just arrived with the 1636 from Bradford-Interchange at Huddersfield on 11 December 2021. *Pip Dunn*

158816	NOR	ANG	HT	52816		57816	
158817	NOR	ANG	HT	52817		57817	
158842	NOR	ANG	HT	52842		57842	
158843	NOR	ANG	HT	52843		57843	
158844	NOR	ANG	HT	52844		57844	
158845	NOR	ANG	HT	52845		57845	
158848	NOR	ANG	HT	52848		57848	
158849	NOR	ANG	HT	52849		57849	
158850	NOR	ANG	HT	52850		57850	
158851	NOR	ANG	HT	52851		57851	
158853	NOR	ANG	HT	52853		57853	
158855	NOR	ANG	HT	52855		57855	
158859	NOR	ANG	HT	52859		57859	
158860	NOR	ANG	HT	52860		57860	*Ian Dewhurst*
158861	NOR	ANG	HT	52861		57861	*Magna Carta 800 Lincoln 2015*
158867	NOR	ANG	NL	52867		57867	
158868	NOR	ANG	NL	52868		57868	
158869	NOR	ANG	NL	52869		57869	
158870	NOR	ANG	NL	52870		57870	
158871	NOR	ANG	NL	52871		57871	
158872	NOR	ANG	NL	52872		57872	

Class 158/9

Former West Yorkshire PTE Metro units.

	Livery	Owner	Depot	DMSL	DMSL
158901	NOR	EVS	NL	52901	57901
158902	NOR	EVS	NL	52902	57902
158903	NOR	EVS	NL	52903	57903
158904	NOR	EVS	NL	52904	57904
158905	NOR	EVS	NL	52905	57905
158906	NOR	EVS	NL	52906	57906
158907	NOR	EVS	NL	52907	57907
158908	NOR	EVS	NL	52908	57908
158909	NOR	EVS	NL	52909	57909
158910	NOR	EVS	NL	52910	57910

Class 170

These Turbostar units released from ScotRail in 2018-19 and work mostly in Yorkshire on services such as the Harrogate Circle and Sheffield to Scarborough.

All are in the EDHQ pool.

	Livery	Owner	Depot	DMSL	MS	DMSL
170453	NOR	POR	NL	50453	56453	79453
170454	NOR	POR	NL	50454	56454	79454
170455	NOR	POR	NL	50455	56455	79455
170456	NOR	POR	NL	50456	56456	79456
170457	NOR	POR	NL	50457	56457	79457
170458	NOR	POR	NL	50458	56458	79458
170459	NOR	POR	NL	50459	56459	79459
170460	NOR	POR	NL	50460	56460	79460
170461	NOR	POR	NL	50461	56461	79461
170472	NOR	POR	NL	50472	56472	79472
170473	NOR	POR	NL	50473	56473	79473
170474	NOR	POR	NL	50474	56474	79474
170475	NOR	POR	NL	50475	56475	79475
170476	NOR	POR	NL	50476	56476	79476
170477	NOR	POR	NL	50477	56477	79477
170478	NOR	POR	NL	50478	56478	79478

170453-457 have a DMCL instead of a DMSL.

A fleet of 16 ex-ScotRail Class 170 Turbostars, released after electrification projects in the Scottish central belt, have been redeployed with Northern. On 11 December 2021, 170473 was at Leeds awaiting its next duty. *Pip Dunn*

Class 195

These CAF diesel units are all now in traffic and work many local and regional services, especially in the North-West and Yorkshire.

All are in the EDHQ pool.

Class 195/0

Two-car sets.

	Livery	Owner	Depot	DMS	DMS
195001	NOR	EVS	NH	101001	103001
195002	NOR	EVS	NH	101002	103002
195003	NOR	EVS	NH	101003	103003
195004	NOR	EVS	NH	101004	103004
195005	NOR	EVS	NH	101005	103005
195006	NOR	EVS	NH	101006	103006
195007	NOR	EVS	NH	101007	103007
195008	NOR	EVS	NH	101008	103008
195009	NOR	EVS	NH	101009	103009
195010	NOR	EVS	NH	101010	103010
195011	NOR	EVS	NH	101011	103011
195012	NOR	EVS	NH	101012	103012
195013	NOR	EVS	NH	101013	103013
195014	NOR	EVS	NH	101014	103014
195015	NOR	EVS	NH	101015	103015

195016	NOR	EVS	NH	101016	103016
195017	NOR	EVS	NH	101017	103017
195018	NOR	EVS	NH	101018	103018
195019	NOR	EVS	NH	101019	103019
195020	NOR	EVS	NH	101020	103020
195021	NOR	EVS	NH	101021	103021
195022	NOR	EVS	NH	101022	103022
195023	NOR	EVS	NH	101023	103023
195024	NOR	EVS	NH	101024	103024
195025	NOR	EVS	NH	101025	103025

Class 195/1

Three-car sets.

	Livery	Owner	Depot	DMS	MS	DMS	
195101	NOR	EVS	NH	101101	102101	103101	
195102	NOR	EVS	NH	101102	102102	103102	
195103	NOR	EVS	NH	101103	102103	103103	
195104	NOR	EVS	NH	101104	102104	103104	*Deva Victrix*
195105	NOR	EVS	NH	101105	102105	103105	
195106	NOR	EVS	NH	101106	102106	103106	
195107	NOR	EVS	NH	101107	102107	103107	
195108	NOR	EVS	NH	101108	102108	103108	
195109	NOR	EVS	NH	101109	102109	103109	*Pride of Cumbria*
195110	NOR	EVS	NH	101110	102110	103110	
195111	NOR	EVS	NH	101111	102111	103111	
195112	NOR	EVS	NH	101112	102112	103112	
195113	NOR	EVS	NH	101113	102113	103113	
195114	NOR	EVS	NH	101114	102114	103114	
195115	NOR	EVS	NH	101115	102115	103115	
195116	NOR	EVS	NH	101116	102116	103116	

The Class 195s are two- or three-car units. One of the latter, 195116, waits to set off with the 1542 Leeds-Chester on 12 February 2022. *Pip Dunn*

195117	NOR	EVS	NH	101117	102117	103117	
195118	NOR	EVS	NH	101118	102118	103118	
195119	NOR	EVS	NH	101119	102119	103119	
195120	NOR	EVS	NH	101120	102120	103120	
195121	NOR	EVS	NH	101121	102121	103121	
195122	NOR	EVS	NH	101122	102122	103122	
195123	NOR	EVS	NH	101123	102123	103123	
195124	NOR	EVS	NH	101124	102124	103124	
195125	NOR	EVS	NH	101125	102125	103125	
195126	NOR	EVS	NH	101126	102126	103126	
195127	NOR	EVS	NH	101127	102127	103127	
195128	NOR	EVS	NH	101128	102128	103128	*Calder Champion*
195129	NOR	EVS	NH	101129	102129	103129	
195130	NOR	EVS	NH	101130	102130	103130	
195131	NOR	EVS	NH	101131	102131	103131	
195132	NOR	EVS	NH	101132	102132	103132	
195133	NOR	EVS	NH	101133	102133	103133	

Some of the Class 319s declared surplus by Thameslink were taken on by Northern after additional electrification of its network was competed. On 12 July 2019, 319370 calls at Preston as it heads to Liverpool. *Pip Dunn*

Class 319

These ex-Thameslink units dual-voltage EMUs, displaced by Class 700s, work Liverpool to Wigan, Blackpool and Manchester Airport.

All are in the EDHQ pool.

	Livery	Owner	Depot	DTS	MS	TS	DTS
319361	NOR	POR	AN	77459	63043	71929	77458
319366	NOR	POR	AN	77469	63048	71934	77468
319367	NOR	POR	AN	77471	63049	71935	77470
319368	NOR	POR	AN	77473	63050	71936	77472
319369	NOR	POR	AN	77475	63051	71937	77474
319370	NOR	POR	AN	77477	63052	71938	77476
319372	NOR	POR	AN	77481	63054	71940	77480
319375	NOR	POR	AN	77487	63057	71943	77486
319378	NOR	POR	AN	77493	63060	71946	77492
319379	NOR	POR	AN	77495	63061	71947	77494
319381	NOR	POR	AN	77973	63093	71979	77974
319383	NOR	POR	AN	77977	63095	71981	77978
319384	NOR	POR	AN	77979	63096	71982	77980
319385	NOR	POR	AN	77981	63097	71983	77982
319386	NOR	POR	AN	77983	63098	71984	77984

Class 323

These ex-BR units, built in 1992-93 by Hunslet, are used out of Manchester to Glossop/Hadfield, Manchester Airport, Crewe and Stoke. Some of the West Midlands Trains Class 323 fleet are expected to join the fleet when displaced by Class 730s.

All are in the EDHQ pool.

	Livery	Owner	Depot	DMS	TS	DMS
323223	NOR	POR	AN	64023	75323	65023
323224	NOR	POR	AN	64024	75324	65024
323225	NOR	POR	AN	64025	75325	65025
323226	NOR	POR	AN	64026	75326	65026
323227	NOR	POR	AN	64027	75327	65027
323228	NOR	POR	AN	64028	75328	65028
323229	NOR	POR	AN	64029	75329	65029
323230	NOR	POR	AN	64030	75330	65030
323231	NOR	POR	AN	64031	75331	65031
323232	NOR	POR	AN	64032	75332	65032
323233	NOR	POR	AN	64033	75333	65033
323234	NOR	POR	AN	64034	75334	65034
323235	NOR	POR	AN	64035	75335	65035
323236	NOR	POR	AN	64036	75336	65036
323237	NOR	POR	AN	64037	75337	65037
323238	NOR	POR	AN	64038	75338	65038
323239	NOR	POR	AN	64039	75339	65039

Class 323s work local trains out of Manchester to the south; 323235 works the 0756 Stoke-on-Trent-Manchester Piccadilly at Stockport on 31 March 2021. *Tom McAtee*

Class 331

These new CAF units are now at work out of Leeds, Manchester and Liverpool on the electrified routes on the Northern Network.

All are in the EDHQ pool.

Class 331/0

Three-car sets.

	Livery	Owner	Depot	DMS	PTS	DTS
331001	NOR	EVS	AN	463001	464001	466001
331002	NOR	EVS	AN	463002	464002	466002
331003	NOR	EVS	AN	463003	464003	466003
331004	NOR	EVS	AN	463004	464004	466004
331005	NOR	EVS	AN	463005	464005	466005
331006	NOR	EVS	AN	463006	464006	466006
331007	NOR	EVS	AN	463007	464007	466007
331008	NOR	EVS	AN	463008	464008	466008
331009	NOR	EVS	AN	463009	464009	466009
331010	NOR	EVS	AN	463010	464010	466010
331011	NOR	EVS	AN	463011	464011	466011
331012	NOR	EVS	AN	463012	464012	466012
331013	NOR	EVS	AN	463013	464013	466013
331014	NOR	EVS	AN	463014	464014	466014
331015	NOR	EVS	AN	463015	464015	466015
331016	NOR	EVS	AN	463016	464016	466016
331017	NOR	EVS	AN	463017	464017	466017
331018	NOR	EVS	AN	463018	464018	466018
331019	NOR	EVS	AN	463019	464019	466019
331020	NOR	EVS	AN	463020	464020	466020
331021	NOR	EVS	AN	463021	464021	466021
331022	NOR	EVS	AN	463022	464022	466022
331023	NOR	EVS	AN	463023	464023	466023
331024	NOR	EVS	AN	463024	464024	466024
331025	NOR	EVS	AN	463025	464025	466025
331026	NOR	EVS	AN	463026	464026	466026
331027	NOR	EVS	AN	463027	464027	466027
331028	NOR	EVS	AN	463028	464028	466028
331029	NOR	EVS	AN	463029	464029	466029
331030	NOR	EVS	AN	463030	464030	466030
331031	NOR	EVS	AN	463031	464031	466031

Class 331/1

Four-car units.

	Livery	Owner	Depot	DMS	PTS	MS	DTS
331101	NOR	EVS	NL	463101	464101	465101	466101
331102	NOR	EVS	NL	463102	464102	465102	466102
331103	NOR	EVS	NL	463103	464103	465103	466103
331104	NOR	EVS	NL	463104	464104	465104	466104
331105	NOR	EVS	NL	463105	464105	465105	466105
331106	NOR	EVS	NL	463106	464106	465106	466106
331107	NOR	EVS	NL	463107	464107	465107	466107
331108	NOR	EVS	NL	463108	464108	465108	466108
331109	NOR	EVS	NL	463109	464109	465109	466109
331110	NOR	EVS	NL	463110	464110	465110	466110
331111	NOR	EVS	NL	463111	464111	465111	466111
331112	NOR	EVS	NL	463112	464112	465112	466112

New CAF DMUs and EMUs have been put into use with Northern in the last three years; 331107 works the 1621 Leeds-Doncaster past Adwick on 11 June 2021. *Stuart West*

Class 333

These Siemens/CAF units were similar in design to the now withdrawn and scrapped Heathrow Express Class 332s. They work Aire Valley line trains from Leeds to Skipton, Bradford and Ilkley.

All are in the EDHQ pool.

	Livery	Owner	Depot	DMS	PTS	MS	DMS
333001	NOR	ANG	NL	78451	74461	74477	78452
333002	NOR	ANG	NL	78453	74462	74478	78454
333003	NOR	ANG	NL	78455	74463	74479	78456
333004	NOR	ANG	NL	78457	74464	74480	78458
333005	NOR	ANG	NL	78459	74465	74481	78460
333006	NOR	ANG	NL	78461	74466	74482	78462
333007	NOR	ANG	NL	78463	74467	74483	78464
333008	NOR	ANG	NL	78465	74468	74484	78466
333009	NOR	ANG	NL	78467	74469	74485	78468
333010	NOR	ANG	NL	78469	74470	74486	78470
333011	NOR	ANG	NL	78471	74471	74487	78472
333012	NOR	ANG	NL	78473	74472	74488	78474
333013	NOR	ANG	NL	78475	74473	74489	78476
333014	NOR	ANG	NL	78477	74474	74490	78478
333015	NOR	ANG	NL	78479	74475	74491	78480
333016	NOR	ANG	NL	78481	74476	74492	78482

Below left: CAF-built Class 333s were the first new electric units ordered for the Aire Valley lines, which were electrified in the 1990s and initially operated with second-hand Class 308s. 333006 stands at Leeds prior to working the 1033 to Ilkley on 11 December 2021. *Pip Dunn*

Below: Class 769 Flex Bi-mode units have entered service with Northern. This is 769442 at Hoscar with the 1531 Southport-Stalybridge on 17 May 2021. *Tom McAtee*

Class 769

Northern has eight bi-mode FLEX units converted from redundant Class 319s for working Liverpool to Stalybridge.

All are in the EDHQ pool.

	Old No.	Livery	Owner	Depot	DMC	MS	TS	DMS
769424	319424	NOR	POR	AN	77337	62914	71795	77336
769431	319431	NOR	POR	AN	77351	62921	71802	77350
769434	319434	NOR	POR	AN	77357	62924	71805	77356
769442	319442	NOR	POR	AN	77373	62932	71813	77372
769448	319448	NOR	POR	AN	77433	62962	71867	77432
769450	319450	NOR	POR	AN	77437	62964	71869	77436
769456	319456	NOR	POR	AN	77449	62970	71875	77448
769458	319458	NOR	POR	AN	77453	62972	71877	77452

ScotRail

Contact details

Website: www.scotrail.co.uk
Twitter: @Scotrail

Key personnel

Managing Director: Alex Hynes
Engineering Director: Willie Marshall

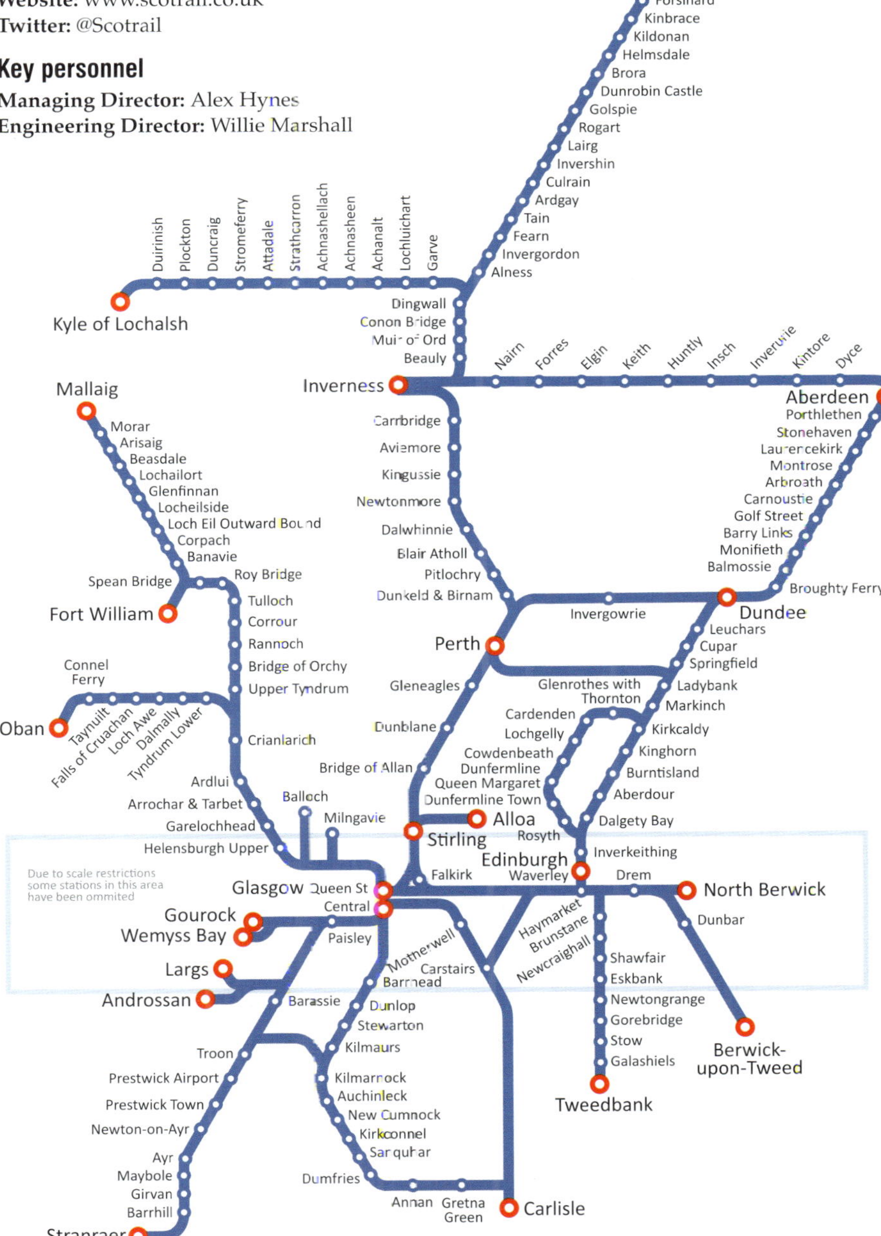

Overview

ScotRail operates all regional and commuter routes in Scotland. Electrification of many Central Belt routes means the EMU fleet north of the border is now a mix of Class 318s dating from 1985-86, Class 320s from 1990 (recently bolstered by 12 ex-London Midland Class 321 units), Alstom Class 334s from 1999-2001, Siemens Class 380s from 2009-10 and Hitachi Class 385s from 2016-18. The elderly Class 314s have been withdrawn, and some spare DMUs were moved to other TOCs.

The EMUs work most services in North Ayrshire, Inverclyde, to Balloch / Helensburgh, Edinburgh-Glasgow via Falkirk, Bathgate, Shotts and Carstairs and from Edinburgh to North Berwick, Dunblane and Dunbar.

Class 156s remain in use on the West Highland and Glasgow South Western routes, and five Class 153 single-car DMUs have been modified to run as luggage- and bike-carrying vehicles for the West Highland line.

Class 158s and 170s work to Tweedbank, the Highland Main Line, Aberdeen to Inverness, the Fife Circle and other commuter routes. The lines north of Inverness are worked by Class 158s. HSTs run with four or five trailers on Glasgow QS / Edinburgh-Inverness / Aberdeen routes.

Most of the investment and enhancement in Scotland has already been completed, but there are plans to reinstate the Levan branch in Fife, and Alloa to Kincardine for passenger use. A longer-term aspiration is the reopening of the remaining part of the Waverley route, the 63 miles from Tweedbank to Carlisle. Other station reopenings are planned across the network.

Class 43

A fleet of ex-GWR HSTs were taken on by ScotRail for fast inter-city operations between seven cities – Glasgow, Edinburgh, Stirling, Perth, Inverness, Dundee and Aberdeen. They run on Glasgow QS-Inverness, Glasgow QS-Aberdeen, Edinburgh-Aberdeen, Edinburgh-Inverness and Inverness-Aberdeen routes.

Fifty-four power cars are leased from Angel Trains and all are in the HAPC pool. Following the fatal accident near Stonehaven on 12 August 2020, 43140 was seriously damaged and scrapped, while 43030 is also withdrawn. Neither has been replaced by any spare loco.

All are in the HAPC pool.

43003	SCT	ANG	HA	
43012	SCT	ANG	HA	
43015	SCT	ANG	HA	
43021	SCT	ANG	HA	
43026	SCT	ANG	HA	
43028	SCT	ANG	HA	
43030	SCT	ANG	ZK (U)	
43031	SCT	ANG	HA	
43032	SCT	ANG	HA	
43033	SCT	ANG	HA	
43034	SCT	ANG	HA	
43035	SCT	ANG	HA	
43036	SCT	ANG	HA	
43037	SCT	ANG	HA	
43124	SCT	ANG	HA	
43125	SCT	ANG	HA	
43126	SCT	ANG	HA	
43127	SCT	ANG	HA	
43128	SCT	ANG	HA	
43129	SCT	ANG	HA	
43130	SCT	ANG	HA	
43131	SCT	ANG	HA	
43132	SCT	ANG	HA	
43133	SCT	ANG	HA	
43134	SCT	ANG	HA	*Gordon Aikmen BEM MND Campaigner 1985-2017*

43135	SCT	ANG	HA
43136	SCT	ANG	HA
43137	SCT	ANG	HA
43138	SCT	ANG	HA
43139	SCT	ANG	HA
43141	SCT	ANG	HA
43142	SCT	ANG	HA
43143	SCT	ANG	HA
43144	SCT	ANG	HA
43145	SCT	ANG	HA
43146	SCT	ANG	HA
43147	SCT	ANG	HA
43148	SCT	ANG	HA
43149	SCT	ANG	HA
43150	SCT	ANG	HA
43151	SCT	ANG	HA
43152	SCT	ANG	HA
43163	SCT	ANG	HA
43164	SCT	ANG	HA
43168	SCT	ANG	HA
43169	SCT	ANG	HA
43175	SCT	ANG	HA
43176	SCT	ANG	HA
43177	SCT	ANG	HA
43179	SCT	ANG	HA
43181	SCT	ANG	HA
43182	SCT	ANG	HA
43183	SCT	ANG	HA

A fleet of HSTs, running with four, or five, Mk 3 trailers are used across the ScotRail inter-city network. On 29 June 2021, passing Forteviot are 43175 and 43028 with the 1041 Glasgow Queen Street-Aberdeen. *Paul Shannon*

Mk 3 trailers

All are in the HAHQ pool.

TGFB –first-class trailer guard with buffet counter

40600	SCT	ANG	IS
40601	SCT	ANG	IS
40602	SCT	ANG	IS
40603	SCT	ANG	IS
40604	SCT	ANG	IS
40605	SCT	ANG	IS
40606	SCT	ANG	IS
40607	SCT	ANG	IS
40608	SCT	ANG	IS
40609	SCT	ANG	IS
40610	SCT	ANG	IS
40611	SCT	ANG	IS
40612	SCT	ANG	IS
40613	SCT	ANG	IS
40614	SCT	ANG	IS
40615	SCT	ANG	IS
40616	SCT	ANG	IS
40617	SCT	ANG	IS
40618	SCT	ANG	IS
40619	SCT	ANG	IS
40620	SCT	ANG	IS
40621	SCT	ANG	IS
40623	SCT	ANG	IS
40624	SCT	ANG	IS
40625	SCT	ANG	IS
40626	SCT	ANG	IS

TS – trailer standard

42004	SCT	ANG	IS
42009	SCT	ANG	IS
42010	SCT	ANG	IS
42012	SCT	ANG	IS
42013	SCT	ANG	IS
42014	SCT	ANG	IS
42019	SCT	ANG	IS
42021	SCT	ANG	IS
42023	SCT	ANG	IS
42029	SCT	ANG	IS
42030	SCT	ANG	IS
42032	SCT	ANG	IS
42033	SCT	ANG	IS
42034	SCT	ANG	IS
42035	SCT	ANG	IS
42045	SCT	ANG	IS
42046	SCT	ANG	IS
42047	SCT	ANG	IS
42054	SCT	ANG	IS
42055	SCT	ANG	IS
42056	SCT	ANG	IS
42072	SCT	ANG	IS
42075	SCT	ANG	IS
42077	SCT	ANG	IS
42078	SCT	ANG	IS
42096	SCT	ANG	IS
42107	SCT	ANG	IS
42129	SCT	ANG	IS
42143	SCT	ANG	IS
42144	SCT	ANG	IS
42183	SCT	ANG	IS
42184	SCT	ANG	IS
42185	SCT	ANG	IS
42200	SCT	ANG	IS
42206	SCT	ANG	IS
42207	SCT	ANG	IS
42209	SCT	ANG	IS
42213	SCT	ANG	IS
42245	SCT	ANG	IS
42250	SCT	ANG	IS
42252	SCT	ANG	IS
42253	SCT	ANG	IS
42207	SCT	ANG	IS
42255	SCT	ANG	IS
42257	SCT	ANG	IS
42259	SCT	ANG	IS
42265	SCT	ANG	IS
42267	SCT	ANG	IS

42268	SCT	ANG	IS
42269	SCT	ANG	IS
42275	SCT	ANG	IS
42276	SCT	ANG	IS
42277	SCT	ANG	IS
42279	SCT	ANG	IS
42280	SCT	ANG	IS
42281	SCT	ANG	IS
42288	SCT	ANG	IS
42289	SCT	ANG	IS
42291	SCT	ANG	IS
42292	SCT	ANG	IS
42293	SCT	ANG	IS
42295	SCT	ANG	IS
42297	SCT	ANG	IS
42299	SCT	ANG	IS
42300	SCT	ANG	IS
42301	SCT	ANG	IS
42325	SCT	ANG	IS
42333	SCT	ANG	IS
42343	SCT	ANG	IS
42345	SCT	ANG	IS
42350	SCT	ANG	IS
42351	SCT	ANG	IS
42360	SCT	ANG	IS
42551	SCT	ANG	IS
42553	SCT	ANG	IS
42555	SCT	ANG	IS
42557	SCT	ANG	IS
42558	SCT	ANG	IS
42559	SCT	ANG	IS
42561	SCT	ANG	IS
42562	SCT	ANG	IS
42567	SCT	ANG	IS
42568	SCT	ANG	IS
42571	SCT	ANG	IS
42574	SCT	ANG	IS
42576	SCT	ANG	IS
42577	SCT	ANG	IS
42578	SCT	ANG	IS
42579	SCT	ANG	IS
42581	SCT	ANG	IS

ScotRail HST formations

As a rule the ScotRail HST vehicle trailers stay in fixed formations, but the spare vehicles may replace vehicles as required for heavy maintenance or because of faults or damage. Power cars will be swapped around as required and are not assigned to any particular set. The 17 of the four-car sets are in the process of being strengthened to five-cars.

Sets HA01-26 are refurbished sets but set HA22 was written off in August 2020.

Set	TGFB	TSD	TS	TS	TSL
HA01	40601	42004		42561	42046
HA02	40602	42292		42562	42045
HA03	40603	42021		42557	42143
HA04	40604	42183		42559	42343
HA05	40605	42345	42029	42034	42184
HA06	40606	42206	43033	42581	42208
HA07	40607	42207		42574	42288
HA08	40608	42055		42571	42019
HA09	40609	42253		42107	42257
HA10	40610	42360	42351	42551	42252
HA11	40611	42267	42023	42325	42301
HA12	40612	42275		42576	42276
HA13	40613	42279		42280	42296
HA14	40614	42012		42245	42013
HA15	40615	42030		42579	42010
HA16	40616	42291		42577	42075
HA17	40617	42295		42558	42250
HA18	40618	42297		42555	42014
HA19	40619	42255		42575	42256
HA20	40620	42200		42568	42129
HA21	40621	42299		42300	42277
HA23	40623	42268		42567	42269
HA24	40624	42265		42553	42293
HA25	40625	42259		42578	42333
HA26	40626	42281		42072	42350
spare		42009	42032	42035	42047
spare		42054	42056	42077	42078
spare		42096	42144	42185	42209
spare		42213			

The spare vehicles will be used to increase more sets to five-cars.

Class 153

These five former GWR/Northern Class 153s have been converted to luggage/bike vehicles but retain some seats. They also have free wi-fi and at-seat power sockets fitted. A refurbished toilet with controlled-emission toilet (CET) tank is also fitted. They work with Class 156s on the West Highland line to Oban to provide room to store skis, rucksacks and bikes.

Each unit can carry up to 20 bikes in custom-designed racks, sporting equipment and large items of luggage, as well as more seats for customers. All have been outshopped in an artistic wrap depicting typical Highland scenery.

All are in the HAHQ pool.

	Livery	Owner	Depot	DMSL
153305	ADV	ANG	CK	52305
153370	ADV	ANG	CK	57370
153373	ADV	ANG	CK	57373
153377	ADV	ANG	CK	57377
153380	ADV	ANG	CK	57380

All these units are in unique promotional liveries with graphics of the Scottish Highlands.

Class 156

These two-car Metro Cammell units built in 1987-89 have been the mainstay of the West Highland and Glasgow South Western routes since 1989, plus they are also used on some local commuter trains in the Glasgow area as required. They briefly worked between Glasgow Queen Street and Edinburgh until displaced by Class 158s in 1990.

156445/446/450/453/456-458/474/476-478/492/493/499/500 have RETB fitted for working on the West Highland Line.

All are in the HAHQ pool.

In 1989 Class 156s were introduced on the West Highland lines and more than 33 years later they are still in use to Mallaig and Oban. On 4 September 2021, 156450 waits to set off with the 1200 Oban-Glasgow QS. *Pip Dunn*

	Livery	Owner	Depot	DMSL	DMS
156430	SCR	ANG	CK	52430	57430
156431	SCR	ANG	CK	52431	57431
156432	SCR	ANG	CK	52432	57432
156433	SCR	ANG	CK	52433	57433
156434	SCR	ANG	CK	52434	57434
156435	SCR	ANG	CK	52435	57435
156436	SCR	ANG	CK	52436	57436
156437	SCR	ANG	CK	52437	57437
156439	SCR	ANG	CK	52439	57439
156442	SCR	ANG	CK	52442	57442
156445	SCR	ANG	CK	52445	57445
156446	SCR	ANG	CK	52446	57446
156450	SCR	ANG	CK	52450	57450
156453	SCR	ANG	CK	52453	57453
156456	SCR	ANG	CK	52456	57456
156457	SCR	ANG	CK	52457	57457
156458	SCR	ANG	CK	52458	57458
156462	SCR	ANG	CK	52462	57462
156467	SCR	ANG	CK	52467	57467
156474	SCR	ANG	CK	52474	57474
156476	SCR	ANG	CK	52476	57476
156477	SCR	ANG	CK	52477	57477
156478	SCR	BRO	CK	52478	57478
156492	SCR	ANG	CK	52492	57492
156493	SCR	ANG	CK	52493	57493
156494	SCR	ANG	CK	52494	57494
156495	SCR	ANG	CK	52495	57495
156499	SCR	ANG	CK	52499	57499
156501	SCR	ANG	CK	52501	57501
156502	SCR	ANG	CK	52502	57502
156503	SCR	ANG	CK	52503	57503
156504	SCR	ANG	CK	52504	57504
156505	SCR	ANG	CK	52505	57505
156506	SCR	ANG	CK	52506	57506
156507	SCR	ANG	CK	52507	57507
156508	SCR	ANG	CK	52508	57508
156509	SCR	ANG	CK	52509	57509
156510	SCR	ANG	CK	52510	57510
156511	SCR	ANG	CK	52511	57511
156512	SCR	ANG	CK	52512	57512
156513	SCR	ANG	CK	52513	57513
156514	SCR	ANG	CK	52514	57514

Class 158

These BREL-built two-car units work across all trains north of Inverness to Wick and Kyle of Lochalsh and also some trains from Inverness to Aberdeen, on the Highland Main line, plus the Borders Railway from Edinburgh to Tweedbank. They also work other local commuter trains as required across Scotland. 158701-736/738-741 are RETB compliant.

All are in the HAHQ pool.

	Livery	Owner	Depot	DMSL	DMSL
158701	SCR	POR	IS	52701	57701
158702	SCR	POR	IS	52702	57702
158703	SCR	POR	IS	52703	57703
158704	SCR	POR	IS	52704	57704
158705	SCR	POR	IS	52705	57705
158706	SCR	POR	IS	52706	57706
158707	SCR	POR	IS	52707	57707
158708	SCR	POR	IS	52708	57708
158709	SCR	POR	IS	52709	57709
158710	SCR	POR	IS	52710	57710
158711	SCR	POR	IS	52711	57711
158712	SCR	POR	IS	52712	57712
158713	SCR	POR	IS	52713	57713
158714	SCR	POR	IS	52714	57714
158715	SCR	POR	IS	52715	57715
158716	SCR	POR	IS	52716	57716
158717	SCR	POR	IS	52717	57717
158718	SCR	POR	IS	52718	57718
158719	SCR	POR	IS	52719	57719
158720	SCR	POR	IS	52720	57720
158721	SCR	POR	IS	52721	57721
158722	SCR	POR	IS	52722	57722
158723	SCR	POR	IS	52723	57723
158724	SCR	POR	IS	52724	57724
158725	SCR	POR	IS	52725	57725
158726	SCR	POR	CK	52726	57726
158727	SCR	POR	CK	52727	57727
158728	SCR	POR	CK	52728	57728
158729	SCR	POR	CK	52729	57729
158730	SCR	POR	CK	52730	57730
158731	SCR	POR	CK	52731	57731
158732	SCR	POR	CK	52732	57732
158733	SCR	POR	CK	52733	57733
158734	SCR	POR	CK	52734	57734
158735	SCR	POR	CK	52735	57735
158736	SCR	POR	CK	52736	57736
158737	SCR	POR	CK	52737	57737
158738	SCR	POR	CK	52738	57738
158739	SCR	POR	CK	52739	57739
158740	SCR	POR	CK	52740	57740
158741	SCR	POR	CK	52741	57741

Class 158s first appeared in Scotland in 1990 and replaced loco-hauled trains on several key inter-city routes. They are regularly seen these days on the new Borders route to Tweedbank, and 158722 passes Falahill on 16 June 2016 heading south with another 158. *John Chalcraft*

Class 170

These three-car Turbostar units were delivered from Bombardier in 1999-2005. Initially the ScotRail fleet comprised 170401-434/450-461/470-478 but units have redeployed over the years to Southern, Northern and East Midland Railway. The remaining Scottish units are used on the Borders Railway, and on other commuter and inter-city routes across Scotland. They are also still used on the Highland Main Line and to Aberdeen, albeit reduced following the introduction of HSTs. They often work in multiple with Class 158s.

All are in the HAHQ pool.

Class 170/3

The four units were initially ordered and used by Hull Trains from 2004 but transferred to Scotland in May 2005.

	Livery	Owner	Depot	DMCL	MS	DMCL
170393	SCR	POR	HA	50393	56393	79393
170394	SCR	POR	HA	50394	56394	79394
170395	SCR	POR	HA	50395	56395	79394
170396	SCR	POR	HA	50396	56396	79396

Class 170/4

The original order for ScotRail Turbostars was 170401-424 built in 1999-2001, and then supplemented by 170425-434 built in 2003-05. Units 170416-420 have transferred to East Midlands Railway and 170421-424 have transferred to Govia Thameslink and been converted to Class 171s.

	Livery	Owner	Depot	DMCL	MS	DMCL
170401	SCR	POR	HA	50401	56401	79401
170402	SCR	POR	HA	50402	56402	79402
170403	SCR	POR	HA	50403	56403	79403
170404	SCR	POR	HA	50404	56404	79404
170405	SCR	POR	HA	50405	56405	79405
170406	SCR	POR	HA	50406	56406	79406
170407	ADV	POR	HA	50407	56407	79407
170408	SCR	POR	HA	50408	56408	79408
170409	SCR	POR	HA	50409	56409	79409
170410	SCR	POR	HA	50410	56410	79410
170411	SCR	POR	HA	50411	56411	79411
170412	SCR	POR	HA	50412	56412	79412
170413	SCR	POR	HA	50413	56413	79413
170414	SCR	POR	HA	50414	56414	79414
170415	SCR	POR	HA	50415	56415	79415
170425	SCR	POR	HA	50425	56425	79425
170426	SCR	POR	HA	50426	56426	79426
170427	SCR	POR	HA	50427	56427	79427
170428	SCR	POR	HA	50428	56428	79428
170429	SCR	POR	HA	50429	56429	79429
170430	SCR	POR	HA	50430	56430	79430
170431	SCR	POR	HA	50431	56431	79431
170432	SCR	POR	HA	50432	56432	79432
170433	SCR	POR	HA	50433	56433	79433
170434	SCR	POR	HA	50434	56434	79434

On 2 June 2021, 170403 passes Dalwhinnie with the 0648 Edinburgh-Inverness. The Turbostar Units, now over 20 years old, share these Highland Main Line duties with HSTs and Class 158s. *Paul Shannon*

Class 170/4

These later orders for ScotRail were 170450-461 built in 2003-05 and 170470-478 built in 2001-05 and were standard class only, so were used on Central Belt local trains. 170453-461/472-478 are now with Northern. 170450-452 now have first-class seats added.

	Livery	Owner	Depot	DMCL	MS	DMCL
170450	SCR	POR	HA	50450	56450	79450
170451	SCR	POR	HA	50451	56451	79451
170452	SCR	POR	HA	50452	56452	79452
170470	SCR	POR	HA	50470	56470	79470
170471	SCR	POR	HA	50471	56471	79471

Class 318

These BREL units, which date from 1985-87, are used on local trains south of the River Clyde to the likes of Lanark, Neilston and Ayrshire. The fleet is being refurbished.

All are in the HAHQ pool.

	Livery	Owner	Depot	DTS	MS	DTS
318250	SCR	EVS	GW	77240	62866	77260
318251	SCR	EVS	GW	77241	62867	77261
318252	SCR	EVS	GW	77242	62868	77262
318253	SCR	EVS	GW	77243	62869	77263
318254	SCR	EVS	GW	77244	62870	77264
318255	SCR	EVS	GW	77245	62871	77265
318256	SCR	EVS	GW	77246	62872	77266
318257	SCR	EVS	GW	77247	62873	77267
318258	SCR	EVS	GW	77248	62874	77268
318259	SCR	EVS	GW	77249	62875	77269
318260	SCR	EVS	GW	77250	62876	77270
318261	SCR	EVS	GW	77251	62877	77271
318262	SCR	EVS	GW	77252	62878	77272
318263	SCR	EVS	GW	77253	62879	77273
318264	SCR	EVS	GW	77254	62880	77274
318265	SCR	EVS	GW	77255	62881	77275
318266	SCR	EVS	GW	77256	62882	77276
318267	SCR	EVS	GW	77257	62883	77277
318268	SCR	EVS	GW	77258	62884	77278
318269	SCR	EVS	GW	77259	62885	77279
318270	SCR	EVS	GW	77288	62890	77289

Class 318s are used on local duties in the Glasgow area; 318260 waits to set off with the 0853 Lanark-Glasgow Central on 26 July 2018. *Pip Dunn*

Class 320

These BREL-built units were delivered to ScotRail in 1990, initially in Strathclyde PTE orange, they work mostly north of the Clyde on local commuter trains. They were joined by 12 Class 320/4s. These were former London Midland Region Class 321s, which were delivered as four-car sets. They were shortened to three-car sets and renumbered from Class 321 to 320 to create a unified fleet.

All are in the HAHQ pool.

A fleet of 22 BREL-built three-car Class 320s were introduced in Scotland in 1988 and have been joined by 12 similar Class 321s – renumbered as Class 320/4s – after their release from local duties in England. 320304 waits at Glasgow Central on 29 May 2019. *Pip Dunn*

Class 320/3

Original three-car Class 320s built for ScotRail.

	Livery	Owner	Depot	DMS	PTS	DMS
320301	SCR	EVS	GW	77899	63021	77921
320302	SCR	EVS	GW	77900	63022	77922
320303	SCR	EVS	GW	77901	63023	77923
320304	SCR	EVS	GW	77902	63024	77924
320305	SCR	EVS	GW	77903	63025	77925
320306	SCR	EVS	GW	77904	63026	77926
320307	SCR	EVS	GW	77905	63027	77927
320308	SCR	EVS	GW	77906	63028	77928
320309	SCR	EVS	GW	77907	63029	77929
320310	SCR	EVS	GW	77908	63030	77930
320311	SCR	EVS	GW	77909	63031	77931
320312	SCR	EVS	GW	77910	63032	77932
320313	SCR	EVS	GW	77911	63033	77933
320314	SCR	EVS	GW	77912	63034	77934
320315	SCR	EVS	GW	77913	63035	77935
320316	SCR	EVS	GW	77914	63036	77936
320317	SCR	EVS	GW	77915	63037	77937
320318	SCR	EVS	GW	77916	63038	77938
320319	SCR	EVS	GW	77917	63039	77939
320320	SCR	EVS	GW	77918	63040	77940
320321	SCR	EVS	GW	77919	63041	77941
320322	SCR	EVS	GW	77920	63042	77942

Class 320/4

Converted from ex-London Midland Class 321/4s reduced from four- to three-car sets.

	Old No.	Livery	Owner	Depot	DTS	MS	DTS
320401	321401	SCR	EVS	GW	78095	63063	77943
320403	321403	SCR	EVS	GW	78097	63065	77945
320404	321404	SCR	EVS	GW	78098	63066	77946
320411	321411	SCR	EVS	GW	78105	63073	77953
320412	321412	SCR	EVS	GW	78106	63074	77954
320413	321413	SCR	EVS	GW	78107	63075	77955
320414	321414	SCR	EVS	GW	78108	63076	77956
320415	321415	SCR	EVS	GW	78109	63077	77957
320416	321416	SCR	EVS	GW	78110	63078	77958
320417	321417	SCR	EVS	GW	78111	63079	77959
320418	321418	SCR	EVS	GW	78112	63080	77962
320420	321420	SCR	EVS	GW	78114	63082	77960

Class 334

These Alstom Juniper 25kV AC three-car units were built at Washwood Heath in 1999-2001 and were the first new EMUs for Scotland after privatisation. They are used on many outer suburban services mostly in the Glasgow area but they do also work to Edinburgh.

All are in the HAHQ pool.

	Livery	Owner	Depot	DMS	PTS	DMS
334001	SCR	EVS	GW	64101	74301	65101
334002	SCR	EVS	GW	64102	74302	65102
334003	SCR	EVS	GW	64103	74303	65103
334004	SCR	EVS	GW	64104	74304	65104
334005	SCR	EVS	GW	64105	74305	65105
334006	ADV	EVS	GW	64106	74306	65106
334007	SCR	EVS	GW	64107	74307	65107
334008	SCR	EVS	GW	64108	74308	65108
334009	SCR	EVS	GW	64109	74309	65109
334010	SCR	EVS	GW	64110	74310	65110
334011	SCR	EVS	GW	64111	74311	65111
334012	SCR	EVS	GW	64112	74312	65112
334013	SCR	EVS	GW	64113	74313	65113
334014	SCR	EVS	GW	64114	74314	65114
334015	SCR	EVS	GW	64115	74315	65115
334016	SCR	EVS	GW	64116	74316	65116
334017	SCR	EVS	GW	64117	74317	65117
334018	SCR	EVS	GW	64118	74318	65118
334019	SCR	EVS	GW	64119	74319	65119
334020	SCR	EVS	GW	64120	74320	65120
334021	SCR	EVS	GW	64121	74321	65121
334022	SCR	EVS	GW	64122	74322	65122
334023	SCR	EVS	GW	64123	74323	65123
334024	SCR	EVS	GW	64124	74324	65124
334025	SCR	EVS	GW	64125	74325	65125
334026	SCR	EVS	GW	64126	74326	65126
334027	SCR	EVS	GW	64127	74327	65127
334028	SCR	EVS	GW	64128	74328	65128
334029	SCR	EVS	GW	64129	74329	65129
334030	SCR	EVS	GW	64130	74330	65130
334031	SCR	EVS	GW	64131	74331	65131
334032	SCR	EVS	GW	64132	74332	65132
334033	SCR	EVS	GW	64133	74333	65133
334034	SCR	EVS	GW	64134	74334	65134
334035	SCR	EVS	GW	64135	74335	65135
334036	SCR	EVS	GW	64136	74336	65136
334037	SCR	EVS	GW	64137	74337	65137
334038	SCR	EVS	GW	64138	74338	65138
334039	SCR	EVS	GW	64139	74339	65139
334040	SCR	EVS	GW	64140	74340	65140

Note: The front three-quarters of 64106 from unit 334006 is wrapped in a Pride rainbow livery. The rest is in SCR livery.

Alstom Class 334 Juniper units were introduced at the turn of the century: 334036 calls at Haymarket on 29 March 2017. *Pip Dunn*

Class 380

Siemens Desiro units built in Krefeld and delivered in 2009-10, they are used on many Inverclyde services and other local trains across the Central Belt.

All are in the HAHQ pool.

Class 380/0

Three-car units.

	Livery	Owner	Depot	DMS	PTS	DMS
380001	SCR	EVS	GW	38501	38601	38701
380002	SCR	EVS	GW	38502	38602	38702
380003	SCR	EVS	GW	38503	38603	38703
380004	SCR	EVS	GW	38504	38604	38704
380005	SCR	EVS	GW	38505	38605	38705
380006	SCR	EVS	GW	38506	38606	38706
380007	SCR	EVS	GW	38507	38607	38707
380008	SCR	EVS	GW	38508	38608	38708
380009	SCR	EVS	GW	38509	38609	38709
380010	SCR	EVS	GW	38510	38610	38710
380011	SCR	EVS	GW	38511	38611	38711
380012	SCR	EVS	GW	38512	38612	38712
380013	SCR	EVS	GW	38513	38613	38713
380014	SCR	EVS	GW	38514	38614	38714
380015	SCR	EVS	GW	38515	38615	38715
380016	SCR	EVS	GW	38516	38616	38716
380017	SCR	EVS	GW	38517	38617	38717
380018	SCR	EVS	GW	38518	38618	38718
380019	SCR	EVS	GW	38519	38619	38719
380020	SCR	EVS	GW	38520	38620	38720
380021	SCR	EVS	GW	38521	38621	38721
380022	SCR	EVS	GW	38522	38622	38722

Class 380/1

Four-car units.

	Livery	Owner	Depot	DMS	PTS	TS	DMS
380101	SCR	EVS	GW	38551	38651	38851	38751
380102	SCR	EVS	GW	38552	38652	38852	38752
380103	SCR	EVS	GW	38553	38653	38853	38753
380104	SCR	EVS	GW	38554	38654	38854	38754
380105	SCR	EVS	GW	38555	38655	38855	38755
380106	SCR	EVS	GW	38556	38656	38856	38756
380107	SCR	EVS	GW	38557	38657	38857	38757
380108	SCR	EVS	GW	38558	38658	38858	38758
380109	SCR	EVS	GW	38559	38659	38859	38759
380110	SCR	EVS	GW	38560	38660	38860	38760
380111	SCR	EVS	GW	38561	38661	38861	38761
380112	SCR	EVS	GW	38562	38662	38862	38762
380113	SCR	EVS	GW	38563	38663	38863	38763
380114	SCR	EVS	GW	38564	38664	38864	38764
380115	SCR	EVS	GW	38565	38665	38865	38765
380116	SCR	EVS	GW	38566	38666	38866	38766

ScotRail has a varied fleet of 25kV AC EMUs, with classes 318, 320, 334, 380 and 385 all in use. Siemens Desiro 380109 passes Cleland on the 1119 Glasgow Central-Edinburgh on 19 November 2020. *Robin Ralston*

Class 385

Hitachi AT200 units delivered in 2016-18, built at Newton Aycliffe and Kasado in Japan. They are used on Glasgow QS-Edinburgh, Glasgow Central-Edinburgh, Glasgow QS-Dunblane and other local commuter workings in the Scottish Central Belt.

All are in the HAHQ pool.

Class 385/0

Hitachi AT200 three-car units.

	Livery	Owner	Depot	DMS	PTS	DMS
385001	SCR	CRL	EC	441001	442001	444001
385002	SCR	CRL	EC	441002	442002	444002
385003	SCR	CRL	EC	441003	442003	444003
385004	SCR	CRL	EC	441004	442004	444004
385005	SCR	CRL	EC	441005	442005	444005
385006	SCR	CRL	EC	441006	442006	444006
385007	SCR	CRL	EC	441007	442007	444007
385008	SCR	CRL	EC	441008	442008	444008
385009	SCR	CRL	EC	441009	442009	444009
385010	SCR	CRL	EC	441010	442010	444010
385011	SCR	CRL	EC	441011	442011	444011
385012	SCR	CRL	EC	441012	442012	444012
385013	SCR	CRL	EC	441013	442013	444013
385014	SCR	CRL	EC	441014	442014	444014
385015	SCR	CRL	EC	441015	442015	444015
385016	SCR	CRL	EC	441016	442016	444016
385017	SCR	CRL	EC	441017	442017	444017
385018	SCR	CRL	EC	441018	442018	444018
385019	SCR	CRL	EC	441019	442019	444019

385020	SCR	CRL	EC	441020	442020	444020
385021	SCR	CRL	EC	441021	442021	444021
385022	SCR	CRL	EC	441022	442022	444022
385023	SCR	CRL	EC	441023	442023	444023
385024	SCR	CRL	EC	441024	442024	444024
385025	SCR	CRL	EC	441025	442025	444025
385026	SCR	CRL	EC	441026	442026	444026
385027	SCR	CRL	EC	441027	442027	444027
385028	SCR	CRL	EC	441028	442028	444028
385029	SCR	CRL	EC	441029	442029	444029
385030	SCR	CRL	EC	441030	442030	444030
385031	SCR	CRL	EC	441031	442031	444031
385032	SCR	CRL	EC	441032	442032	444032
385033	SCR	CRL	EC	441033	442033	444033
385034	SCR	CRL	EC	441034	442034	444034
385035	SCR	CRL	EC	441035	442035	444035
385036	SCR	CRL	EC	441036	442036	444036
385037	SCR	CRL	EC	441037	442037	444037
385038	SCR	CRL	EC	441038	442038	444038
385039	SCR	CRL	EC	441039	442039	444039
385040	SCR	CRL	EC	441040	442040	444040
385041	SCR	CRL	EC	441041	442041	444041
385042	SCR	CRL	EC	441042	442042	444042
385043	SCR	CRL	EC	441043	442043	444043
385044	SCR	CRL	EC	441044	442044	444044
385045	SCR	CRL	EC	441045	442045	444045
385046	SCR	CRL	EC	441046	442046	444046

Hitachi AT200 units – Class 385s – are the newest EMUs in use with ScotRail. On 20 March 2019, 385020 passes Princes Street Gardens as it arrives at Edinburgh. *Rob France*

Class 385/1

Hitachi AT200 four-car units.

	Livery	Owner	Depot	DMS	PTS	TS	DMS
385101	SCR	CRL	EC	441101	442101	443101	444101
385101	SCR	CRL	EC	441101	442101	443101	444101
385102	SCR	CRL	EC	441102	442102	443102	444102
385103	SCR	CRL	EC	441103	442103	443103	444103
385104	SCR	CRL	EC	441104	442104	443104	444104
385105	SCR	CRL	EC	441105	442105	443105	444105
385106	SCR	CRL	EC	441106	442106	443106	444106
385107	SCR	CRL	EC	441107	442107	443107	444107
385108	SCR	CRL	EC	441108	442108	443108	444108
385109	SCR	CRL	EC	441109	442109	443109	444109
385110	SCR	CRL	EC	441110	442110	443110	444110
385111	SCR	CRL	EC	441111	442111	443111	444111
385112	SCR	CRL	EC	441112	442112	443112	444112
385113	SCR	CRL	EC	441113	442113	443113	444113
385114	SCR	CRL	EC	441114	442114	443114	444114
385115	SCR	CRL	EC	441115	442115	443115	444115
385116	SCR	CRL	EC	441116	442116	443116	444116
385117	SCR	CRL	EC	441117	442117	443117	444117
385118	SCR	CRL	EC	441118	442118	443118	444118
385119	SCR	CRL	EC	441119	442119	443119	444119
385120	SCR	CRL	EC	441120	442120	443120	444120
385121	SCR	CRL	EC	441121	442121	443121	444121
385122	SCR	CRL	EC	441122	442122	443122	444122
385123	SCR	CRL	EC	441123	442123	443123	444123
385124	SCR	CRL	EC	441124	442124	443124	444124

Southeastern

Contact details

Website: www.southeasternrailway.co.uk
Twitter: @Se_Railway

Key personnel

Managing Director: Steve White
Engineering Director: Mark Johnson

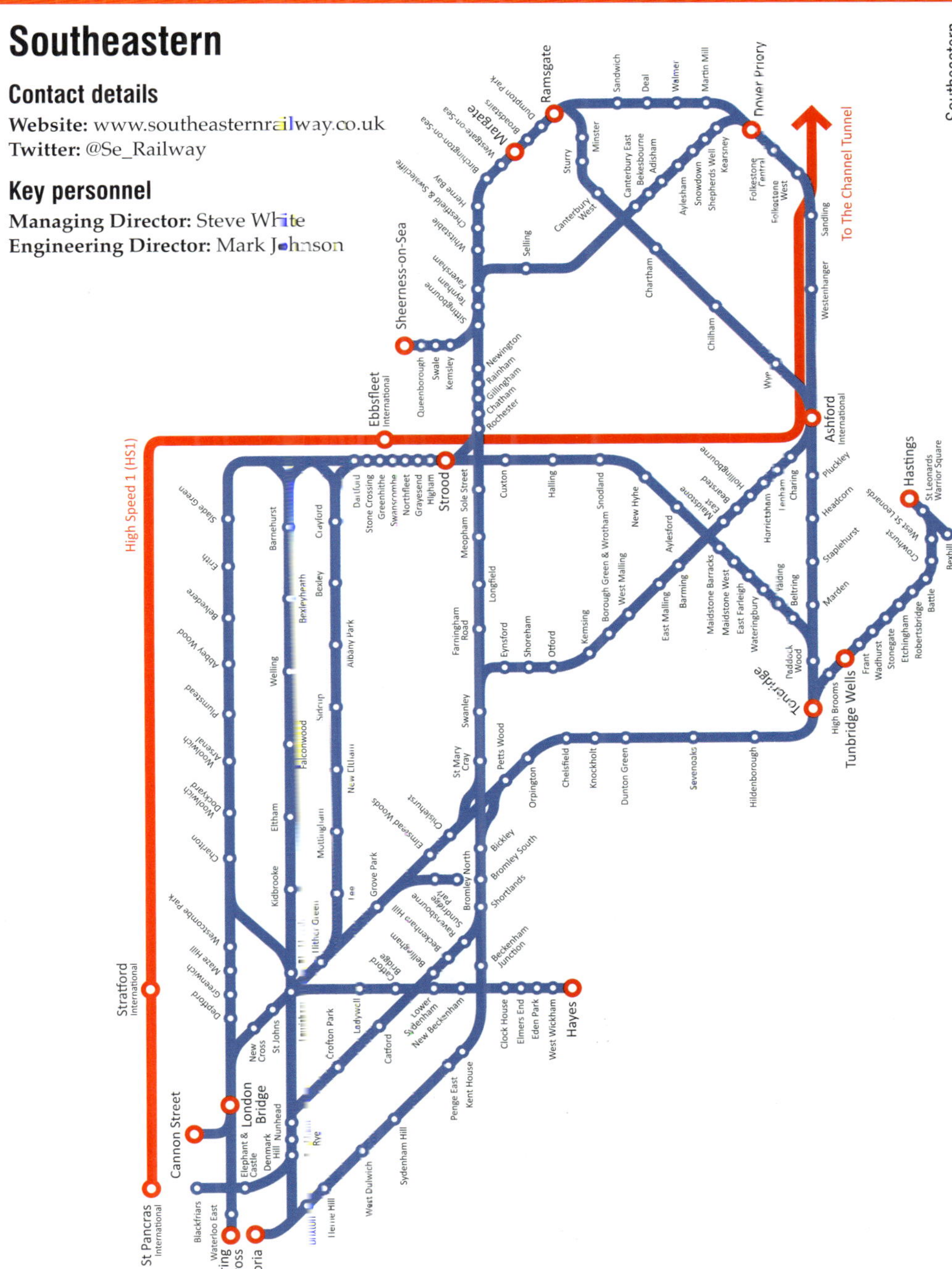

Overview

Southeastern operates commuter trains across the South-East in Kent and East Sussex. It replaced most of its all-electric fleet in the late 1990s and early 2000s and as well as the Class 465/466 Networker units it inherited from BR days, it now has a mostly Electrostar fleet, including the Class 376 metro-style trains. It also operates high-speed domestic trains on HS1 using 29 six-car Class 395 Hitachi Javelin units.

It was due to receive the 30 five-car Siemens Desiro City Class 707 units from South Western Railway when they were due to be replaced by Class 701s in 2021 but delays in the new Bombardier units has seen 12 units remain with SWR until further notice. Class 465 units are now being returned to their leasing companies and stored pending disposal.

Class 375

These Bombardier Electrostar units work the majority of Southeastern's outer suburban commuter services. Most are four-car units, although the 375/3s are three-car sets.

Eversholt has awarded a £10m contract to carry out modifications to its fleet of 112 Class 375s, including charging points, new lights, improved LED external lights and bio-reactor toilets.

All are in the HUHQ pool.

Class 375/5

750V DC Express units.

	Livery	Owner	Depot	DMS	TS	DMC
375301	SEB	EVS	RM	67921	74351	67931
375302	SEB	EVS	RM	67922	74352	67932
375303	SEB	EVS	RM	67923	74353	67933
375304	SEB	EVS	RM	67924	74354	67934
375305	SEB	EVS	RM	67925	74355	67935
375306	SEB	EVS	RM	67926	74356	67936
375307	SEB	EVS	RM	67927	74357	67937
375308	SEB	EVS	RM	67928	74358	67938
375309	SEB	EVS	RM	67929	74359	67939
375310	SEB	EVS	RM	67930	74360	67940

Class 375/6

Dual-voltage 750V DC and 25kV AC units.

	Livery	Owner	Depot	DMS	MC	PTS	DMS
375601	SEB	EVS	RM	67801	74251	74201	67851
375602	SEB	EVS	RM	67802	74252	74202	67852
375603	SEB	EVS	RM	67803	74253	74203	67853
375604	SEB	EVS	RM	67804	74254	74204	67854
375605	SEB	EVS	RM	67805	74255	74205	67855
375606	SEB	EVS	RM	67806	74256	74206	67856
375607	SEB	EVS	RM	67807	74257	74207	67857

Southeastern is a fully electric operation with a mix of 750V DC units. Bombardier Electrostar 375607 works the 1040 Charing Cross-Dover Priory past Knockholt on 17 February 2022. *David Staines*

375608	SEB	EVS	RM	67808	74258	74208	67858	
375609	SEB	EVS	RM	67809	74259	74209	67859	
375610	SEB	EVS	RM	67810	74260	74210	67860	
375611	SEB	EVS	RM	67811	74261	74211	67861	
375612	SEB	EVS	RM	67812	74262	74212	67862	
375613	SEB	EVS	RM	67813	74263	74213	67863	
375614	SEB	EVS	RM	67814	74264	74214	67864	
375615	SEB	EVS	RM	67815	74265	74215	67865	
375616	SEB	EVS	RM	67816	74266	74216	67866	
375617	SEB	EVS	RM	67817	74267	74217	67867	
375618	SEB	EVS	RM	67818	74268	74218	67868	
375619	SEB	EVS	RM	67819	74269	74219	67869	*Driver John Neve*
375620	SEB	EVS	RM	67820	74270	74220	67870	
375621	SEB	EVS	RM	67821	74271	74221	67871	
375622	SEB	EVS	RM	67822	74272	74222	67872	
375623	SEB	EVS	RM	67823	74273	74223	67873	*Hospice in the Weald*
375624	SEB	EVS	RM	67824	74274	74224	67874	
375625	SEB	EVS	RM	67825	74275	74225	67875	
375626	SEB	EVS	RM	67826	74276	74226	67876	
375627	SEB	EVS	RM	67827	74277	74227	67877	
375628	SEB	EVS	RM	67828	74278	74228	67878	
375629	SEB	EVS	RM	67829	74279	74229	67879	
375630	SEB	EVS	RM	67830	74280	74230	67880	

Class 375/7

750V DC express units with limited first-class seats.

	Livery	Owner	Depot	DMS	MC	TS	DMS	
375701	SEB	EVS	RM	67831	74281	74231	67881	*Kent Air Ambulance Explorer*
375702	SEB	EVS	RM	67832	74282	74232	67882	
375703	SEB	EVS	RM	67833	74283	74233	67883	
375704	SEB	EVS	RM	67834	74284	74234	67884	
375705	SEB	EVS	RM	67835	74285	74235	67885	
375706	SEB	EVS	RM	67836	74286	74236	67886	
375707	SEB	EVS	RM	67837	74287	74237	67887	
375708	SEB	EVS	RM	67838	74288	74238	67888	
375709	SEB	EVS	RM	67839	74289	74239	67889	
375710	SEB	EVS	RM	67840	74290	74240	67890	*Rochester Castle*
375711	SEB	EVS	RM	67841	74291	74241	67891	
375712	SEB	EVS	RM	67842	74292	74242	67892	
375713	SEB	EVS	RM	67843	74293	74243	67893	
375714	SEB	EVS	RM	67844	74294	74244	67894	*Rochester Cathedral*
375715	SEB	EVS	RM	67845	74295	74245	67895	

Class 375/8

750V DC express units with limited first-class seats.

	Livery	Owner	Depot	DMS	MC	TS	DMS
375801	SEB	EVS	RM	73301	79001	78201	73701
375802	SEB	EVS	RM	73302	79002	78202	73702
375803	SEB	EVS	RM	73303	79003	78203	73703
375804	SEB	EVS	RM	73304	79004	78204	73704
375805	SEB	EVS	RM	73305	79005	78205	73705
375806	SEB	EVS	RM	73306	79006	78206	73706
375807	SEB	EVS	RM	73307	79007	78207	73707
375808	SEB	EVS	RM	73308	79008	78208	73708
375809	SEB	EVS	RM	73309	79009	78209	73709
375810	SEB	EVS	RM	73310	79010	78210	73710

375811	SEB	EVS	RM	73311	79011	78211	73711	
375812	SEB	EVS	RM	73312	79012	78212	73712	
375813	SEB	EVS	RM	73313	79013	78213	73713	
375814	SEB	EVS	RM	73314	79014	78214	73714	
375815	SEB	EVS	RM	73315	79015	78215	73715	
375816	SEB	EVS	RM	73316	79016	78216	73716	
375817	SEB	EVS	RM	73317	79017	78217	73717	
375818	SEB	EVS	RM	73318	79018	78218	73718	
375819	SEB	EVS	RM	73319	79019	78219	73719	
375820	SEB	EVS	RM	73320	79020	78220	73720	
375821	SEB	EVS	RM	73321	79021	78221	73721	
375822	SEB	EVS	RM	73322	79022	78222	73722	
375823	SEB	EVS	RM	73323	79023	78223	73723	*Ashford Proudly Served by Rail for 175 years*
375824	SEB	EVS	RM	73324	79024	78224	73724	
375825	SEB	EVS	RM	73325	79025	78225	73725	
375826	SEB	EVS	RM	73326	79026	78226	73726	
375827	SEB	EVS	RM	73327	79027	78227	73727	
375828	SEB	EVS	RM	73328	79028	78228	73728	
375829	SEB	EVS	RM	73329	79029	78229	73729	
375830	SEB	EVS	RM	73330	79030	78230	73730	

Class 375/9

750V DC outer suburban units with first class.

	Livery	Owner	Depot	DMC	MS	TS	DMC
375901	SEB	EVS	RM	73331	79031	79061	73731
375902	SEB	EVS	RM	73332	79032	79062	73732
375903	SEB	EVS	RM	73333	79033	79063	73733
375904	SEB	EVS	RM	73334	79034	79064	73734
375905	SEB	EVS	RM	73335	79035	79065	73735
375906	SEB	EVS	RM	73336	79036	79066	73736
375907	SEB	EVS	RM	73337	79037	79067	73737
375908	SEB	EVS	RM	73338	79038	79068	73738
375909	SEB	EVS	RM	73339	79039	79069	73739
375910	SEB	EVS	RM	73340	79040	79070	73740
375911	SEB	EVS	RM	73341	79041	79071	73741
375912	SEB	EVS	RM	73342	79042	79072	73742
375913	SEB	EVS	RM	73343	79043	79073	73743
375914	SEB	EVS	RM	73344	79044	79074	73744
375915	SEB	EVS	RM	73345	79045	79075	73745
375916	SEB	EVS	RM	73346	79046	79076	73746
375917	SEB	EVS	RM	73347	79047	79077	73747
375918	SEB	EVS	RM	73348	79048	79078	73748
375919	SEB	EVS	RM	73349	79049	79079	73749
375920	SEB	EVS	RM	73350	79050	79080	73750
375921	SEB	EVS	RM	73351	79051	79081	73751
375922	SEB	EVS	RM	73352	79052	79082	73752
375923	SEB	EVS	RM	73353	79053	79083	73753
375924	SEB	EVS	RM	73354	79054	79084	73754
375925	SEB	EVS	RM	73355	79055	79085	73755
375926	SEB	EVS	RM	73356	79056	79086	73756
375927	SEB	EVS	RM	73357	79057	79087	73757

Class 376

These five-car Bombardier Electrostar metro-style trains are used on inner-suburban commuter services only as they have no toilets. They also do not have corridor connections.

All are in the HUHQ pool.

The Class 376s are a basic metro train with few frills for passengers, and not even any toilets! 376001 works the 1034 Charing Cross-Sevenoaks past Polhill on 17 February 2022. *David Staines*

	Livery	Owner	Depot	DMS	MS	TS	MS	DMS	
376001	SEW	EVS	SG	61101	63301	64301	63501	61601	*Alan Doggett*
376002	SEW	EVS	SG	61102	63302	64302	63502	61602	
376003	SEW	EVS	SG	61103	63303	64303	63503	61603	
376004	SEW	EVS	SG	61104	63304	64304	63504	61604	
376005	SEW	EVS	SG	61105	63305	64305	63505	61605	
376006	SEW	EVS	SG	61106	63306	64306	63506	61606	
376007	SEW	EVS	SG	61107	63307	64307	63507	61607	
376008	SEW	EVS	SG	61108	63308	64308	63508	61608	
376009	SEW	EVS	SG	61109	63309	64309	63509	61609	
376010	SEW	EVS	SG	61110	63310	64310	63510	61610	
376011	SEW	EVS	SG	61111	63311	64311	63511	61611	
376012	SEW	EVS	SG	61112	63312	64312	63512	61612	
376013	SEW	EVS	SG	61113	63313	64313	63513	61613	
376014	SEW	EVS	SG	61114	63314	64314	63514	61614	
376015	SEW	EVS	SG	61115	63315	64315	63515	61615	
376016	SEW	EVS	SG	61116	63316	64316	63516	61616	
376017	SEW	EVS	SG	61117	63317	64317	63517	61617	
376018	SEW	EVS	SG	61118	63318	64318	63518	61618	
376019	SEW	EVS	SG	61119	63319	64319	63519	61619	
376020	SEW	EVS	SG	61120	63320	64320	63520	61620	
376021	SEW	EVS	SG	61121	63321	64321	63521	61621	
376022	SEW	EVS	SG	61122	63322	64322	63522	61622	
376023	SEW	EVS	SG	61123	63323	64323	63523	61623	
376024	SEW	EVS	SG	61124	63324	64324	63524	61624	
376025	SEW	EVS	SG	61125	63325	64325	63525	61625	
376026	SEW	EVS	SG	61126	63326	64326	63526	61626	
376027	SEW	EVS	SG	61127	63327	64327	63527	61627	
376028	SEW	EVS	SG	61128	63328	64328	63528	61628	
376029	SEW	EVS	SG	61129	63329	64329	63529	61629	
376030	SEW	EVS	SG	61130	63330	64330	63530	61630	
376031	SEW	EVS	SG	61131	63331	64331	63531	61631	
376032	SEW	EVS	SG	61132	63332	64332	63532	61632	
376033	SEW	EVS	SG	61133	63333	64333	63533	61633	
376034	SEW	EVS	SG	61134	63334	64334	63534	61634	
376035	SEW	EVS	SG	61135	63335	64335	63535	61635	
376036	SEW	EVS	SG	61136	63336	64336	63536	61636	

Class 377

These units are essentially the same as the Class 375s. When ordered Southeastern classed its Electrostar as Class 375s and Southern opted for 377s. These dual-voltage units were ordered by Southern, then sub-leased to First Capital Connect before moving to Southeastern when displaced by Class 700s.

All are in the HUHQ pool.

	Livery	Owner	Depot	DMC	MS	PTS	DMS
377163	SOU	POR	RM	78563	77163	78963	78763
377164	SOU	POR	RM	78564	77164	78964	78764
377501	FIR	POR	RM	73501	75901	74901	73601
377502	FIR	POR	RM	73502	75902	74902	73602
377503	FIR	POR	RM	73503	75903	74903	73603
377504	FIR	POR	RM	73504	75904	74904	73604
377505	FIR	POR	RM	73505	75905	74905	73605
377506	FIR	POR	RM	73506	75906	74906	73606
377507	FIR	POR	RM	73507	75907	74907	73607
377508	FIR	POR	RM	73508	75908	74908	73608
377509	FIR	POR	RM	73509	75909	74909	73609
377510	FIR	POR	RM	73510	75910	74910	73610
377511	FIR	POR	RM	73511	75911	74911	73611
377512	FIR	POR	RM	73512	75912	74912	73612
377513	FIR	POR	RM	73513	75913	74913	73613
377514	FIR	POR	RM	73514	75914	74914	73614
377515	FIR	POR	RM	73515	75915	74915	73615
377516	FIR	POR	RM	73516	75916	74916	73616
377517	FIR	POR	RM	73517	75917	74917	73617
377518	FIR	POR	RM	73518	75918	74918	73618
377519	FIR	POR	RM	73519	75919	74919	73619
377520	FIR	POR	RM	73520	75920	74920	73620
377521	FIR	POR	RM	73521	75921	74921	73621
377522	FIR	POR	RM	73522	75922	74922	73622
377523	FIR	POR	RM	73523	75923	74923	73623

Southeastern's 377514 leads one of the TOC's two Class 377/1s, 377164, at Victoria on 13 January 2019. *Rob France*

Class 395

These six-car Hitachi-built trains, called Javelins, work on the HS1 route from St Pancras to Ashford and Dover.

All are in the HUHQ pool.

	Livery	Owner	Depot	PDTS	MS	MS	MS	MS	PDTS	
395001	SEB	EVS	AD	39011	39012	39013	39014	39015	39016	*Dame Kelly Holmes*
395002	SEB	EVS	AD	39021	39022	39023	39024	39025	39026	*Sebastian Coe*
395003	SEB	EVS	AD	39031	39032	39033	39034	39035	39036	*Sir Steve Redgrave*
395004	SEB	EVS	AD	39041	39042	39043	39044	39045	39046	*Sir Chris Hoy*
395005	SEB	EVS	AD	39051	39052	39053	39054	39055	39056	*Dame Tanni Grey-Thompson*
395006	SEB	EVS	AD	39061	39062	39063	39064	39065	39066	*Daley Thompson*
395007	SEB	EVS	AD	39071	39072	39073	39074	39075	39076	*Steve Backley*
395008	SEB	EVS	AD	39081	39082	39083	39084	39085	39086	*Ben Ainslie*
395009	SEB	EVS	AD	39091	39092	39093	39094	39095	39096	*Rebecca Adlington*
395010	SEB	EVS	AD	39101	39102	39103	39104	39105	39106	*Duncan Goodhew*
395011	SEB	EVS	AD	39111	39112	39113	39114	39115	39116	*Katherine Grainger*
395012	SEB	EVS	AD	39121	39122	39123	39124	39125	39126	
395013	SEB	EVS	AD	39131	39132	39133	39134	39135	39136	*Hornby Visitor Centre*
395014	SEB	EVS	AD	39141	39142	39143	39144	39145	39146	*Dina Asher-Smith*
395015	SEB	EVS	AD	39151	39152	39153	39154	39155	39156	
395016	SEB	EVS	AD	39161	39162	39163	39164	39165	39166	
395017	SEB	EVS	AD	39171	39172	39173	39174	39175	39176	
395018	SEB	EVS	AD	39181	39182	39183	39184	39185	39186	*The Victory Javelin*
395019	SEB	EVS	AD	39191	39192	39193	39194	39195	39196	*Jessica Ennis*
395020	SEB	EVS	AD	39201	39202	39203	39204	39205	39206	*Jason Kenny*
395021	SEB	EVS	AD	39211	39212	39213	39214	39215	39216	*Ed Clancy MBE*
395022	SEB	EVS	AD	39221	39222	39223	39224	39225	39226	*Alistair Brownlee*
395023	SEB	EVS	AD	39231	39232	39233	39234	39235	39236	*Ellie Simmonds*
395024	SEB	EVS	AD	39241	39242	39243	39244	39245	39246	*Jonnie Peacock*
395025	SEB	EVS	AD	39251	39252	39253	39254	39255	39256	*Victoria Pendleton*
395026	SEB	EVS	AD	39261	39262	39263	39264	39265	39266	*Marc Woods*
395027	SEB	EVS	AD	39271	39272	39273	39274	39275	39276	*Hannah Cockroft*
395028	SEB	EVS	AD	39281	39282	39283	39284	39285	39286	*Laura Trott*
395029	SEB	EVS	AD	39291	39292	39293	39294	39295	39296	*David Weir*

The High Speed 1 line from St Pancras to Kent is worked by Hitachi Class 395 Javelin six-car units operated by SET. The 2037 St Pancras-Margate passes Singlewell loops on 21 June 2018. *David Staines*

Class 465

The Class 465s are four-car BREL Networker units built in the early 1990s and work many commuter trains out of London. They are now being withdrawn.

All are in the HUHQ pool.

	Livery	Owner	Depot	DMS	TS	TS	DMS
465001	SEW	EVS	SG	64759	72028	72029	64809
465002	SEW	EVS	SG	64760	72030	72031	64810
465003	SEW	EVS	SG	64761	72032	72033	64811
465004	SEW	EVS	SG	64762	72034	72035	64812
465005	SEW	EVS	SG	64763	72036	72037	64813
465006	SEW	EVS	SG	64764	72038	72039	64814
465007	SEW	EVS	SG	64765	72040	72041	64815
465008	SEW	EVS	SG	64766	72042	72043	64816
465009	SEW	EVS	SG	64767	72044	72045	64817
465010	SEW	EVS	WK (S)	64768	72046	72047	64818
465011	SEW	EVS	SG	64769	72048	72049	64819
465012	SEW	EVS	SG	64770	72050	72051	64820
465013	SEW	EVS	SG	64771	72052	72053	64821
465014	SEW	EVS	SG	64772	72054	72055	64822
465015	SEW	EVS	SG	64773	72056	72057	64823
465016	SEW	EVS	SG	64774	72058	72059	64824
465017	SEW	EVS	SG	64775	72060	72061	64825
465018	SEW	EVS	SG	64776	72062	72063	64826
465019	SEW	EVS	WK (S)	64777	72064	72065	64827
465020	SEW	EVS	SG	64778	72066	72067	64828
465021	SEW	EVS	SG	64779	72068	72069	64829
465022	SEW	EVS	SG	64780	72070	72071	64830
465023	SEW	EVS	SG	64781	72072	72073	64831
465024	SEW	EVS	SG	64782	72074	72075	64832
465025	SEW	EVS	SG	64783	72076	72077	64833
465026	SEW	EVS	SG	64784	72078	72079	64834
465027	SEW	EVS	SG	64785	72080	72081	64835
465028	SEW	EVS	SG	64786	72082	72083	64836
465029	SEW	EVS	SG	64787	72084	72085	64837
465030	SEW	EVS	SG	64788	72086	72087	64838
465031	SEW	EVS	SG	64789	72088	72089	64839
465032	SEW	EVS	SG	64790	72090	72091	64840
465033	SEW	EVS	SG	64791	72092	72093	64841
465034	SEW	EVS	SG	64792	72094	72095	64842
465035	SEW	EVS	SG	64793	72096	72097	64843
465036	SEW	EVS	SG	64794	72098	72099	64844
465037	SEW	EVS	SG	64795	72100	72101	64845
465038	SEW	EVS	SG	64796	72102	72103	64846
465039	SEW	EVS	SG	64797	72104	72105	64847
465040	SEW	EVS	SG	64798	72106	72107	64848
465041	SEW	EVS	SG	64799	72108	72109	64849
465042	SEW	EVS	SG	64800	72110	72111	64850
465043	SEW	EVS	SG	64801	72112	72113	64851
465044	SEW	EVS	SG	64802	72114	72115	64852
465045	SEW	EVS	SG	64803	72116	72117	64853
465046	SEW	EVS	SG	64804	72118	72119	64854
465047	SEW	EVS	SG	64805	72120	72121	64855
465048	SEW	EVS	SG	64806	72122	72123	64856
465049	SEW	EVS	SG	64807	72124	72125	64857
465050	SEW	EVS	SG	64808	72126	72127	64858
465151	SEW	EVS	SG	65800	72900	72901	65847
465152	SEW	EVS	SG	65801	72902	72903	65848
465153	SEW	EVS	SG	65802	72904	72905	65849

465154	SEW	EVS	SG	65803	72906	72907	65850
465155	SEW	EVS	SG	65804	72908	72909	65851
465156	SEW	EVS	SG	65805	72910	72911	65852
465157	SEW	EVS	SG	65806	72912	72913	65853
465158	SEW	EVS	SG	65807	72914	72915	65854
465159	SEW	EVS	SG	65808	72916	72917	65855
465160	SEW	EVS	SG	65809	72918	72919	65856
465161	SEW	EVS	SG	65810	72920	72921	65857
465162	SEW	EVS	SG	65811	72922	72923	65858
465163	SEW	EVS	SG	65812	72924	72925	65859
465164	SEW	EVS	SG	65813	72926	72927	65860
465165	SEW	EVS	SG	65814	72928	72929	65861
465166	SEW	EVS	SG	65815	72930	72931	65862
465167	SEW	EVS	SG	65816	72932	72933	65863
465168	SEW	EVS	SG	65817	72934	72935	65864
465169	SEW	EVS	SG	65818	72936	72937	65865
465170	SEW	EVS	SG	65819	72938	72939	65866
465171	SEW	EVS	SG	65820	72940	72941	65867
465172	SEW	EVS	SG	65821	72942	72943	65868
465173	SEW	EVS	SG	65822	72944	72945	65869
465174	SEW	EVS	SG	65823	72946	72947	65870
465175	SEW	EVS	SG	65824	72948	72949	65871
465176	SEW	EVS	SG	65825	72950	72951	65872
465177	SEW	EVS	SG	65826	72952	72953	65873
465178	SEW	EVS	SG	65827	72954	72955	65874
465179	SEW	EVS	SG	65828	72956	72957	65875
465180	SEW	EVS	SG	65829	72958	72959	65876
465181	SEW	EVS	SG	65830	72960	72961	65877
465182	SEW	EVS	SG	65831	72962	72963	65878
465183	SEW	EVS	SG	65832	72964	72965	65879
465184	SEW	EVS	SG	65833	72966	72967	65880
465185	SEW	EVS	SG	65834	72968	72969	65881
465186	SEW	EVS	SG	65835	72970	72971	65882
465187	SEW	EVS	SG	65836	72972	72973	65883
465188	SEW	EVS	SG	65837	72974	72975	65884
465189	SEW	EVS	SG	65838	72976	72977	65885
465190	SEW	EVS	SG	65839	72978	72979	65886
465191	SEW	EVS	SG	65840	72980	72981	65887
465192	SEW	EVS	SG	65841	72982	72983	65888
465193	SEW	EVS	SG	65842	72984	72985	65889
465194	SEW	EVS	SG	65843	72986	72987	65890
465195	SEW	EVS	SG	65844	72988	72989	65891
465196	SEW	EVS	SG	65845	72990	72991	65892
465197	SEW	EVS	SG	65846	72992	72993	65893
465235	SEW	ANG	EY (S)	65734	72787	72788	65784
465236	SEW	ANG	EY (S)	65735	72789	72790	65785
465237	SEW	ANG	EY (S)	65736	72791	72792	65786
465238	SEW	ANG	EY (S)	65737	72793	72794	65787
465239	SEW	ANG	EY (S)	65738	72795	72796	65788
465240	SEW	ANG	EY (S)	65739	72797	72798	65789
465241	SEW	ANG	WK (S)	65740	72799	72800	65790
465242	SEW	ANG	WK (S)	65741	72801	72802	65791
465243	SEW	ANG	EY (S)	65742	72803	72804	65792
465244	SEW	ANG	EY (S)	65743	72805	72806	65793
465245	SEW	ANG	EY (S)	65744	72807	72808	65794
465246	SEW	ANG	EY (S)	65745	72809	72810	65795
465247	SEW	ANG	WK (S)	65746	72811	72812	65796
465248	SEW	ANG	EY (S)	65747	72813	72814	65797
465249	SEW	ANG	EY (S)	65748	72815	72816	65798
465250	SEW	ANG	EY (S)	65749	72817	72818	65799

Class 465/9

Refurbished units converted from Class 465/2s in 2005 and with some first-class seats added.

	Old No.	Livery	Owner	Depot	DMC	TS	TS	DMC
465901	465201	SEW	ANG	SG	65700	72719	72720	65750
465902	465202	SEW	ANG	SG	65701	72721	72722	65751
465903	465203	SEW	ANG	SG	65702	72723	72724	65752
465904	465204	SEW	ANG	SG	65703	72725	72726	65753
465905	465205	SEW	ANG	SG	65704	72727	72728	65754
465906	465206	SEW	ANG	SG	65705	72729	72730	65755
465907	465207	SEW	ANG	SG	65706	72731	72732	65756
465908	465208	SEW	ANG	SG	65707	72733	72734	65757
465909	465209	SEW	ANG	SG	65708	72735	72736	65758
465910	465210	SEW	ANG	SG	65709	72737	72738	65759
465911	465211	SEW	ANG	SG	65710	72739	72740	65760
465912	465212	SEW	ANG	SG	65711	72741	72742	65761
465913	465213	SEW	ANG	SG	65712	72743	72744	65762
465914	465214	SEW	ANG	SG	65713	72745	72746	65763
465915	465215	SEW	ANG	SG	65714	72747	72748	65764
465916	465216	SEW	ANG	SG	65715	72749	72750	65765
465917	465217	SEW	ANG	SG	65716	72751	72752	65766
465918	465218	SEW	ANG	SG	65717	72753	72754	65767
465919	465219	SEW	ANG	SG	65718	72755	72756	65768
465920	465220	SEW	ANG	SG	65719	72757	72758	65769
465921	465221	SEW	ANG	SG	65720	72759	72760	65770
465922	465222	SEW	ANG	SG	65721	72761	72762	65771
465923	465223	SEW	ANG	SG	65722	72763	72764	65772
465924	465224	SEW	ANG	SG	65723	72765	72766	65773
465925	465225	SEW	ANG	SG	65724	72767	72768	65774
465926	465226	SEW	ANG	SG	65725	72769	72770	65775
465927	465227	SEW	ANG	SG	65726	72771	72772	65776
465928	465228	SEW	ANG	SG	65727	72773	72774	65777
465929	465229	SEW	ANG	SG	65728	72775	72776	65778
465930	465230	SEW	ANG	SG	65729	72777	72778	65779
465931	465231	SEW	ANG	SG	65730	72779	72780	65780
465932	465232	SEW	ANG	SG	65731	72781	72782	65781
465933	465233	SEW	ANG	SG	65732	72783	72784	65782
465934	465234	SEW	ANG	SG	65733	72785	72786	65783

Southeastern inherited all the ex-BR Networker Class 465/466 EMUs, although the first withdrawals are now starting to take place. On 13 January 2019, 465933 and 465022 stop at London Bridge. *Rob France*

SET is to take the fleet of 30 Class 707s that were ordered by South West Trains and so far 18 have transferred over, with the other dozen held back by SWR due to delays in Class 701s coming into use. Now in SET Blue, 707010/009 work the 1120 Cannon Street-Orpington through Petts Wood on 27 September 2021; this was the first passenger diagram on the first day in service with the TOC. *David Staines*

Class 466

Two-car Networker units for inner-suburban commuter services, they are usually added to eight-car Class 465 sets to strengthen to ten-cars. They are also now being withdrawn.

All are in the HUHQ pool.

	Livery	Owner	Depot	DMS	DTS
466001	SEW	ANG	SG	64860	78312
466002	SEW	ANG	SG	64861	78313
466003	SEW	ANG	SG	64862	78314
466004	SEW	ANG	WK (S)	64863	78315
466005	SEW	ANG	SG	64864	78316
466006	SEW	ANG	SG	64865	78317
466007	SEW	ANG	SG	64866	78318
466008	SEW	ANG	SG	64867	78319
466009	SEW	ANG	SG	64868	78320
466010	SEW	ANG	WK (S)	64869	78321
466011	SEW	ANG	SG	64870	78322
466012	SEW	ANG	SG	64871	78323
466013	SEW	ANG	SG	64872	78324
466014	SEW	ANG	SG	64873	78325
466015	SEW	ANG	SG	64874	78326
466016	SEW	ANG	WK (S)	64875	78327
466017	SEW	ANG	SG	64876	78328
466018	SEW	ANG	SG	64877	78329
466019	SEW	ANG	SG	64878	78330
466020	SEW	ANG	SG	64879	78331
466021	SEW	ANG	SG	64880	78332
466022	SEW	ANG	SG	64881	78333
466023	SEW	ANG	SG	64882	78334
466024	SEW	ANG	WK (S)	64883	78335
466025	SEW	ANG	SG	64884	78336
466026	SEW	ANG	SG	64885	78337
466027	SEW	ANG	SG	64886	78338
466028	SEW	ANG	SG	64887	78339
466029	SEW	ANG	SG	64888	78340
466030	SEW	ANG	SG	64889	78341
466031	SEW	ANG	SG	64890	78342
466032	SEW	ANG	SG	64891	78343
466033	SEW	ANG	SG	64892	78344
466034	SEW	ANG	SG	64893	78345
466035	SEW	ANG	SG	64894	78346
466036	SEW	ANG	SG	64895	78347
466037	SEW	ANG	SG	64896	78348
466038	SEW	ANG	SG	64897	78349
466039	SEW	ANG	SG	64898	78350
466040	SEW	ANG	SG	64899	78351
466041	SEW	ANG	SG	64900	78352
466042	SEW	ANG	SG	64901	78353
466043	SEW	ANG	WK (S)	64902	78354

Class 707

All 30 five-car units were ordered for use by South West Trains, then transferred to South Western Railway, although SWR soon relinquished the lease on the units to transfer to SE.

All should have transferred by now, but 12 units have been retained by SWR to cover for a shortage of stock caused by delays to the Class 701s. They are branded as CityBeam and have initially have been put to work on local trains from Cannon Street and Charing Cross to Maidstone, Dartford, Hayes, Orpington and Sevenoaks. The SE units are in the HUHQ pool.

	Livery	Owner	Depot	DMS	TS	TS	TS	DMS	
707001	SEB	ANG	SG	421001	422001	423001	424001	425001	
707002	SEB	ANG	SG	421002	422002	423002	424002	425002	
707003	SWM	ANG	SG	421003	422003	423003	424003	425003	
707004	SWM	ANG	SG	421004	422004	423004	424004	425004	
707005	SEB	ANG	SG	421005	422005	423005	424005	425005	*Rt Hon James Brokenshire MP*
707006	SEB	ANG	SG	421006	422006	423006	424006	425006	
707007	SEB	ANG	SG	421007	422007	423007	424007	425007	
707008	SEB	ANG	SG	421008	422008	423008	424008	425008	
707009	SEB	ANG	SG	421009	422009	423009	424009	425009	
707010	SEB	ANG	SG	421010	422010	423010	424010	425010	
707011	SEB	ANG	SG	421011	422011	423011	424011	425011	
707012	SEB	ANG	SG	421012	422012	423012	424012	425012	
707013	SEB	ANG	SG	421013	422013	423013	424013	425013	
707025	SEB	ANG	SG	421025	422025	423025	424025	425025	
707026	SEB	ANG	SG	421026	422026	423026	424026	425026	
707027	SEB	ANG	SG	421027	422027	423027	424027	425027	
707028	SEB	ANG	SG	421028	422028	423028	424028	425028	
707029	SEB	ANG	SG	421029	422029	423029	424029	425029	

South Western Railway

Contact details

Website: www.southwesternrailway.com
Twitter: @SW_Help
Key personnel
Managing Director: Claire Mann
Engineering Director: Neil Drury

Overview

South Western Railway is the old South West Trains franchise. It has a mostly EMU fleet, although a fleet of DMUs are used for the Waterloo-Salisbury-Exeter line and Bristol TM-Portsmouth cross-country workings.

A massive order for Class 701 units was placed when the new franchise was let in 2017, which will allow for older EMUs of Classes 455/456 to be withdrawn. All 456s have now been dispensed with. It will also allow SWR to return its brand-new Class 707s to their ROSCOs; which are moving for the Southeastern operation. All 30 units should have moved by 2021, but delays in deliveries of the 701s has meant 12 have been held back by SWR.

Class 73

A single ex-BR Class 73 Electro diesel is retained as a Thunderbird/stock move loco.

73235	73135	POR	SWU	BM

Class 158

Ex-BR two-car units released from Transpennine Express in 2007, they are used for local services in the Wessex area.

All are in the HYHQ pool.

	Old no.	Livery	Owner	Depot	DMCL	DMSL
158880	158737	SWW	POR	SA	52737	57737
158881	158742	SWW	POR	SA	52742	57742
158882	158743	SWW	POR	SA	52743	57743
158883	158744	SWW	POR	SA	52744	57744
158884	158772	SWW	POR	SA	52772	57772
158885	158775	SWW	POR	SA	52775	57775
158886	158779	SWW	POR	SA	52779	57779
158887	158781	SWR	POR	SA	52781	57781
158888	158802	SWR	POR	SA	52802	57802
158890	158814	SWR	POR	SA	52814	57814

Class 159

This fleet of three-car trains was built by BREL for Network SouthEast in 1992-93 to replace loco-hauled trains on the Exeter St David's-Waterloo line. They are essentially the same as a Class 158. The fleet has been expanded with additional Class 158s – some being renumbered as Class 159/1s – added to the fleet to provide additional services and extra capacity.

All are in the HYHQ pool.

Class 159/0

The 22 Class 159/0s are ex-BR units from 1992-93 purposely built for the Exeter St David's-Waterloo route.

	Livery	Owner	Depot	DMCL	MSL	DMSL
159001	SWR	POR	SA	52873	58718	57873
159002	SWR	POR	SA	52874	58719	57874
159003	SWR	POR	SA	52875	58720	57875
159004	SWR	POR	SA	52876	58721	57876
159005	SWR	POR	SA	52877	58722	57877
159006	SWR	POR	SA	52878	58723	57878
159007	SWR	POR	SA	52879	58724	57879
159008	SWR	POR	SA	52880	58725	57880
159009	SWR	POR	SA	52881	58726	57881
159010	SWR	POR	SA	52882	58727	57882
159011	SWR	POR	SA	52883	58728	57883
159012	SWR	POR	SA	52884	58729	57884
159013	SWR	POR	SA	52885	58730	57885
159014	SWR	POR	SA	52886	58731	57886
159015	SWR	POR	SA	52887	58732	57887
159016	SWR	POR	SA	52888	58733	57888
159017	SWR	POR	SA	52889	58734	57889

London
Waterloo
Vauxhall
Queenstown Road
Clapham
Junction
Earlsfield
Wimbledon
Raynes Park
Wandsworth Town
Putney
Barnes
Barnes Bridge
Chiswick
Kew Bridge
Mortlake
North Sheen
Richmond
St Margaret's
Twickenham
Brentford
Syon Lane
Isleworth
Hounslow
Whitton
Strawberry Hill
Teddington
Hampton Wick
Kingston
Norbiton
New Malden
Fulwell
Hampton
Kempton Park
Sunbury
Upper Halliford
Shepperton
Feltham
Ashford
Staines
Datchet
Sunnymeads
Wraysbury
Windsor &
Eton Riverside
Egham
Virginia Water
Longcross
Sunningdale
Ascot
Martins Heron
Bracknell
Wokingham
Winnersh
Winnersh Triangle
Earley
Reading
Motspur Park
Worcester Park
Stoneleigh
Ewell West
Epsom
Ashtead
Leatherhead
Box Hill &
Westhumble
Dorking
Berrylands
Surbiton
Malden Manor
Tolworth
Chessington North
Chessington South
Hampton Court
Thames Ditton
Esher
Hersham
Walton-on-Thames
Hinchley Wood
Claygate
Oxshott
Cobham &
Stoke D'Abernon
Bookham
Effingham
Junction
Horsley
Clandon
London Road
Weybridge
West Byfleet
Byfleet &
New Haw
Addlestone
Chertsey
Woking
Worplesdon
Guildford
Farncombe
Godalming
Milford
Witley
Haslemere
Liphook
Liss
Petersfield
Rowlands Castle
Havant
Bedhampton
Hilsea
Fratton
Portsmouth & Southsea
Portsmouth
Harbour
Wanborough
Ash Vale
Ash
Aldershot
Farnham
Bentley
Alton
Brookwood
Frimley
Camberley
Bagshot
Farnborough
Fleet
Winchfield
Hook
Basingstoke
Micheldever
Winchester
Shawford
Eastleigh
Hedge End
Botley
Portchester
Fareham
Cosham
Swanwick
Bursledon
Hamble
Netley
Sholing
Woolston
Bitterne
Southampton
Central
Southampton
Airport Parkway
Swaythling
St Denys
Chandler's Ford
Romsey
Millbrook
Redbridge
Totton
Ashurst New Forrest
Beaulieu Road
Brockenhurst
Lymington Town
Lymington
Pier
Sway
New Milton
Hinton Admiral
Christchurch
Pokesdown
Bournemouth
Branksome
Parkstone
Poole
Hamworthy
Holton Heath
Wareham
Wool
Moreton
Dorchester
South
Upwey
Weymouth
Ryde
Pier Head
Ryde Esplanade
Ryde St John's Road
Smallbrook
Junction
Brading
Sandown
Lake
Shanklin
Isle of Wight
Overton
Whitchurch
Andover
Grateley
Salisbury
Dean
Mottisfont &
Dunbridge
Warminster
Westbury
Tisbury
Gillingham
Templecombe
Sherborne
Yeovil
Junction
Frome
Bruton
Castle Cary
Yeovil Pen Mill
Trowbridge
Bradford-on-Avon
Bath Spa
Oldfield Park
Keynsham
Bristol
Temple Meads
Crewkerne
Axminster
Honiton
Feniton
Whimple
Cranbrook
Pinhoe
Exeter Central
Exeter
St David's

The Class 159 fleet has been the mainstay of the Exeter St David's-Waterloo line for three decades now. 159009, in the new SWR livery, passes Berkeley Marsh, just west of Frome, with a Yeovil Pen Mill-Waterloo, via Westbury, train on 13 September 2020. *Mark Pike*

159018	SWR	POR	SA	52890	58735	57890
159019	SWR	POR	SA	52891	58736	57891
159020	SWR	POR	SA	52892	58737	57892
159021	SWR	POR	SA	52893	58738	57893
159022	SWR	POR	SA	52894	58739	57894

Class 159/1

The SWT diesel fleet was increased with eight additional ex-Transpennine Class 158 three-car sets dating from 1990, which were refurbished and renumbered as Class 159/1s in 2006-07. They replaced Class170/3 Turbostars and work alongside the Class 159/0s. 159102 was damaged in an incident near Salisbury in November 2021 and is stored pending repairs.

	Old No.	Livery	Owner	Depot	DMCL	MSL	DMSL
159101	158800	SWW	POR	SA	52800	58717	57800
159102	158803	SWW	POR	LM (U)	52803	58703	57803
159103	158804	SWW	POR	SA	52804	58704	57804
159104	158805	SWW	POR	SA	52805	58705	57805
159105	158807	SWW	POR	SA	52807	58707	57807
159106	158809	SWW	POR	SA	52809	58709	57809
159107	158811	SWW	POR	SA	52811	58711	57811
159108	158801	SWW	POR	SA	52801	58701	57801

Class 444

Siemens Desiro five-car inter-city units ordered by SWT and delivered in 2003-04 from Krefeld/Vienna. All are in the HYHQ pool.

	Livery	Owner	Depot	DMS	TS	TS	TS	DMC	
444001	SWW	ANG	NT	63801	67101	67151	67201	63851	*Naomi House*
444002	SWW	ANG	NT	63802	67102	67152	67202	63852	
444003	SWW	ANG	NT	63803	67103	67153	67203	63853	
444004	SWW	ANG	NT	63804	67104	67154	67204	63854	
444005	SWR	ANG	NT	63805	67105	67155	67205	63855	

Siemens Desiro units are the mainstay of the SWR fleet with 127 four-car Class 450s and 45 five-car Class 444s. One of the latter, 444021 works the 1203 Weymouth-Waterloo at Bournemouth on 21 September 2020. *Tom McAtee*

444006	SWR	ANG	NT	63806	67106	67156	67206	63856	
444007	SWR	ANG	NT	63807	67107	67157	67207	63857	
444008	SWW	ANG	NT	63808	67108	67158	67208	63858	
444009	SWW	ANG	NT	63809	67109	67159	67209	63859	
444010	SWW	ANG	NT	63810	67110	67160	67210	63860	
444011	SWR	ANG	NT	63811	67111	67161	67211	63861	
444012	SWW	ANG	NT	63812	67112	67162	67212	63862	*Destination Weymouth*
444013	SWW	ANG	NT	63813	67113	67163	67213	63863	
444014	SWW	ANG	NT	63814	67114	67164	67214	63864	
444015	SWR	ANG	NT	63815	67115	67165	67215	63865	
444016	SWW	ANG	NT	63816	67116	67166	67216	63866	
444017	SWR	ANG	NT	63817	67117	67167	67217	63867	
444018	SWR	ANG	NT	63818	67118	67168	67218	63868	*The Fab 444*
444019	SWR	ANG	NT	63819	67119	67169	67219	63869	
444020	SWR	ANG	NT	63820	67120	67170	67220	63870	
444021	SWR	ANG	NT	63821	67121	67171	67221	63871	
444022	SWW	ANG	NT	63822	67122	67172	67222	63872	
444023	SWW	ANG	NT	63823	67123	67173	67223	63873	
444024	SWW	ANG	NT	63824	67124	67174	67224	63874	
444025	SWW	ANG	NT	63825	67125	67175	67225	63875	
444026	SWW	ANG	NT	63826	67126	67176	67226	63876	
444027	SWW	ANG	NT	63827	67127	67177	67227	63877	
444028	SWW	ANG	NT	63828	67128	67178	67228	63878	
444029	SWR	ANG	NT	63829	67129	67179	67229	63879	
444030	SWW	ANG	NT	63830	67130	67180	67230	63880	
444031	SWW	ANG	NT	63831	67131	67181	67231	63881	
444032	SWW	ANG	NT	63832	67132	67182	67232	63882	
444033	SWW	ANG	NT	63833	67133	67183	67233	63883	
444034	SWW	ANG	NT	63834	67134	67184	67234	63884	
444035	SWW	ANG	NT	63835	67135	67185	67235	63885	
444036	SWW	ANG	NT	63836	67136	67186	67236	63886	
444037	SWW	ANG	NT	63837	67137	67187	67237	63887	
444038	SWR	ANG	NT	63838	67138	67188	67238	63888	*South Western Railway*
444039	SWW	ANG	NT	63839	67139	67189	67239	63889	
444040	SWO	ANG	NT	63840	67140	67190	67240	63890	*The D-Day Story Portsmouth*
444041	SWW	ANG	NT	63841	67141	67191	67241	63891	
444042	SWR	ANG	NT	63842	67142	67192	67242	63892	
444043	SWR	ANG	NT	63843	67143	67193	67243	63893	
444044	SWW	ANG	NT	63844	67144	67194	67244	63894	
444045	SWW	ANG	NT	63845	67145	67195	67245	63895	

444019 has Pride graphics on its cabs.

Class 450

Four-car Siemens Desiro EMUs built in 2002-06, they are used on semi-fast and outer suburban workings. Those numbered in the 450/5 series had revised seat layouts but are now back with their original set numbers.

All are in the HYHQ pool.

	Old No.	Livery	Owner	Depot	DMC	TS	TS	DMC	
450001		SWB	ANG	NT	63201	64201	68101	63601	
450002		SWB	ANG	NT	63202	64202	68102	63602	
450003		SWB	ANG	NT	63203	64203	68103	63603	
450004		SWB	ANG	NT	63204	64204	68104	63604	
450005		SWB	ANG	NT	63205	64205	68105	63605	
450006		SWR	ANG	NT	63206	64206	68106	63606	
450007		SWR	ANG	NT	63207	64207	68107	63607	
450008		SWB	ANG	NT	63208	64208	68108	63608	
450009		SWR	ANG	NT	63209	64209	68109	63609	
450010		SWR	ANG	NT	63210	64210	68110	63610	
450011		SWR	ANG	NT	63211	64211	68111	63611	
450012		SWB	ANG	NT	63212	64212	68112	63612	
450013		SWR	ANG	NT	63213	64213	68113	63613	*Desiro*
450014		SWR	ANG	NT	63214	64214	68114	63614	
450015		SWB	ANG	NT	63215	64215	68115	63615	
450016		SWR	ANG	NT	63216	64216	68116	63616	
450017		SWR	ANG	NT	63217	64217	68117	63617	
450018		SWB	ANG	NT	63218	64218	68118	63618	
450019		SWB	ANG	NT	63219	64219	68119	63619	
450020		SWB	ANG	NT	63220	64220	68120	63620	
450021		SWR	ANG	NT	63221	64221	68121	63621	
450022		SWR	ANG	NT	63222	64222	68122	63622	
450023		SWB	ANG	NT	63223	64223	68123	63623	
450024		SWR	ANG	NT	63224	64224	68124	63624	
450025		SWB	ANG	NT	63225	64225	68125	63625	
450026		SWB	ANG	NT	63226	64226	68126	63626	
450027		SWB	ANG	NT	63227	64227	68127	63627	
450028		SWB	ANG	NT	63228	64228	68128	63628	
450029		SWR	ANG	NT	63229	64229	68129	63629	
450030		SWR	ANG	NT	63230	64230	68130	63630	
450031		SWR	ANG	NT	63231	64231	68131	63631	
450032		SWR	ANG	NT	63232	64232	68132	63632	
450033		SWB	ANG	NT	63233	64233	68133	63633	
450034		SWB	ANG	NT	63234	64234	68134	63634	
450035		SWB	ANG	NT	63235	64235	68135	63635	
450036		SWR	ANG	NT	63236	64236	68136	63636	
450037		SWB	ANG	NT	63237	64237	68137	63637	
450038		SWB	ANG	NT	63238	64238	68138	63638	

The Class 450 Desiro units are used on local and semi-fast services across much of the SWR's electrified network. 450036 leads the 0850 Winchester-Bournemouth away from Eastleigh on 24 September 2021. *Mark Pike*

450039		SWB	ANG	NT	63239	64239	68139	63639	
450040		SWB	ANG	NT	63240	64240	68140	63640	
450041		SWB	ANG	NT	63241	64241	68141	63641	
450042		SWR	ANG	NT	63242	64242	68142	63642	*Treloar College*
450043	450543	SWR	ANG	NT	63243	64243	68143	63643	
450044	450544	SWB	ANG	NT	63244	64244	68144	63644	
450045	450545	SWB	ANG	NT	63245	64245	68145	63645	
450046	450546	SWB	ANG	NT	63246	64246	68146	63646	
450047	450547	SWB	ANG	NT	63247	64247	68147	63647	
450048	450548	SWB	ANG	NT	63248	64248	68148	63648	
450049	450549	SWB	ANG	NT	63249	64249	68149	63649	
450050	450550	SWB	ANG	NT	63250	64250	68150	63650	
450051	450551	SWB	ANG	NT	63251	64251	68151	63651	
450052	450552	SWB	ANG	NT	63252	64252	68152	63652	
450053	450553	SWB	ANG	NT	63253	64253	68153	63653	
450054	450554	SWR	ANG	NT	63254	64254	68154	63654	
450055	450555	SWR	ANG	NT	63255	64255	68155	63655	
450056	450556	SWB	ANG	NT	63256	64256	68156	63656	
450057	450557	SWR	ANG	NT	63257	64257	68157	63657	
450058	450558	SWB	ANG	NT	63258	64258	68158	63658	
450059	450559	SWB	ANG	NT	63259	64259	68159	63659	
450060	450560	SWB	ANG	NT	63260	64260	68160	63660	
450061	450561	SWR	ANG	NT	63261	64261	68161	63661	
450062	450562	SWB	ANG	NT	63262	64262	68162	63662	
450063	450563	SWB	ANG	NT	63263	64263	68163	63663	
450064	450564	SWB	ANG	NT	63264	64264	68164	63664	
450065	450565	SWB	ANG	NT	63265	64265	68165	63665	
450066	450566	SWR	ANG	NT	63266	64266	68166	63666	
450067	450567	ADV	ANG	NT	63267	64267	68167	63667	
450068	450568	SWB	ANG	NT	63268	64268	68168	63668	
450069	450569	SWB	ANG	NT	63269	64269	68169	63669	
450070	450570	SWB	ANG	NT	63270	64270	68170	63670	
450071		SWB	ANG	NT	63271	64271	68171	63671	
450072		SWB	ANG	NT	63272	64272	68172	63672	
450073		SWB	ANG	NT	63273	64273	68173	63673	
450074		SWB	ANG	NT	63274	64274	68174	63674	
450075		SWB	ANG	NT	63275	64275	68175	63675	
450076		SWB	ANG	NT	63276	64276	68176	63676	
450077		SWB	ANG	NT	63277	64277	68177	63677	
450078		SWB	ANG	NT	63278	64278	68178	63678	
450079		SWB	ANG	NT	63279	64279	68179	63679	
450080		SWB	ANG	NT	63280	64280	68180	63680	
450081		SWR	ANG	NT	63281	64281	68181	63681	
450082		SWB	ANG	NT	63282	64282	68182	63682	
450083		SWB	ANG	NT	63283	64283	68183	63683	
450084		SWB	ANG	NT	63284	64284	68184	63684	
450085		SWB	ANG	NT	63285	64285	68185	63685	
450086		SWB	ANG	NT	63286	64286	68186	63686	
450087		SWB	ANG	NT	63287	64287	68187	63687	
450088		SWB	ANG	NT	63288	64288	68188	63688	
450089		SWR	ANG	NT	63289	64289	68189	63689	
450090		SWB	ANG	NT	63290	64290	68190	63690	
450091		SWB	ANG	NT	63291	64291	68191	63691	
450092		SWB	ANG	NT	63292	64292	68192	63692	
450093		SWB	ANG	NT	63293	64293	68193	63693	
450094		SWB	ANG	NT	63294	64294	68194	63694	
450095		SWB	ANG	NT	63295	64295	68195	63695	
450096		SWR	ANG	NT	63296	64296	68196	63696	

450097	SWB	ANG	NT	63297	64297	68197	63697	
450098	SWB	ANG	NT	63298	64298	68198	63698	
450099	SWB	ANG	NT	63299	64299	68199	63699	
450100	SWB	ANG	NT	63300	64300	68200	63700	
450101	SWR	ANG	NT	63701	66851	66801	63751	
450102	SWB	ANG	NT	63702	66852	66802	63752	
450103	SWR	ANG	NT	63703	66853	66803	63753	
450104	SWB	ANG	NT	63704	66854	66804	63754	
450105	SWB	ANG	NT	63705	66855	66805	63755	
450106	SWR	ANG	NT	63706	66856	66806	63756	
450107	SWR	ANG	NT	63707	66857	66807	63757	
450108	SWR	ANG	NT	63708	66858	66808	63758	
450109	SWB	ANG	NT	63709	66859	66809	63759	
450110	SWB	ANG	NT	63710	66860	66810	63760	
450111	SWO	ANG	NT	63901	66921	66901	63921	
450112	SWB	ANG	NT	63902	66922	66902	63922	
450113	SWB	ANG	NT	63903	66923	66903	63923	
450114	SWB	ANG	NT	63904	66924	66904	63924	*Fairbridge Investing in the Future*
450115	SWB	ANG	NT	63905	66925	66905	63925	
450116	SWB	ANG	NT	63906	66926	66906	63926	
450117	SWR	ANG	NT	63907	66927	66907	63927	
450118	SWB	ANG	NT	63908	66928	66908	63928	
450119	SWB	ANG	NT	63909	66929	66909	63929	
450120	SWB	ANG	NT	63910	66930	66910	63930	
450121	SWB	ANG	NT	63911	66931	66911	63931	
450122	SWR	ANG	NT	63912	66932	66912	63932	
450123	SWR	ANG	NT	63913	66933	66913	63933	
450124	SWB	ANG	NT	63914	66934	66914	63934	
450125	SWB	ANG	NT	63915	66935	66915	63935	
450126	SWB	ANG	NT	63916	66936	66916	63936	
450127	SWB	ANG	NT	63917	66937	66917	63937	*Dave Gunson*

450067 carries a Key Workers livery on its two driving cars.

Class 455

These BREL units are the oldest EMUs in the SWR fleet, dating from 1982-85, and are used on many inner-suburban services from Waterloo. They are now being withdrawn and more will be laid up and inevitably scrapped once the new Class 701s finally enter traffic. The TS vehicles in 5701-50 were previously in Class 508 units and date from 1979.

All are in the HYHQ pool.

	Livery	Owner	Depot	DTS	MS	TS	DTS
5701	SWM	POR	WD	77727	62783	71545	77728
5702	SWM	POR	WD	77729	62784	71547	77730
5703	SWM	POR	WD	77731	62785	71540	77732
5704	SWM	POR	LM (S)	77733	62786	71548	77734
5705	SWM	POR	WD	77735	62787	71565	77736
5706	SWM	POR	WD	77737	62788	71534	77738
5707	SWM	POR	WD	77739	62789	71536	77740
5708	SWM	POR	SH (S)	77741	62790	71560	77742
5709	SWM	POR	WD	77743	62791	71532	77744
5710	SWM	POR	WD	77745	62792	71566	77746
5711	SWM	POR	WD	77747	62793	71542	77748
5712	SWM	POR	WD	77749	62794	71546	77750
5713	SWM	POR	WD	77751	62795	71567	77752
5714	SWM	POR	WD	77753	62796	71539	77754
5715	SWM	POR	WD	77755	62797	71535	77756
5716	SWM	POR	WD	77757	62798	71564	77758
5717	SWM	POR	WD	77759	62799	71528	77760

5718	SWM	POR	WD	77761	62800	71557	77762
5719	SWM	POR	WD	77763	62801	71558	77764
5720	SWM	POR	WD	77765	62802	71568	77766
5721	SWM	POR	WD	77767	62803	71553	77768
5722	SWM	POR	WD	77769	62804	71533	77770
5723	SWM	POR	WD	77771	62805	71526	77772
5724	SWM	POR	WD	77773	62806	71561	77774
5725	SWM	POR	WD	77775	62807	71541	77776
5726	SWM	POR	LM (S)	77777	62808	71556	77778
5727	SWM	POR	WD	77779	62809	71562	77780
5728	SWM	POR	WD	77781	62810	71527	77782
5729	SWM	POR	WD	77783	62811	71550	77784
5730	SWM	POR	WD	77785	62812	71551	77786
5731	SWM	POR	WD	77787	62813	71555	77788
5732	SWM	POR	WD	77789	62814	71552	77790
5733	SWM	POR	WD	77791	62815	71549	77792
5734	SWM	POR	WD	77793	62816	71531	77794
5735	SWM	POR	WD	77795	62817	71563	77796
5736	SWM	POR	LM (S)	77797	62818	71554	77798
5737	SWM	POR	WD	77799	62819	71544	77800
5738	SWM	POR	WD	77801	62820	71529	77802
5739	SWM	POR	WD	77803	62821	71537	77804
5740	SWM	POR	WD	77805	62822	71530	77806
5741	SWM	POR	WD	77807	62823	71559	77808
5742	SWM	POR	WD	77809	62824	71543	77810
5750	SWM	POR	WD	77811	62825	71538	77812
5847	SWM	POR	LM (S)	77671	62755	71683	77672
5848	SWM	POR	WD	77673	62756	71684	77674
5849	SWM	POR	WD	77675	62757	71685	77676
5850	SWM	POR	WD	77677	62758	71686	77678
5851	SWM	POR	WD	77679	62759	71687	77680
5852	SWM	POR	WD	77681	62760	71688	77682
5853	SWM	POR	WD	77683	62761	71689	77684
5854	SWM	POR	WD	77685	62762	71690	77686
5855	SWM	POR	WD (S)	77687	62763	71691	77688
5856	SWM	POR	WD	77689	62764	71692	77690
5857	SWM	POR	WD	77691	62765	71693	77692
5858	SWM	POR	WD	77693	62766	71694	77694
5859	SWM	POR	WD	77695	62767	71695	77696
5860	SWM	POR	WD	77697	62768	71696	77698
5861	SWM	POR	WD	77699	62769	71697	77700
5862	SWM	POR	WD	77701	62770	71698	77702
5863	SWM	POR	WD	77703	62771	71699	77704
5864	SWM	POR	WD	77705	62772	71700	77706
5865	SWM	POR	WD	77707	62773	71701	77708
5866	SWM	POR	WD	77709	62774	71702	77710
5867	SWM	POR	WD	77711	62775	71703	77712
5868	SWM	POR	WD	77713	62776	71704	77714
5869	SWM	POR	WD	77715	62777	71705	77716
5870	SWM	POR	WD	77717	62778	71706	77718
5871	SWM	POR	WD	77719	62779	71707	77720
5872	SWM	POR	WD	77721	62780	71708	77722
5873	SWM	POR	WD	77723	62781	71709	77724
5874	SWM	POR	WD	77725	62782	71710	77726
5901	SWM	POR	WD	77813	62826	71714	77814
5902	SWM	POR	WD	77815	62827	71715	77816
5903	SWM	POR	WD	77817	62828	71716	77818
5904	SWM	POR	WD	77819	62829	71717	77820
5905	SWM	POR	WD	77821	62830	71725	77822

Class 455s are now starting to be withdrawn by South Western Railway. On 12 July 2021, 5731 passes Shawford on one of the semi-regular maintenance runs from Wimbledon Depot to Bournemouth TRSMD. *Mark Pike*

5906	SWM	POR	WD	77823	62831	71719	77824
5907	SWM	POR	WD	77825	62832	71720	77826
5908	SWM	POR	WD	77827	62833	71721	77828
5909	SWM	POR	WD	77829	62834	71722	77830
5910	SWM	POR	CJ (S)	77831	62835	71723	77832
5911	SWM	POR	WD	77833	62836	71724	77834
5912	SWM	POR	WD	77835	62837	67400	77836
5913	SWM	POR	WD	77837	67301	71726	77838
5914	SWM	POR	WD	77839	62839	71727	77840
5915	SWM	POR	WD	77841	62840	71728	77842
5916	SWM	POR	WD	77843	62841	71729	77844
5917	SWM	POR	WD	77845	62842	71730	77846
5918	SWM	POR	WD	77847	62843	71732	77848
5919	SWM	POR	WD	77849	62844	71718	77850
5920	SWM	POR	WD	77851	62845	71733	77852

Class 458

These Alstom Juniper units were ordered when the South West Trains franchise was first let in 1996 and were delivered from 1998-2000. They had a troubled start with reliability issues. They were initially four-cars but in 2013 the eight eight-car Class 460s units that had been built for Gatwick Express were disbanded, and six of them became five-car Class 458s – 458531-536, and each of the original 30 Class 458/5s had an additional vehicle added to make them five-cars sets. This project was completed in 2016. Four 'left over' Class 460 vehicles – all driving cars – were scrapped at the end of the project.

The 458s were due to be redundant with SWR once the Class 701 fleet is fully accepted into traffic but will now be retained. 30 will be reduced to four-car sets.

All are in the HYHQ pool.

	Livery	Owner	Depot	DMC	TS	TS	MS	DMC
458501	SWB	POR	WD	67601	74431	74001	74101	67701
458502	SWB	POR	WD	67602	74421	74002	74102	67702
458503	SWB	POR	WD	67603	74441	74003	74103	67703
458504	SWB	POR	WD	67604	74451	74004	74104	67704
458505	SWB	POR	WD (S)	67605	74425	74005	74105	67705
458506	SWB	POR	WD (S)	67606	74436	74006	74106	67706
458507	SWB	POR	LM (S)	67607	74428	74007	74107	67707
458508	SWB	POR	WD	67608	74433	74008	74108	67708
458509	SWB	POR	WD	67609	74452	74009	74109	67709
458510	SWB	POR	WD	67610	74405	74010	74110	67710
458511	SWB	POR	WD	67611	74435	74011	74111	67711
458512	SWB	POR	WD	67612	74427	74012	74112	67712
458513	SWB	POR	WD	67613	74437	74013	74113	67713
458514	SWB	POR	WD	67614	74407	74014	74114	67714
458515	SWB	POR	WD	67615	74404	74015	74115	67715
458516	SWB	POR	WD	67616	74406	74016	74116	67716
458517	SWB	POR	WD	67617	74426	74017	74117	67717
458518	SWB	POR	WD	67618	74432	74018	74118	67718
458519	SWB	POR	WD	67619	74403	74019	74119	67719
458520	SWB	POR	WD	67620	74401	74020	74120	67720
458521	SWB	POR	WD	67621	74438	74021	74121	67721
458522	SWB	POR	WD	67622	74424	74022	74122	67722
458523	SWB	POR	WD	67623	74434	74023	74123	67723
458524	SWB	POR	WD	67624	74402	74024	74124	67724
458525	SWB	POR	WD	67625	74422	74025	74125	67725
458526	SWB	POR	WD	67626	74442	74026	74126	67726
458527	SWB	POR	WD	67627	74412	74027	74127	67727
458528	SWB	POR	WD	67628	74408	74028	74128	67728
458529	SWB	POR	WD	67629	74423	74029	74129	67729

An order for 30 four-car Class 458s was delivered to SWT by Alstom in late 1990s, and they were later extended to five-cars, with six additional five-car sets formed using vehicles from similar ex-Gatwick Express Class 460 EMUs. On 11 January 2019 458525 and 458515 were next to 458827 waiting to leave the former international platforms at Waterloo. *Rob France*

458530	SWB	POR	WD	67630	74411	74030	74130	67730
458531	SWB	POR	WD	67913	74418	74446	74458	67912
458532	SWB	POR	WD	67904	74417	74447	74457	67905
458533	SWB	POR	WD	67917	74413	74443	74453	67916
458534	SWB	POR	WD	67914	74414	74444	74454	67918
458535	SWB	POR	WD	67915	74415	74445	74455	67911
458536	SWB	POR	WD	67903	74416	74448	74456	67902

Class 484

Five VivaRail Class 484 EMUs have been delivered to the Island Line for use on the line from Ryde to Shanklin. They were converted using ex-London Underground D78 stock dating from 1978 and have replaced the dated 1938-built Class 483s. The first units were unveiled in August 2020 but moved to the island in early 2021 and entered service in November.

The Class 483s have been disposed, some going for scrap others for preservation.

	Livery	Owner	Depot	DMS	DMS
484001	SWR	LOM	RY	131	231
484002	SWR	LOM	RY	132	232
484003	SWR	LOM	RY	133	233
484004	SWR	LOM	RY	134	234
484005	SWR	LOM	RY	135	235

The Island Line's fleet of trains was replaced in 2021 with VivaRail Class 484s. Prior to their transfer to the island, 484003 and 484002 arrive with the 0915 Eastleigh-Fareham test run on 14 July 2021. *Marr Pike*

Class 701

New Bombardier Aventra five- and ten-car units – branded as Arterio by SWR – to replace the elderly Class 455 and 456 units as well as the brand new Class 707s, the latter of which will move to Southeastern.

Several of the units have now being delivered and are undergoing main-line testing but they are woefully late into traffic and this has led to the retention of older trains and the inability of SWR to give up all 30 of its 707s to Southeastern.

They will be used on the Reading, Windsor and West London suburban routes. All are expected to be in the HYHQ pool.

	Livery	Owner	Depot	DMS	MS	TS	MS	MS	MS	MS	TS	MS	DMS
701001	SWR	ROC	HQ	480001	481001	482001	483001	484001	485001	486001	487001	488001	489001
701002	SWR	ROC	WD (Q)	480002	481002	482002	483002	484002	485002	486002	487002	488002	489002
701003	SWR	ROC	WD (Q)	480003	481003	482003	483003	484003	485003	486003	487003	488003	489003
701004	SWR	ROC	WD (Q)	480004	481004	482004	483004	484004	485004	486004	487004	488004	489004
701005	SWR	ROC	WD (Q)	480005	481005	482005	483005	484005	485005	486005	487005	488005	489005
701006	SWR	ROC	WD (Q)	480006	481006	482006	483006	484006	485006	486006	487006	488006	489006
701007	SWR	ROC	WD (Q)	480007	481007	482007	483007	484007	485007	486007	487007	488007	489007
701008	SWR	ROC	WD (Q)	480008	481008	482008	483008	484008	485008	486008	487008	488008	489008
701009	SWR	ROC	WD (Q)	480009	481009	482009	483009	484009	485009	486009	487009	488009	489009
701010	SWR	ROC	WD (Q)	480010	481010	482010	483010	484010	485010	486010	487010	488010	489010
701011	SWR	ROC	WD (Q)	480011	481011	482011	483011	484011	485011	486011	487011	488011	489011
701012	SWR	ROC	WD (Q)	480012	481012	482012	483012	484012	485012	486012	487012	488012	489012
701013	SWR	ROC	WD (Q)	480013	481013	482013	483013	484013	485013	486013	487013	488013	489013
701014	SWR	ROC	WD (Q)	480014	481014	482014	483014	484014	485014	486014	487014	488014	489014
701015	SWR	ROC	WD (Q)	480015	481015	482015	483015	484015	485015	486015	487015	488015	489015
701016	SWR	ROC	ZG (Q)	480016	481016	482016	483016	484016	485016	486016	487016	488016	489016
701017	SWR	ROC	ZG (Q)	480017	481017	482017	483017	484017	485017	486017	487017	488017	489017
701018	SWR	ROC	WD (Q)	480018	481018	482018	483018	484018	485018	486018	487018	488018	489018
701019	SWR	ROC	WD (Q)	480019	481019	482019	483019	484019	485019	486019	487019	488019	489019
701020	SWR	ROC	WD (Q)	480020	481020	482020	483020	484020	485020	486020	487020	488020	489020
701021	SWR	ROC	WD (Q)	480021	481021	482021	483021	484021	485021	486021	487021	488021	489021
701022	SWR	ROC	WD (Q)	480022	481022	482022	483022	484022	485022	486022	487022	488022	489022
701023	SWR	ROC	WD (Q)	480023	481023	482023	483023	484023	485023	486023	487023	488023	489023
701024	SWR	ROC	WD (Q)	480024	481024	482024	483024	484024	485024	486024	487024	488024	489024
701025	SWR	ROC	ZG (Q)	480025	481025	482025	483025	484025	485025	486025	487025	488025	489025
701026	SWR	ROC	ZD (S)	480026	481026	482026	483026	484026	485026	486026	487026	488026	489026

It is more than three years since production of the Class 701 Aventra units started at Bombardier's Derby works, yet none have entered traffic and the delay has only been felt less severely because of the pandemic. Brand new 701028 passes Eastleigh with a Waterloo to Eastleigh Depot mileage accumulation run on 5 July 2021. *Mark Pike*

701027	SWR	ROC	ZD (Q)	480027	481027	482027	483027	484027	485027	486027	487027	488027	489027
701028	SWR	ROC	WD (Q)	480028	481028	482028	483028	484028	485028	486028	487028	488028	489028
701029	SWR	ROC	WD (Q)	480029	481029	482029	483029	484029	485029	486029	487029	488029	489029
701030	SWR	ROC	WD (Q)	480030	481030	482030	483030	484030	485030	486030	487030	488030	489030
701031	SWR	ROC	WD (Q)	480031	481031	482031	483031	484031	485031	486031	487031	488031	489031
701032	SWR	ROC	WD (Q)	480032	481032	482032	483032	484032	485032	486032	487032	488032	489032
701033	SWR	ROC	WD (Q)	480033	481033	482033	483033	484033	485033	486033	487033	488033	489033
701034	SWR	ROC	ZD (Q)	480034	481034	482034	483034	484034	485034	486034	487034	488034	489034
701035	SWR	ROC	WD (Q)	480035	481035	482035	483035	484035	485035	486035	487035	488035	489035
701036	SWR	ROC	WD (Q)	480036	481036	482036	483036	484036	485036	486036	487036	488036	489036
701037	SWR	ROC	WD (Q)	480037	481037	482037	483037	484037	485037	486037	487037	488037	489037
701038	SWR	ROC	WK (S)	480038	481038	482038	483038	484038	485038	486038	487038	488038	489038
701039	SWR	ROC	HQ	480039	481039	482039	483039	484039	485039	486039	487039	488039	489039
701040	SWR	ROC	WK (S)	480040	481040	482040	483040	484040	485040	486040	487040	488040	489040
701041	SWR	ROC		480041	481041	482041	483041	484041	485041	486041	487041	488041	489041
701042	SWR	ROC		480042	481042	482042	483042	484042	485042	486042	487042	488042	489042
701043	SWR	ROC		480043	481043	482043	483043	484043	485043	486043	487043	488043	489043
701044	SWR	ROC		480044	481044	482044	483044	484044	485044	486044	487044	488044	489044
701045	SWR	ROC		480045	481045	482045	483045	484045	485045	486045	487045	488045	489045
701046	SWR	ROC		480046	481046	482046	483046	484046	485046	486046	487046	488046	489046
701047	SWR	ROC		480047	481047	482047	483047	484047	485047	486047	487047	488047	489047
701048	SWR	ROC		480048	481048	482048	483048	484048	485048	486048	487048	488048	489048
701049	SWR	ROC		480049	481049	482049	483049	484049	485049	486049	487049	488049	489049
701050	SWR	ROC		480050	481050	482050	483050	484050	485050	486050	487050	488050	489050
701051	SWR	ROC		480051	481051	482051	483051	484051	485051	486051	487051	488051	489051
701052	SWR	ROC		480052	481052	482052	483052	484052	485052	486052	487052	488052	489052
701053	SWR	ROC		480053	481053	482053	483053	484053	485053	486053	487053	488053	489053
701054	SWR	ROC		480054	481054	482054	483054	484054	485054	486054	487054	488054	489054
701055	SWR	ROC		480055	481055	482055	483055	484055	485055	486055	487055	488055	489055
701056	SWR	ROC		480056	481056	482056	483056	484056	485056	486056	487056	488056	489056
701057	SWR	ROC		480057	481057	482057	483057	484057	485057	486057	487057	488057	489057
701058	SWR	ROC		480058	481058	482058	483058	484058	485058	486058	487058	488058	489058
701059	SWR	ROC		480059	481059	482059	483059	484059	485059	486059	487059	488059	489059
701060	SWR	ROC		480060	481060	482060	483060	484060	485060	486060	487060	488060	489060

Class 701/5

All are expected to be in the HYHQ pool.

	Livery	Owner	Depot	DMS	MS	TS	MS	DMS
701501	SWR	ROC	OD (Q)	480101	481101	482101	483101	484101
701502	SWR	ROC	OD (Q)	480102	481102	482102	483102	484102
701503	SWR	ROC		480103	481103	482103	483103	484103
701504	SWR	ROC		480104	481104	482104	483104	484104
701505	SWR	ROC		480105	481105	482105	483105	484105
701506	SWR	ROC	ZD (Q)	480106	481106	482106	483106	484106
701507	SWR	ROC		480107	481107	482107	483107	484107
701508	SWR	ROC		480108	481108	482108	483108	484108
701509	SWR	ROC		480109	481109	482109	483109	484109
701510	SWR	ROC		480110	481110	482110	483110	484110
701511	SWR	ROC		480111	481111	482111	483111	484111
701512	SWR	ROC		480112	481112	482112	483112	484112
701513	SWR	ROC		480113	481113	482113	483113	484113
701514	SWR	ROC		480114	481114	482114	483114	484114
701515	SWR	ROC		480115	481115	482115	483115	484115
701516	SWR	ROC		480116	481116	482116	483116	484116
701517	SWR	ROC		480117	481117	482117	483117	484117
701518	SWR	ROC		480118	481118	482118	483118	484118
701519	SWR	ROC		480119	481119	482119	483119	484119

701520	SWR	ROC		480120	481120	482120	483120	484120
701521	SWR	ROC		480121	481121	482121	483121	484121
701522	SWR	ROC		480122	481122	482122	483122	484122
701523	SWR	ROC		480123	481123	482123	483123	484123
701524	SWR	ROC		480124	481124	482124	483124	484124
701525	SWR	ROC		480125	481125	482125	483125	484125
701526	SWR	ROC		480126	481126	482126	483126	484126
701527	SWR	ROC		480127	481127	482127	483127	484127
701528	SWR	ROC		480128	481128	482128	483128	484128
701529	SWR	ROC		480129	481129	482129	483129	484129
701530	SWR	ROC		480130	481130	482130	483130	484130

Class 707

These Siemens Desiro City inner-suburban EMUs units were ordered by South West Trains in 2014 and entered traffic as recently as 2016 but are due to be replaced by new Class 701s and the entire fleet of 30 units were due to move to the Southeastern operation.

However, delays of the 701s entering traffic – they are three years behind schedule – mean these 12 units are remaining with SWR until at least August 2022.

All are in the HYHQ pool.

	Livery	Owner	Depot	DMS	TS	TS	TS	DMS
707014	SWM	ANG	WD	421014	422014	423014	424014	425014
707015	SWM	ANG	WD	421015	422015	423015	424015	425015
707016	SWM	ANG	WD	421016	422016	423016	424016	425016
707017	SWM	ANG	WD	421017	422017	423017	424017	425017
707018	SWM	ANG	WD	421018	422018	423018	424018	425018
707019	SWM	ANG	WD	421019	422019	423019	424019	425019
707020	SWM	ANG	WD	421020	422020	423020	424020	425020
707021	SWM	ANG	WD	421021	422021	423021	424021	425021
707022	SWM	ANG	WD	421022	422022	423022	424022	425022
707023	SWM	ANG	WD	421023	422023	423023	424023	425023
707024	SWM	ANG	WD	421024	422024	423024	424024	425024
707030	SWM	ANG	WD	421030	422030	423030	424030	425030

The Class 707s were ordered by South West Trains, but were then unwanted by new franchisee South Western Railway and reassigned to SET. However, delays in 701 deliveries mean a dozen will remain with SWR for the foreseeable. One of the units that has since left the franchise, 707004, snakes its way past Nine Elms soon after leaving with a Waterloo to Waterloo circular train on 13 August 2019. *Mark Pike*

Transpennine Express

Contact details

Website: www.tpexpress.co.uk
Twitter: @TPExpressTrains

Key personnel

Interim Managing Director: Matthew Golton
Fleet Director: Paul Staples

Overview

The First Group-operated franchise runs from Liverpool Lime Street to Scarborough, Middlesbrough and Newcastle, from Manchester Piccadilly to Cleethorpes and Hull, via the Hope Valley and also from Manchester Airport to Edinburgh and Glasgow Central via the WCML and from Liverpool to Edinburgh via the ECML.

The TPE fleet has gone through a complete overhaul in recent times. Of the 51 three-car Class 185s on lease, 15 were due to be returned to their ROSCO in December 2020 but delays in getting all its new fleets into traffic postponed that and all remain in use with the operator.

DRS Class 68s are hired by TPE to work with CAF MK 5A coaches on the Liverpool-Scarborough route, although their introduction has been beset with problems, and more recently complaints over the noise of the locos. 68025 *Superb* rests at Scarborough on 5 March 2022. *Pip Dunn*

In their place have come 13 sets of Mk 5A CAF coaches hauled by Stadler Class 68s. The locos are provided by Direct Rail Services, of which 68019-032 are in TPE livery and 68033/034 are 'spare locos' that remain in DRS colours. The locos remain owned and maintained by DRS and 13 a day (with 11 in traffic) will be necessary if the full Mk 5A timetable is finally introduced.

The Nova 3 Mk 5A sets were planned for the Liverpool-Scarborough route and also to Middlesbrough. The latter has not happened and due to noise complaints from Scarborough residents, their use to the town is much curtailed. TPE is now training drivers on the locos and stock for introduction on some turns from Manchester to Cleethorpes via the Hope Valley. Any locos not used by TPE are free to be used by DRS if required.

The reign of Class 68s may be short-lived as First Group has issued an invitation to tender for a fleet of 15-30 new locos (with an option for five more for its GWR 'sleeper' operation) to replace the 68s, as it wants bi-mode locomotives.

A dozen brand new Nova 2 five-car Class 397 Civity units from CAF work on the Manchester Airport to Glasgow Central and Edinburgh route, while the TOC now runs Liverpool to Edinburgh trains via Leeds, York and the ECML using Bi-Mode Hitcahi AT300 Class 802 Nova 1s.

Class 185

These three-car Siemens Desiro units have been the mainstay of TPE's workings since their introduction in 2005-06.

All are in the EAHQ pool.

	Livery	Owner	Depot	DMCL	MSL	DMS
185101	TPE	EVS	AK	51101	53101	54101
185102	TPE	EVS	AK	51102	53102	54102
185103	TPE	EVS	AK	51103	53103	54103
185104	TPE	EVS	AK	51104	53104	54104
185105	TPE	EVS	AK	51105	53105	54105
185106	TPE	EVS	AK	51106	53106	54106
185107	TPE	EVS	AK	51107	53107	54107
185108	TPE	EVS	AK	51108	53108	54108
185109	TPE	EVS	AK	51109	53109	54109
185110	TPE	EVS	AK	51110	53110	54110
185111	TPE	EVS	AK	51111	53111	54111
185112	TPE	EVS	AK	51112	53112	54112
185113	TPE	EVS	AK	51113	53113	54113
185114	TPE	EVS	AK	51114	53114	54114
185115	TPE	EVS	AK	51115	53115	54115
185116	TPE	EVS	AK	51116	53116	54116
185117	TPE	EVS	AK	51117	53117	54117
185118	TPE	EVS	AK	51118	53118	54118
185119	TPE	EVS	AK	51119	53119	54119
185120	TPE	EVS	AK	51120	53120	54120
185121	TPE	EVS	AK	51121	53121	54121
185122	TPE	EVS	AK	51122	53122	54122
185123	TPE	EVS	AK	51123	53123	54123

Transpennine Express runs train from Manchester to Cleethorpes via Sheffield using Class 185s. On 12 June 2021, 185120 was stabled at Sheffield. *Stuart West*

185124	TPE	EVS	AK	51124	53124	54124
185125	TPE	EVS	AK	51125	53125	54125
185126	TPE	EVS	AK	51126	53126	54126
185127	TPE	EVS	AK	51127	53127	54127
185128	TPE	EVS	AK	51128	53128	54128
185129	TPE	EVS	AK	51129	53129	54129
185130	TPE	EVS	AK	51130	53130	54130
185131	TPE	EVS	AK	51131	53131	54131
185132	TPE	EVS	AK	51132	53132	54132
185133	TPE	EVS	AK	51133	53133	54133
185134	TPE	EVS	AK	51134	53134	54134
185135	TPE	EVS	AK	51135	53135	54135
185136	TPE	EVS	AK	51136	53136	54136
185137	TPE	EVS	AK	51137	53137	54137
185138	TPE	EVS	AK	51138	53138	54138
185139	TPE	EVS	AK	51139	53139	54139
185140	TPE	EVS	AK	51140	53140	54140
185141	TPE	EVS	AK	51141	53141	54141
185142	TPE	EVS	AK	51142	53142	54142
185143	TPE	EVS	AK	51143	53143	54143
185144	TPE	EVS	AK	51144	53144	54144
185145	TPE	EVS	AK	51145	53145	54145
185146	TPE	EVS	AK	51146	53146	54146
185147	TPE	EVS	AK	51147	53147	54147
185148	TPE	EVS	AK	51148	53148	54148
185149	TPE	EVS	AK	51149	53149	54149
185150	TPE	EVS	AK	51150	53150	54150
185151	TPE	EVS	AK	51151	53151	54151

Class 397

These Nova 2 12 five-car CAF Civity 25kV units have replaced Class 350s and work all Manchester Airport/Liverpool to Glasgow/Edinburgh services.

All are in the TPEC pool.

	Livery	Owner	Depot	DMF	PTS	MS	PTS	DMS
397001	TPE	EVS	MA	471001	472001	473001	474001	475001
397002	TPE	EVS	MA	471002	472002	473002	474002	475002
397003	TPE	EVS	MA	471003	472003	473003	474003	475003
397004	TPE	EVS	MA	471004	472004	473004	474004	475004
397005	TPE	EVS	MA	471005	472005	473005	474005	475005
397006	TPE	EVS	MA	471006	472006	473006	474006	475006
397007	TPE	EVS	MA	471007	472007	473007	474007	475007
397008	TPE	EVS	MA	471008	472008	473008	474008	475008
397009	TPE	EVS	MA	471009	472009	473009	474009	475009
397010	TPE	EVS	MA	471010	472010	473010	474010	475010
397011	TPE	EVS	MA	471011	472011	473011	474011	475011
397012	TPE	EVS	MA	471012	472012	473012	474012	475012

In 2007, TPE started to run through trains to Edinburgh and Glasgow Central along the West Coast Main Line. It started with Class 185 DMUs, then purpose-built Class 350/4 EMUs before introducing CAF Civity Class 397s in 2018. On 25 January 2021, 397008 passes Auchengray with the 1012 Edinburgh Manchester Airport. *Robin Ralston*

Class 802/2

Nineteen Nova 1 Hitachi bi-mode AT300 units are used on Liverpool-Edinburgh (via York) and Manchester Airport to Newcastle routes.

All are in the EAHQ pool.

	Livery	Owner	Depot	PDTS	MS	MS	MC	PDTF
802201	TPE	ANG	DN	831201	832201	833201	834201	835201
802202	TPE	ANG	DN	831202	832202	833202	834202	835202
802203	TPE	ANG	DN	831203	832203	833203	834203	835203
802204	TPE	ANG	DN	831204	832204	833204	834204	835204
802205	TPE	ANG	DN	831205	832205	833205	834205	835205
802206	TPE	ANG	DN	831206	832206	833206	834206	835206
802207	TPE	ANG	DN	831207	832207	833207	834207	835207
802208	TPE	ANG	DN	831208	832208	833208	834208	835208
802209	TPE	ANG	DN	831209	832209	833209	834209	835209
802210	TPE	ANG	DN	831210	832210	833210	834210	835210
802211	TPE	ANG	DN	831211	832211	833211	834211	835211
802212	TPE	ANG	DN	831212	832212	833212	834212	835212
802213	TPE	ANG	DN	831213	832213	833213	834213	835213
802214	TPE	ANG	DN	831214	832214	833214	834214	835214
802215	TPE	ANG	DN	831215	832215	833215	834215	835215
802216	TPE	ANG	DN	831216	832216	833216	834216	835216
802217	TPE	ANG	DN	831217	832217	833217	834217	835217
802218	TPE	ANG	DN	831218	832218	833218	834218	835218
802219	TPE	ANG	DN	831219	832219	833219	834219	835219

TPE also has a fleet of Hitachi Class 802 bi-mode units, which run from Liverpool to Newcastle and on to Edinburgh, using the 25kV AC wires from York. 802205 pauses at Leeds while working the 1554 Liverpool-Newcastle on 11 December 2021. *Pip Dunn*

Mk 5A coaches

TPE has 66 Mk 5A coaches built by CAF in Spain to form 13 full sets of five-car loco-hauled sets comprising a Driving Open Brake Standard (of which 14 have been delivered), three Open Standards and an Open First next to the loco.

These Nova 3 trains work in push-pull mode on the Liverpool-Scarborough route but will also be deployed on some Manchester-Cleethorpes trains when all sets are in traffic. They have been heavily delayed into traffic and the full 11 sets a day is some way off happening.

All are in the EAHQ pool.

Open first

11501	TPE	BEA	MA
11502	TPE	BEA	MA
11503	TPE	BEA	MA
11504	TPE	BEA	MA
11505	TPE	BEA	MA
11506	TPE	BEA	MA
11507	TPE	BEA	MA
11508	TPE	BEA	MA
11509	TPE	BEA	MA
11510	TPE	BEA	MA
11511	TPE	BEA	MA
11512	TPE	BEA	MA
11513	TPE	BEA	MA

Open standard

12701	TPE	BEA	MA
12702	TPE	BEA	MA
12703	TPE	BEA	MA
12704	TPE	BEA	MA
12705	TPE	BEA	MA
12706	TPE	BEA	MA
12707	TPE	BEA	MA
12708	TPE	BEA	MA
12709	TPE	BEA	MA
12710	TPE	BEA	MA
12711	TPE	BEA	MA
12712	TPE	BEA	MA
12713	TPE	BEA	MA
12714	TPE	BEA	MA
12715	TPE	BEA	MA
12716	TPE	BEA	MA
12717	TPE	BEA	MA
12718	TPE	BEA	MA
12719	TPE	BEA	MA
12720	TPE	BEA	MA
12721	TPE	BEA	MA
12722	TPE	BEA	MA
12723	TPE	BEA	MA
12724	TPE	BEA	MA
12725	TPE	BEA	MA
12726	TPE	BEA	MA
12727	TPE	BEA	MA
12728	TPE	BEA	MA
12729	TPE	BEA	MA
12730	TPE	BEA	MA
12731	TPE	BEA	MA
12732	TPE	BEA	MA
12733	TPE	BEA	MA
12734	TPE	BEA	MA
12735	TPE	BEA	MA
12736	TPE	BEA	MA
12737	TPE	BEA	MA
12738	TPE	BEA	MA
12739	TPE	BEA	MA

Driving open brake standard

12801	TPE	BEA	MA
12802	TPE	BEA	MA
12803	TPE	BEA	MA
12804	TPE	BEA	MA
12805	TPE	BEA	MA
12806	TPE	BEA	MA
12807	TPE	BEA	MA
12808	TPE	BEA	MA
12809	TPE	BEA	MA
12810	TPE	BEA	MA
12811	TPE	BEA	MA
12812	TPE	BEA	MA
12813	TPE	BEA	MA
12814	TPE	BEA	MA

On 5 March 2022, TPE driving open brake standard 12811 stands at Scarborough prior to leading the 1534 to Manchester, powered by 68020 *Reliance*. *Pip Dunn*

Transport for Wales/Trafnidiaeth Cymru

Contact details

Website: www.tfwrail.wales or www.trctrenau.cymru
Twitter: @TfWrail

Key personnel

Managing Director: Jan Chaudhry-van de Velde
Engineering Director: Ryan Williams

Overview

Transport for Wales is the name given to what was the Wales & Borders franchise previously run by Arriva under the name Arriva Trains Wales. It was awarded to Keolis Amey Operations but was brought under direct Welsh Government control in February 2021.

It inherited a mix of ex-BR rolling stock, and in recent times has eliminated its Class 142/143. TfW also had two loco-hauled sets of Mk 3 coaches that were used from Holyhead to Cardiff to Manchester using DB Cargo Class 67s but now ex-LNER Mk 4s have replaced them on the Welsh Marches route, with three sets in use, and TfW has just agreed to take four more sets, which will run until December 2028. 67008/010/012-015/017/022/025 have been prepared by DBC to work with these coaches and so far 67008/014/017/022/025 have been painted in TfW livery (see page 265). Other Class 67s cannot work in push-pull mode with Mk 4s unless modified, with an additional jumper cable required.

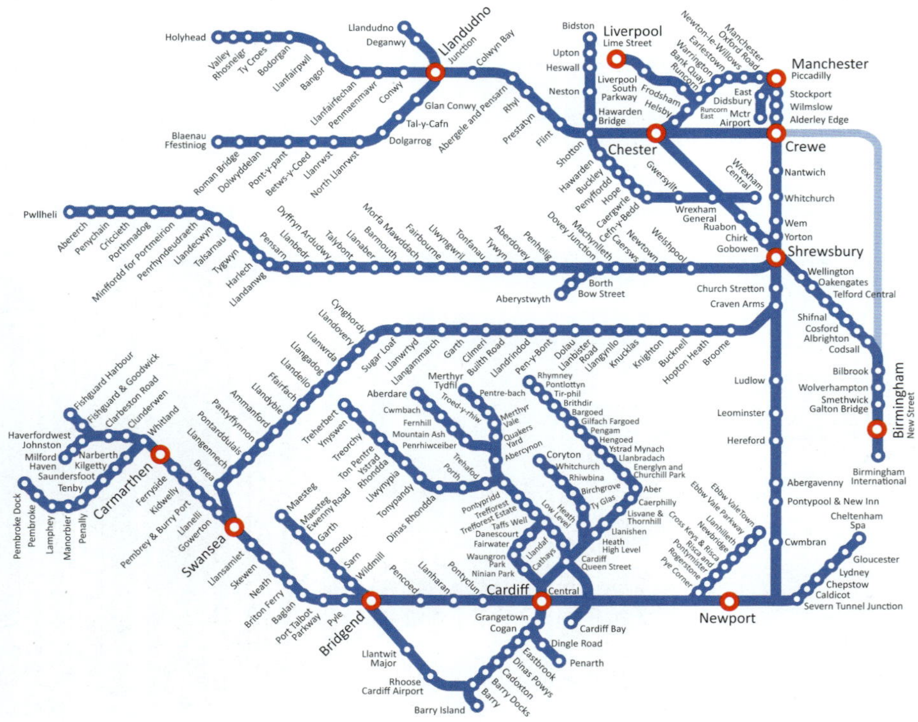

Class 67 are hired by TfW for its loco-hauled operation, and nine locos have been modified – instantly noticeable by the additional jumper socket on the cab front to the top right of the multiple working cable. 67015 shows off this modification as it waits to leave with the 1712 Cardiff Central-Holyhead on 3 August 2021. *Stuart West*

TfW is starting a massive introduction of new train fleets, with Class 197 and 231 units now being delivered and Class 398 and 756 units being constructed. They will replace Class 150/153/158/175 fleets. Additionally, five VivaRail Class 230 and eight FLEX Class 769s have also been introduced by the TOC. The Class 230s are used on the Wrexham Central to Bidston line, while the Class 769s have started work on the Valley Lines.

TfW has acquired the Pullman Rail operation from the Colas Rail group, which includes the former EWS depot at Cardiff Canton.

Class 150

These share duties with Class 153/158/170 units but also work Crewe to Chester, Chester to Liverpool, Carmarthen/Swansea to Shrewsbury/Crewe, Blaenau Ffestiniog to Llandudno, Wrexham Central to Bidston (due to be replaced by Class 230s), Fishguard Harbour/Pembroke Dock to Swansea, Swansea to Cardiff Central, Maesteg to Cheltenham Spa and Fishguard Harbour to Gloucester, all duties they share with other TfW units.

All are in the HLHQ pool.

	Livery	Owner	Depot	DMSL	DMS
150208	ARV	POR	CF	52208	57208
150213	ARV	POR	CF	52213	57213
150217	TFW	POR	CF	52217	57217
150227	TFW	POR	CF	52227	57227
150229	ARV	POR	CF	52229	57229
150230	ARV	POR	CF	52230	57230
150231	ARV	POR	CF	52231	57231
150235	TFW	POR	CF	52235	57235
150236	ARV	POR	CF	52236	57236
150237	TFW	POR	CF	52237	57237
150240	ARV	POR	CF	52240	57240
150241	ARV	POR	CF	52241	57241
150242	ARV	POR	CF	52242	57242
150245	ARV	POR	CF	52245	57245
150250	ARV	POR	CF	52250	57250
150251	ARV	POR	CF	52251	57251
150252	ARV	POR	CF	52252	57252
150253	ARV	POR	CF	52253	57253
150254	ARV	POR	CF	52254	57254
150255	ARV	POR	CF	52255	57255
150256	ARV	POR	CF	52256	57256

Built in the mid-1980s, 150256 works the 1402 Rhymney-Penarth, calling at Cardiff Central on 29 July 2021. As a rule, these units are used on local trains. *Stuart West*

150257	ARV	POR	CF	52257	57257
150258	ARV	POR	CF	52258	57258
150259	TFW	POR	CF	52259	57259
150260	ARV	POR	CF	52260	57260
150262	ARV	POR	CF	52262	57262
150264	ARV	POR	CF	52264	57264
150267	ARV	POR	CF	52267	57267
150278	ARV	POR	CF	52278	57278
150279	ARV	POR	CF	52279	57279
150280	ARV	POR	CF	52280	57280
150281	ARV	POR	CF	52281	57281
150282	TFW	POR	CF	52282	57282
150283	TFW	POR	CF	52283	57283
150284	TFW	POR	CF	52284	57284
150285	ARV	POR	CF	52285	57285

Class 153

The fleet has recently been enlarged with surplus units released by East Midlands Railway and they work pretty much the same duties as 150s.

The Class 153/9 are former East Midlands Railways and Greater Anglia units that have undergone interior refurbishment and as they do not have compliant toilets as a rule operate in pairs with 153/3s on long journeys, though they can operate on their own on short turns such as the Llandudno Junction to Llandudno shuttles.

All are in the HLHQ pool.

	Livery	Owner	Depot	DMSL
153303	TFW	TFW	CF	52303
153312	TFW	TFW	CF	52312
153320	TFW	POR	CF	52320
153323	TFW	POR	CF	52323
153325	TFW	POR	CF	52325
153327	TFW	TFW	DF	52327
153329	TFW	POR	CF	52329
153333	TFW	POR	CF	52333
153353	TFW	TFW	CF	57353
153361	TFW	POR	CF	57361
153362	TFW	TFW	CF	57362
153367	TFW	POR	CF	57367
153369	TFW	POR	CF	57369

153374		TFW	TFW	CF	57362
153906	153306	TFW	POR	CF	52306
153909	153309	TFW	POR	CF	52309
153910	153310	TFW	POR	CF	52310
153913	153313	TFW	POR	CF	52313
153914	153314	TFW	POR	CF	52314
153918	153318	EMT	TFW	CF	52321
153921	153321	TFW	POR	CF	52321
153922	153322	TFW	POR	CF	52322
153926	153326	TFW	POR	CF	52326
153935	153335	TFW	POR	CF	52335
153968	153368	EMT	TFW	CF	57368
153972	153372	TFW	TFW	CF	57372
153982	153382	EMT	TFW	CF	57382

Several ex-Greater Anglia and East Midlands Railway Class 153s have been taken on by TfW. Some have been renumbered as 153/9s to show their toilets are non-compliant and so locked out of use. Therefore they work either in multiple with compliant DMU types or on short trips such as this; 153926 arrives with the 1508 Llandudno Junction-Llandudno on 20 November 2021. *Pip Dunn*

Class 158

These units can be fitted with plug-in ERTMS in-cab signalling equipment, meaning they are the only TfW trains that can work Cambrian line duties west of Shrewsbury to Aberystwyth and Pwllheli, trains that usually start at Birmingham International.

They are also used on Birmingham International to Holyhead, Manchester Airport to Llandudno, Crewe to Chester, Chester to Liverpool and sometimes on the South West Wales routes. They have recently been refurbished.

All are in the HLHQ pool.

	Livery	Owner	Depot	DMSL	DMSL
158818	TFW	ANG	MN	52818	57818
158819	TFW	ANG	MN	52819	57819
158820	TFW	ANG	MN	52820	57820
158821	TFW	ANG	MN	52821	57821
158822	TFW	ANG	MN	52822	57822
158823	TFW	ANG	MN	52823	57823
158824	TFW	ANG	MN	52824	57824
158825	TFW	ANG	MN	52825	57825
158826	TFW	ANG	MN	52826	57826
158827	TFW	ANG	MN	52827	57827
158828	TFW	ANG	MN	52828	57828
158829	TFW	ANG	MN	52829	57829
158830	TFW	ANG	MN	52830	57830
158831	TFW	ANG	MN	52831	57831
158832	TFW	ANG	MN	52832	57832
158833	TFW	ANG	MN	52833	57833
158834	TFW	ANG	MN	52834	57834
158835	TFW	ANG	MN	52835	57835
158836	TFW	ANG	MN	52836	57836
158837	TFW	ANG	MN	52837	57837
158838	TFW	ANG	MN	52838	57838
158839	TFW	ANG	MN	52839	57839
158840	TFW	ANG	MN	52840	57840
158841	TFW	ANG	MN	52841	57841

Transport for Wales has a fleet of ex-BR DMUs of Classes 150, 153, 158 supplemented by some post-privatisation Class 170s and 175s, but is now undergoing a massive fleet overhaul. 158840 works the 0831 Manchester Piccadilly-Bridgend, calling at Stockport on 31 March 2021. *Tom McAtee*

Class 170/2

TfW took all of the 12 Class 170 Turbostars displaced by Greater Anglia. The eight three-car and three two-car sets are now used mostly on Maesteg to Cheltenham Spa and Bridgend to Ebbw Vale Town routes. They should all move to EMR.

All are in the HLHQ pool.

	Livery	Owner	Depot	DMCL	MSL	DMSL
170201	GAW	POR	CF	50201	56201	79201
170202	GAW	POR	CF	50202	56202	79202
170203	GAW	POR	CF	50203	56203	79203
170204	GAW	POR	CF	50204	56204	79204
170205	GAW	POR	CF	50205	56205	79205
170206	GAW	POR	CF	50206	56206	79206
170207	GAW	POR	CF	50207	56207	79207
170208	GAW	POR	CF	50208	56208	79208
170270	GAW	POR	CF	50270		79270
170271	GAW	POR	CF	50271		79271
170272	GAW	POR	CF	50272		79272

TfW took all 12 Class 170/2 Turbostars released by Greater Anglia, although this unit, 170273, has since left the fleet and moved to East Midlands Railway and the others should follow. It was at Cardiff Central on 19 December 2019 working the 1307 Bridgend-Ebbw Vale Town. *Stuart West*

Class 175

Alstom two- and three-car units originally ordered by First North Western, they are used mostly on the Welsh Marches, North Wales Coast and South West Wales routes. The fleet has recently been refurbished.

All are in the HLHQ pool.

Class 175/0

Original two-car sets.

	Livery	Owner	Depot	DMSL	DMSL
175001	TFW	ANG	CH	50701	79701
175002	TFW	ANG	CH	50702	79702
175003	TFW	ANG	CH	50703	79703
175004	TFW	ANG	CH	50759	79759
175005	TFW	ANG	CH	50705	79751
175006	TFW	ANG	CH	50706	79765
175007	TFW	ANG	CH	50707	79707
175008	TFW	ANG	CH	50708	79708
175009	TFW	ANG	CH	50709	79709
175010	TFW	ANG	CH	50710	79710
175011	TFW	ANG	CH	50711	79711

First North Western ordered the Class 175 Coradia DMUs from Alstom in the late 1990s and they entered traffic in 2000. They then passed to Arriva Trains Wales and now TfW. 175006 passes Wistanstow, near Shrewsbury, working the 0931 Manchester Piccadilly-Carmarthen on 20 October 2021. *Glen Batten*

Class 175/1

Original three-car sets.

	Livery	Owner	Depot	DMSL	MSL	DMSL
175101	TFW	ANG	CH	50751	56751	79704
175102	TFW	ANG	CH	50702	56752	79752
175103	TFW	ANG	CH	50753	56753	79753
175104	TFW	ANG	CH	50754	56754	79754
175105	TFW	ANG	CH	50755	56755	79755
175106	TFW	ANG	CH	50756	56756	79756
175107	TFW	ANG	CH	50757	56757	79757
175108	TFW	ANG	CH	50758	56758	79758
175109	TFW	ANG	CH	50704	56759	79759
175110	TFW	ANG	CH	50760	56760	79760
175111	TFW	ANG	CH	50761	56761	79761
175112	TFW	ANG	CH	50762	56762	79762
175113	TFW	ANG	CH	50763	56763	79763
175114	TFW	ANG	CH	50764	56764	79764
175115	TFW	ANG	CH	50765	56765	79706
175116	TFW	ANG	CH	50766	56766	79766

Class 197

Transport for Wales has several new fleets of train on order, the first of which are now being delivered for testing. They will replace Class 150, 153, 158 and 175s. 197022-041 will be fitted with ERTMS to operate on the Cambrian lines and replace similarly fitted Class 158s. The units have been approved for use by the ORR.

Active units will be in the HLHQ pool.

	Livery	Owner	Depot	DMS	DMS
197001	TFW	SMBC	CL (Q)	131001	133001
197002	TFW	SMBC	CL (Q)	131002	133002
197003	TFW	SMBC	CL (Q)	131003	133003
197004	TFW	SMBC	CL (Q)	131004	133004
197005	TFW	SMBC		131005	133005
197006	TFW	SMBC	DO (S)	131006	133006
197007	TFW	SMBC	DO (S)	131007	133007
197008	TFW	SMBC	DO (S)	131008	133008
197009	TFW	SMBC		131009	133009
197010	TFW	SMBC	DO (S)	131010	133010
197011	TFW	SMBC		131011	133011
197012	TFW	SMBC		131012	133012
197013	TFW	SMBC		131013	133013
197014	TFW	SMBC		131014	133014
197015	TFW	SMBC		131015	133015
197016	TFW	SMBC		131016	133016
197017	TFW	SMBC		131017	133017
197018	TFW	SMBC		131018	133018
197019	TFW	SMBC		131019	133019
197020	TFW	SMBC		131020	133020
197021	TFW	SMBC		131021	133021
197022	TFW	SMBC		131022	133022
197023	TFW	SMBC		131023	133023
197024	TFW	SMBC		131024	133024
197025	TFW	SMBC		131025	133025
197026	TFW	SMBC		131026	133026
197027	TFW	SMBC		131027	133027
197028	TFW	SMBC		131028	133028
197029	TFW	SMBC		131029	133029
197030	TFW	SMBC		131030	133030
197031	TFW	SMBC		131031	133031
197032	TFW	SMBC		131032	133032
197033	TFW	SMBC		131033	133033
197034	TFW	SMBC		131034	133034
197035	TFW	SMBC		131035	133035
197036	TFW	SMBC		131036	133036
197037	TFW	SMBC		131037	133037
197038	TFW	SMBC		131038	133038
197039	TFW	SMBC		131039	133039
197040	TFW	SMBC		131040	133040
197041	TFW	SMBC		131041	133041
197042	TFW	SMBC		131042	133042
197043	TFW	SMBC		131043	133043
197044	TFW	SMBC		131044	133044
197045	TFW	SMBC		131045	133045
197046	TFW	SMBC		131046	133046
197047	TFW	SMBC		131047	133047
197048	TFW	SMBC		131048	133048
197049	TFW	SMBC		131049	133049
197050	TFW	SMBC		131050	133050
197051	TFW	SMBC		131051	133051

Class 197/1

Units 197113-126 have first-class seating and will be used on the Swansea-Manchester route. 197101-112 are standard-class-only units.

	Livery	Owner	Depot	DMS	MS	DMS,DMF
197101	TFW	SMBC	CL (Q)	131101	132101	133101
197102	TFW	SMBC	CL (Q)	131102	132102	133102
197103	TFW	SMBC	DO (S)	131103	132103	133103
197104	TFW	SMBC		131104	132104	133104
197105	TFW	SMBC		131105	132105	133105
197106	TFW	SMBC		131106	132106	133106
197107	TFW	SMBC		131107	132107	133107
197108	TFW	SMBC		131108	132108	133108
197109	TFW	SMBC		131109	132109	133109
197110	TFW	SMBC		131110	132110	133110
197111	TFW	SMBC		131111	132111	133111
197112	TFW	SMBC		131112	132112	133112
197113	TFW	SMBC		131113	132113	133113
197114	TFW	SMBC		131114	132114	133114
197115	TFW	SMBC		131115	132115	133115
197116	TFW	SMBC		131116	132116	133116

The new order for TfW will be Class 197s, which will replace Class 158s and 175s on longer-distance journeys. Passing Christleton near Chester is 197101 on the 1309 Llandudno-Crewe Carriage Sidings training trip on 8 October 2021. *Paul Shannon*

197117	TFW	SMBC		131117	132117	133117
197118	TFW	SMBC		131118	132118	133118
197119	TFW	SMBC		131119	132119	133119
197120	TFW	SMBC		131120	132120	133120
197121	TFW	SMBC		131121	132121	133121
197122	TFW	SMBC		131122	132122	133122
197123	TFW	SMBC		131123	132123	133123
197124	TFW	SMBC		131124	132124	133124
197125	TFW	SMBC		131125	132125	133125
197126	TFW	SMBC		131126	132126	133126

Class 230s

VivaRail diesel-battery hybrid units converted from former London Underground D78 stock for use on the Bidston to Wrexham route.

	Livery	Owner	Depot	DMS	TS	DMS
230006	TFW	TFW	BD	300006	300206	300106
230007	TFW	TFW	BD	300007	300207	300107
230008	TFW	TFW	BD	300008	300208	300108
230009	TFW	TFW	BD	300009	300209	300109
230010	TFW	TFW	BD	300010	300210	300110

TfW has five VivaRail Class 230 DMUs, modified from old London Underground stock. They will be used on the line from New Brighton to Wrexham. On 11 August 2020, 230006 passes Chester on the 0756 Birkenhead North-Chester DMU Depot test run. *Paul Shannon*

Stadler is building 11 four-car Class 231 FLIRT DMUs for TfW, which like the Greater Anglia Class 755s have a fifth 'vehicle' for the engines, Unlike the 755s, they are diesel units only. The first two units arrived in late 2021 and are now on test. 231004 stands on Canton depot shortly after arrival. *TfW*

Class 231

New four-car DMUs built by Stadler for use Ebbw Vale and Maesteg line, The first two sets, 231002/004, were delivered in November 2021. They feature a centre vehicle that houses the engine, which technically makes them five-cars.

	Livery	Owner	Depot	DMS	TS	PP	TS	DMS
231001	TFW	SMBC	CF (Q)	381001	381201	381401	381301	381101
231002	TFW	SMBC	CF (Q)	381002	381202	381402	381302	381102
231003	TFW	SMBC	CF (Q)	381003	381203	381403	381303	381103
231004	TFW	SMBC	CF (Q)	381004	381204	381404	381304	381104
231005	TFW	SMBC		381005	381205	381405	381305	381105
231006	TFW	SMBC		381006	381206	381406	381306	381106
231007	TFW	SMBC		381007	381207	381407	381307	381107
231008	TFW	SMBC		381008	381208	381408	381308	381108
231009	TFW	SMBC		381009	381209	381409	381309	381109
231010	TFW	SMBC		381010	381210	381410	381310	381110
231011	TFW	SMBC		381011	381211	381411	381311	381111

Class 398

Stadler City Link bi-mode battery and 25kV AC metro-tram units on order. The trains will be maintained at a new purpose-built depot at Taff Wells, which is currently being constructed.

	Livery	Owner	Depot	DMS	MS	DMS
398001	TFW	SMBC		999051	999151	999251
398002	TFW	SMBC		999052	999152	999252
398003	TFW	SMBC		999053	999153	999253
398004	TFW	SMBC		999054	999154	999254
398005	TFW	SMBC		999055	999155	999255
398006	TFW	SMBC		999056	999156	999256
398007	TFW	SMBC		999057	999157	999257
398008	TFW	SMBC		999058	999158	999258
398009	TFW	SMBC		999059	999159	999259
398010	TFW	SMBC		999060	999160	999260
398011	TFW	SMBC		999061	999161	999261
398012	TFW	SMBC		999062	999162	999262
398013	TFW	SMBC		999063	999163	999263
398014	TFW	SMBC		999064	999164	999264
398015	TFW	SMBC		999065	999165	999265

398016	TFW	SMBC		999066	999166	999266
398017	TFW	SMBC		999067	999167	999267
398018	TFW	SMBC		999068	999168	999268
398019	TFW	SMBC		999069	999169	999269
398020	TFW	SMBC		999070	999170	999270
398021	TFW	SMBC		999071	999171	999271
398022	TFW	SMBC		999072	999172	999272
398023	TFW	SMBC		999073	999173	999273
398024	TFW	SMBC		999074	999174	999274
398025	TFW	SMBC		999075	999175	999275
398026	TFW	SMBC		999076	999176	999276
398027	TFW	SMBC		999077	999177	999277
398028	TFW	SMBC		999078	999178	999278
398029	TFW	SMBC		999079	999179	999279
398030	TFW	SMBC		999080	999180	999280
398031	TFW	SMBC		999081	999181	999281
398032	TFW	SMBC		999082	999182	999282
398033	TFW	SMBC		999083	999183	999283
398034	TFW	SMBC		999084	999184	999284
398035	TFW	SMBC		999085	999185	999285
398036	TFW	SMBC		999086	999186	999286

Class 756/0

These Stadler FLIRT tri-mode units with diesel, battery and 25kV AC capability will work on the Valley Lines and help replace Class 150/153 and 769 units. They are similar to the Class 755s in use with Greater Anglia. They are technically three-car sets with an additional vehicle for the diesel engines.

	Livery	Owner	Depot	DMS	PP	PTS	DMS
756001	TFW	SMBC		911001	971001	981001	912001
756002	TFW	SMBC		911002	971002	981002	912002
756003	TFW	SMBC		911003	971003	981003	912003
756004	TFW	SMBC		911004	971004	981004	912004
756005	TFW	SMBC		911005	971005	981005	912005
756006	TFW	SMBC		911006	971006	981006	912006
756007	TFW	SMBC		911007	971007	981007	912007

Class 756/1

Like the Class 756/0s but with four passenger vehicles.

	Livery	Owner	Depot	DMS	PTS	PP	PTS	DMS
756101	TFW	SMBC		911101	961101	971101	981101	912101
756102	TFW	SMBC		911102	961102	971102	981102	912102
756103	TFW	SMBC		911103	961103	971103	981103	912103
756104	TFW	SMBC		911104	961104	971104	981104	912104
756105	TFW	SMBC		911105	961105	971105	981105	912105
756106	TFW	SMBC		911106	961106	971106	981106	912106
756107	TFW	SMBC		911107	961107	971107	981107	912107
756108	TFW	SMBC		911108	961108	971108	981108	912108
756109	TFW	SMBC		911109	961109	971109	981109	912109
756110	TFW	SMBC		911110	961110	971110	981110	912110
756111	TFW	SMBC		911111	961111	971111	981111	912111
756112	TFW	SMBC		911112	961112	971112	981112	912112
756113	TFW	SMBC		911113	961113	971113	981113	912113
756114	TFW	SMBC		911114	961114	971114	981114	912114
756115	TFW	SMBC		911115	961115	971115	981115	912115
756116	TFW	SMBC		911116	961116	971116	981116	912116
756117	TFW	SMBC		911117	961117	971117	981117	912117

Class 769

All of these ex-Class 319 EMUs converted to bi-mode operation are now in traffic. All are in the HLHQ pool.

TfW has nine Class 769 FLEX units and these have finally entered traffic after long delays in their commissioning. 769003 waits to work the 1207 Penarth-Bargoed on 29 July 2021. *Stuart West*

	Old No.	Livery	Owner	Depot	DMC	MS	TS	DMS
769002	319002	TFW	POR	CF	77293	62892	71773	77292
769003	319003	TFW	POR	CF	77295	62893	71774	77294
769006	319006	TFW	POR	CF	77301	62896	71777	77300
769007	319007	TFW	POR	CF	77303	62897	71778	77302
769008	319008	TFW	POR	CF	77305	62898	71779	77304
769421	319421	TFW	POR	CF	77331	62911	71792	77330
769426	319426	TFW	POR	LB (U)	77341	62916	71796	77340
769445	319445	TFW	POR	CF	77379	62935	71816	77378
769452	319452	TFW	POR	CF	77441	62966	71871	77440

Loco-hauled coaches

Seven sets of ex-LNER Mk 4 coaches have been taken on by TfW for push-pull operation with DB Cargo Class 67s. Three initial sets are now in use on the Welsh Marches route, but recently four additional sets were acquired that had originally been earmarked for Grand Central's Blackpool-Euston Open Access service that did not come to fruition because of the pandemic.

The initial three sets replaced two sets of Mk 3 coaches. The Class 67s are those in the WAWC pool, listed on page 265.

All are in the HLHQ pool.

Mk 4 kitchen buffet standard

Vehicle	Livery	Owner	Depot
10301	GCR	TFW	EH (S)
10305	LNE	TFW	CF
10312	LNE	TFW	CF
10318	GCR	TFW	CF
10321	GCR	TFW	ZG (S)
10325	LNE	TFW	CF
10328	LNE	TFW	CF
10330	GCR	TFW	CB (S)

Mk 4 open first

Vehicle	Livery	Owner	Depot
11319	GCR	TFW	CF
11320	GCR	TFW	ZG (S)
11321	GCR	TFW	CB (S)
11322	GCR	TFW	EH (S)
11323	LNE	TFW	CF
11324	LNE	TFW	CF
11325	LNE	TFW	CF

Mk 4 open standard (end)

Vehicle	Livery	Owner	depot
12210	GCR	TFW	EH (S)
12211	GCR	TFW	CF
12215	LNE	TFW	CF
12217	LNE	TFW	CF
12219	LNE	TFW	CF
12222	GCR	TFW	CB (S)
12224	GCR	TFW	ZG (S)
12225	LNE	TFW	CF

Ex-LNER MK 4 open first coach 11323 retains its old Virgin livery but is expected to be repainted into TfW colours in the fullness of time. It was at Cardiff Central on 3 August 2021. *Stuart West*

Mk 4 open standard (disabled)

Vehicle	Livery	Owner	Depot
12304	LNE	TFW	CF
12310	GCR	TFW	CF
12315	LNE	TFW	CF
12316	GCR	TFW	EH (S)
12318	LNE	TFW	CF
12319	LNE	TFW	CF
12322	LNE	TFW	CF
12323	GCR	TFW	CB (S)
12324	LNE	TFW	CF
12326	GCR	TFW	ZG (S)

Mk 4 open standard

Vehicle	Livery	Owner	Depot
12434	GCR	TFW	CF
12446	LNE	TFW	CF
12447	LNE	TFW	CF
12452	GCR	TFW	EH (S)
12454	LNE	TFW	CF
12461	GCR	TFW	CF
12477	GCR	TFW	ZG (S)

Mk 4 DVT

Vehicle	Livery	Owner	Depot
82200	GCR	TFW	ZG (S)
82201	GCR	TFW	CF
82216	ADV	TFW	CF
82220	LNE	TFW	CF
82226	ADV	TFW	CF
82227	GCR	TFW	CB (S)
82229	ADV	TFW	CF
82230	GCR	TFW	EH (S)
82216 is in Hope House Children's Hospice livery.			
82226 is in Alzheimer's Society livery.			
82229 is in RNLI black livery.			

TfW uses loco haulage on a few trains on the Holyhead to Cardiff route, with the Mk 3 coaches having recently been replaced by spare ex-LNER Mk 4s. Only modified Class 67s from DB Cargo can work with the coaches, currently 67008/010/013-015/017/022/025. On 3 August 2021, DVT 82226 departs Cardiff Central on the rear of the 1712 to Holyhead. *Stuart West*

Transport for Wales Mk 4 formations

	DVT	OS	FOD	KBS	OS	TSOT
HD01	82226	12454	11323	10325		12225
HD02	82229	12447	11324	10328		12219
HD03	82216	12446	11325	10312		12217
GC01	82201	12434	11319	10318	12310	12211
GC02	82200	12477	11320	10321	12326	12224
GC03	82227	12461	11321	10330	12323	12222
GC04	82230	12452	11322	10301	12316	12210
spare	82220	12324	12315	10305	12034	12215

West Midlands Trains

Contact details

Websites: www.westmidlandsrailway.co.uk (WMR); www.londonnorthwesternrailway.co.uk (LNWR)
Twitter: @WestMidRailway or @LNRailway

Key personnel

Managing Director: Ian McConnell
Engineering Director: Zena Dent

Overview

West Midlands Trains is one franchise but split into two distinct parts, each having their own fleets and livery. It was created from the old London Midland Railway franchise, which was previously part of the Silverlink operations.

London North Western Railway (LNWR) runs the Euston to Liverpool semi-fast trains (via Northampton) as well as the Bedford to Bletchley and Watford Junction to St Albans Abbey branches and West Midlands Railway (WMR), which covers most of the local and commuter trains in, and emanating from, Birmingham and extending as far as Leamington Spa, Stratford-on-Avon, Worcester, Shrewsbury, Rugeley and the Cross-City line from Redditch / Bromsgrove to Lichfield.

LNWR trains are predominately dark green, while WMR uses a gold and purple colour scheme. Both parts retain a mixed fleet of both ex-BR rolling stock and early post-privatisation trains but is also now taking deliveries of its CAF Class 196 diesel units (for WMR) and Class 730 EMUs for both LNWR and WMR.

They will allow its older fleets to be replaced – either redeployed or returned to their ROSCOs for probable disposal. In the last year, all its remaining Class 153s have been withdrawn.

Class 08

The TOC has two ex-BR Class 08 shunters, one based at Tyseley and one based at Soho.

Both are in the EJLO pool.

08616	LMR	LON	TS	*Tyseley 100*
08805	RSR	LON	SO	*Robin Jones 40 Years Service*

Class 139

These single-car units are part of the WMR fleet built by Parry People Mover and work exclusively on the self-contained ¾-mile Stourbridge Junction to Stourbridge Town branch. The units are currently being tested to run on biomethane.

Both are in the EJHQ pool.

	Livery	Owner	Depot	DMM
139001	WMG	POR	SJ	39001
139002	WMG	POR	SJ	39002

Class 170

All the Class 170 Turbostar units, initially ordered by Central Trains, are to move to East Midlands Railway, hence several units have been repainted into EMR purple despite currently remaining with WMR.

All are in the EJHQ pool.

On 6 November 2020, 170513 arrives at the delightful Great Malvern Station with a train for Hereford. These turns will be handed over to Class 196s soon. *Jack Boskett*

	Old No.	Livery	Owner	Depot	DMSL	DMSL
170501		EMP	POR	TS	50501	79501
170502		EMP	POR	TS	50502	79502
170504		EMP	POR	TS	50504	79504
170505		WMG	POR	TS	50505	79505
170506		EMP	POR	TS	50506	79506
170507		EMP	POR	TS	50507	79507
170508		EMP	POR	TS	50508	79508
170509		EMP	POR	TS	50509	79509
170510		EMP	POR	TS	50510	79510
170512		EMP	POR	TS	50512	79512
170513		EMP	POR	TS	50513	79513
170514		EMP	POR	TS	50514	79514
170516		WMG	POR	TS	50516	79516
170533	170633	WMG	POR	TS	50633	79633
170535	170635	WMG	POR	TS	50635	79635

Class 172

A development of the Class 170 Turbostar units, these differ by having corridor connections to allow them to run in multiple formations with one guard. All are in the EJHQ pool.

Class 172/0

These two-car units were previously ordered and used by London Overground for the Gospel Oak-Barking line, but have now transferred to WMR and are used on the Nuneaton to Leamington Spa route as well as being used to strengthen other WMR services. Unlike other 172s, they do not have corridor connections.

	Livery	Owner	Depot	DMSL	DMS
172001	WMG	ANG	TS	59311	59411
172002	WMG	ANG	TS	59312	59412
172003	WMG	ANG	TS	59313	59413
172004	WMG	ANG	TS	59314	59414
172005	WMG	ANG	TS	59315	59415
172006	WMG	ANG	TS	59316	59416
172007	WMG	ANG	TS	59317	59417
172008	WMG	ANG	TS	59318	59418

Class 172/1

These two-car Turbostar units date from 2009-10. They are Chiltern units on sub-lease to West Midlands Trains.

All are in the HOHQ pool.

	Livery	Owner	Depot	DMSL	DMS
172101	CRO	ANG	TS	59111	59211
172102	CRO	ANG	TS	59112	59212
172103	CRO	ANG	TS	59113	59213
172104	CRO	ANG	TS	59114	59214

Class 172/2

These two-car units, along with the three-car Class 172/3s, were ordered by LMR. They have corridor connections and work local trains from Leamington Spa and Stratford-upon-Avon to Stourbridge Junction and Worcester, via Birmingham Snow Hill.

	Livery	Owner	Depot	DMSL	DMS
172211	WMG	POR	TS	50211	79211
172212	WMG	POR	TS	50212	79212
172213	WMG	POR	TS	50213	79213
172214	WMG	POR	TS	50214	79214
172215	WMG	POR	TS	50215	79215
172216	WMG	POR	TS	50216	79216
172217	WMG	POR	TS	50217	79217
172218	WMG	POR	TS	50218	79218
172219	WMG	POR	TS	50219	79219
172220	WMG	POR	TS	50220	79220
172221	WMG	POR	TS	50221	79221
172222	WMG	POR	TS	50222	79222

On 22 March 2021, Turbostar 172218 arrives at Whitlocks End with a train from Birmingham Snow Hill to Stratford-upon-Avon. Most of the stations on the Shakespeare line have rows of flower beds, which are looked after by volunteers. *Jack Boskett*

Class 172/3

Like the 172/2s, these three-car units have corridor connections and the two fleets work side by side.

	Livery	Owner	Depot	DMSL	MS	DMS
172331	WMG	POR	TS	50331	56631	79331
172332	WMG	POR	TS	50332	56632	79332
172333	WMG	POR	TS	50333	56633	79333
172334	WMG	POR	TS	50334	56634	79334
172335	WMG	POR	TS	50335	56635	79335
172336	WMG	POR	TS	50336	56636	79336
172337	WMG	POR	TS	50337	56637	79337
172338	WMG	POR	TS	50338	56638	79338
172339	WMG	POR	TS	50339	56639	79339
172340	WMG	POR	TS	50340	56640	79340
172341	WMG	POR	TS	50341	56641	79341
172342	WMG	POR	TS	50342	56642	79342
172343	WMG	POR	TS	50343	56643	79343
172344	WMG	POR	TS	50344	56644	79344
172345	WMG	POR	TS	50345	56645	79345

Class 196

These new CAF Civity DMUs will work the Birmingham NS-Hereford/Shrewsbury and Snow Hill line services and replace Class 170s. The first units are now on test and were expected to enter traffic in the summer of 2022.

Class 196/0

These two-car units should start to enter traffic in 2022.

Active units will be in the EJHQ pool.

	Livery	Owner	Depot	DMS	DMS
196001	WMG	COR	TS (Q)	121001	124001
196002	WMG	COR	DO (S)	121002	124002
196003	WMG	COR	TS (Q)	121003	124003
196004	WMG	COR	TS (Q)	121004	124004
196005	WMG	COR	TS (Q)	121005	124005
196006	WMG	COR	BY (S)	121006	124006
196007	WMG	COR	BY (S)	121007	124007
196008	WMG	COR	TS (Q)	121008	124008
196009	WMG	COR	TS (Q)	121009	124009
196010	WMG	COR	BY (S)	121010	124010
196011	WMG	COR	BY (S)	121011	124011
196012	WMG	COR	TS (Q)	121012	124012

Class 196/1

The first of these four-car CAF units are now on test.

The new CAF Class 196s were about to go into service as the rail guide closed for press. 196104 rests on Tyseley. *WMR*

	Livery	Owner	Depot	DMS	MS	MS	DMS
196101	WMG	COR	WS (Q)	121101	122101	123101	124101
196102	WMG	COR	TS (Q)	121102	122102	123102	124102
196103	WMG	COR	TS (Q)	121103	122103	123103	124103
196104	WMG	COR	TS (Q)	121104	122104	123104	124104
196105	WMG	COR	TS (Q)	121105	122105	123105	124105
196106	WMG	COR	LM (S)	121106	122106	123106	124106
196107	WMG	COR	WS (Q)	121107	122107	123107	124107
196108	WMG	COR	LM (S)	121108	122108	123108	124108
196109	WMG	COR	LM (S)	121109	122109	123109	124109
196110	WMG	COR	TS (Q)	121110	122110	123110	124110
196111	WMG	COR	DO (S)	121111	122111	123111	124111
196112	WMG	COR	TS (Q)	121112	122112	123112	124112
196113	WMG	COR	TS (Q)	121113	122113	123113	124113
196114	WMG	COR	TS (Q)	121114	122114	123114	124114

Class 230

LNWR has three VivaRail Class 230 DMUs made using London Underground D78 Stock bodies. They run a two-car sets on the Bedford-Bletchley Marston Vale line.

All are in the EJHQ pool.

West Midlands Trains is split into two – West Midland Railways for local trains emanating from Birmingham and London North Western for trains on and around the WCML. The latter includes the Bedford-Bletchley route, which is now worked by VivaRail three-car Class 230 DMUs. On 19 December 2019, 230003 calls at Millbrook. *Rob France*

	Livery	Owner	Depot	DMS	DMS
230003	LMR	VIV	BY	300003	300103
230004	LMR	VIV	BY	300004	300104
230005	LMR	VIV	BY	300005	300105

Class 319

LMR was another TOC to take some surplus Class 319s when displaced from Thameslink. It uses the sets on the Watford Junction to St Albans Abbey branch and also on additional WCML workings from Euston to Northampton. They will be replaced from 2023 when Class 730/1s come on stream and the first have been stood down.

All are in the EJHQ pool.

	Livery	Owner	Depot	DTS	MS	TS	DTS
319005	WHI	POR	NN	77299	62895	71776	77298
319012	WHI	POR	NN	77313	62902	71783	77312
319013	LMR	POR	BU (S)	77315	62903	71784	77314
319214	WHI	POR	NN	77317	62904	71785	77316
319215	WHI	POR	NN	77319	62905	71786	77318
319216	LMR	POR	BU (S)	77321	62906	71787	77320
319217	WHI	POR	NN	77323	62907	71788	77322
319218	WHI	POR	BU (S)	77325	62908	71789	77324
319219	WHI	POR	NN	77327	62909	71790	77326
319220	WHI	POR	NN	77329	62910	71791	77328
319429	LMR	POR	NN	77347	62919	71800	77346
319433	LMR	POR	NN	77355	62923	71804	77354
319457	LMR	POR	NN	77451	62971	71876	77450
319460	LMR	POR	BU (S)	77457	62974	71879	77456

Class 323

These WMR units are used mostly on the Cross-City line from Bromsgrove and Redditch to Lichfield via Birmingham NS and Sutton Coldfield. They also work Walsall-Wolverhampton via Aston and the Chase line from Birmingham NS to Rugeley, which was recently electrified north of Walsall. When new Class 730/0s enter traffic, many of the fleet will be returned to their ROSCO, but indications are 17 should move to Northern. The fleet is steadily being named with stick-on 'nameplates' of stations on the Cross-City route.

All are in the EJHQ pool.

	Livery	Owner	Depot	DMS	TS	DMS	
323201	WMG	POR	SO	64001	75301	65001	*Duddeston*
323202	WMG	POR	DO	64002	75302	65002	*Butlers Lane*
323203	WMG	POR	SO	64003	75303	65003	*Aston*
323204	WMG	POR	SO	64004	75304	65004	*Selly Oak*
323205	WMG	POR	SO	64005	75305	65005	*Blake Street*
323206	WMG	POR	SO	64006	75306	65006	*Barnt Green*
323207	WMG	POR	SO	64007	75307	65007	*Bournville*
323208	WMG	POR	SO	64008	75308	65008	*Five Ways*
323209	WMG	POR	SO	64009	75309	65009	*New Street*
323210	WMG	POR	SO	64010	75310	65010	*Shenstone*
323211	WMG	POR	SO	64011	75311	65011	*Four Oaks*
323212	WMG	POR	SO	64012	75312	65012	*Bromsgrove*
323213	WMG	POR	SO	64013	75313	65013	*Sutton Coldfield*
323214	WMG	POR	SO	64014	75314	65014	*Wylde Green*
323215	WMG	POR	SO	64015	75315	65015	*Gravelly Hill*
323216	WMG	POR	SO	64016	75316	65016	*University*
323217	WMG	POR	SO	64017	75317	65017	*Chester Road*
323218	WMG	POR	SO	64018	75318	65018	*Lichfield City*
323219	WMG	POR	SO	64019	75319	65019	*King's Norton*
323220	WMG	POR	SO	64020	75320	65020	*Lichfield Trent Valley*
323221	WMG	POR	SO	64021	75321	65021	*Northfield*
323222	WMG	POR	SO	64022	75322	65022	*Redditch*
323240	WMG	POR	SO	64040	75340	65040	*Erdington*
323241	WMG	POR	SO	64041	75341	65041	*Dave Pomroy 323 Fleet Engineer 40 Years Service*
323242	WMG	POR	SO	64042	75342	65042	*Alvechurch*
323243	WMG	POR	SO	64043	75343	65043	*Longbridge*

Class 350

The mainstay of the LNWR fleet, which work most trains from Euston to Birmingham NS and Birmingham NS to Liverpool plus local services in the West Midlands. They sometimes work alongside 323s on the Chase line.

All are in the EJHQ pool.

Class 350/0

Units originally ordered by the Strategic Rail Authority but transferred to London Midland.

	Livery	Owner	Depot	DMC	TC	PTS	DMS
350101	LNW	ANG	NN	63761	66811	66861	63711
350102	LNW	ANG	NN	63762	66812	66862	63712
350103	LNW	ANG	NN	63763	66813	66863	63713
350104	LNW	ANG	NN	63764	66814	66864	63714
350105	LNW	ANG	NN	63765	66815	66865	63715
350106	LNW	ANG	NN	63766	66816	66866	63716
350107	LNW	ANG	NN	63767	66817	66867	63717
350108	ADV	ANG	NN	63768	66818	66868	63718
350109	LNW	ANG	NN	63769	66819	66869	63719
350110	LNW	ANG	NN	63770	66820	66870	63720
350111	LNW	ANG	NN	63771	66821	66871	63721
350112	LNW	ANG	NN	63772	66822	66872	63722
350113	LNW	ANG	NN	63773	66823	66873	63723
350114	LNW	ANG	NN	63774	66824	66874	63724
350115	LNW	ANG	NN	63775	66825	66875	63725
350116	LNW	ANG	NN	63776	66826	66876	63726
350117	LNW	ANG	NN	63777	66827	66877	63727
350118	LNW	ANG	NN	63778	66828	66878	63728
350119	LNW	ANG	NN	63779	66829	66879	63729

350120	LNW	ANG	NN	63780	66830	66880	63730
350121	LNW	ANG	NN	63781	66831	66881	63731
350122	LNW	ANG	NN	63782	66832	66882	63732
350123	LNW	ANG	NN	63783	66833	66883	63733
350124	LNW	ANG	NN	63784	66834	66884	63734
350125	LNW	ANG	NN	63785	66835	66885	63735
350126	LNW	ANG	NN	63786	66836	66886	63736
350127	LNW	ANG	NN	63787	66837	66887	63737
350128	LNW	ANG	NN	63788	66838	66888	63738
350129	LNW	ANG	NN	63789	66839	66889	63739
350130	LMR	ANG	NN	63790	66840	66890	63740

350108 has anti-trespass rail safety livery on 63768.

Class 350/2

The units were due to return to their ROSCOs, but this has been delayed.

	Livery	Owner	Depot	DMC	TC	PTS	DMS
350231	LNW	POR	NN	61431	65231	67531	61531
350232	LNW	POR	NN	61432	65232	67532	61532
350233	LMR	POR	NN	61433	65233	67533	61533
350234	LNW	POR	NN	61434	65234	67534	61534
350235	ADV	POR	NN	61435	65235	67535	61535
350236	LMR	POR	NN	61436	65236	67536	61536
350237	LMR	POR	NN	61437	65237	67537	61537
350238	LMR	POR	NN	61438	65238	67538	61538
350239	LNW	POR	NN	61439	65239	67539	61539
350240	LNW	POR	NN	61440	65240	67540	61540
350241	LMR	POR	NN	61441	65241	67541	61541
350242	LMR	POR	NN	61442	65242	67542	61542
350243	LMR	POR	NN	61443	65243	67543	61543
350244	LNW	POR	NN	61444	65244	67544	61544
350245	LNW	POR	NN	61445	65245	67545	61545
350246	LMR	POR	NN	61446	65246	67546	61546
350247	LMR	POR	NN	61447	65247	67547	61547
350248	LMR	POR	NN	61448	65248	67548	61548
350249	LMR	POR	NN	61449	65249	67549	61549
350250	LMR	POR	NN	61450	65250	67550	61550
350251	LMR	POR	NN	61451	65251	67551	61551
350252	LNW	POR	NN	61452	65252	67552	61552
350253	LNW	POR	NN	61453	65253	67553	61553
350254	LNW	POR	NN	61454	65254	67554	61554
350255	LMR	POR	NN	61455	65255	67555	61555
350256	LMR	POR	NN	61456	65256	67556	61556
350257	LNW	POR	NN	61457	65257	67557	61557
350258	LNW	POR	NN	61458	65258	67558	61558
350259	LNW	POR	NN	61459	65259	67559	61559
350260	LMR	POR	NN	61460	65260	67560	61560
350261	LMR	POR	NN	61461	65261	67561	61561
350262	LNW	POR	NN	61462	65262	67562	61562
350263	LNW	POR	NN	61463	65263	67563	61563
350264	LMR	POR	NN	61464	65264	67564	61564
350265	LMR	POR	NN	61465	65265	67565	61565
350266	LMR	POR	NN	61466	65266	67566	61566
350267	LNW	POR	NN	61467	65267	67567	61567

Note: 350235 is in LMR livery apart from vehicle 65235, which is in a BTP promotional livery.

The LNW operation from Euston to Birmingham and Liverpool is currently worked by a fleet of 87 four-car Class 350s, usually running as eight- or 12-car sets. On 11 September 2021, 350369 and 350255 pass South Kenton with the 1314 Birmingham NS-Euston. The lead unit is in the current livery, the rear unit is still in the old London Midland livery. *Paul Shannon*

Class 350/3

	Livery	Owner	Depot	DMC	TC	PTS	DMS
350368	LNW	ANG	NN	60141	60511	60651	60151
350369	LNW	ANG	NN	60142	60512	60652	60152
350370	LNW	ANG	NN	60143	60513	60653	60153
350371	LNW	ANG	NN	60144	60514	60654	60154
350372	LNW	ANG	NN	60145	60515	60655	60155
350373	LNW	ANG	NN	60146	60516	60656	60156
350374	LNW	ANG	NN	60147	60517	60657	60157
350375	LNW	ANG	NN	60148	60518	60658	60158
350376	LNW	ANG	NN	60149	60519	60659	60159
350377	LNW	ANG	NN	60150	60520	60660	60160

Class 350/4

Ex-Transpennine Express units.

	Livery	Owner	Depot	DMC	TC	PTS	DMS
350401	LNW	ANG	NN	60691	60901	60941	60671
350402	LNW	ANG	NN	60692	60902	60942	60672
350403	LNW	ANG	NN	60693	60903	60943	60673
350404	LNW	ANG	NN	60694	60904	60944	60674
350405	LNW	ANG	NN	60695	60905	60945	60675
350406	LNW	ANG	NN	60696	60906	60946	60676
350407	LNW	ANG	NN	60697	60907	60947	60677
350408	LNW	ANG	NN	60698	60908	60948	60678
350409	LNW	ANG	NN	60699	60909	60949	60679
350410	LNW	ANG	NN	60700	60910	60950	60680

Class 730

These Bombardier Aventra units are now being built, and the first units are being tested. The three-car Class 730/0s will work on the Cross-City line and also from Wolverhampton to Birmingham NS and Walsall.

A fleet of 81 Class 730 EMUs is being constructed for WMT; 36 three-cars for West Midlands Railway while LNW will have 29 four-car and 16 five-car sets. The first units are currently undergoing testing. On 26 January 2022, 730005 passes Great Bridgeford. *Jack Boskett*

Class 730/0

Three-car units.

	Livery	Owner	Depot	DMS	PMS	DMS
730001	WMG	COR	OD (Q)	490001	491001	492001
730002	WMG	COR	SO (Q)	490002	491002	492002
730003	WMG	COR	OY (Q)	490003	491003	492003
730004	WMG	COR	OD (Q)	490004	491004	492004
730005	WMG	COR	OY (Q)	490005	491005	492005
730006	WMG	COR	OY (Q)	490006	491006	492006
730007	WMG	COR		490007	491007	492007
730008	WMG	COR		490008	491008	492008
730009	WMG	COR		490009	491009	492009
730010	WMG	COR		490010	491010	492010
730011	WMG	COR		490011	491011	492011
730012	WMG	COR		490012	491012	492012
730013	WMG	COR		490013	491013	492013
730014	WMG	COR		490014	491014	492014
730015	WMG	COR		490015	491015	492015
730016	WMG	COR		490016	491016	492016
730017	WMG	COR		490017	491017	492017
730018	WMG	COR		490018	491018	492018
730019	WMG	COR		490019	491019	492019
730020	WMG	COR		490020	491020	492020
730021	WMG	COR		490021	491021	492021
730022	WMG	COR		490022	491022	492022
730023	WMG	COR		490023	491023	492023
730024	WMG	COR		490024	491024	492024
730025	WMG	COR		490025	491025	492025
730026	WMG	COR		490026	491026	492026
730027	WMG	COR		490027	491027	492027
730028	WMG	COR		490028	491028	492028
730029	WMG	COR		490029	491029	492029
730030	WMG	COR		490030	491030	492030
730031	WMG	COR		490031	491031	492031
730032	WMG	COR		490032	491032	492032
730033	WMG	COR		490033	491033	492033
730034	WMG	COR		490034	491034	492034
730035	WMG	COR		490035	491035	492035
730036	WMG	COR		490036	491036	492036

Class 730/1

These five-car units will be used for the outer-suburban LNWR operations from Euston to Birmingham NS.

	Livery	Owner	Depot	DMS	MS	PMS	MS	DMS
730101	LNW	COR	OY (Q)	490101	491101	492101	493101	494101
730102	LNW	COR	OD (Q)	490102	491102	492102	493102	494102
730103	LNW	COR	OY (Q)	490103	491103	492103	493103	494103
730104	LNW	COR		490104	491104	492104	493104	494104
730105	LNW	COR		490105	491105	492105	493105	494105
730106	LNW	COR		490106	491106	492106	493106	494106
730107	LNW	COR		490107	491107	492107	493107	494107
730108	LNW	COR		490108	491108	492108	493108	494108
730109	LNW	COR		490109	491109	492109	493109	494109
730110	LNW	COR		490110	491110	492110	493110	494110
730111	LNW	COR		490111	491111	492111	493111	494111
730112	LNW	COR		490112	491112	492112	493112	494112
730113	LNW	COR		490113	491113	492113	493113	494113
730114	LNW	COR		490114	491114	492114	493114	494114
730115	LNW	COR		490115	491115	492115	493115	494115
730116	LNW	COR		490116	491116	492116	493116	494116
730117	LNW	COR		490117	491117	492117	493117	494117
730118	LNW	COR		490118	491118	492118	493118	494118
730119	LNW	COR		490119	491119	492119	493119	494119
730120	LNW	COR		490120	491120	492120	493120	494120
730121	LNW	COR		490121	491121	492121	493121	494121
730122	LNW	COR		490122	491122	492122	493122	494122
730123	LNW	COR		490123	491123	492123	493123	494123
730124	LNW	COR		490124	491124	492124	493124	494124
730125	LNW	COR		490125	491125	492125	493125	494125
730126	LNW	COR		490126	491126	492126	493126	494126
730127	LNW	COR		490127	491127	492127	493127	494127
730128	LNW	COR		490128	491128	492128	493128	494128
730129	LNW	COR		490129	491129	492129	493129	494129

Class 730/2

These five-car units will be used for the longer-distance LNWR operations from Euston to Crewe.

	Livery	Owner	Depot	DMS	MS	PMS	MS	DMS
730201	LNW	COR		490201	491201	492201	493201	494201
730202	LNW	COR		490202	491202	492202	493202	494202
730203	LNW	COR		490203	491203	492203	493203	494203
730204	LNW	COR		490204	491204	492204	493204	494204
730205	LNW	COR		490205	491205	492205	493205	494205
730206	LNW	COR		490206	491206	492206	493206	494206
730207	LNW	COR		490207	491207	492207	493207	494207
730208	LNW	COR		490208	491208	492208	493208	494208
730209	LNW	COR		490209	491209	492209	493209	494209
730210	LNW	COR		490210	491210	492210	493210	494210
730211	LNW	COR		490211	491211	492211	493211	494211
730212	LNW	COR		490212	491212	492212	493212	494212
730213	LNW	COR		490213	491213	492213	493213	494213
730214	LNW	COR		490214	491214	492214	493214	494214
730215	LNW	COR		490215	491215	492215	493215	494215
730216	LNW	COR		490216	491216	492216	493216	494216

Open Access Operators Overview

When the railways were privatised in the mid-1990s, the existing passenger operations – which had already been split into three sectors – InterCity, Regional Railways and Network SouthEast, were split into 25 – mostly regional – franchises, which were then let for pre-defined periods.

Part of the cornerstone of privatisation, however, was to allow entrepreneurs and free enterprise to run their own operations. This meant – theoretically – anyone could have access to the railway to operate their own trains, providing they meet certain criteria.

This has led to a number of what are called Open Access Operators – privately funded companies who operate their own trains, take their own risk and enjoy their own profits. It's not as simple as that, of course, but that is the theory.

For a start, OA operators are not allowed to stop at certain stations if they will seriously impinge on the revenue of the established franchised TOCs, but likewise they can exploit what is known as ORCATs raiding, where they get a percentage of all fares on certain types of ticket from certain stations on certain routes regardless of whether passengers actually use their trains.

Open Access successes

The first OA operator was Heathrow Express, which started operating on the Paddington to Heathrow Airport line in 1998 and has been doing so ever since. It ordered new trains – interestingly, since scrapped and replaced by EMUs it hires from GWR. HEx has also been incredibly expensive, and Paddington is not the best station for connectivity but the service has been a success and continues to be so.

In September 2000, Hull Trains, set up by Renaissance Railways, part funded by GB Railways, started its Hull-King's Cross operation running with hired in Anglia Railways three-car Class 170 Turbostars for three return trains a day. Now own by First Group, Hull Trains has grown and, after periods using new-build Class 222/1s and then Class 180 Adelante units, now uses brand new five-car Class 802/3 Hitachi AT300 bi-mode units, of the same design as the Azuma sets use by LNER.

In December 2007 Grand Central started its Sunderland to King's Cross operation and expanded to start running from Bradford in 2010. It started operations using HSTs, then added Class 180 DMUs to its fleet for the Bradford trains. It has since dispensed with its HSTs, which moved to East Midlands Trains and have since been sold and now has a fleet of ten Class 180s.

GC was looking to start a Blackpool North to Euston service using ex-LNER Mk 4 coaches and hired DB Cargo Class 90s, but the plan was shelved due to the pandemic.

However, on the East Coast a third OA operator did indeed launch in the pandemic; and in October 2021 First East Coast – under the brand name Lumo – launched, using standard-class only Class 803 Hitachi AT300 EMUs. Lumo is aiming to offer a cheaper alternative to flying between Edinburgh and King's Cross. However, it can only stop its trains at Morpeth, Newcastle and Stevenage.

Open Access failures

At the same time GC was granted operating rights, Wrexham, Shropshire & Marylebone Railway started operations in April 2008 from Wrexham to Marylebone via Shrewsbury. It saw a gap in the market for Shrewsbury to London through trains, a service Virgin had ended. However, Virgin complained and the WSMR operation was not allowed to stop at Birmingham New Street or Coventry, so instead called at Tame Bridge Parkway and Birmingham International.

WSMR, which was another Renaissance Railways venture, was sold totally to Deutsche Bahn in 2009 and used DB Schenker Class 67s and Mk 3 stock, with DVTs. When the WSMR service ended in January 2011 having failed to meet its business case expectations, the locos and stock transferred to Chiltern Railways, which was also own by DB Regio.

Grand Central

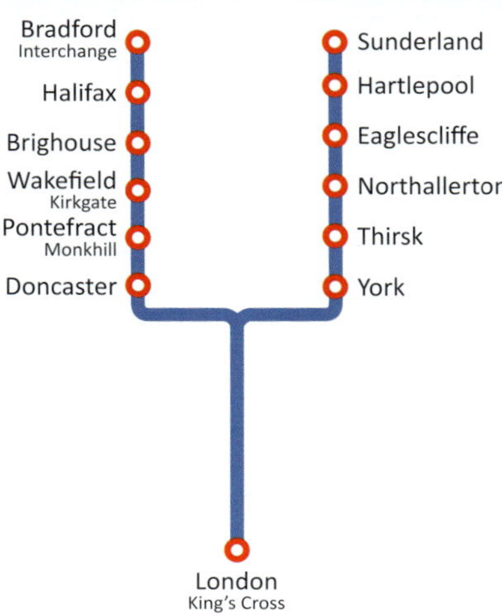

Contact details

Website: www.grandcentralrail.com
Twitter: @GC_Rail

Key personnel

Managing Director: Sean English

Overview

Grand Central's first route was launched in December 2007 from King's Cross to Sunderland, calling at York, Thirsk, Northallerton, Eaglescliffe and Hartlepool.

In May 2010, a new service from Bradford Interchange to King's Cross was launched, serving Halifax, Brighouse, Wakefield, Pontefract and Doncaster, with stops at Mirfield added in 2011 and Low Moor in 2017.

Despite running 'under the wires' for the majority of its routes, diesel trains are used because the section from Northallerton to Sunderland and from Bradford to Hare Park Junction are not electrified.

After starting its operations with three HSTs, expansion saw GC acquire ex-FGW Class 180s to run new services. It has since relinquished its HSTs to concentrate on a wholly-Class 180 fleet. All the trains received a full interior and exterior refurbishment in 2018-19.

Class 180

Ten five-car Class 180s DMUs, built by Alstom in 2000-01, form the Grand Central fleet for its ECML operations. All are based at Heaton. They work from Sunderland and Bradford to King's Cross.

DMSL 59112 from unit 180112 has been converted to run on duel fuel – a combination of diesel and liquified natural gas. It has a green bodyside stripe instead of the normal orange.

All are in the ECGC pool.

	Livery	Owner	Depot	DMSL	MFL	MSL	MSLRB	DMSL	
180101	GCR	ANG	HT	50901	54901	55901	56901	59901	
180102	GCR	ANG	HT	50902	54902	55902	56902	59902	
180103	GCR	ANG	HT	50903	54903	55903	56903	59903	
180104	GCR	ANG	HT	50904	54904	55904	56904	59904	
180105	GCR	ANG	HT	50905	54905	55905	56905	59905	*The Yorkshire Artist Ashley Jackson*
180106	GCR	ANG	HT	50906	54906	55906	56906	59906	
180107	GCR	ANG	HT	50907	54907	55907	56907	59907	*Hart of the North*
180108	GCR	ANG	HT	50908	54908	55908	56908	59908	*William Shakespeare*
180112	GCR	ANG	HT	50912	54912	55912	56912	59912	*James Herriot Celebrating 100 years 1916-2016*
180114	GCR	ANG	HT	50914	54914	55914	56914	59914	*Kirkgate Calling*

On 25 November 2021, 180114 waits to leave with the 1457 King's Cross-Bradford Interchange. There are ten of these Adelante units in the GC fleet and they replaced HSTs as well as allowing more services. *Pip Dunn*

Heathrow Express

Contact details

Website: www.heathrowexpress.com
Twitter: @HeathrowExpress

Key personnel

Managing Director: Sophie Chapman

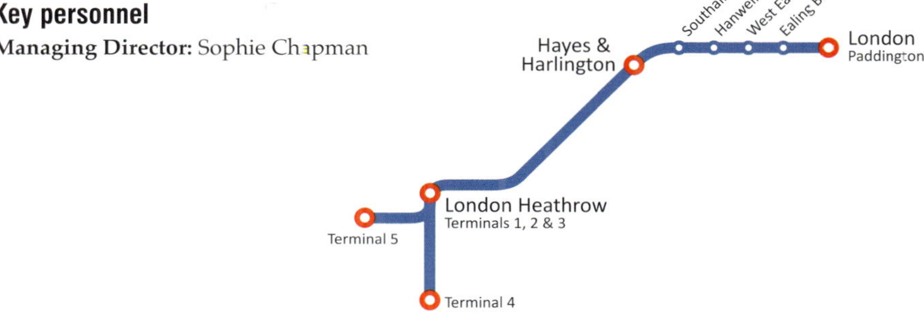

Overview

The first Open Access operator, launched in 1998, HEx operates a regular non-stop Paddington to Heathrow Airport shuttle. It later introduced the Heathrow Connect, which stopped at some intermediate stations on the GWML from Airport Junction to Paddington.

It started with a fleet of 14 Class 334 four-car units, five of which were increased to five-cars. It then acquired five four-car Class 360/2 Siemens Desiro units for Connect service.

In 2021 the 332s and 360s were dispensed with, the former going for scrap and the latter initially going into store before being taken on by ROG.

A dozen GWR Class 387/1s have now been hired and modified for use on HEx, which operationally will now be managed by GWR. The trains, 387130-141, have been reliveried in HEx colours and have been modified internally to include first-class accommodation (in the DMS(A)), higher-speed wi-fi and have additional luggage racks as well as on-board entertainment.

	Livery	Owner	Depot	DMS (A)	MS	PTS	DMS (B)	
387130	HEX	POR	RG	421130	422130	423130	424130	
387131	HEX	POR	RG	421131	422131	423131	424131	
387132	HEX	POR	RG	421132	422132	423132	424132	
387133	HEX	POR	RG	421133	422133	423133	424133	*Tokyo*
387134	HEX	POR	RG	421134	422134	423134	424134	*Barcelona*
387135	HEX	POR	RG	421135	422135	423135	424135	*Rome*
387136	HEX	POR	RG	421136	422136	423136	424136	*Paris*
387137	HEX	POR	RG	421137	422137	423137	424137	
387138	HEX	POR	RG	421138	422138	423138	424138	
387139	HEX	POR	RG	421139	422139	423139	424139	*Dublin*
387140	HEX	POR	RG	421140	422140	423140	424140	*London*
387141	HEX	POR	RG	421141	422141	423141	424141	

Hull Trains

Contact details

Website: www.hulltrains.co.uk
Twitter: @Hull_Trains

Key personnel

Managing Director: David Gibson
Production Director: Louise Mendham

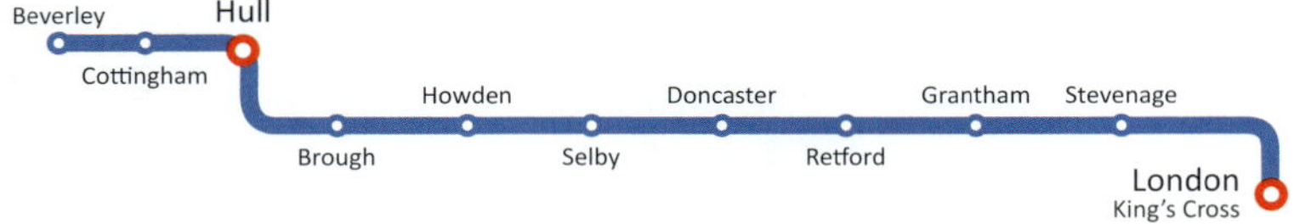

Overview

Hull Trains was the second Open Access operator and has been running trains between Hull and King's Cross since September 2000. It initially used hired Anglia Railways Class 170 Turbostars before acquiring its own four three-car Class 170/3s in 2004. These were replaced by four four-car Class 222/1s in 2005.

These were moved to East Midlands Trains in 2008-09 and replaced by four five-car Class 180 Adelante units. These have recently been replaced by five brand new five-car Class 802/3 bi-mode Hitachi AT300 multiple units.

HT has used a hired GWR HST as cover in previous years, and also a Class 86 and Mk 3 coaches for Doncaster to King's Cross workings when short of units. Some trains start at Beverley.

Class 802

All are in the PFHQ pool.

	Livery	Owner	Depot	PDTS	MS	MS	MC	PDTF
802301	HUL	ANG	DN	831301	832301	833301	834301	835301
802302	HUL	ANG	DN	831302	832302	833302	834302	835302
802303	HUL	ANG	DN	831303	832303	833303	834303	835303
802304	HUL	ANG	DN	831304	832304	833304	834304	835304
802305	HUL	ANG	DN	831305	832305	833305	834305	835305

Hull Trains has steadily built up its customer base and now runs brand new trains with five five-car Bi-Mode Class 802/3 units. On 28 September 2021, 802303 approaches Hull with the 0948 from King's Cross. *Robin Ralston*

Lumo

Contact details

Website: www.firstgroupplc.com

Key personnel

Managing Director: TBA

Overview

Trading as Lumo, East Coast Trains is a new Open Access operation run by First Group on the Edinburgh-King's Cross ECML that started in October 2021. While it competes with LNER across the whole route, plus also Transpennine Express and CrossCountry trains on the section north of Newcastle, its aim is to offer an alternative to internal air journeys from Scotland to London.

Its five trains a day call at Morpeth, Newcastle and Stevenage. ECT says the 'average' fare between Edinburgh and King's Cross is just £25.

The Lumo operation uses five brand new five-car Hitachi AT300 Class 803 EMUs. The units were assembled at Newton Aycliffe using bodyshells shipped from Japan and, unlike the LNER BMMUs, are electric only and do not have diesel engines, but batteries will keep the train supply live in the event of a failure. They only have standard-class seating.

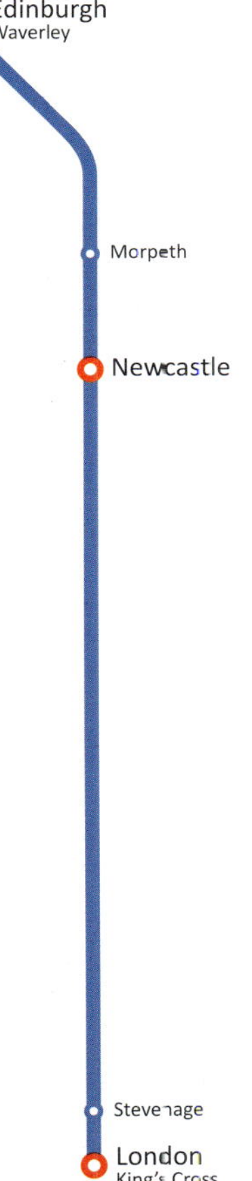

Class 803

	Livery	Owner	Depot	PDTS	MS	MS	MC	PDTF
803001	LUM	BEA	EC	841001	842001	843001	844001	845001
803002	LUM	BEA	EC	841002	842002	843002	844002	845002
803003	LUM	BEA	EC	841003	842003	843003	844003	845003
803004	LUM	BEA	EC	841004	842004	843004	844004	845004
803005	LUM	BEA	EC	841005	842005	843005	844005	845005

Lumo is one of the few Open Access Operators to start its service with brand new trains. It ordered five five-car Class 803/0 EMUs, and on 25 November 2021, 803001 waits to set off with the 1436 King's Cross-Edinburgh. *Pip Dunn*

Channel Tunnel Operations Overview

The Channel Tunnel opened in 1994 linking Cheriton in the UK with Coquelles in France and is the only fixed connection between the UK and mainland Europe. It is over 31 miles long and features three tunnels, two for trains and a service tunnel in between.

It is owned and managed by Getlink – which retains the Eurotunnel trading name – and they also operate the lorry-carrying and car-carrying shuttle trains. International passenger trains using the tunnel are operated by Eurostar.

Other operators, such as Deutsche Bahn and more recently Renfe, have expressed interest in operating through the Channel Tunnel and ran a promotional train to St Pancras in 2010, but as yet no operations are planned.

Eurostar

Contact details

Website: www.eurostar.com
Twitter: @EurostarUK

Key personnel

Chief Executive Officer: Jacques Damas

Overview

Eurostar is – currently – the only operator of long-distance passenger trains through the Channel Tunnel and operates out of St Pancras International in London via the HS1 High Speed line to the Tunnel and then on into France, Belgium and now the Netherlands. Trains run to Paris Gard du Nord, Bruxelles Midi and now Amsterdam Centraal.

The fleet is made up of a mix of older refurbished Class 373 units, which date from the mid-1990s, and new Class 374 Siemens Velaro e320 units delivered in 2012-17.

The 373s were owned by Eurostar, UK or SNCF but used indiscriminately. However, some 373/3s were used for French domestic services but have since been withdrawn and scrapped. One Class 373 is in use with SNCF majority-owned operator Thalys and used for that company's 'Izy' low-cost service between Paris and Brussels, while other sets have been sent for scrap at EMR in Kingsbury.

Each 373 runs with a driving power car at the head of a rake of eight vehicles all coupled via articulated bogies, and they run in pairs, essentially permanently coupled, to create a 20-coach train; a power car then five standard-class coaches, a buffet and three first-class coaches in one half of the train and the same in reverse in the second half so each train has two buffets, six first and ten standard-class coaches between the two power cars. There is a spare power car, 3999, which can be used to replace any long-term defective sister vehicle.

Eurostar and Thalys merged in 2021 forming a new company still branded Eurostar both companies are majority owned by French State Railways SNCF with part shares owned by Belgium's SNCB; Eurostar also has private shareholders, while Dutch operator NS is a part shareholder in Thalys.

Class 08

Sole shunter based at Temple Mills. It is in the GPSS pool.

08948	EUK	EUK	TI

Class 373

These units only carry the last four digits of their numbers on the power car cabside; they carry the full European Vehicle Number on each vehicle. They are based at Temple Mills depot in East London and Le Landy in Paris. Most of these sets were mothballed during the pandemic but have since been returned to traffic. These sets have been refurbished and any original unrefurbished units have been disposed of.

3007	ESB	EUK	TI
3008	ESB	EUK	TI
3015	ESB	EUK	TI
3016	ESB	EUK	TI
3205	ESB	SNC	LY
3206	ESB	SNC	LY
3209	ESB	SNC	LY
3210	ESB	SNC	LY
3211	ESB	SNC	LY
3212	ESB	SNC	LY
3213	LZY	SNC	LY
3214	ESO	SNC	LY (S)
3215	ESO	SNC	TI (U)
3216	ESO	SNC	TI (U)
3217	ESO	SNC	TI (U)
3218	ESO	SNC	TI (U)
3219	ESB	SNC	LY
3220	ESB	SNC	LY
3221	ESB	SNC	LY
3222	ESB	SNC	LY
3223	ESO	SNC	LY (S)
3224	LZY	SNC	LY
3229	ESB	SNC	LY
3230	ESB	SNC	LY
9999	ESB	SNC	LY

Opposite: One of the original Class 373 Eurostar sets, led by power car 3212, rises along the gradient through to Stratford International Station with a London-bound train on 4 December 2015. *Jack Boskett*

Class 374

New Siemens Velaro e320 trains delivered in 2012-17 to replace unrefurbished Class 373s and provide new capacity. They have distributed power as opposed to power cars.

4001	ESB	EUK	TI
4002	ESB	EUK	TI
4003	ESB	EUK	TI
4004	ESB	EUK	TI
4005	ESB	EUK	TI
4006	ESB	EUK	TI
4007	ESB	EUK	TI
4008	ESB	EUK	TI
4009	ESB	EUK	TI
4010	ESB	EUK	TI
4011	ESB	EUK	TI
4012	ESB	EUK	TI
4013	ESB	EUK	TI
4014	ESB	EUK	TI
4015	ESB	EUK	TI
4016	ESB	EUK	TI
4017	ESB	EUK	TI
4018	ESB	EUK	TI
4019	ESB	EUK	TI
4020	ESB	EUK	TI
4021	ESB	EUK	TI
4022	ESB	EUK	TI
4023	ESB	EUK	TI
4024	ESB	EUK	TI
4025	ESB	EUK	TI
4026	ESB	EUK	TI
4027	ESB	EUK	TI
4028	ESB	EUK	TI
4029	ESB	EUK	TI
4030	ESB	EUK	TI
4031	ESB	EUK	TI
4032	ESB	EUK	TI
4033	ESB	EUK	TI
4034	ESB	EUK	TI

Barrier Vehicles

96380	86386, 6380	EUK	Mk 1
96381	86187, 6381	EUK	Mk 1
96383	86664, 6383	EUK	Mk 1
96384	86955, 6384	EUK	Mk 1

Class 374 Eurostar sets wait in the platforms at St Pancras having arrived from France on 9 September 2021. *Jack Boskett*

Eurotunnel

Contact details

Website: www.eurotunnel.com
Twitter: @LeShuttle

Key personnel

CEO: Jacques Gounon

Overview

The Channel Tunnel is marketed as Eurotunnel but is managed and operated by Getlink (Groupe Eurotunnel – link) and operates tourist shuttles – for cars, caravans, and motorhomes – and freight shuttles for trucks and large vans. The latter are naturally heavier so they use dedicated freight locos.

Each freight shuttle uses a loco, a club car, two 16-wagon rakes – to give 32 wagons, a club car and a loco – and are 745m long. The first vehicle will drive on from a platform via a ramp and then drive the length of the empty train so it will be the first off at the other end. Subsequent vehicles follow it and the last one off will have to drive the length of the train to disembark, and will be the last truck off the train. The trains run at a maximum of 140kmh (87mph), whereas Eurostar trains can run at 160kmh (100mph) in the Tunnel.

Shuttle locos

The shuttle locos were built by Brush from 1993-2003. The initial fleet was 38 7,500hp locos built in 1992-94, for both freight shuttles (trucks) and tourist shuttles (cars). Loco 9030 was written off in the 1996 Tunnel fire and 9040 was built as a replacement for it in 1998.

9101-113 were built in 1998-2001 as dedicated freight locos at 7,500hp supplemented by 9701-07 in 2001/02 at 9,400hp. The 91xx locos were then uprated to 9,400hp and renumbered as 97xx locos. Between 2004-12, 24 of the original locos from the 90xx series were uprated to 9,400hp power output and renumbered as 98xx, the last two digits remaining the same.

90xx locos

Original Bo-Bo-Bo locos built by Brush 1992-94, many have now been upgraded to higher power for use of freight shuttles. The remaining original locos tend to work tourist shuttles only.

No.	Livery	Owner	Depot	
9005	EUT	EUT	CO	*Jessye Norman*
9007	EUT	EUT	CO	*Dame Joan Sutherland*
9011	EUT	EUT	CO	*José Van Dam*
9013	EUT	EUT	CO	*Maria Callas*
9015	EUT	EUT	CO	*Lötschberg 1913*
9018	EUT	EUT	CO	*Wilhelmenia Fernandez*
9022	EUT	EUT	CO	*Dame Janet Baker*
9024	EUT	EUT	CO	*Gotthard 1882*
9026	EUT	EUT	CO	*Furkatunnel 1982*
9029	EUT	EUT	CO	*Thomas Allen*
9033	EUT	EUT	CO	*Monserrat Caballe*
9036	EUT	EUT	CO	*Alain Foundary*
9037	EUT	EUT	CO	

97xx locos

Dedicated freight locos, these 9,400hp locos – of which 9711-13 were originally 7,500hp – work alongside 98xx locos.

No.	Old No.	Livery	Owner	Depot
9701		EUT	EUT	CO
9702		EUT	EUT	CO
9703		EUT	EUT	CO
9704		EUT	EUT	CO
9705		EUT	EUT	CO
9706		EUT	EUT	CO
9707		EUT	EUT	CO
9711	9101	EUT	EUT	CO
9712	9102	EUT	EUT	CO
9713	9103	EUT	EUT	CO
9714	9104	EUT	EUT	CO
9715	9105	EUT	EUT	CO
9716	9106	EUT	EUT	CO
9717	9107	EUT	EUT	CO
9718	9108	EUT	EUT	CO
9719	9109	EUT	EUT	CO
9720	9110	EUT	EUT	CO
9721	9111	EUT	EUT	CO
9722	9112	EUT	EUT	CO
9723	9113	EUT	EUT	CO

98xx locos

These were originally 90xx locos but rebuilt with higher power for working freight shuttles.

No.	Old No.	Livery	Owner	Depot	
9801	9001	EUT	EUT	CO	*Lesley Garrett*
9802	9002	EUT	EUT	CO	*Stuart Burrows*
9803	9003	EUT	EUT	CO	*Benjamin Luxon*
9804	9004	EUT	EUT	CO	
9806	9006	EUT	EUT	CO	*Regine Crespin*
9808	9008	EUT	EUT	CO	*Elisabeth Soderstrom*
9809	9009	EUT	EUT	CO	
9810	9010	EUT	EUT	CO	
9812	9012	EUT	EUT	CO	
9814	9014	EUT	EUT	CO	*Lucia Popp*
9816	9016	EUT	EUT	CO	
9819	9019	EUT	EUT	CO	*Maria Ewing*
9820	9020	EUT	EUT	CO	*Nicolai Ghiarov*
9821	9021	EUT	EUT	CO	
9823	9023	EUT	EUT	CO	*Dame Elisabeth Legge-Schwarzkoff*
9825	9025	EUT	EUT	CO	
9827	9027	EUT	EUT	CO	*Barbara Hendricks*
9828	9028	EUT	EUT	CO	
9831	9031	EUT	EUT	CO	
9832	9032	EUT	EUT	CO	*Renata Tebaldi*
9834	9034	EUT	EUT	CO	*Mirella Freni*
9835	9035	EUT	EUT	CO	*Nocolai Gedda*
9838	9038	EUT	EUT	CO	*Hildegard Behrens*
9840	9040	EUT	EUT	CO	

Recovery and maintenance locos

Getlink has three bespoke fleets of locos used for recovering failed trains or working maintenance trains. The MaK (Maschinenbau Kiel) Bo-Bo locos can run on the HS1 route through to St Pancras if required, although only 0001-0005 have TVM signalling, meaning any of the 0006-0010 fleet can only be used if in multiple with one of the earlier locos.

Because of their ability to appear on HS1, these ten locos have TOPS numbers in the Class 21/9 series. The MaK Bo-Bos, built in 1991-92, are 1,275hp locos and used to operate any maintenance trains plus act as Thunderbirds to recover any stricken trains.

TOPS no.	ET No.	Livery	Owner	Depot
21901	0001	EUY	EUT	CT
21902	0002	EUY	EUT	CT
21903	0003	EUY	EUT	CT
21904	0004	EUY	EUT	CT
21905	0005	EUY	EUT	CT

MaK Bo-Bo locos built in 1990-91 for Dutch operator NS but rebuilt for Channel Tunnel use in 2011 or 2016, these 1,580hp locos are also used to operate any maintenance trains plus act as Thunderbirds to recover any stricken trains. They also act as shunters if required.

TOPS no.	ET No.	Old NS no.	Livery	Owner	Depot
21906	0006	6456	EUY	EUT	CT
21907	0007	6457	EUY	EUT	CT
21908	0008	6450	EUY	EUT	CT
21909	0009	6451	EUY	EUT	CT
21910	0010	6447	EUY	EUT	CT

These small 230hp 0-4-0 locos were built by Hunslet Engine Company in Leeds in 1990 at 900mm gauge but in 1993-94 were rebuilt by Schöma in Germany to standard gauge. They are used as shunters and maintenance inspection locos.

ET no.	Livery	Owner	Depot	Name
0031	EUY	EUT	CT	*Frances*
0032	EUY	EUT	CT	*Elisabeth*
0033	EUY	EUT	CT	*Silke*
0034	EUY	EUT	CT	*Amanda*
0035	EUY	EUT	CT	*Mary*
0036	EUY	EUT	CT	*Laurence*
0037	EUY	EUT	CT	*Lydie*
0038	EUY	EUT	CT	*Jenny*
0039	EUY	EUT	CT	*Pacita*
0040	EUY	EUT	CT	*Jill*
0041	EUY	EUT	CT	*Kim*
0042	EUY	EUT	CT	*Nicole*

Channel Tunnel freight operators

Freight trains throughout the Channel Tunnel are operated by either DB Cargo or GB Railfreight. Some Class 92s owned by both companies are fitted with TVM430 signalling to allow them to also operate on HS1 and through the Channel Tunnel.

The DBC fleet in the WFBC pool passed for HS1 use are 92011/015/019/036/041/042 (with withdrawn 92009/016/031 also TVM430 compatible), while GBRf has 92010/018/028/038/043 in its GBST pool (with 92023/032/044 also TVM430 compatible). These locos are listed in their relevant owner's sections.

Eurotunnel MaK locos 0002 and 0001 run light from Singlewell to Dollands Moor, and pass Cuxton on 4 June 2021. These locos were built in 1991 and are rated at 1,275hp. *David Staines*

Locomotive Services Limited

Contact details

Website: www.lsltoc.co.uk
Twitter: @CharterRail

Key personnel

Managing Director: Tony Bush

Overview

Locomotive Services Limited is a nationwide Train Operating Company for charter trains and is owned by businessman Jeremy Hosking.

It is one of five TOCs that are capable of operating steam alongside West Coast Railways, DB Cargo, Vintage Trains and SLC Operations, although the latter does not do so.

LSL promotes its own series of charters through its Saphos Trains, Midland Pullman, InterCity and Rail Charter Services banners as well as running the Staycation Express on the Settle and Carlisle line, which was an HST operation in 2021 but has been cacelled for 2022. HSTs also run the Midland Pullman, which has the power cars and Mk 3 trailers repainted into the Blue Pullman livery of the 1960s.

LSL owns several steam, diesel and electric locos as well as two Class 121 single-car DMUs. It has recently expanded its loco fleet by acquiring Class 20s and 57s from DRS.

The year 2022 should hopefully see it return both 45118 and 55022 to the main line again. It owns examples of Classes 20, 37, 45, 47, 55, 73, 86, 87, 90 and HST, also has hire arrangements for Class 40 D213 and has recently added 89001 to its fleet on a hire arrangement.

Included in its steam fleet – either owned or under its custody – are 46100 *Royal Scot*, 70000 *Britannia*, 6024 *King Edward I*, 60532 *Blue Peter*, 34046 *Braunton*, 35027 *Port Line* and 35022 *Holland America Line*, although not all are currently in traffic.

Class 08

08483	LSLS	GRW	LSL	CD	*Bungle*
08631	MBDL	BRW	LSL	WO	*George*
08737		GWS	LSL	SO	
08780	LSLO	GRW	LSL	CD	*Zippy*

Class 20

The Class 20/0s are ex-HNRC locos acquired in 2020, while the two Class 20/3s are ex-DRS locos that joined the fleet in 2021.

	Old No.					
20096		LSLO	GYP	LSL	CD	
20107		LSLO	GYP	LSL	CD	*Jocelyn Feilding 1940-2020*
20302	20084	LSLO	DRC	DRS	CD (S)	
20305	20095	LSLO	DRC	DRS	CD (S)	

Locomotive Services is a nationwide Train Operating Company specialising in charter trains, hauled by steam, diesel or electric traction. One of its four Class 20s, D8107 (20107), passes Dent with the 1438 Appleby-Skipton 'Staycation Express' trip on 12 August 2020. *Tom McAtee*

Class 37

37688 is on long-term hire.

	Old No.					
37521	37117	LSLO	GYP	LSL	CD	
37667	37151	LSLO	GYP	LSL	CD	*Flopsie*
37688	37205	MBDL	TLA	D05	CD	

LSL has three Class 37s in its fleet, one of which is on long-term hire. One of the pair it owns is D6851 *Flopsie* (37667), which was at Llandudno on 20 November 2021. *Pip Dunn*

Class 40

This loco is owned by Shaun Wright and on a long-term hire deal with LSL. The company has also hired preserved 40145 from the Class 40 Preservation Society, but this has returned to Bury.

40013	LSLO	GYP	SHW	CD	*Andania*

Class 43

These ex-EMR HST Power Cars, along with rakes of Mk 3 trailers, are used on the Midland Pullman and Staycation Express turns. Two are painted in Blue Pullman livery and two in Rail Charter Services green, while 43049 is in InterCity Executive colours. All are in the SBXL pool but expected to move into an LSL pool.

43046	BPU	LSL	CD	*Geoff Drury 1930-1999*
43047	EMW	LSL	CD	
43049	ICE	LSL	CD	*Neville Hill*
43055	BPU	LSL	CD	
43058	RCS	LSL	CD	
43059	RCS	LSL	CD	
43083	EMW	LSL	CD	

LSL has taken two full HST sets with a pool of six VP185 ex-EMR Power cars. One set and two power cars have been painted in the Nanking Blue Pullman livery. 43046/055 pass Bruton with the 0636 Peterborough to Kingswear 'Devon Pullman' on 30 August 2021. *Mark Pike*

Class 45

This loco is undergoing contract repairs at Barrow Hill and it is hoped it will return to traffic in the latter part of 2022.

45118	LSLS	BRB	LSL	BH (U)	*Royal Artilleryman*

Class 47

The mainstay of the LSL diesel fleet, 47712/828 are on long-term hire while 47811/816 are spares donors only. The company also owns 47841, which is a static exhibit at Margate (see page 370). 47853 carries its old 47614 number but has not been renumbered on TOPS.

Unless specific traction is requested, Class 47s provide the power for the majority of LSL's diesel-hauled charter operations. This is D1924 *Crewe Diesel Depot* (47810) at Crewe on the rear of a charter to Scarborough hauled by Class 40 D213 *Andania* on 6 October 2018. *Pip Dunn*

47501		LSLO	GYP	LSL	CD	*Craftsman*
47593	47272, 47673, 47790	LSLO	BLL	LSL	CD	*Galloway Princess*
47712	47505	LSLO	SCO	CDP	CD	*Lady Diana Spencer*
47805	47257, 47650	LSLO	GYP	LSL	CD	*Roger Hosking MA 1925-2013*
47810	47247, 47655	LSLO	GYP	LSL	CD	*Crewe Diesel Depot*
47811	47128, 47656	DHLT	FPG	LSL	CD (U)	
47816	47066, 47661	DHLT	FPG	LSL	CD (U)	
47828	47266, 47629	LSLO	ICS	D05	CD	
47853	47141, 47614	LSLO	BRB	LSL	CD	

Class 55

Main-line registered Deltic, the loco is undergoing repairs to return it to use and is expected to be back in service in late 2022. The company also owns D9016 (55016), which is a static exhibit at Margate (see page 371).

55022	LSLS	BRB	LSL	CD (U)	*Royal Scots Grey*

Class 57

Former Virgin Trains and DRS locos bought in 2021-22. All need major expenditure before they can return to use.

57004	47347	XSDP	DRC	LSL	LT (S)
57302	47251, 47589, 47827	XSDP	DRC	LSL	ZG (S)
57311	47032, 47662, 47817	XSDP	DRC	LSL	KM (S)

Class 73

Electro-diesel bought from preservation with a view to be returned to the main line. The company also owns 73002, which is to be restored as a static exhibit at Margate but is currently at Eastleigh Works (see page 371).

	Old No.				
73001	73901	MBED	BRB	LSL	CD (S)

Class 86

Ex-ACLG main-line registered locomotive repainted into InterCity Swallow livery.

	Old No.					
86101	86201	LSLO	ICS	LSL	CD	*Sir William A Stanier FRS*

Class 87

Like 86101, also bought from the ACLG for main-line use and repainted into InterCity colours.

87002	LSLO	ICS	LSL	CD	*Royal Sovereign*

There are examples of Classes 86, 87 and 90s in the LSL AC electric fleet, with 89001 set to join the operation in 2022. On 26 January 2022, 87002 *Royal Sovereign* works north through Rugeley Trent Valley with a Euston-Glasgow Central Intercity railtour. *Jack Boskett*

Class 89

In December 2021 the AC Loco Group announced 89001 was moving to the LSL stable and would be main-line registered and expected to be completed sometime in 2022.

89001	LSLS	ICS	ACL	CD (S)	*Avocet*

Class 90

Two ex-Greater Anglia locos bought from Porterbrook and both repainted into InterCity colours.

90001	LSLO	ICS	LSL	CD	*Royal Scot*
90002	LSLO	ICS	LSL	CD	*Wolf of Badenoch*

Class 121

Ex-Chiltern Railways Single car DMU 55034 was acquired for main-line use as a route learner, although it has not been used as such. 55022 moved to Crewe in February 2022.

55022		BRB	LSL	CD (S)
55034	121034	GYP	LSL	CD

Class 142

				DMS	DMSL
142003	MPT	LSL	ZG	55544	55594
142007	NBL	LSL	ZG	55548	55598
142014	NBL	LSL	ZG	55555	55605

Coaches

Not all these vehicles are currently in traffic.

No.	Old No.	Livery	Type	Name
1203	3291	CAC	Mk 2 RFO	
1211	3305	PUL	Mk 2 RFO	*Car No. 1211*
1658		BRB	Mk 1 RBR	
1659		CAC	Mk 1 RBR	
1679		GRC	Mk 1 RBR	
1696		SRG	Mk 1 RBR	
1863		CHC	Mk 1 RMB	
1883	80021	CAC	Mk 1 Bar	
1954		BRM	Mk 1 RUK	
3045		CHC	Mk 1 FO	
3051		CHC	Mk 1 FO	
3060		CAC	Mk 1 FO	
3091		CHC	Mk 1 FO	
3100		CAC	Mk 1 FO	
3107		CAC	Mk 1 FO	
3112		CAC	Mk 1 FO	
3122		CAC	Mk 1 FO	
3125		CAC	Mk 1 FO	
3140		CAC	Mk 1 FO	
3148		CAC	Mk 1 FO	
3188		PUL	Mk 2 FO	*Cadir Idris*
3223		BCM	Mk 2 FO	*Diamond*
3229		PUL	Mk 2 FO	*Snowdon*
3231		PUL	Mk 2 FO	*Ben Cruachan*
3240		BCM	Mk 2 FO	*Sapphire*
3277		ANG	Mk 2 FO	
3295		ANG	Mk 2 FO	
3312		PUL	Mk 2 FO	*Helvellyn*
3330		CAC	Mk 2 FO	
3344		PUL	Mk 2 FO	*Scafell*
3348		PUL	Mk 2 FO	*Ingleborough*
3384		PUL	Mk 2 FO	*Pen-Y-Ghent*
3426		PUL	Mk 2 FO	*Ben Nevis*
3438		PUL	Mk 2 FO	*Ben Lomond*
5054		CHC	Mk 1 TSO	
5366		CAC	Mk 2 TSO	
5797		ICS	Mk 2 TSO	
5912		CAC	Mk 2 TSO	
5991		CAC	Mk 2 TSO	
6311	92911	CAC	Mk 1 BGGV	
6412	3168	BRM	Mk 2 SO	
6702		SCP	Mk 2 RLO	
6704		SCP	Mk 2 RLO	

Mk 2f FO 3312 *Helvellyn* stands at Peterborough on 17 July 2021 as part of a Potters Bar-Scarborough charter for Statesman rail. *Stuart West*

6705		CAC	Mk 2 RLO	
6706		CAC	Mk 2 RLO	
6707		CAL	Mk 2 RLO	
6708		PUL	Mk 2 RLO	*Mount Helicon*
9804		SCP	Mk 2 BUO	
9809		SCP	Mk 2 BUO	
9479		PUL	Mk 2 BSO	*Car No. 9479*
10411		ICS	Mk 3 TSOB	
10416		ICS	Mk 3 TSOB	
10504		SCP	Mk 3 TSOB	
10513		SCP	Mk 3 TSOB	
10519		SCP	Mk 3 TSOB	
10520		CAC	Mk 3 TSOB	
10648		SCP	Mk 3 TSOB	
10650		SCP	Mk 3 TSOB	
10675		SCP	Mk 3 TSOB	
10683		SCP	Mk 3 TSOB	
10688		SCP	Mk 3 TSOB	
11068		ICS	Mk 3 FO	
11070		ICS	Mk 3 FO	
11075		ANG	Mk 3 FO	
11076		ANG	Mk 3 FO	
11077		ANG	Mk 3 FO	
11087		ANG	Mk 3 FO	
11091		ANG	Mk 3 FO	
11098		ANG	Mk 3 FO	
12111		ANG	Mk 3 TSO	
12171		ICS	Mk 3 TSO	
13227		CAC	Mk 1 FK	
13508		UND	Mk 2 FK	
16204		UND	Mk 1 CK	
17056		PUL	Mk 2 BFK	
17080		PUL	Mk 2 BFK	*Car No. 17080*
17159	14159	CAC	Mk 2 BFK	
21268		BRM	Mk 1 BCK	
35511	17130	BRM	Mk 2 BFK	
80043	1680	PUL	Mk 1 RBR	*Kitchen Car*
80044	1659	CAC	Mk 1 RBR	
82127		ICS	Mk 3 DVT	
82139		ICS	Mk 3 DVT	
93568		BRG	Mk 1 GUV	
94538		BRM	Mk 1 NQA	
99241	35449	BRM	Mk 1 TSO	
99993	5067	CAC	Mk 1 RKF	*Club Car*

LSL has Mk 1 coach 6311 converted for use as a generator car to allow train supply to coaches when not hauled by electric heat or steam heat locos. It passes Craigenhill on the 1312 Carstairs to Glenfinnan special on 5 March 2021. *Robin Ralston*

Support coaches

The following coaches are support vehicles for LSL's steam fleet and now used in the general fleet.

17013	99130	CHC	Mk 1 BFK
35317		CAC	Mk 1 BSK
35333		CHC	Mk 1 BSK
35451		CAC	Mk 1 BSK
35461		CHC	Mk 1 BSK
35465		CAC	Mk 1 BSK
35511	17130	BRM	Mk 1 BFK

Loram

Contact details

Website: www.loram.co.uk
Twitter: @LoramUK

Key personnel

Managing Director: Richard Kelly
International Business Development Director: Andrew Watson

Overview

Loram bought the Derby-based company Rail Vehicle Engineering Ltd (RVEL) in 2016 and now looks after the Network Rail Class 37 and 73 and Infrastructure Monitoring vehicle fleets, and the maintenance of the New Measurement Trains.

It also offers a range of maintenance to other third-party companies, including heavy 'F-' and 'G-exam' overhauls, vehicle life extensions, modifications and routine maintenance. Recent work has been 'F-exams' on DRS Class 37s and 57s and on NR's four Class 97/3s.

The company also owns C2101, a production Rail Grinder, which is contracted for five years to Network Rail. It is also currently involved in major re-power programmes for alternative propulsion for ROSCOs, TOCs and technology OEMs.

Loram – which is an American-owned company – has a freight train operating licence that it uses for moving and operating infrastructure equipment. It often hires Class 20s from Michael Owen.

The company has its own drivers and staff, but works closely with FOCs for drivers on routes or traction it is not familiar with. It has no aspirations to operate passenger or freight trains.

5919	LOM	Mk 2 TSO
6046	LOM	Mk 2 TSO

Loram now has a Train Operating Licence but does not own any locos, and hires in as required. 20189/205 pass Dawlish Warren on 19 August 2020 with the 1118 Derby RTC-Plymouth driver training run. *Tony Christie*

SLC Operations

Contact details

Website: www.slcoperations.com

Key personnel

Managing Director: Cath Bellamy

Overview

A new TOC created in 2020 to assist with driver training, vehicle commissioning and delivering rolling stock between sites. It started operations with a nationwide non-passenger licence but has recently gained a passenger licence and will work some charter trains. It has a base at King's Norton, Birmingham, where it sometimes stables locos it has on hire.

SLC Operations has been set up by three shareholders: Cath Bellamy and Adrian Shooter, ex-Vintage Trains, with Ian Walters, the managing director of SLC Rail limited, a consultancy company with ten years' experience in the UK rail sector. SLC Operations is a wholly owned subsidiary of SLC Rail.

The company will support Mr Shooter's VivaRail business in mileage accumulation on Class 230 and 484 multiple units converted from former London Underground stock, with the DMUs in service with London North Western Railway and Transport for Wales and the electric units working on the Isle of Wight for South Western Railway.

Twenty-four staff are currently on the payroll, 12 of which who are currently part time.

SLC undertook the driver training for SWR crews on Class 484 units for the Isle of Wight, and provides route learning and driver training for third parties. It also is able to move rolling stock of many types for other companies. It has worked with Class 20s from Michael Owen and recently a Class 50 from Hanson & Hall, which does not have an operating licence.

It has recently agreed a deal with Northumbria Rail to operate ex-EMR Power cars 43045/060 (which had been used briefly by Colas) for training and route learning roles, while 43048/089, preserved by the 125 Group, have also been used by the company.

43045	SBXL	EMB	POR	HQ		43060	SBXL	EMB	POR	HQ	
43048	SBXL	EMB	125	HQ	*TCB Miller*	43089	SBXL	EMB	125	HQ	

Another new, but growing, train operator is SLC Operations. It too hires in locos from third parties and often uses Class 20s and Hanson & Hall's 50008 *Thunderer*. On 11 February 2022, the Class 50 passes Evesham with two ex-Cargowaggon vehicles acting as barrier vehicles to King's Norton, having worked a train to Long Marston the previous day. *Jack Boskett*

Vintage Trains

Contact details

Website: www.vintagetrains.co.uk
Twitter: @Vintage_Trains

Key personnel

Chairman: Michael Whitehouse

Vintage Trains operates mostly vacuum-braked trains, so it often hires pairs of Class 20s to work excursions. On 4 August 2021, 20227 and 20189 approach Swineshead, between Boston and Sleaford, with the 1825 Skegness-Tyseley charter. This route was regularly worked by Class 20s in 1974-93. *Pip Dunn*

Overview

Vintage Trains is based at Tyseley Locomotives Works and recently gained its own Train Operating Licence after years of using West Coast Railways to run its charter trains.

It owns a handful of steam locos and a single Class 47, 47773. It also owns a Class 08. It often hires in Class 20s from 20189 Ltd and other locos from main-line preservation groups. It has recently acquired three Class 144 railbuses.

Class 08

Depot shunter at Tyseley Locomotive Works.

08021	BLK	TLW	TM

Class 47

An ex-EWS loco, this dual-braked 47 is the main support loco for VTL's steam operations. It is in the MBDL pool.

47773	47541	GYP	VIN	TM

The only diesel loco owned by Vintage Trains is 47773. On 29 July 2021, the loco was working an evening circular trip from Tyseley to Leicester, where the train is seen. *Jack Boskett*

Class 144

Ex-Northern Rail Railbuses bought from Angel Trains.

144014	NBL	VIN	TM	55814	55850	55837
144019	NBL	VIN	TM	55819	55855	55842
144023	NBL	VIN	TM	55821	55859	55846

Coaches

Pullman FK 99361 *Eagle* stands at Stratford-upon-Avon on the 1252 charter to Birmingham Snow Hill, the 'Shakespeare Express', on 8 August 2021. *Pip Dunn*

No.	Old No.	Livery	type	depot	Name
99349	349	PUL	Pullman SP	TS	*CAR No 349*
99353	353	PUL	Pullman SP	TS	*CAR No 353*
99361	311	PUL	Pullman FK	TS	*EAGLE*
5157		CHC	Mk 2 TSO	TS	
5177		CHC	Mk 2 TSO	TS	
5191		CHC	Mk 2 TSO	TS	
5198		CHC	Mk 2 TSO	TS	
5212		CHC	Mk 2 TSO	TS	
9101	9398	CHC	Mk 2 BFK	TS	
17018		CHC	Mk 1 BFK	TS	
35470		CHC	Mk 1 BSK	TS	

W9101 is a Mk 2 BFK that is often used as a staff coach on Vintage Trains excursions. It was at Stratford-upon-Avon on 8 August 2021. *Pip Dunn*

West Coast Railways

Contact details

Website: www.westcoastrailways.co.uk
Twitter: @westcoastrail

Key personnel

Chairman: David Smith
Managing Director: Patricia Marshall

Overview

West Coast Railways is an independent train operator, best known for its steam operations including the annual summer Fort William-Mallaig 'Jacobite', which it started to run in 1998. It has its main base at Carnforth but also operates out of Southall in west London.

In 2000 it gained a nationwide operating licence and started to grow its fleet of diesel locos. It now owns Class 33s, 37s, 47s and 57s and has a single 25kV AC electric loco, 86401. However, it also operates other loco types, steam diesel and electric, owned by other companies.

Modern traction-wise, the firm started by acquiring a 47 and then a 57 from Porterbrook and has since gone on to grow to typically about 30 locos, although not all are in traffic at the same time. It has a large fleet of Mk 1 and 2 coaches.

It owns the Northern Belle Mk 2 set, plus the Queen of Scots Pullman, although the latter is only used very occasionally for special events.

Class 03

These locos were inherited when WCR acquired Steamtown, apart from 03084, which is owned by Chris Beet but based at Carnforth.

03084		GWS	CRB	CS
03196		GNY	WCR	CS
D2381		GNY	WCR	CS (U)

Class 08

08678 was inherited when WCR acquired Steamtown, while the other two locos were bought from EWS in 2010.

08418		EWS	WCR	CS
08485		BRW	WCR	CS
08678	AWCX	WCR	WCR	CS

Class 33

Ex-DRS locos acquired in 2005, 33030 has never run for WCR and is used as a source of spares. These locos have no train heating. 33025/207 are dual braked.

33025	AWCA	WCR	WCR	CS	
33029	AWCA	WCR	WCR	CS	
33030	AWCX	DRU	WCR	CS (U)	
33207	AWCA	WCR	WCR	CS	*Jim Martin*

WCR has three Class 33s, although they lack train heat and this loco, 33029, is air-braked only. The Crompton passes Black Moss just outside Windermere with the 1130 from Oxenholme on 22 June 2018. On the rear is 57316. *Graham Roose*

Class 37

Locos mostly acquired directly, or indirectly from EWS. 37165/517 have never run for WCR and are used as a source of spares along with 37712.

37165	37374		CCT	WCR	CS (U)	
37516	37086	AWCA	WCR	WCR	CS	*Loch Laidon*
37517	37018		LHO	WCR	CS (U)	
37518	37076	AWCA	WCR	WCR	CS	*Fort William/An Gearasdan*
37668	37257	AWCA	WCR	WCR	CS	
37669	37129	AWCA	WCR	WCR	CS	
37676	37126	AWCA	WCR	WCR	CS	*Loch Rannoch*
37685	37234	AWCA	WCR	WCR	CS	*Loch Arkaig*
37706	37016	AWCA	WCR	WCR	CS	
37712	37102	AWCX	WCR	WCR	CS (U)	

WCR's fleet of seven Class 37s are used regularly on charters, stock moves and occasionally freight. On 22 June 2019, 37669 passes Lambrigg, near Oxenholme, with the 0850 Fort William Yard-Carnforth Steamtown ECS move. *Rob France*

Class 47

Locos acquired mostly from Porterbrook, EWS, FM Rail, DRS and Rail Operations Group. 47194/368/492/526/768/776 have never run for WCR and are used as a source of spares. 47237/245/746/760/772/776/802/804 are dual braked.

47194	AWCX	RFD	WCR	CS (U)
47237	AWCA	WCR	WCR	CS

WCR built its business up using second-hand Class 47s, which in the main have proved pretty reliable. 47746 *Chris Fudge*, with 47237 dead in train, passes Upton Noble, near Bruton, with a Saltburn Railtours charter from Saltburn to Torquay on 16 July 2021. *Glen Batten*

47245		AWCA	WCR	WCR	CS	*VE Day 75th Anniversary*
47270		AWCA	BRB	WCR	CS	
47355		AWCX	FRG	WCR	CS (U)	
47368			TTG	WCR	CS (U)	
47492		AWCX	RES	WCR	CS (U)	
47526			BLL	WCR	CS (U)	
47746	47160, 47605	AWCA	WCR	WCR	CS	*Chris Fudge 29.7.70-22.6.10*
47760	47036, 47562, 47672	AWCA	WCR	WCR	BU (U)	
47768	47490	AWCX	UND	WCR	CS (U)	
47772	47537	AWCA	WCR	WCR	CS	*Carnforth TMD*
47776	47181, 47578	AWCX	RES	WCR	CS (U)	
47786	47138, 47607, 47821	AWCA	WCR	WCR	BU (U)	*Roy Castle OBE*
47787	47163, 47610, 47823	AWCX	WCR	WCR	CS (U)	
47802	47552	AWCA	WCR	WCR	CS	
47804	47265, 47591, 47792	AWCA	WCR	WCR	CS	
47812	47239, 47657	AWCA	ROU	WCR	CS	
47813	47129, 47658	AWCA	ROU	WCR	CS	
47815	47155, 47660	AWCA	GYP	WCR	CS	*Great Western*
47826	47274, 47637	AWCA	WCR	WCR	CS	
47832	47031, 47560	AWCA	WCR	WCR	CS	
47848	47068, 47632	AWCA	WCR	WCR	CS	
47851	47064, 47639	AWCA	WCR	WCR	CS	
47854	47271, 47604, 47674	AWCA	WCR	WCR	CS	*Diamond Jubilee*

Class 57

Locos acquired from Porterbrook, Advenza and DRS. 57005 has never run for WCR and is used as a source of spares. WCR's 57/0s have been returned to 95mph locos.

57001	47356	AWCA	WCR	WCR	CS	
57005	47350	AWCX	ADZ	WCR	CS (U)	
57006	47187	AWCA	WCR	WCR	CS	
57008	47060	XDSP	DRC	WCR	CS (U)	
57009	47079	XSDP	DRC	WCR	LT (S)	
57010	47231	XSDP	DRN	WCR	LT (S)	
57011	47329	XDSP	DRC	WCR	CS (U)	
57012	47204	XDSP	DRC	WCR	CS (U)	
57313	47371	AWCA	NOB	WCR	CS	*Scarborough Castle*
57314	47372	AWCA	WCR	WCR	CS	*Conwy Castle*

WCR has found second-hand Class 57s in plentiful supply, which give it an extra level of reliability over the old Sulzer engines. 57313 *Scarborough Castle* and 57601 *Windsor Castle* have both been repainted into Pullman livery to match the Northern Belle luxury train, which WCR now owns. They work the 0948 Carnforth-Carnforth test run past Wilpshire on 18 May 2021. *Tom McAtee*

57315	47234	AWCA	WCR	WCR	CS (U)	
57316	47290	AWCA	WCR	WCR	CS	*Alnwick Castle*
57601	47165, 47590, 47825	AWCA	NOB	WCR	CS	*Windsor Castle*

Class 86

Acquired for the ACLG, this loco works alongside preserved 86259.

86401	86001	AWCA	CAL	WCR	CS	*Mons Meg*

Coaches

WCR also owns many scrap vehicles, which it retains for spare parts and they are not currently in traffic so are not listed here.

No.	Other No.	Livery	Type	Depot	Name
1666*		WCR	Mk 1 RBR	CS	
1861		WCR	Mk 1 RMB	CS	
1961		WCR	Mk 1 RBR	CS	
3058		WCR	Mk 1 FO	CS	*Florence*

WCR has a fleet of Mk 1 and Mk 2 coaches, plus several pre-nationalisation vehicles as well. Mk 1 FO 3058 *Florence* was at Bognor Regis on 24 March 2018.

No.	Other No.	Livery	Type	Depot	Name
3093		WCR	Mk 1 FO	CS	*Florence*
3136		WCR	Mk 1 FO	CS	*Diana*
3143		WCR	Mk 1 FO	CS	*Patricia*
3313		WCR	Mk 2 FO	CS	
3326		WCR	Mk 2 FO	CS	
3350		WCR	Mk 2 FO	CS	
3352		WCR	Mk 2 FO	CS	
3392		WCR	Mk 2 FO	CS	
3395		WCR	Mk 2 FO	CS	
3431		WCR	Mk 2 FO	CS	

West Coast Railways

4854		WCR	Mk 1 TSO	CS	
4905		WCR	Mk 1 TSO	CS	
4940		WCR	Mk 1 TSO	CS	
4951		WCR	Mk 1 TSO	CS	
4960		WCR	Mk 1 TSO	CS	
4973		WCR	Mk 1 TSO	CS	
4984		WCR	Mk 1 TSO	CS	
4994		WCR	Mk 1 TSO	CS	
5032		WCR	Mk 1 TSO	CS	
5035		WCR	Mk 1 TSO	CS	
5171		WCR	Mk 2 TSO	CS	
5200		WCR	Mk 2 TSO	CS	
5216		WCR	Mk 2 TSO	CS	
5222		WCR	Mk 2 TSO	CS	
5229		WCR	Mk 2 TSO	CS	
5236		WCR	Mk 2 TSO	CS	
5237		WCR	Mk 2 TSO	CS	
5239		WCR	Mk 2 TSO	CS	
5249		WCR	Mk 2 TSO	CS	
5278		WCR	Mk 2 TSO	CS	
5419		WCR	Mk 2 TSO	CS	
5487		WCR	Mk 2 TSO	CS	
5903		WCR	Mk 2 TSO	CS (U)	
6021		WCR	Mk 2 TSO	CS	
6022		WCR	Mk 2 TSO	CS	
6103		WCR	Mk 2 TSO	CS	
6115		WCR	Mk 2 TSO	CS	
6312	81023	WCR	Mk 1 BG	CS	
6528		WCR	Mk 2 TSOT	CS	
9104		WCR	Mk 2 BSOT	CS	
9391		WCR	Mk 2 BSO	CS	
9392		WCR	Mk 2 BSO	CS	
9493		WCR	Mk 2 BSO	CS	
13320		WCR	Mk 1 FO	CS	*Anna*
13440		WCR	Mk 2 FK	CS	
21266		WCR	Mk 1 BCK	CS	
95402	326	PUL	Mk 1 FP	CS	*Emerald*
99025	325	PUL	Mk 1 FP	CS	*Amber*
99035	35322	WCR	Mk 1 BSK	CS	
99121	3105	WCR	Mk 1 FO	CS	*Julia*
99122	3106	WCR	Mk 1 FO	CS	*Alexandra*
99125	3113	WCR	Mk 1 FO	CS	*Jessica*
99127	3117	WCR	Mk 1 FO	CS	*Christina*
99128	3130	WCR	Mk 1 FO	CS	*Pamela*
99304	21256	WCR	Mk 1 BCK	CS	
99311	1882	WCR	Mk 1 RMB	CS	
99316	13321	WCR	Mk 1 RBR	CS	
99318	4912	WCR	Mk 1 TSO	CS	
99326	4954	WCR	Mk 1 TSO	CS	
99327	5044	WCR	Mk 1 TSO	CS	
99328	5033	WCR	Mk 1 TSO	CS	
99329	4931	WCR	Mk 1 TSO	CS	
99347	347	WCR	Mk 1 SP	CS	
99348	348	PUL	Mk 1 SP	CS	*Topaz*
99350	350	PUL	Mk 1 SP	CS	*Tanzanite*
99351	351	PUL	Mk 1 SP	CS	*Sapphire*
99352	352	PUL	Mk 1 SP	CS	*Amethyst*
99354	354	PUL	Mk 1 BAR	CS	*The Hadrian Bar*

99371	3128	WCR	Mk 1 FO	CS	*Victoria*
99680	17102	WCR	Mk 2 BFK	CS	
99710	18767	WCR	Mk 1 SK	CS	
99716	25808	WCR	Mk 1 SK	CS	
999506		WCR	Inspection Saloon	CS	
99678	504	PUL	Mk 2 PFK	CS	*Ullswater*
99679	506	PUL	Mk 2 PFK	CS	*Windermere*
99670	546	PUL	Mk 2 PFP	CS	*City of Manchester*
99671	548	PUL	Mk 2 PFP	CS	*Grasmere*
99672	549	PUL	Mk 2 PFP	CS	*Bassenthwaite*
99673	550	PUL	Mk 2 PFP	CS	*Rydal Water*
99674	551	PUL	Mk 2 PFP	CS	*Buttermere*
99675	552	PUL	Mk 2 PFP	CS	*Ennerdale Water*
99676	553	PUL	Mk 2 PFP	CS	*Crummock Water*
99677	586	PUL	Mk 2 PFB	CS	*Derwent Water*
99722	18806	PUL	Mk 1 TSO	CS	
99723	35459	BRM	Mk 1 BSK	CS	
1566		NOB	Mk 1 RKB	CS	
1953		NOB	Mk 1 RU	CS	
3174		NOB	Mk 2 FO	CS	*Glamis*
3182		NOB	Mk 2 FO	CS	*Warwick*
3232		BRG	Mk 2 FO	CS (U)	
3247		NOB	Mk 2 FO	CS	*Chatsworth*
3267		NOB	Mk 2 FO	CS	*Belvoir*
3273		NOB	Mk 2 FO	CS	*Alnwick*
3275		NOB	Mk 2 FO	CS	*Harlech*
3362		NOB	Mk 2 FO	CS	
17167		NOB	Mk 2 BFK	CS	*Mow Cop*
10729		NOB	Mk 3 SLE	CS	
10734		NOB	Mk 3 SLE	CS	*Balmoral*
159		LWR		CS	
807		TEK		CS	
9004		NOB		CS	
99886	35407	LWR	Mk 1 BSK	CS	

Note: 1666 is owned by Steam Dreams.

Support coaches

These are usually used to support WCR's steam locos and as a rule are not used in its normal charter trains, although being brake vehicles would not preclude that.

17025		BRM	Mk 1 BFK
35447		BRM	Mk 1 BSK
35518	17097	BGC	Mk 1 BFK
80204	35297	BRM	Mk 1 BSK
80217	35299	BRM	Mk 1 BSK
80220	35276	BRM	Mk 1 BSK
99035	35322	BRM	Mk 1 BSK
99312	35463	BRM	Mk 1 BSK

Northern Ireland Railways

Contact details

Website: www.translink.co.uk
Twitter: @nirailways

Key personnel

General Manger Rail Services Operations: Richard Knox
Head of Rail Fleet Engineering: Richard Noble

Overview

Unlike the passenger operations in England, Scotland and Wales, those in Northern Ireland were not privatised in the mid-1990s. There are only a handful of routes in the country, and those that remain emanate mostly from Belfast.

The trains are operated by Translink NI Railways, part of the national transport operator Translink. Unlike the British mainland, the track gauge is not the standard 4ft 8½in, but the same 5ft 3in gauge as used in the Republic of Ireland.

Through InterCity cross-border services between Belfast and Dublin are jointly operated by Translink NI Railways and Iarnród Éireann using 201 Class GM diesel locos in push-pull mode and De-Dietrich-built coaching stock. Translink NI Railways owns two of these GM locos, 8208/8209, which are used on these services on a common user basis with certain dedicated Iarnród Éireann-owned locos.

The routes exclusively operated by Translink NI Railways are Belfast-Derry/Londonderry, Coleraine-Portrush, Belfast-Newry, Bangor and Larne.

The majority of the Translink NI Railways fleet now consists of just two classes of CAF-built diesel railcars following extensive fleet replacement in the last two decades to allow elderly trains to be withdrawn. These consist of 23 three-car 3000 class sets and 20 three-car 4000 class sets.

In December 2018, 21 additional 4000 class intermediate vehicles were ordered from CAF at a cost of £50m. These have been used to lengthen seven of the 20 Class 4000 units (4014-4020) to six-car sets. The first of these extra vehicles were delivered in 2021, with the remainder expected to be added to their sets in 2022.

111 Class

These three GM 2,475hp Co-Co locos, built in 1980-84, are the same as IÉ's 071 Class locos, and are used on infrastructure trains on the Translink NI Railways network.

8111	*Great Northern*
8112	*Northern Counties*
8113	*Belfast & County Down*

201 Class

These two GM 3200hp Co-Co locos are used alongside similar locos owned and operated by IÉ. They normally work cross-border 'Enterprise' services between Belfast Lanyon Place and Dublin Connolly with De-Dietrich-built trainsets, although they can also occasionally be found working internal passenger and freight services within the Republic of Ireland.

8208	*River Lagan*
8209	

3000 Class

These three-car railcar sets were built by CAF in Spain in 2004-05 and can operate at up to 90mph. They are used along with the 4000 class on all internal services within Northern Ireland and occasionally on cross-border 'Enterprise' services to Dublin Connolly. Sets 3001-3006 are specially fitted with IÉ CAWS and train radios for this purpose.

They replaced the old 80 Class DEMUs, although a handful of the former were retained as Sandite units until 2017.

Set No.	DM1	M	DM2
3001	3301	3501	3401
3002	3302	3502	3402
3003	3303	3503	3403
3004	3304	3504	3404
3005	3305	3505	3405
3006	3306	3506	3406
3007	3307	3507	3407
3008	3308	3508	3408
3009	3309	3509	3409
3010	3310	3510	3410
3011	3311	3511	3411
3012	3312	3512	3412
3013	3313	3513	3413
3014	3314	3514	3414
3015	3315	3515	3415
3016	3316	3516	3416
3017	3317	3517	3417
3018	3318	3518	3418
3019	3319	3519	3419
3020	3320	3520	3420
3021	3321	3521	3421
3022	3322	3522	3422
3023	3323	3523	3423

4000 class

Visually the same as the older 3000 Class units, these CAF-built units were delivered in 2010-11. They differ by having an alternative driveline. They are used along with the 3000 class on all internal services within Northern Ireland.

These trains were ordered to replace the 450 Class DEMUs and to provide extra capacity on local services.

Set No.	DM1	M3	M4	M5	M6	DM2
4001	4301	4501				4401
4002	4302	4502				4402
4003	4303	4503				4403
4004	4304	4504				4404
4005	4305	4505				4405
4006	4306	4506				4406
4007	4307	4507				4407
4008	4308	4508				4408
4009	4309	4509				4409
4010	4310	4510				4410
4011	4311	4511				4411
4012	4312	4512				4412
4013	4313	4513				4413
4014	4314	4514	4614	4714	4814	4414
4015	4315	4515	4615	4715	4815	4415
4016	4316	4516				4416
4017	4317	4517	4617	4717	4817	4417
4018	4318	4518				4418
4019	4319	4519	4619	4719	4819	4419
4020	4320	4520				4420

Sets 4014 to 4020 are in the process of being extended to six-car sets with the addition of three extra intermediate vehicles currently being delivered, these being M4 type vehicles 4614 to 4620, M5 type vehicles 4714 to 4720 and M6 type vehicles 4814 to 4820.

Irish Railways – Iarnród Éireann

Ballina
Sligo
Collooney
Ballymote
To Belfast (NIR)
Dundalk
Foxford
Boyle
Dromod
Carrick on Shannon
Manulla Junction
Edgeworthstown
Westport
Castlebar
Longford
Drogheda
Castlerea
Claremorris
Ballyhaunis
Mullingar
Roscommon
Enfield
Dunboyne/M3
Kilcock
Athlone
Maynooth
Dublin Connolly
Oranmore
Athenry
Attymon
Woodlawn
Clara
Clonsilla
Galway
Tullamore
Monasterevin
Dublin Pearse
Ballinasloe
Dublin Heuston
Dun Laoghaire
Kildare
Craughwell
Cloughjordan
Portarlington
Roscrea
Newbridge
Bray
Greystones
Portlaoise
Ardrahan
Athy
Kilcoole
Rathdrum
Nenagh
Ballybrophy
Birdhill
Wicklow
Gort
Castleconnell
Templemore
Carlow
Arklow
Ennis
Limerick
Thurles
Muine Bheag
Gorey
Kilkenny
Sixmilebridge
Enniscorthy
Charleville
Limerick Junction
Thomastown
Tralee
Tipperary
Clonmel
Wexford
Cahir
Rosslare Strand
Farranfore
Millstreet
Mallow
Glounthaune
Carrick on Suir
Waterford
Rosslare Europort
Killarney
Rathmore
Banteer
Midleton
Cork
Cobh

Showing the challenge of expanding services on the DART network – the 1120 Dublin-Belfast Enterprise races through Howth Junction on its way north on 20 February 2018. The plan is to extend the DART north to Drogheda and at the same time go hourly with the Enterprise. *Tony Miles*

Overview

The main railway network of the Republic of Ireland is operated by Iarnród Éireann (IÉ). The last two decades has seen a massive change in the IÉ fleet, with traditional loco-hauled passenger trains, some being predominately steam-heated using steam generator vans, being replaced by newer diesel railcar sets and air-conditioned loco-hauled coaches operating in push-pull mode.

The only loco-hauled operations now are the cross-border 'Enterprise' trains between Dublin Connolly and Belfast Lanyon Place using a fleet of De-Dietrich-built coaches (owned jointly with Translink NI Railways), and most of the InterCity services between Dublin Heuston and Cork using CAF-built Mk 4 coaches.

Newer 22000 and 29000 diesel railcar sets operate most of the other regional and InterCity routes, while local trains in the Cork and Limerick areas are operated by the older 2600 and 2800 diesel railcars. The electrified Dublin Area Rapid Transit system (DART) uses a fleet of 8100/8300 and 8500/8600 EMUs.

There are currently just under 1,700 miles of 5ft 3in gauge railway in the island of Ireland and most of it is diesel-only, apart from the DART system, which is electrified using 1,500V DC overhead.

Locomotives

There are two main loco types currently in service on IÉ; the older, 18-strong, 2,475hp Co-Co GM 071 class, dating from 1976, which are used for freight and infrastructure work only. These locomotives no longer have any booked passenger work except for the occasional special working.

The newer locos are the 32-strong, 3,200hp Co-Co GM 201 Class, which date from 1994-95. Most of these locos are equipped for push-pull operation and were fitted with head-end power (HEP) – a version of electric train supply – but this is no longer used and has been isolated.

There are also five withdrawn 950hp Bo-Bo GM 141 Class locomotives, which were built in 1962, still owned by IÉ. These are 144/147/162/171/177. All are stored at Inchicore Works.

201 Class

Thirty-four locos, numbered 201-234, were built for use in Ireland; two (8208/8209) are owned by Translink NI Railways, while the rest are owned by IÉ.

Locos 201-205/210-214 have been withdrawn from traffic and have been heavily robbed for spares. They were unable to operate in push-pull mode with the Mk 4 and De-Dietrich coaches, which was part of the reason why they were laid up, coupled with a significant decline in railfreight.

They could conceivably be returned to traffic, albeit only with major investment and a significant increase in railfreight, which is unlikely.

224 has also been withdrawn after cracking its chassis, while 230 is also stored out of use following significant fire damage. The remainder, 206-207/215-223/226-229/231-234, are in traffic.

206/207/227/228/231/233 (and formerly 230), together with Translink NI Railways 8208/8209, are modified to operate with the De-Dietrich coaches for the InterCity 'Enterprise' trains between Dublin and Belfast. These locos are fitted with Translink NI Railways TPWS and train radios to enable them to operate north of the border. There are also four former IÉ Mk 3 generator vans, now owned by Translink NI Railways, which were introduced in September 2012 to work with the De-Dietrich coaches to improve reliability of service and reduce maintenance costs.

The other operational 201s are mainly used in push-pull mode with the CAF-built Mk 4 coaches on passenger services between Dublin Heuston and Cork, but can also work the limited freight services such as liner trains between Dublin Port and Ballina, timber trains between Ballina/Westport and Waterford, and certain infrastructure trains. Those locos that normally work on the cross-border 'Enterprise' services can also work with the Mk 4s.

The locos have a top speed of 100mph, and all were intended to be named after rivers, with the 32 IÉ locos having nameplates in both Gaelic and English, although not all carry the correct plates.

The 201 Class 231 heads the 1120 Dublin Connolly-Belfast Enterprise train through Clontarf Road station, next to the Fairview DART Depot, on 1 May 2019. *Tony Miles*

Loco	Livery	Status	Name (English)	Name (Gaelic)
201	Orange (IÉ plug logo)	Stored unserviceable	*River Shannon*	*Abhainn na Sionnaine*
202	Orange (IÉ plug logo)	Stored unserviceable	*River Lee*	*Abhainn na Laoi*
203	Orange (IÉ plug logo)	Stored unserviceable	*River Corrib*	*Abhainn na Coiribe*
204	Orange (IÉ plug logo)	Stored unserviceable	*River Barrow*	*Abhainn na Bearú*
205	Orange (IÉ plug logo)	Stored unserviceable	*River Nore*	*Abhainn na Feoire*
206	Enterprise			*Abhainn na Life*
207	Enterprise		*River Boyne*	*Abhainn na Bóinne*
8208	Enterprise		*River Lagan*	
8209	Enterprise			
210	Orange (IÉ plug logo)	Stored unserviceable	*River Erne*	*Abhainn na Héirne*
211	Orange (IÉ plug logo)	Stored unserviceable	*River Suck*	*Abhainn na Suca*
212	Orange (IÉ plug logo)	Stored unserviceable	*River Slaney*	*Abhainn na Sláine*
213	Orange (IÉ plug logo)	Stored unserviceable	*River Moy*	*Abhainn na Muaidhe*
214	Orange (IÉ plug logo)	Stored unserviceable	*River Brosna*	*Abhainn na Brosnaí*
215	InterCity (Current logo)		*River Avonmore*	*An Abhainn Mhór*
216	Belmond Grand Hibernian		*River Dodder*	*Abhainn na Dothra*
217	InterCity (Current logo)		*River Flesk*	*Abhainn na Fleisce*
218	InterCity (Current logo)		*River Garavogue*	*Abhainn na Garbhóige*
219	InterCity (Current logo)		*River Tolka*	
220	InterCity (Current logo)		*River Blackwater*	*An Abhainn Dhubh*
221	InterCity (Current logo)		*River Fealge*	*Abhainn na Feilge*
222	InterCity (Current logo)		*River Dargle*	
223	InterCity (Current logo)		*River Anner*	*Abhainn na Hainnire*
224	InterCity (Current logo)	Withdrawn		*Abhainn na Féile*
225	InterCity (Current logo)		*River Deel*	*Abhainn na Daoile*
226	InterCity (Current logo)			*Abhainn na Siuire*
227	Enterprise		*River Laune*	
228	Enterprise			*An Abhainn Bhuí*
229	InterCity (Current logo)		*River Maine*	*Abhainn na Mainge*
230	Old Enterprise	Stored unserviceable		*Abhainn na Bandan*
231	Common-User Livery			*Abhainn na Maighe*
232	InterCity (Current logo)		*River Cummeragh*	
233	Common-User Livery		*River Clare*	*Abhainn na Chláir*
234	InterCity (Current logo)		*River Aherlow*	

With the Croagh Patrick mountain in the background, General Motors 071 Class locomotive 078 sets off with the 1105 Westport-Waterford timber train on 16 May 2016. *Graham Roose*

071 Class

An older GM-built class of locos dating from 1976. These 2,475hp locos used to work most of the express passenger trains with Mk 2 and Mk 3 coaches but have now been relegated to freight and infrastructure use only. They have a maximum speed of 90mph.

They are the same design as the 8111 Class owned by Translink NI Railways.

loco	Livery	Name
071	Supertrain	
072	Slate Grey	
073	Original Irish Rail	
074	Slate Grey	
075	Slate Grey	
076	Slate Grey	
077	Slate Grey	
078	Slate Grey	
079	Slate Grey	
080	Slate Grey	
081	Slate Grey	
082	Slate Grey	*Cumann la Ninneal Tóirí – The Institution of Engineers of Ireland*
083	Slate Grey	
084	Slate Grey	
085	Slate Grey	
086	Slate Grey	
087	Slate Grey	
088	Slate Grey	

DART Fleet

The Dublin Area Rapid Transit, or DART, provides the majority of local commuter rail services centred on Dublin Connolly, extending to Malahide and Howth on the north side of the City and Greystones to the south. It uses a mixed fleet of 1,500V DC overhead EMUs.

8100/8300 Class EMU

These are the original German-built two-car sets introduced when the DART was first constructed, dating from 1983. There were originally 40 sets but two were withdrawn following fire damage in 2001, leaving 38 sets left in traffic.

They operate local services between Greystones/Bray to Malahide and Howth, formed into two-, four-, six- or eight-car sets. The fleet underwent a major refurbishment programme in 2005-07.

IÉ also has five Alstom-built 8200/8400 series two-car sets, but these have all been withdrawn from traffic and are stored at Inchicore Works. These are formed 8201+8401, 8202+8402, 8203 +8405, 8204+8404 and 8205+8403.

DMS	DTS										
8101	8301	8107	8307	8115	8315	8122	8322	8129	8329	8137	8337
8102	8302	8108	8308	8116	8316	8123	8323	8130	8330	8138	8338
8103	8303	8109	8311	8117	8317	8124	8324	8131	8331	8139	8339
8104	8304	8111	8311	8118	8318	8125	8325	8132	8332	8140	8340
8105	8305	8112	8312	8119	8319	8126	8326	8133	8333		
8106	8306	8113	8313	8120	8320	8127	8327	8134	8334		
		8114	8314	8121	8321	8128	8328	8135	8335		

Class 8500/8600 Class EMU

These are three types of Japanese-built four-car EMU sets: the 8500, 8510 and 8520 sets, which were built in 2000, 2001 and 2003-04 respectively.

They operate local services between Greystones/Bray to Malahide and Howth as either four- or eight-car trains. They have a maximum speed of 63mph.

DTS	PMS	PMS	DTS								
8601	8501	8502	8602	8613	8513	8514	8614	8629	8529	8530	8630
8603	8503	8504	8604	8615	8515	8516	8616	8631	8531	8532	8632
8605	8505	8506	8606	8621	8521	8522	8622	8633	8533	8534	8634
8607	8507	8508	8608	8623	8523	8524	8624	8635	8535	8536	8636
8611	8511	8512	8612	8625	8525	8526	8626	8637	8537	8538	8638
				8627	8527	8528	8628	8639	8539	8540	8640

An Irish Rail DART set comprising three 8500 Class sets led by 8633 head towards Malahide at Howth Junction on 1 May 2019. These units were made by the Tokyu Car Corporation of Japan and entered service between July 2001 and 2004. *Tony Miles*

Sets 8621+8521+8522+8622 and 8625+8525+8526+8626 are currently running in a semi-permanent 8-piece 'hybrid' set formed 8621+8521+8526+8626+8625+8525+8522+8622, as the two driving cars from set 25/26 have been fitted with a new cab signalling system currently being developed/introduced on parts of the Iarnród Éireann network.

These two test cars, 8625 and 8626, can only be used as leading driving cabs on sections of lines currently fitted with the new system. When not required for testing, they are coupled 'face to face' in the centre of the 'hybrid' set to enable the sets to be used in passenger service.

Iarnród Éireann has 41 new intermediate diesel railcar vehicles on order to strengthen the existing 22000 Class. These are due to enter traffic from 2022. It is also looking to acquire up to 600 electric or battery-electric hybrid vehicles to replace the older DART EMU fleet, and to expand services.

Diesel Railcar Fleet

29000 Class

There are 29 of these four-car sets built in 2002-05 and they work in the Greater Dublin Area out to Longford, Maynooth, Dundalk, Drogheda, Arklow and Gorey

They run as four- or eight-car sets and have a maximum speed of 80mph.

Set No.	DM1	MDT	MT	DM2
29001	29101	29201	29301	29401
29002	29102	29202	29302	29402
29003	29103	29203	29303	29403
29004	29104	29204	29304	29404
29005	29105	29205	29305	29405
29006	29106	29206	29306	29406
29007	29107	29207	29307	29407
29008	29108	29208	29308	29408
29009	29109	29209	29309	29409
29010	29110	29210	29310	29410
29011	29111	29211	29311	29411
29012	29112	29212	29312	29412
29013	29113	29213	29313	29413
29014	29114	29214	29314	29414
29015	29115	29215	29315	29415
29016	29116	29216	29316	29416
29017	29117	29217	29317	29417
29018	29118	29218	29318	29418
29019	29119	29219	29319	29419
29020	29120	29220	29320	29420
29021	29121	29221	29321	29421
29022	29122	29222	29322	29422
29023	29123	29223	29323	29423
29024	29124	29224	29324	29424
29025	29125	29225	29325	29425
29026	29126	29226	29326	29426
29027	29127	29227	29327	29427
29028	29128	29228	29328	29428
29029	29129	29229	29329	29429

Sets 29022 and 29029 are currently running in a semi-permanent eight-piece 'hybrid' set formed 29129+29222+29322+29422+29122+29229+29329+29429 as the two driving cars from set 29022 have been fitted with a new cab signalling system currently being developed/introduced on parts of the Iarnród Éireann network.

These two test cars, 29122 and 29422, can only be used as leading driving cabs on sections of lines currently fitted with the new system. When not required for testing, they are coupled 'face to face' in the centre of the 'hybrid' set to enable the sets to be used in passenger service.

2800 DMU

These 80mph two-car units, built by the Tokyu Car Corporation in Japan in 2000, predominately operate in the Limerick area on local services to Ballybrophy via Nenagh, Ennis and Galway, and also on the Ballina to Manulla Junction branch.

DC1	DC2
2801	2802
2803	2804
2805	2806
2807	2808
2809	2810
2811	2812
2813	2814
2815	2816
2817	2818
2819	2820

2600 DMU

These 75mph two-car units, built by the Tokyu Car Corporation in Japan in 1993, mainly operate local trains out of Cork to Cobh, Midleton and Mallow. Occasionally they also operate on the Tralee line.

These run as two-, four-, six- or eight-car trains and have a maximum speed of 75mph.

DC1	DC2	
2601	2602	
2603	2604	
2605	2616	
2607	2608	
2609		Stored
2611	2612	
2613	2610	
2615	2606	
2617	2614	

DC1 Vehicle 2609 was specially modified to operate with a spare M2 2700 class vehicle, 2716, as a 'hybrid' two-car unit. Following the withdrawal of the rest of the 2700 class, this vehicle is now stored out of use along with them.

IÉ also has 25 Alstom-built 2700 Class railcar vehicles (12 × M1, 13 × M2), plus two single-car 2750 Class railcar vehicles, all of which are now withdrawn. They are stored at two locations, Inchicore Works and Cork. These are formed 2701+2702, 2703+2704, 2705+2706, 2707+2708, 2709+2710, 2711+2712, 2713+2714, 2715+2724, 2716+2609, 2717+2718, 2719+2722, 2720+2721, 2723+2726, 2751, 2753.

InterCity Rotem Fleet

The 22000 Class are a 100mph InterCity railcar, formed into a mix of three-, four- and five-car sets and were built by Rotem in 2007-12.

These sets operate on all InterCity routes from Dublin, including to Cork, Limerick, Tralee, Waterford, Galway, Westport, Sligo and Rosslare Europort, as well as commuter trains to Newbridge, Kildare, Portlaoise, Athlone and Carlow. Certain sets, 22001 to 22006, are also equipped to operate cross-border 'Enterprise' trains to Belfast.

Set No.	A2	B	A3
22001	22201	22401	22301
22002	22202	22402	22302
22003	22203	22403	22303
22004	22204	22404	22304
22005	22205	22405	22305
22006	22206	22406	22306
22007	22207	22407	22307
22008	22208	22408	22308
22009	22209	22409	22309
22010	22210	22410	22310

Set No.	A2	B	B	A3
22011	22211	22811	22411	22311
22012	22212	22812	22412	22312
22013	22213	22813	22413	22313
22014	22214	22814	22414	22314
22015	22215	22815	22415	22315
22016	22216	22816	22416	22316
22017	22217	22817	22417	22317
22018	22218	22818	22418	22318
22019	22219	22819	22419	22319
22020	22220	22820	22420	22320
22021	22221	22821	22421	22321
22022	22222	22822	22422	22322
22023	22223	22823	22423	22323
22024	22224	22824	22424	22324
22025	22225	22825	22425	22325
22026	22226	22826	22426	22326
22027	22227	22827	22427	22327
22028	22228	22828	22428	22328
22029	22229	22829	22429	22329
22030	22230	22830	22430	22330

Rotem 22000 Class DMU, led by 22327, approaches Westport with the 0908 Athlone-Westport on 16 May 2016. *Graham Roose*

Set No.	A1	B	B1	B	A3
22031	22131	22731	22631	22431	22331
22032	22132	22732	22632	22432	22332
22033	22133	22733	22633	22433	22333
22034	22134	22734	22634	22434	22334
22035	22135	22735	22635	22435	22335
22036	22136	22736	22636	22436	22336
22037	22137	22737	22637	22437	22337
22038	22138	22738	22638	22438	22338
22039	22139	22739	22639	22439	22339
22040	22140	22740	22640	22440	22340

Set No.	A2	B	B	A3
22041	22241	22541	22441	22341
22042	22242	22542	22442	22342
22043	22243	22543	22443	22343
22044	22244	22544	22444	22344
22045	22245	22545	22445	22345

Set No.	A2	B	A3
22046	22246	22446	22346
22047	22247	22447	22347
22048	22248	22448	22348
22049	22249	22449	22349
22050	22250	22450	22350
22051	22251	22451	22351
22052	22252	22452	22352
22053	22253	22453	22353
22054	22254	22454	22354
22055	22255	22455	22355
22056	22256	22456	22356
22057	22257	22457	22357
22058	22258	22458	22358
22059	22259	22459	22359
22060	22260	22460	22360
22061	22261	22461	22361
22062	22262	22462	22362
22063	22263	22463	22363

Sets 22001 and 22002 are currently running in a semi-permanent eight-piece 'hybrid' set formed 22302+22402+22201+22301+22401+22202 as the two driving cars from set 22001 have been fitted with a new cab signalling system currently being developed / introduced on parts of the Iarnród Éireann network.

These two test cars, 22201 and 22301, can only be used as leading driving cabs on sections of lines currently fitted with the new system. When not required for testing, they are coupled 'face to face' in the centre of the 'hybrid' set to enable the sets to be used in passenger service.

Mk 4

These CAF-built loco-hauled coaches operate in push-pull mode with 201 Class locos. There is a mixture of first and standard coaches, plus driving trailer / generator vans and buffet cars. They were built between 2004 and 2005.

They usually work as eight-coach formations between Dublin Heuston and Cork with five standard-class coaches, a buffet car, one first-class coach and a DVT.

Driving brake generator coaches

4001	4005
4002	4006
4003	4007
4004	4008

IÉ Mk 4 Driving Van trailer 4007 leads a Mk 4 set through Kildare on the 1020 Cork-Dublin Heuston on 29 April 2016. *Tony Miles*

Standard-class open coaches

4101	4105	4109	4113	4117	4121	4125	4129	4133	4137	4141
4102	4106	4110	4114	4118	4122	4126	4130	4134	4138	4142
4103	4107	4111	4115	4119	4123	4127	4131	4135	4139	4143
4104	4108	4112	4116	4120	4124	4128	4132	4136	4140	

Coaches 4105/10/15/20/25/30/35/140 are connector coaches with gangways at only one end. The non-gangway end is coupled immediately next to the locomotive.

First-class open coaches

4201	4203	4205	4207
4202	4204	4206	4208

Standard-class buffet coaches

4401	4403	4405	4407
4402	4404	4406	4408

De Dietrich (Enterprise)

These loco-hauled coaches date from 1996 and operate in push-pull mode with certain modified 201 Class locos on the Dublin to Belfast 'Enterprise' service, a joint Iarnród Éireann and Translink NI Railways venture. The fleet is jointly owned by the two rail companies, odd-numbered vehicles by Iarnród Éireann and even-numbered ones by Translink NI Railways, except for the four former Mk 3 generator cars, which are owned by Translink NI Railways.

They run as eight-coach sets, with a similar formation as the Mk 4 sets.

Driving trailer brake open first

9001	9003
9002	9004

Open first coaches

9101	9103
9102	9104

Open standard coaches

9201	9205	9209	9213
9202	9206	9210	9214
9203	9207	9211	9215
9204	9208	9212	9216

Catering coaches

9401	9403
9402	9404

Mk 3 generator coaches

	Old IÉ no.
9602	7604
9604	7605
9606	7608
9608	7613

The London Underground is the oldest metro system in the world and transports millions of passengers every day. S7 set 21552 arrives at London Paddington on 2 July 2021 as it heads to Edgware Road on the Circle line. *Jack Boskett*

London Underground

The London Underground is understandably the UK's busiest and best-known metro system, with as many as 5 million journeys made each day. It comprises 11 lines covering 250 miles and 270 stations. It includes a number of 'sub-surface' lines, which head out into the suburbs and are far from underground on these sections.

Bakerloo line

Map Colour: Brown
Routes: Harrow & Wealdstone to Elephant and Castle via Willesden Junction, Paddington, Marylebone, Piccadilly Circus, Waterloo
Mileage: 14.4 miles
Number of stations: 25
Trains: 36 × seven-car sets built in 1972
Seven-car sets formed of a four-car single-ended set coupled to a three-car single-ended set; the former tend to be the south end of the train. The trains comprise a driving motor, two unpowered trailer cars and a middle-motor car on the four-car sets and an unmanned driving motor, trailer and a driving motor car on the three-car sets. Each motor car has four Brush traction motors.

Trams, Metros and Light Rail

DM	T	T	MM	UNDM	T	DM
3231	4231	4331	3331	3431	4531	3531
3232	4232	4332	3332	3432	4532	3532
3233	4233	4333	3333	3433	4533	3533
3234	4234	4334	3334	3434	4534	3534
3235	4235	4335	3335	3435	4535	3535
3236	4236	4336	3336	3436	4536	3536
3237	4237	4337	3337	3437	4537	3537
3238	4238	4338	3338	3438	4538	3538
3239	4239	4339	3339	3439	4539	3539
3240	4240	4340	3340	3440	4540	3540
3241	4241	4341	3341	3441	4541	3541
3242	4242	4342	3342	3442	4542	3542
3243	4243	4343	3343	3443	4543	3543
3244	4244	4344	3344	3444	4544	3544
3245	4245	4345	3345	3445	4545	3545
3246	4246	4346	3346	3446	4546	3546
3247	4247	4347	3347	3447	4547	3547
3248	4248	4348	3348	3448	4548	3548
3249	4249	4349	3349	3449	4549	3549
3250	4250	4350	3350	3450	4550	3550
3251	4251	4351	3351	3451	4551	3551
3252	4252	4352	3352	3452	4552	3552
3253	4253	4353	3353	3453	4553	3553
3254	4254	4354	3354	3454	4554	3554
3255	4255	4355	3355	3455	4555	3555
3256	4256	4356	3356	3456	4556	3556
3257	4257	4357	3357	3457	4557	3557
3258	4258	4358	3358	3458	4558	3558
3259	4259	4359	3359	3459	4559	3559
3260	4260	4360	3360	3460	4560	3560
3261	4261	4361	3361	3461	4561	3561
3262	4262	4362	3362	3462	4562	3562
3263	4263	4363	3363	3463	4563	3563
3264	4264	4364	3364	3464	4564	3564
3265	4265	4365	3365	3465	4565	3565
3266	4266	4366	3366	3466	4566	3566
3267	4267	4367	3367	3467	4567	3567
3299	4299	4399	3399			

Central Line

Map Colour: Red

Routes: West Ruislip/Ealing Broadway to Epping via Greenford, White City, Oxford Circus, Bank, Liverpool St, Stratford and Leytonstone; loop line from Leytonstone to Woodford via Grange Hill

Mileage: 46

Number of stations: 49

Trains: 85 × eight-car sets built in 1992 – they are essentially two-car sets that run in formations of eight vehicles.

A+B two-car sets

DM	NDM
91001	92001
91003	92003
91005	92005
91007	92007
91009	92009
91011	92011
91013	92013
91015	92015
91017	92017
91019	92019
91021	92021
91023	92023
91025	92025
91027	92027
91029	92029
91031	92031
91033	92033
91035	92035
91037	92037
91039	92039
91041	92041
91043	92043
91045	92045
91047	92047
91049	92049
91051	92051
91053	92053
91055	92055
91057	92057
91059	92059
91061	92061
91063	92063
91065	92065
91067	92067
91069	92069
91071	92071
91073	92073
91075	92075
91077	92077
91079	92079
91081	92081
91083	92083
91085	92085
91087	92087
91089	92089
91091	92091
91093	92093
91095	92095
91097	92097
91099	92099
91101	92101
91103	92103
91105	92105
91107	92107
91109	92109
91111	92111
91113	92113
91115	92115
91117	92117
91119	92119
91121	92121
91123	92123
91125	92125
91127	92127
91129	92129
91131	92131
91133	92133
91135	92135
91137	92137
91139	92139
91141	92141
91143	92143
91145	92145
91147	92147
91149	92149
91151	92151
91153	92153
91155	92155
91157	92157
91159	92159
91161	92161
91163	92163
91165	92165
91167	92167
91169	92169
91171	92171
91173	92173
91175	92175
91177	92177
91179	92179
91181	92181
91183	92183
91185	92185
91187	92187
91189	92189
91191	92191
91193	92193
91195	92195
91197	92197
91199	92199
91201	92201
91203	92203
91205	92205
91207	92207
91209	92209
91211	92211
91213	92213
91215	92215
91217	92217
91219	92219
91221	92221
91223	92223
91225	92225
91227	92227
91229	92229
91231	92231
91233	92233
91235	92235
91237	92237
91239	92239
91241	92241
91243	92243
91245	92245
91247	92247
91249	92249
91251	92251
91253	92253
91255	92255
91257	92257
91259	92259
91261	92261
91263	92263
91265	92265
91267	92267
91269	92269
91271	92271
91273	92273
91275	92275
91277	92277
91279	92279
91281	92281
91283	92283
91285	92285
91287	92287
91289	92289
91291	92291
91293	92293
91295	92295
91297	92297

91299	92299
91301	92301
91303	92303
91305	92305
91307	92307
91309	92309
91311	92311
91313	92313
91315	92315
91317	92317
91319	92319
91321	92321
91323	92323
91325	92325
91327	92327
91329	92329
91331	92331
91333	92333
91335	92335
91337	92337
91339	92339
91341	92341
91343	92343
91345	92345
91347	92347
91349	92349

B+C two-car sets

NDM	NDM
92002	93002
92004	93004
92006	93006
92008	93008
92010	93010
92012	93012
92014	93014
92016	93016
92018	93018
92020	93020
92022	93022
92024	93024
92026	93026
92028	93028
92030	93030
92032	93032
92034	93034
92036	93036
92038	93038
92040	93040
92042	93042
92044	93044
92046	93046
92048	93048
92050	93050
92052	93052
92054	93054
92056	93056
92058	93058
92060	93060
92062	93062
92064	93064
92066	93066
92068	93068
92070	93070
92072	93072
92074	93074
92076	93076
92078	93078
92080	93080
92082	93082
92084	93084
92086	93086
92088	93088
92090	93090
92092	93092
92094	93094
92096	93096
92098	93098
92100	93100
92102	93102
92104	93104
92106	93106
92108	93108
92110	93110
92112	93112
92114	93114
92116	93116
92118	93118
92120	93120
92122	93122
92124	93124
92126	93126
92128	93128
92130	93130
92132	93132
92134	93134
92136	93136
92138	93138
92140	93140
92142	93142
92144	93144
92146	93146
92148	93148
92150	93150
92152	93152
92154	93154
92156	93156
92158	93158
92160	93160
92162	93162
92164	93164
92166	93166
92168	93168
92170	93170
92172	93172
92174	93174
92176	93176
92178	93178
92180	93180
92182	93182
92184	93184
92186	93186
92188	93188
92190	93190
92192	93192
92194	93194
92196	93196
92198	93198
92200	93200
92202	93202
92204	93204
92206	93206
92208	93208
92210	93210
92212	93212
92214	93214
92216	93216
92218	93218
92220	93220
92222	93222
92224	93224
92226	93226
92228	93228
92230	93230
92232	93232
92234	93234
92236	93236
92238	93238
92240	93240
92242	93242
92244	93244
92246	93246
92248	93248
92250	93250
92252	93252
92254	93254
92256	93256
92258	93258
92260	93260
92262	93262
92264	93264
92266	93266

B+D two-car sets

NDM	NDM
92402	93402
92404	93404
92406	93406
92408	93408
92410	93410
92412	93412
92414	93414
92416	93416
92418	93418
92420	93420
92422	93422
92424	93424
92426	93426
92428	93428
92430	93430
92432	93432
92434	93434
92436	93436
92438	93438
92440	93440
92442	93442
92444	93444
92446	93446
92448	93448
92450	93450
92452	93452
92454	93454
92456	93456
92458	93458
92460	93460
92462	93462
92464	93464

Jubilee Line

Map Colour: silver
Routes: Stanmore to Stratford via Neasden, Baker St, Green Park, Waterloo, London Bridge, Canary Wharf and Canning Town
Mileage: 22.5
Number of stations: 27
Trains: 63 × seven-car sets built in 1996
These trains run as seven-car sets using three-car and four-car single-ended units coupled together.

Four-car single-ended units

DM	T	T	UNDM
96001	96201	96601	96401
96003	96203	96603	96403
96005	96205	96605	96405
96007	96207	96607	96407
96009	96209	96609	96409
96011	96211	96611	96411
96013	96213	96613	96413
96015	96215	96615	96415
96017	96217	96617	96417
96019	96219	96619	96419
96021	96221	96621	96421
96023	96223	96623	96423
96025	96225	96625	96425
96027	96227	96627	96427
96029	96229	96629	96429
96031	96231	96631	96431
96033	96233	96633	96433
96035	96235	96635	96435
96037	96237	96637	96437
96039	96239	96639	96439

96041	96241	96641	96441
96043	96243	96643	96443
96045	96245	96645	96445
96047	96247	96647	96447
96049	96249	96649	96449
96051	96251	96651	96451
96053	96253	96653	96453
96055	96255	96655	96455
96057	96257	96657	96457
96059	96259	96659	96459
96061	96261	96661	96461
96063	96263	96663	96463
96065	96265	96665	96465
96067	96267	96667	96467
96069	96269	96669	96469
96071	96271	96671	96471
96073	96273	96673	96473
96075	96275	96675	96475
96077	96277	96677	96477
96079	96279	96679	96479
96081	96281	96681	96481
96083	96283	96683	96483
96085	96285	96685	96485
96087	96287	96687	96487
96089	96289	96689	96489
96091	96291	96691	96491
96093	96293	96693	96493
96095	96295	96695	96495
96097	96297	96697	96497
96098	96298	96698	96498
96099	96299	96699	96499
96101	96301	96701	96501
96103	96303	96703	96503
96105	96305	96705	96505
96107	96307	96707	96507
96109	96309	96709	96509
96111	96311	96711	96511
96113	96313	96713	96513
96115	96315	96715	96515
96117	96317	96717	96517
96119	96319	96719	96519
96121	96321	96721	96521
96123	96323	96723	96523
96125	96325	96725	96525

Three-car single-ended units

UNDM	T	DM
96402	96202	96002
96404	96204	96004
96406	96206	96006
96408	96208	96008
96410	96210	96010
96412	96212	96012
96414	96214	96014
96416	96216	96016
96418	96218	96018
96420	96220	96020
96422	96222	96022
96424	96224	96024
96426	96226	96026
96428	96228	96028
96430	96230	96030
96432	96232	96032
96434	96234	96034
96436	96236	96036
96438	96238	96038
96440	96240	96040
96442	96242	96042
96444	96244	96044
96446	96246	96046
96448	96248	96048
96450	96250	96050
96452	96252	96052
96454	96254	96054
96456	96256	96056
96458	96258	96058
96460	96260	96060
96462	96262	96062
96464	96264	96064
96466	96266	96066
96468	96268	96068
96470	96270	96070
96472	96272	96072
96474	96274	96074
96476	96276	96076
96478	96278	96078
96480	96880	96080
96482	96882	96082
96484	96884	96084
96486	96886	96086
96488	96888	96088
96490	96890	96090
96492	96892	96092
96494	96894	96094
96496	96896	96096
96498	96898	96098
96500	96900	96100
96502	96902	96102
96504	96904	96104
96506	96906	96106
96508	96908	96108
96510	96910	96110
96512	96912	96112
96514	96914	96114
96516	96916	96116
96518	96918	96118
96520	96320	96120
96522	96322	96122
96524	96324	96124
96526	96326	96126

Northern Line

Map Colour: Black

Routes: Edgware to Morden via Brent Cross, Camden Town, Euston, Tottenham Court Road, Waterloo, Stockwell, Balham and Colliers Wood; branches from Camden Town to High Barnet via Finchley Central, Finchley Central to Mill Hill East, Kennington to Battersea Power Station and Camden Town to Kennington via King's Cross St Pancras, Bank and London Bridge

Mileage: 42

Number of stations: 34

Trains: 106 × six-car sets built 1995

DM	T	UNDM	UNDM	T	DM
51501	52501	53501	53701	52701	51701
51502	52502	53502	53503	52503	51503
51504	52504	53504	53505	52505	51505
51506	52506	53506	53507	52507	51507
51508	52508	53508	53702	52702	51702
51509	52509	53509	53510	52510	51510
51511	52511	53511	53512	52512	51512
51513	52513	53513	53514	52514	51514
51515	52515	53515	53703	52703	51703
51516	52516	53516	53517	52517	51517
51518	52518	53518	53519	52519	51519
51520	52520	53520	53704	52704	51704
51521	52521	53521	53522	52522	51522
51523	52523	53523	53524	52524	51524
51525	52525	53525	53705	52705	51705
51526	52526	53526	53527	52527	51527
51528	52528	53528	53529	52529	51529
51530	52530	53530	53531	52531	51531
51532	52532	53532	53706	52706	51706
51533	52533	53533	53534	52534	51534
51535	52535	53535	53536	52536	51536
51537	52537	53537	53538	52538	51538
51539	52539	53539	53707	52707	51707
51540	52540	53540	53541	52541	51541
51542	52542	53542	53543	52543	51543

51544	52544	53544	53545	52545	51545
51546	52546	53546	53708	52708	51708
51547	52547	53547	53548	52548	51548
51549	52549	53549	53550	52550	51550
51551	52551	53551	53711	52711	51711
51553	52553	53553	53709	52709	51709
51554	52554	53554	53555	52555	51555
51556	52556	53556	53557	52557	51557
51558	52558	53558	53559	52559	51559
51560	52560	53560	53710	52710	51710
51561	52561	53561	53562	52562	51562
51563	52563	53563	53564	52564	51564
51565	52565	53565	53566	52566	51566
51567	52567	53567	53552	52552	51551
51568	52568	53568	53569	52569	51569
51570	52570	53570	53571	52571	51571
51572	52572	53572	53573	52573	51573
51574	52574	53574	53712	52712	51712
51575	52575	53575	53576	52576	51576
51577	52577	53577	53578	52578	51578
51579	52579	53579	53580	52580	51580
51581	52581	53581	53713	52713	51713
51582	52582	53582	53583	52583	51583
51584	52584	53584	53585	52585	51585
51586	52586	53586	53587	52587	51587
51588	52588	53588	53714	52714	51714
51589	52589	53589	53590	52590	51590
51591	52591	53591	53592	52592	51592
51593	52593	53593	53594	52594	51594
51595	52595	53595	53715	52715	51715
51596	52596	53596	53597	52597	51597
51598	52598	53598	53599	52599	51599
51600	52600	53600	53601	52601	51601
51602	52602	53602	53716	52716	51716
51603	52603	53603	53604	52604	51604
51605	52605	53605	53606	52606	51606
51607	52607	53607	53608	52608	51608
51609	52609	53609	53717	52717	51717
51610	52610	53610	53611	52611	51611
51612	52612	53612	53613	52613	51613
51614	52614	53614	53615	52615	51615

51616	52616	53616	53718	52718	51718
51617	52617	53617	53618	52618	51618
51619	52619	53619	53620	52620	51620
51621	52621	53621	53622	52622	51622
51623	52623	53623	53719	52719	51719
51624	52624	53624	53625	52625	51625
51626	52626	53626	53627	52627	51627
51628	52628	53628	53629	52629	51629
51630	52630	53630	53720	52720	51720
51631	52631	53631	53632	52632	51632
51633	52633	53633	53634	52634	51634
51635	52635	53635	53636	52636	51636
51636	52636	53636	53721	52721	51721
51638	52638	53638	53639	52639	51639
51640	52640	53640	53641	52641	51641
51642	52642	53642	53643	52643	51643
51644	52644	53644	53722	52722	51722
51645	52645	53645	53646	52646	51646
51647	52647	53647	53648	52648	51648
51649	52649	53649	53650	52650	51650
51651	52651	53651	53723	52723	51723
51652	52652	53652	53653	52653	51653
51654	52654	53654	53655	52655	51655
51656	52656	53656	53657	52657	51657
51658	52658	53658	53724	52724	51724
51659	52659	53659	53660	52660	51660
51661	52661	53661	53662	52662	51662
51663	52663	53663	53664	52664	51664
51665	52665	53665	53725	52725	51725
51666	52666	53666	53667	52667	51667
51668	52668	53668	53669	52669	51669
51670	52670	53670	53671	52671	51671
51672	52672	53672	53726	52726	51726
51673	52673	53673	53674	52674	51674
51675	52675	53675	53676	52676	51676
51677	52677	53677	53678	52678	51678
51679	52679	53679	53680	52680	51680
51681	52681	53681	53682	52682	51682
51683	52683	53683	53684	52684	51684
51685	52685	53685	53686	52686	51686

Piccadilly Line

Map Colour: dark blue

Routes: Cockfosters to Uxbridge via Finsbury Park, King's Cross St Pancras, Leicester Square, Earls Court, Acton Town, Rayners Lane, and Ruislip; branches from Acton Town to Heathrow terminals 2/3 and 5, plus a loop from Hatton Cross to Terminals 23.

Mileage: 44

Number of stations: 53

Trains: 86 × six-car sets built 1973.

From 2025, a new generation trains for the Piccadilly line will begin to replace the 1970s fleet. Ninety-four 'Inspiro London' sets will also offer more capacity with a more frequent and reliable service. They have Wider doors and walk-through, air-conditioned carriages with improved access from the platform to train for a more comfortable and accessible journey. They are designed with sustainability in mind, reducing energy consumption by 20 per cent.

Three-car single-ended units

DM	T	UNDM
100	500	300
101	501	301
102	502	302
103	503	303
104	504	304
105	505	305
106	506	306
107	507	307
108	508	308
109	509	309
110	510	310
111	511	311
112	512	312
113	513	313
115	515	315
116	516	316
117	517	317
118	518	318
119	519	319
120	520	320
121	521	321
122	522	322
123	523	323
124	524	324
125	525	325
126	526	326
127	527	327
128	528	328
129	529	329
130	530	330
131	531	331
132	532	332
133	533	333
134	534	334
135	535	335
136	536	336
137	537	337
138	538	338
139	539	339
140	540	340
141	541	341
142	542	342
143	543	343
144	544	344
145	545	345
146	546	346
147	547	347
148	548	348
149	549	349
150	550	350
151	551	351
152	552	352
153	553	353
154	554	354
155	555	355
156	556	356
157	557	357
158	558	358
159	559	359
160	560	360
161	561	361
162	562	362
163	563	363
164	564	364
165	565	365
167	567	367
168	568	368
169	569	369
170	570	370
171	571	371
172	572	372
173	573	373
174	574	374
175	575	375
176	576	376
177	577	377
178	578	378
179	579	379
180	580	380
181	581	381
182	582	382
183	583	383
184	584	384
185	585	385
186	586	386
187	587	387
188	588	388
189	589	389
190	590	390
191	591	391
192	592	392
193	593	393
194	594	394
195	595	395
196	596	396
197	597	397
198	598	398
199	599	399
200	600	400
201	601	401
202	602	402
203	603	403
205	605	405
206	606	406
207	607	407
208	608	408
209	609	409
210	610	410
211	611	411
212	612	412
213	613	413
214	614	414
215	615	415
216	616	416
217	617	417
218	618	418
219	619	419
220	620	420
221	621	421
222	622	422
223	623	423
224	624	424
225	625	425
226	626	426
227	627	427
228	628	428
229	629	429
230	630	430
231	631	431
232	632	432
233	633	433
234	634	434
235	635	435
236	636	436
237	637	437
238	638	438
239	639	439
240	640	440
241	641	441
242	642	442
243	643	443
244	644	444
245	645	445
246	646	446
247	647	447
248	648	448
249	649	449
250	650	450
251	651	451
252	652	452
253	653	453
566	366	

Three-car double-ended units

DM	T	DM
854	654	855
856	656	857
858	658	859
860	660	861
862	662	863
864	664	865
866	666	867
868	668	869
870	670	871
872	672	873
874	674	875
876	676	877
878	678	879
880	680	881
882	682	883
884	684	885
886	686	887
890	690	891
892	692	893
894	694	895
896	696	897

Victoria Line

Map Colour: Light blue

Routes: Walthamstow Central to Brixton via Finsbury Park, King's Cross St Pancras, Euston, Green Park, Victoria and Vauxhall

Mileage: 13

Number of stations: 16

Trains: 47 × eight-car sets built 2009

DM	T	NDM	UNDM	UNDM	NDM	T	DM
11001	12001	13001	14001	14002	13002	12002	11002
11003	12003	13003	14003	14004	13004	12004	11004
11005	12005	13005	14005	14006	13006	12006	11006
11007	12007	13007	14007	14008	13008	12008	11008
11009	12009	13009	14009	14010	13010	12010	11010
11011	12011	13011	14011	14012	13012	12012	11012
11013	12013	13013	14013	14014	13014	12014	11014
11015	12015	13015	14015	14016	13016	12016	11016
11017	12017	13017	14017	14018	13018	12018	11018
11019	12019	13019	14019	14020	13020	12020	11020
11021	12021	13021	14021	14022	13022	12022	11022
11023	12023	13023	14023	14024	13024	12024	11024
11025	12025	13025	14025	14026	13026	12026	11026
11027	12027	13027	14027	14028	13028	12028	11028
11029	12029	13029	14029	14030	13030	12030	11030
11031	12031	13031	14031	14032	13032	12032	11032
11033	12033	13033	14033	14034	13034	12034	11034
11035	12035	13035	14035	14036	13036	12036	11036
11037	12037	13037	14037	14038	13038	12038	11038
11039	12039	13039	14039	14040	13040	12040	11040
11041	12041	13041	14041	14042	13042	12042	11042
11043	12043	13043	14043	14044	13044	12044	11044
11045	12045	13045	14045	14046	13046	12046	11046
11047	12047	13047	14047	14048	13048	12048	11048
11049	12049	13049	14049	14050	13050	12050	11050
11051	12051	13051	14051	14052	13052	12052	11052
11053	12053	13053	14053	14054	13054	12054	11054
11055	12055	13055	14055	14056	13056	12056	11056
11057	12057	13057	14057	14058	13058	12058	11058
11059	12059	13059	14059	14060	13060	12060	11060
11061	12061	13061	14061	14062	13062	12062	11062
11063	12063	13063	14063	14064	13064	12064	11064
11065	12065	13065	14065	14066	13066	12066	11066
11067	12067	13067	14067	14068	13068	12068	11068
11069	12069	13069	14069	14070	13070	12070	11070
11071	12071	13071	14071	14072	13072	12072	11072
11073	12073	13073	14073	14074	13074	12074	11074
11075	12075	13075	14075	14076	13076	12076	11076
11077	12077	13077	14077	14078	13078	12078	11078
11079	12079	13079	14079	14080	13080	12080	11080
11081	12081	13081	14081	14082	13082	12082	11082
11083	12083	13083	14083	14084	13084	12084	11084
11085	12085	13085	14085	14086	13086	12086	11086
11087	12087	13087	14087	14088	13088	12088	11088
11089	12089	13089	14089	14090	13090	12090	11090
11091	12091	13091	14091	14092	13092	12092	11092
11093	12093	13093	14093	14094	13094	12094	11094

The Piccadilly Line has ordered 94 nine-car Siemens Inspiro trains for delivery from 2025.

Waterloo & City Line

Map Colour: Turquoise
Routes: Bank to Waterloo
Mileage: 1.5
Number of stations: 2
Trains: 5 × four-car sets built 1992. Each train runs in a 655xx+675xx+675xx+655xx formation.

Two-car sets

DM	M
65501	67501
65503	67503
65505	67505
65507	67507
65509	67509

Two-car sets

M	DM
67502	65502
67504	65504
67506	65506
67508	65508
67510	65510

Sub-surface Lines Stock

A fleet of 132 seven-car and 80 eight-car sets work the lines known as the sub-surface lines – the Circle, District, Hammersmith & City and Metropolitan lines.

Circle Line

Map Colour: Yellow
Routes: Edgware Road to Hammersmith via Paddington, Victoria, Embankment, Cannon St, Liverpool St, King's Cross St Pancras, Baker St, Edgware Road and Latimer Road
Mileage: 17
Number of stations: 36
Trains: 132 × seven-car sets built 2008-12. These trains are shared between District and Hammersmith & City lines

District Line

Map Colour: Green
Routes: Ealing Broadway to Upminster via Turnham Green, Earls Court, Victoria, Embankment. Tower Hill, Bow Road, West Ham, Barking and Dagenham East; Edgware Road to Wimbledon; branches from Kensington Olympia to Earls Court and Turnham Green to Richmond
Mileage: 40
Number of stations: 60
Trains: 132 × seven-car sets built 2008-12. These trains are shared between Circle and Hammersmith & City lines

Hammersmith & City Line

Map Colour: Pink
Routes: Hammersmith to Barking, via Paddington, Euston Square, King's Cross St Pancras, Liverpool St, Bow Road and West Ham
Mileage: 15.8
Number of stations: 29
Trains: 132 × seven-car sets built 2008-12. These trains are shared between District and Circle lines

Trams, Metros and Light Rail

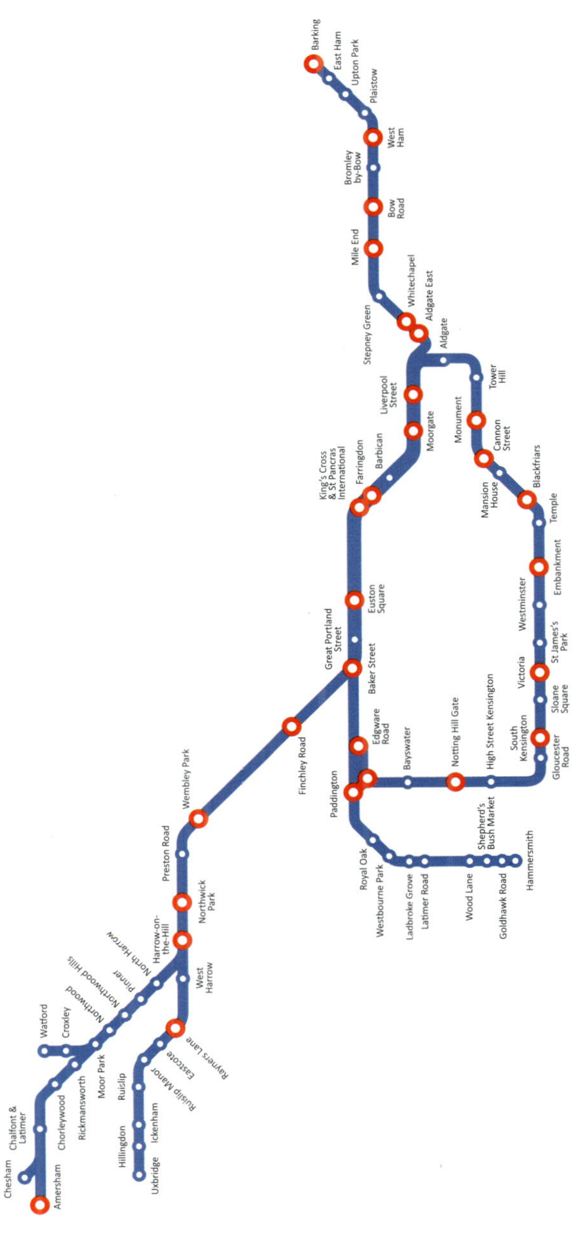

Trams, Metros and Light Rail

Metropolitan Line

Map Colour: Maroon
Routes: Amersham to Aldgate via Rickmansworth, Harrow-on-the-Hill, Wembley park, Baker St, King's Cross St Pancras and Liverpool Street; branches from Chalfont & Latimer to Chesham, Moor Park to Watford, Harrow-on-the-Hill to Uxbridge
Mileage: 42
Number of stations: 34
Trains: 60 × eight-car sets built 2008-12

Sub-surface stock S-stock fleets

S8 eight-car sets

DM	M	M	MS	MS	M	M	DM
21001	22001	23001	24001	24002	25002	22002	21002
21003	22003	23003	24003	24004	25004	22004	21004
21005	22005	23005	24005	24006	25006	22006	21006
21007	22007	23007	24007	24008	25008	22008	21008
21009	22009	23009	24009	24010	25010	22010	21010
21011	22011	23011	24011	24012	25012	22012	21012
21013	22013	23013	24013	24014	25014	22014	21014
21015	22015	23015	24015	24016	25016	22016	21016
21017	22017	23017	24017	24018	25018	22018	21018
21019	22019	23019	24019	24020	25020	22020	21020
21021	22021	23021	24021	24022	25022	22022	21022
21023	22023	23023	24023	24024	25024	22024	21024
21025	22025	23025	24025	24026	25026	22026	21026
21027	22027	23027	24027	24028	25028	22028	21028
21029	22029	23029	24029	24030	25030	22030	21030
21031	22031	23031	24031	24032	25032	22032	21032
21033	22033	23033	24033	24034	25034	22034	21034
21035	22035	23035	24035	24036	25036	22036	21036
21037	22037	23037	24037	24038	25038	22038	21038
21039	22039	23039	24039	24040	25040	22040	21040
21041	22041	23041	24041	24042	25042	22042	21042
21043	22043	23043	24043	24044	25044	22044	21044
21045	22045	23045	24045	24046	25046	22046	21046
21047	22047	23047	24047	24048	25048	22048	21048
21049	22049	23049	24049	24050	25050	22050	21050
21051	22051	23051	24051	24052	25052	22052	21052
21053	22053	23053	24053	24054	25054	22054	21054
21055	22055	23055	24055	24056	25056	22056	21056
21057	22057	23057	24057	24058	23058	22058	21058
21059	22059	23059	24059	24060	23060	22060	21060
21061	22061	23061	24061	24062	23062	22062	21062
21063	22063	23063	24063	24064	23064	22064	21064
21065	22065	23065	24065	24066	23066	22066	21066
21067	22067	23067	24067	24068	23068	22068	21068
21069	22069	23069	24069	24070	23070	22070	21070
21071	22071	23071	24071	24072	23072	22072	21072
21073	22073	23073	24073	24074	23074	22074	21074
21075	22075	23075	24075	24076	23076	22076	21076
21077	22077	23077	24077	24078	23078	22078	21078
21079	22079	23079	24079	24080	23080	22080	21080
21081	22081	23081	24081	24082	23082	22082	21082
21083	22083	23083	24083	24084	23084	22084	21084

21085	22085	23085	24085	24086	23086	22086	21086
21087	22087	23087	24087	24088	23088	22088	21088
21089	22089	23089	24089	24090	23090	22090	21090
21091	22091	23091	24091	24092	23092	22092	21092
21093	22093	23093	24093	24094	23094	22094	21094
21095	22095	23095	24095	24096	23096	22096	21096
21097	22097	23097	24097	24098	23098	22098	21098
21099	22099	23099	24099	24100	23100	22100	21100
21101	22101	23101	24101	24102	23102	22102	21102
21103	22103	23103	24103	24104	23104	22104	21104
21105	22105	23105	24105	24106	23106	22106	21106
21107	22107	23107	24107	24108	23108	22108	21108
21109	22109	23109	24109	24110	23110	22110	21110
21111	22111	23111	24111	24112	23112	22112	21112
21113	22113	23113	24113	24114	23114	22114	21114
21115	22115	23115	24115	24116	23116	22116	21116

S7 – seven-car sets (two are S7+1, eight-car sets)

DM	M	M	MS	MS	M	M	DM
21301	22301		24301	24302	25302	22302	21302
21303	22303		24303	24304	25304	22304	21304
21305	22305		24305	24306	25306	22306	21306
21307	22307		24307	24308	25308	22308	21308
21309	22309		24309	24310	25310	22310	21310
21311	22311		24311	24312	25312	22312	21312
21313	22313		24313	24314	25314	22314	21314
21315	22315		24315	24316	25316	22316	21316
21317	22317		24317	24318	25318	22318	21318
21319	22319		24319	24320	25320	22320	21320
21321	22321		24321	24322	25322	22322	21322
21323	22323	25384	24323	24324	25324	22324	21324
21325	22325		24325	24326	25326	22326	21326
21327	22327	25386	24327	24328	25328	22328	21328
21329	22329		24329	24330	25330	22330	21330
21331	22331		24331	24332	25332	22332	21332
21333	22333		24333	24334	25334	22334	21334
21335	22335		24335	24336	25336	22336	21336
21337	22337		24337	24338	25338	22338	21338
21339	22339		24339	24340	25340	22340	21340
21341	22341		24341	24342	25342	22342	21342
21343	22343		24343	24344	25344	22344	21344
21345	22345		24345	24346	25346	22346	21346
21347	22347		24347	24348	25348	22348	21348
21349	22349		24349	24350	25350	22350	21350
21351	22351		24351	24352	25352	22352	21352
21353	22353		24353	24354	25354	22354	21354
21355	22355		24355	24356	25356	22356	21356
21357	22357		24357	24358	25358	22358	21358
21359	22359		24359	24360	25360	22360	21360
21361	22361		24361	24362	25362	22362	21362
21363	22363		24363	24364	25364	22364	21364
21365	22365		24365	24366	25366	22366	21366
21367	22367		24367	24368	25368	22368	21368
21369	22369		24369	24370	25370	22370	21370
21371	22371		24371	24372	25372	22372	21372
21373	22373		24373	24374	25374	22374	21374
21375	22375		24375	24376	25376	22376	21376

21377	22377		24377	24378	25378	22378	21378
21379	22379		24379	24380	25380	22380	21380
21381	22381		24381	24382	25382	22382	21382
21383	22383		24383	24384	23384	22384	21384
21385	22385		24385	24386	23386	22386	21386
21387	22387		24387	24388	23388	22388	21388
21389	22389		24389	24390	23390	22390	21390
21391	22391		24391	24392	23392	22392	21392
21393	22393		24393	24394	23394	22394	21394
21395	22395		24395	24396	23396	22396	21396
21397	22397		24397	24398	23398	22398	21398
21399	22399		24399	24400	23400	22400	21400
21401	22401		24401	24402	23402	22402	21402
21403	22403		24403	24404	23404	22404	21404
21405	22405		24405	24406	23406	22406	21406
21407	22407		24407	24408	23408	22408	21408
21409	22409		24409	24410	23410	22410	21410
21411	22411		24411	24412	23412	22412	21412
21413	22413		24413	24414	23414	22414	21414
21415	22415		24415	24416	23416	22416	21416
21417	22417		24417	24418	23418	22418	21418
21419	22419		24419	24420	23420	22420	21420
21421	22421		24421	24422	23422	22422	21422
21423	22423		24423	24424	23424	22424	21424
21425	22425		24425	24426	23426	22426	21426
21427	22427		24427	24428	23428	22428	21428
21429	22429		24429	24430	23430	22440	21430
21431	22441		24431	24432	23432	22442	21432
21433	22443		24433	24434	23434	22444	21434
21435	22445		24435	24436	23436	22446	21436
21437	22447		24437	24438	23438	22448	21438
21439	22449		24439	24440	23440	22440	21440
21441	22441		24441	24442	23442	22442	21442
21443	22443		24443	24444	23444	22444	21444
21445	22445		24445	24446	23446	22446	21446
21447	22447		24447	24448	23448	22448	21448
21449	22449		24449	24450	23450	22450	21450
21451	22451		24451	24452	23452	22452	21452
21453	22453		24453	24454	23454	22454	21454
21455	22455		24455	24456	23456	22456	21456
21457	22457		24457	24458	23458	22458	21458
21459	22459		24459	24460	23460	22460	21460
21461	22461		24461	24462	23462	22462	21462
21463	22463		24463	24464	23464	22464	21464
21465	22465		24465	24466	23466	22466	21466
21467	22467		24467	24468	23468	22468	21468
21469	22469		24469	24470	23470	22470	21470
21471	22471		24471	24472	23472	22472	21472
21473	22473		24473	24474	23474	22474	21474
21475	22475		24475	24476	23476	22476	21476
21477	22477		24477	24478	23478	22478	21478
21479	22479		24479	24480	23480	22480	21480
21481	22481		24481	24482	23482	22482	21482
21483	22483		24483	24484	23484	22484	21484
21485	22485		24485	24486	23486	22486	21486
21487	22487		24487	24488	23488	22488	21488
21489	22489		24489	24490	23490	22490	21490
21491	22491		24491	24492	23492	22492	21492

21493	22493	24493	24494	23494	22494	21494
21495	22495	24495	24496	23496	22496	21496
21497	22497	24497	24498	23498	22498	21498
21499	22499	24499	24400	25500	22500	21500
21501	22501	24501	24502	23502	22502	21502
21503	22503	24503	24504	23504	22504	21504
21505	22505	24505	24506	23506	22506	21506
21507	22507	24507	24508	23508	22508	21508
21509	22509	24509	24510	23510	22510	21510
21511	22511	24511	24512	23512	22512	21512
21513	22513	24513	24514	23514	22514	21514
21515	22515	24515	24516	23516	22516	21516
21517	22517	24517	24518	23518	22518	21518
21519	22519	24519	24520	23520	22520	21520
21521	22521	24521	24522	23522	22522	21522
21523	22523	24523	24524	23524	22524	21524
21525	22525	24525	24526	23526	22526	21526
21527	22527	24527	24528	23528	22528	21528
21529	22529	24529	24530	23530	22540	21530
21531	22541	24531	24532	23532	22542	21532
21533	22543	24533	24534	23534	22544	21534
21535	22545	24535	24536	23536	22546	21536
21537	22547	24537	24538	23538	22548	21538
21539	22549	24539	24540	23540	22540	21540
21541	22541	24541	24542	23542	22542	21542
21543	22543	24543	24544	23544	22544	21544
21545	22545	24545	24546	23546	22546	21546
21547	22547	24547	24548	23548	22548	21548
21549	22549	24549	24550	23550	22550	21550
21551	22551	24551	24552	23552	22552	21552
21553	22553	24553	24554	23554	22554	21554
21555	22555	24555	24556	23556	22556	21556
21557	22557	24557	24558	23558	22558	21558
21559	22559	24559	24560	23560	22560	21560
21561	22561	24561	24562	23562	22562	21562
21563	22563	24563	24564	23564	22564	21564
21565	22565	24565	24566	23566	22566	21566
21567	22567	24567	24568	23568	22568	21568

Blackpool & Fleetwood Tramway

Routes: Fleetwood to Starr Gate via Broadwater, Bispham, Blackpool Tower and Burlington Road West
Mileage: 11
Number of stations: 38
Trains: 18 × Bombardier Flexity five-section trams

Dating from 1885, this is one of the oldest tramway systems anywhere in the world and comprises one main route. It is electrified using 600V DC overhead line.

It was recently modernised to transition from a heritage tourist attraction to a viable, modern light rail system. Part of that upgrade saw delivery of a fleet of German-built Bombardier Flexity trams dating from 2011-12. They are numbered 001-018 and are referred to as the 'A' Fleet.

The system also has nine 'balloon' double-deck single car trams dating from the 1930s; the 'B fleet'. While these do not meet modern standards, they have exemptions for use but are typically only used in the high summer or autumn Illuminations seasons. They are numbered in the 700 series, although some are in store.

A fleet of vintage trams, some as old as 1901, are also allowed to be used on special workings and private operations, such as for weddings and functions. This is the 'C fleet'.

All the new trams are maintained at a new purpose-built depot at Starr Gate at the southern end of the line, while the original Rigby Road depot is retained for the maintenance and storage of the holder heritage trams.

Flexity2 – 'A' fleet

Built: 2010-12 by Bombardier, Bautzen, Germany

No.	Livery	Name
001	Merlin Attractions (advert)	
002	Purple/white	*Alderman E.E. Wynne*
003	Pretty Little Thing (advert)	
004	Purple/white	
005	Purple/white	
006	Purple/white	
007	Purple/white	*Alan Whitbread*
008	Purple/white	
009	Purple/white	
010	Purple/white	
011	Purple/white	
012	Purple/white	
013	Purple/white	
014	Purple/white	
015	Purple/white	
016	Pretty Little Thing (advert)	
017	Purple/white	
018	Purple/white	

The Blackpool & Fleetway tram system may be most famous for its heritage trams, some of which are still in use, but it also boasts new modem trams for commuters and leisure passengers alike. Bombardier Flexity A2 tram 011 stands in the shadow of the iconic Blackpool tower on 15 May 2019. *Tony Miles*

Modified Balloon Cars – 'B' Fleet

Built: 1934-35 by English Electric, Preston

No.	Livery	Name	Year Last Operated
700	Purple/white		2020
707	GRC		2020
709	Sealife Centre (advert)		2013
711	Purple/white	*Ray Roberts*	2020
713	Houndshill Shopping Centre (advert)		2020
718	GRC		2020
719	Purple/white	*Donna's Dream House*	2019
720	Walls Ice Cream (advert)		2011
724	Lyndene Hotel (advert)		2013

Heritage Trams – 'C' Fleet

Note: those marked as stored/display have not run for several years. They may be operational and some have been loaned to Crich or Beamish and operated at those sites.

Tram	Livery	Name	Notes
OMO 8	Plum/custard		Stored/display
Lytham 43	Blue/white		Stored/display
Bolton 66	Maroon/cream		2020
Standard 143	Red/white/teak		2019
Standard 147	GRC		2020
Boat 227	Red/white	*Charlie Cairoli*	2020
Boat 230	GRC	*George Formby OBE*	Stored/display
Brush 259	Works green		Stored/display)
Twin Car 272+T2	Cream		Stored/display
Railcoach 279	GRC		Stored/display
Brush 290	GRC		Stored/display
Coronation 304	Cream/green		Stored/display
Boat 600	GRC	*Duchess of Cornwall*	2020
Brush 621	GRC		2020
Brush 625	GRC		Stored/display
Brush 631	GRC		2020
Brush 632	GRC		Stored/display
Brush 634	GRC		Stored/display
Centenary 642	Travelcard (advert)		2020
Centenary 645	Red/white	Stored/display	
Centenary 648	GRC		2020
Coronation 660	Cream/green		Stored/display
Coronation 663	GRC		Stored/display
Twin Car 671+681	green/yellow		Stored/display
Twin Car 675+685	GRC		Stored/display
Twin Car 676+686	GRC		Stored/display
Ex-Towing Railcoach 680	GRC		2020
Balloon 701	Red/white		2020
Balloon 703	Sunderland red/white		Stored/display
Balloon 704	Eclipse at the Globe, Pleasure Beach (advert)		Stored/display
Open Top Balloon 706	GRC	*Princess Alice*	Stored/display
Balloon 708	GRC		Stored/display
Balloon 715	GRC		2020
Balloon 717	GRC	*Walter Luff*	2020
Balloon 723	GRC		2020
Balloon 726	HM Coastguard (advert)		2010
Illuminated Rocket 732	Green		Stored/display
Illuminated Western Train 733+734	Red/green/yellow		2019
Blackpool Hovertram 735	White/black		Stored/display
Illuminated Frigate 736	White	*HMS Blackpool*	2020
Illuminated Trawler 737	Blue/red/black/white	*Cevic*	2020
Jubilee 761	Wynsors World of Shoes (advert)		Stored/display
Halle 902	Red & White		Stored/display
Glasgow 1016			Stored/display

Engineering Vehicles

No.	Livery	Purpose
260	Works yellow	Rail Crane
750	Yellow	Reel Wagon
754	Yellow	Overhead Line Car
938	Yellow	Road/Rail Vehicle
939	Yellow/orange	Road/Rail Vehicle
Crab	Yellow	Shunter

Docklands Light Railway

Docklands Light Railway first opened in 1987 as a fully automated operation and in the intervening 35 years has expanded with lots of new routes and new trains. Units 12 and 14 pass Westferry on 16 January 2019. *Rob France*

Routes: Tower Gateway / Bank to Stratford, Beckton, Stratford-Woolwich Arsenal, Canary Wharf-Lewisham
Mileage: 24
Number of stations: 45
Trains: 94 × two-section trains

Famous for using driverless trains, the 750V DC Docklands Light Railway is another light rail system that has organically grown since it first opened in 1987.

The initial network was from Tower Gateway to Island Gardens, in the Isle of Dogs, with a branch from Poplar to Stratford. In 1991 a short spur to serve Bank station in the City of London also opened.

The first major extension came in 1994 with Poplar to Beckton opening. The next new line, built just in time for the Millennium, was from Island Gardens to Lewisham, which included a new tunnel under the Thames and the resiting and closure of the original Island Gardens station.

Canning Town to King George V opened in December 2005 and this line was extended – again under the Thames – to Woolwich Arsenal in January 2009. The last expansion was from Stratford International to Canning Town.

The initial 21 trains, numbered 01-21, were two-car units built by BN in Belgium, which could run in pairs. These were withdrawn in 1991-95 (as they could not work to Bank) and sold for use in Germany.

Trams, Metros and Light Rail

They were replaced by 23 new trains, 22-44, and then supplemented by another 47 units in 1992-95. Then, as further expansion of the network dictated further new trains, another 24 units, 92-99 and 01-16, were added from 2002-03 built by Bombardier. Finally, 55 more units were delivered in 2007-10, also built by Bombardier, and numbered 101-155. These newest units were built in Germany, with the rest built in Belgium. The trains are maintained at depots at Poplar and Beckton. All are in DLR red livery with blue waves.

Class B90

Built: 1991-92 by BN Construction Brugge, Belgium

22	24	26	28	30	32	34	36	38	40	42	44
23	25	27	29	31	33	35	37	39	41	43	

These vehicles are due to be replaced by 43 new CAF vehicles from 2023.

Class B92

Built: 1992-95 by BN Construction, Brugge, Belgium

45	49	53	57	61	65	69	73	77	81	85	89
46	50	54	58	62	66	70	74	78	82	86	90
47	51	55	59	63	67	71	75	79	83	87	91
48	52	56	60	64	68	72	76	80	84	88	

These vehicles are due to be replaced by 43 new CAF vehicles from 2023.

Class B2K

Built: 2001 by Bombardier Transportation, Brugge, Belgium

01	03	05	07	09	11	13	15	92	94	96	98
02	04	06	08	10	12	14	16	93	95	97	99

These vehicles are due to be replaced by 43 new CAF vehicles from 2023.

Class B2007

Built: 2007-08 by Bombardier Transportation, Bautzen, Germany

101	103	105	107	109	111	113	115	117	119	121	123
102	104	106	108	110	112	114	116	118	120	122	124

Class B2009

Built: 2009 by Bombardier Transportation, Bautzen, Germany

125	128	131	134	137	140	143	146	149	152	155
126	129	132	135	138	141	144	147	150	153	
127	130	133	136	139	142	145	148	151	154	

Works locos

The DLR also has a 1979 GEC four-wheel diesel hydraulic shunter numbered 994 and named *Kevin Keaney*, a battery electric works loco numbered 993 and named *Kylie* and a crane trolley numbered 992.

New CAF trains

Forty-three new walkthrough trains are on order from CAF for entry into service from 2023. Numbering and further details are awaited.

London Tramlink

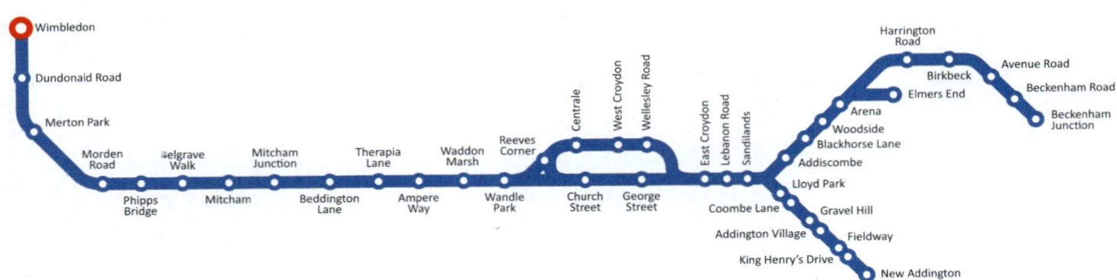

Routes: Wimbledon to New Addington via Mitcham, East Croydon and Sandlands; branches to Elmers End and Beckenham Junction
Mileage: 17.5
Number of stations: 39
Trains: 24 × Bombardier Flexity Swift and 12 x Stadler Variobahn

This system, which first opened in May 2000, uses many former heavy rail lines converted to tram operation, including Wimbledon to West Croydon and Elmers End to Addiscombe. In addition to their conversions, there were extensions into the centre of Croydon and a wholly new line from East Croydon to New Addington, and also from Arena to Beckenham Junction.

The system uses 750V DC overhead lines. It has its only depot at Therapia Lane, midway on the Wimbledon branch, and this maintains the fleet of 24 Bombardier Flexity Swift three-section trams that were delivered to launch the service, and now a new fleet of 12 Stadler Variobahn five-section trams that date from 2011-16. One tram, 2551, was involved in a series accident in 2016 when it fell on its side when the driver took a corner too fast. It is stored and unlikely to return to service.

CR4000s

Built: 1998-99 Bombardier-Wien Schienfahrzeuge, Austria

All in lime green, blue & white livery apart from 2554 in an advertising livery.

No.	Notes
2530	
2531	
2532	
2533	
2534	
2535	Named *Stephen Parascandolo 1980-2007*
2536	
2537	
2538	
2539	
2540	
2541	
2542	
2543	
2544	
2545	
2546	
2547	
2548	
2549	
2550	
2551	Stored after derailment damage
2552	
2553	

Variobahn

Built: 2011-12 (2554-59) and 2015-16 (2560-65) by Stadler, Germany

No.			
2554	2557	2561	2565
2555	2558	2562	
2556	2559	2563	
	2560	2564	

London Trams, marketed as Tramlink, is the metro system operating in Croydon, with a mix of new lines and ex-heavy rail routes. 2541 passes Lloyd Park in Croydon on 28 October 2020. *Rob France*

Edinburgh Trams

Routes: Edinburgh Airport-York Place via Edinburgh Park, Haymarket and Princes Street
Mileage: 8.5
Number of stations: 16
Trains: 27 × CAF Urbos 3

One of the newer tram systems, and a bespoke operation, it opened on 31 May 2014, running one just one route from Edinburgh Airport in the west, through the city centre and out to York Place in the east.

It uses 750V DC overheads and has a fleet of 27 seven-section CAF Urbos 3 trams, maintained at the system's one depot at Gogar.

The system is being extended from York Place to Newhaven, while there is a proposal to build a second branch from Haymarket to Granton. Longer term, a line from Ingliston to Newbridge is a possibility.

Built: 2009-12 by CAF, Spain

No.	Livery
251	White/black/grey/madder
252	CR Smith (advert)
253	CR Smith (advert)
254	Edinburgh Park (advert)
255	Forever Edinburgh (advert)
256	CR Smith (advert)
257	Ray Harryhausen (advert)
258	CR Smith (advert)
259	CR Smith (advert)
260	Johnnie Walker (advert)
261	White/black/grey/madder
262	White/black/grey/madder
263	White/black/grey/madder
264	Johnnie Walker (advert)
265	CR Smith (advert)
266	CR Smith (advert)
267	White/black/grey/madder
268	Edinburgh Park (advert)
269	White/black/grey/madder
270	White/black/grey/madder
271	Parabola My Edinburgh Park (advert)
272	John Lewis (advert)
273	White/black/grey/madder
274	Parabola (advert)
275	Scottish Rugby (advert)
276	Parabola (advert)
277	Parabola My Edinburgh Park (advert)

Edinburgh's tram network opened in 2014 and has new routes being constructed and proposed. CAF set 270 heads east after calling at Ingliston Park & Ride station on 27 August 2021. *Pip Dunn*

Glasgow Underground

Routes: Buchannan St to Buchannan St via Govan (circular route)
Mileage: 6.5
Number of stations: 15
Trains: 41 vehicles running in two- and three-car sets

This 600V DC third rail underground system opened way back in 1896 and uses a non-standard 4ft track gauge. It is known as the clockwork orange as the trains, in bright orange livery, run on the system's circular route.

The current fleet is varied. It has 33 power cars, which run in pairs, dating from 1977-79 and entered traffic in 1980. These were supplemented by eight centre trailer cars from 1992 to allow three-car trains to run.

However, all are on the cusp of being replaced by 17 new four-car Stadler trains, which will be able to be operated driverless.

Driving motors

DM	Livery
101	Maroon/cream
102	Orange
103	Orange
104	Orange
105	Orange
106	Orange
107	Orange
108	Orange
109	Orange
110	Orange
111	Orange
112	Orange
113	Orange
114	Orange
115	Orange
116	Orange
117	Orange
118	Orange
119	Orange
120	Orange
121	Orange
122	Orange
123	Orange
124	Orange
125	Orange
126	Orange
127	Orange
128	Orange
129	Orange
130	Orange/white
131	Orange
132	Orange
133	Orange

Trailer vehicles

No.	Livery
201	Orange
202	Orange
203	Orange
204	Orange
205	Orange
206	Orange
207	Orange
208	Orange

One of the soon-to-be-replaced Glasgow Underground units calls at Glasgow West Street Station on 22 March 2019. *Rob France*

Manchester Metrolink

Routes: Bury to Altrincham via Manchester Victoria, Old Trafford and Timperley, Rochdale to East Didsbury, via Milnrow, Oldham, Failsworth, Manchester Victoria, Chorlton and West Didsbury; Eccles to Ashton-under-Lyne via Salford Quays, Manchester Piccadilly and Droylesden; Manchester Victoria to Manchester Airport; Trafford Centre to Cornbrook
Mileage: 63
Number of stations: 99
Trains: 120 two-car articulated Bombardier M5000 series 750V DC overhead trams
The Manchester Metrolink is a glowing example of how a good modern light rail system can flourish.

The first line opened in April 1992 running from Bury to Altrincham, mostly using old BR lines. It ran into the centre of Manchester and served both of the city's main stations at Victoria and Piccadilly, but the sections from Queens Road to Bury and Old Trafford to Altrincham were ex-BR lines. The Bury line was 1,200V DC third rail, while the new tram system was 750V DC overhead.

In 2000, a new branch from Manchester to Eccles was added, followed by a spur to Media City in in 2010.

In 2013, the transformation of the Dean Lane to Rochdale, via Oldham, was converted from heavy rail to Metrolink with an extension into Rochdale town centre. At the same time, a new branch in the east to Droylsden was added and this was extended to Ashton-under-Lyne in 2014. Also, a short branch from Old Trafford to Chorlton opened in 2013 and was extended to East Didsbury the following year, as was a new line from Chorlton to Manchester Airport.

A second crossing through the city centre opened in 2017, while in 2020 a new line to the Trafford centre brought the network's total mileage up to 63.

The original fleet of trams have been replaced and now the fleet comprises 120 3000 series Flexity Swift two-section trams delivered progressively from 2009 to 2016, with additional orders made as new lines opened. The fleet is maintained at the Queens Road depot close to Victoria.

M5000s

Built: 2008-14 by Bombardier, Bautzen (Germany) AND Vienna (Austria)

No.	Livery	Name
3001	Silver/yellow	
3002	Silver/yellow	
3003	Silver/yellow	
3004	Silver/yellow	
3005	Silver/yellow	
3006	Silver/yellow	
3007	Silver/yellow	
3008	Silver/yellow	
3009	Silver/yellow	*Coronation Street 50th Anniversary 1960-2010*
3010	Silver/yellow	
3011	Silver/yellow	
3012	Silver/yellow	
3013	Silver/yellow	
3014	Silver/yellow	
3015	Silver/yellow	
3016	Silver/yellow	
3017	Silver/yellow	
3018	Silver/yellow	
3019	Silver/yellow	
3020	Silver/yellow	*Lancashire Fusilier*
3021	Silver/yellow	
3022	Spirit of Manchester Silver/yellow	*Spirit of Manchester*
3023	Silver/yellow	
3024	Silver/yellow	

3025	Silver/yellow
3026	Silver/yellow
3027	Silver/yellow
3028	Silver/yellow
3029	Silver/yellow
3030	Silver/yellow
3031	Silver/yellow
3032	Silver/yellow
3033	Silver/yellow
3034	Silver/yellow
3035	Silver/yellow
3036	Silver/yellow
3037	Silver/yellow
3038	Silver/yellow
3039	Silver/yellow
3040	Silver/yellow (Bee Network vinyls)
3041	Silver/yellow
3042	Silver/yellow
3043	Silver/yellow
3044	Silver/yellow
3045	Silver/yellow
3046	Silver/yellow
3047	Silver/yellow
3048	Silver/yellow
3049	Silver/yellow
3050	Silver/yellow
3051	Silver/yellow
3052	Silver/yellow
3053	Silver/yellow
3054	Silver/yellow
3055	Silver/yellow
3056	Silver/yellow
3057	Silver/yellow
3058	Silver/yellow
3059	Silver/yellow
3060	Silver/yellow
3061	Silver/yellow
3062	Silver/yellow (Bee Network Vinyls)
3063	Silver/yellow
3064	Silver/yellow
3065	Silver/yellow
3066	Clean Air Greater Manchester Advert
3067	Silver/yellow
3068	Silver/yellow
3069	Silver/yellow
3070	Silver/yellow
3071	Silver/yellow
3072	Silver/yellow
3073	Silver/yellow
3074	Silver/yellow
3075	Trafford Park Advert
3076	NOCCO advert
3077	Silver/yellow
3078	Silver/yellow
3079	Silver/yellow
3080	Silver/yellow
3081	Silver/yellow
3082	Silver/yellow

3083	Silver/yellow	
3084	Silver/yellow	
3085	Silver/yellow	
3086	Silver/yellow	
3087	Silver/yellow	
3088	Silver/yellow	
3089	Silver/yellow (Bee Network Vinyls)	
3090	Silver/yellow	
3091	Silver/yellow	
3092	Silver/yellow	
3093	Silver/yellow	
3094	Silver/yellow	
3095	Silver/yellow	
3096	Silver/yellow	
3097	Silver/yellow	
3098	Silver/yellow	*Gracie Fields*
3099	Silver/yellow	
3100	Silver/yellow	
3101	Silver/yellow	
3102	Silver/yellow	
3103	Silver/yellow	
3104	Silver/yellow	
3105	Silver/yellow	

The Manchester Metrolink has been a phenomenal success in the last three decades, with continual extensions of the network over the years. Set 3015 heads to East Didsbury and crosses with 3118 on its way to Ashton at Manchester St Peter's Square on 11 December 2020. *Tom McAtee*

3106	Silver/yellow		
3107	Silver/yellow		
3108	Silver/yellow		
3109	Silver/yellow		
3110	Silver/yellow		
3111	Silver/yellow		
3112	Silver/yellow		
3113	Silver/yellow		
3114	Silver/yellow		
3115	Robin Advert		
3116	Silver/yellow		
3117	Lifebuoy Soap Advert		
3118	Silver/yellow		
3119	Silver/yellow		
3120	Silver/yellow		
3121	Silver/yellow		
3122	Silver/yellow		
3123	Silver/yellow		
3124	Silver/yellow		
3125	Silver/yellow	Under construction	
3126	Silver/yellow		
3127	Silver/yellow		
3128	Silver/yellow		
3129	Silver/yellow		
3130	Silver/yellow		
3131	Silver/yellow		
3132	Silver/yellow		
3133	Silver/yellow		
3134	Silver/yellow		
3135	Silver/yellow		
3136	Silver/yellow		
3137	*Under construction for delivery 2022*		
3138	*Under construction for delivery 2022*		
3139	*Under construction for delivery 2022*		
3140	*Under construction for delivery 2022*		
3141	*Under construction for delivery 2022*		
3142	*Under construction for delivery 2022*		
3143	*Under construction for delivery 2022*		
3144	*Under construction for delivery 2022*		
3145	*Under construction for delivery 2022*		
3146	*Under construction for delivery 2022*		
3147	*Under construction for delivery 2022*		

Manx Electric Railway

The Isle of Man has its own network operated by trams, running from Douglas Pier in the south to Ramsey in the north. It runs on 3ft gauge track.

There are two depots at Derby Castle and one at Laxey. In total there is 17 miles of route and the line is electrified using a 550V DC overhead trolley wire system.

The network provides a valuable transport system for locals, but is essentially a tourist attraction and that is reflected in the age of the vehicles used on the network, which date mostly from 1893-1906, although the newest vehicles dates from 1930.

Saloons

Built: 1893 by GF Milnes, Birkenhead

No.	Livery	Usual Location	Year Last Operated
1	Douglas & Laxey Coast Electric Tramways red/white	Derby Castle	2019
2	Manx Electric Railway red/white	Derby Castle	2019

Tunnel Cars

Built: 1894 by GF Milnes, Birkenhead

No.	Livery	Usual Location	Year Last Operated
5	Manx Electric Railway red/white	Derby Castle	2020
6	Manx Electric Railway red/white	Derby Castle	2019
7	Douglas & Laxey Electric Railway blue/white/teak	Derby Castle	2020
9	Manx Electric Railway red/white	Derby Castle	2020

Crossbench Open Cars

Built: 1898 by GF Milnes, Birkenhead

No.	Livery	Usual Location	Year Last Operated
14	Douglas & Ramsey Electric Tramway Maroon	Derby Castle	2019
15	Manx Electric Railway red/white	Derby Castle	1973
16	Manx Electric Railway red/white	Derby Castle	2020
17	Manx Electric Railway red/white	Derby Castle	1973
18	Manx Electric Railway red/white	Derby Castle	2000

Winter Saloons

Built: 1898 by GF Milnes, Birkenhead

No.	Livery	Usual Location	Year Last Operated
19	Douglas, Laxey & Ramsey Electric Tramway cream/teak	Derby Castle	2020
20	Manx Electric Railway red/white	Derby Castle	2020
21	Manx Electric Railway green/white	Derby Castle	2018
22	Manx Electric Railway red/white	Derby Castle	2020

Freight Locomotive

Built: 1900 by Manx Electric Railway, Derby Castle

No.	Livery	Usual Location	Year Last Operated
23	Manx Electric Railway green/grey	Laxey	1993

Paddlebox Crossbench Open Cars

Built: 1898 by GF Milnes, Birkenhead

No.	Livery	Usual Location	Year Last Operated
25	Undercoat red/white	Laxey	1998
26	Manx Electric Railway red/white	Laxey	2009
27	Isle of Man Railways yellow/white	Laxey	2003

Crossbench Open Cars

Built: 1904 by ERTCC, Preston

No.	Livery	Usual Location	Year Last Operated
28	Manx Electric Railway red / white	Laxey	1970
29	Manx Electric Railway red / white	Derby Castle	1979
30	Manx Electric Railway red / white	Laxey	1971
31	Manx Electric Railway red / white	Laxey	2002

Crossbench Open Cars

Built: 1906 by UECC, Preston

No.	Livery	Usual Location	Year Last Operated
32	Manx Electric Railway green / white	Derby Castle	2020
33	Manx Electric Railway red / white	Derby Castle	2020

Replica Locomotive

Built: 1995 by Manx Electric Railway, Derby Castle

No.	Livery	Usual Location	Year Last Operated
34	Manx Electric Railway yellow / red	Derby Castle	2019

Open Trailers

Built: 1894 by GF Milnes, Birkenhead

No.	Livery	Usual Location	Year Last Operated
36	Manx Electric Railway red / white	Laxey	1971
37	Manx Electric Railway red / white	Derby Castle	2020

Open Trailers

Built: 1930 by English Electric, Preston

No.	Livery	Usual Location	Year Last Operated
40	Manx Electric Railway red / white	Derby Castle	2019
41	Manx Electric Railway red / white	Derby Castle	2020
44	Manx Electric Railway red / white	Derby Castle	2019

Open Trailers

Built: 1903 by GF Milnes, Hadley

No.	Livery	Usual Location	Year Last Operated
42	Manx Electric Railway red / white	Derby Castle	2019
43	Manx Electric Railway red / white	Derby Castle	2019

Open Trailers

Built: 1899 by GF Milnes, Birkenhead

No.	Livery	Usual Location	Year Last Operated
46	Manx Electric Railway red / white	Derby Castle	2020
47	Manx Electric Railway red / white	Derby Castle	2020
48	Manx Electric Railway blue / white	Derby Castle	2019

Open Trailers

Built: 1893 by GF Milnes, Birkenhead

No.	Livery	Usual Location	Year Last Operated
49	Douglas & Laxey Coast Electric Tramway maroon/cream	Derby Castle	2019
50	Manx Electric Railway red/white	Laxey	1978
51	Maroon/teak	Derby Castle	2019
53	Manx Electric Railway red/white	Derby Castle	1978
54	Manx Electric Railway red/white	Derby Castle	2019

Open Trailers (as built)

Built: 1904 by GF Milnes, Birkenhead

No.	Livery	Usual Location	Year Last Operated
55	Isle of Man Railways red/white	Derby Castle	1997
56	Manx Electric Railway red/white	Derby Castle	2020

Enclosed Saloon Trailers

Built: 1904 by GF Milnes, Birkenhead

No.	Livery	Usual Location	Year Last Operated
57	Manx Electric Railway red/white	Derby Castle	2019
58	Manx Electric Railway red/white	Derby Castle	2019

Special Saloon

Built: 1895 by GF Milnes, Birkenhead

No.	Livery	Usual Location	Year Last Operated
59	Douglas & Laxey Coast Electric Tramway blue/white	Derby Castle	2020

Open Trailer

Built: 1896 by GF Milnes, Birkenhead

No.	Livery	Usual Location	Year Last Operated
60	Manx Electric Railway red/white	Derby Castle	2019

Open Trailers

Built: 1906 by UECC, Preston

No.	Livery	Usual Location	Year Last Operated
61	Manx Electric Railway red/white	Derby Castle	2019
62	Manx Electric Railway green/white	Derby Castle	2020

Nottingham Express Transit

Routes: Hucknall to Clifton South via Bulwell, Trent University, Nottingham Station and Clifton Centre; Toton Lane to Nottingham via Beeston and University of Nottingham; branch from Phoenix Park to Highbury Vale

Mileage: 20

Number of stations: 50

Trains: 37 × five-section 750V DC overhead trams

Another of the newer, but flourishing, tram systems is that in Nottingham. Phase 1 was from Nottingham Station to Hucknall in the north, with a short spur from Highbury Vale to Phoenix Park. It opened in March 2004.

Phase two was two new branches, one to the south to Clifton via Ruddington Lane, which opened in July 2015, and another to the west to Chilwell via the university and hospital. This line was then extended to Toton Lane, a park and ride station near the M1. It opened in August 2015.

The system uses two fleets of five-section trams. The Bombardier Incentro units were delivered for the opening of Phase 1 and built at nearby Derby. When additional trams were needed, NET opted for Spanish-built Alstom Citadis 302s. All are maintained at the Wilson Street Depot.

No.	Livery	Name
201	Revised Silver/dark green/black	*Torvill and Dean*
202	Silver/dark green/black	*D H Lawrence*
203	Revised Silver/dark green/black	*Bendigo Thompson*
204	Revised Silver/dark green/black	*Erica Beardsmore*
205	Revised Silver/dark green/black	*Lord Byron*
206	Revised Silver/dark green/black	*Angela Alcock*
207	Revised Silver/dark green/black	*Mavis Worthington*
208	Revised Silver/dark green/black	*Dinah Minton*
209	Nottingham Express Transit promotions (advert)	*Sid Standard*
210	Revised Silver/dark green/black	*Sir Jesse Boot*
211	Revised Silver/dark green/black	*Robin Hood*
212	Revised Silver/dark green/black	*William Booth*
213	e-on (advert)	*Mary Potter*
214	Silver/dark green/black	*Dennis McCarthy MBE*
215	Silver/dark green/black	*Brian Clough*
216	Silver/dark green/black	*Dame Laura Knight*
217	Silver/dark green/black	*Carl Froch MBE*
218	Silver/dark green/black	*Jim Taylor*
219	Silver/dark green/black	*Alan Sillitoe*
220	Silver/dark green/black	*Sophie Robson*
221	Silver/dark green/black	*Stephen Lowe*
222	Silver/dark green/black	*David S Stewart OBE*
223	Silver/dark green/black	*Colin Slater MBE*
224	Silver/dark green/black	*Vicky McClure*
225	Silver/dark green/black	*Doug Scott CBE*
226	Silver/dark green/black	*Jimmy Sirrel and Jack Wheeler*
227	Silver/dark green/black	*Sir Peter Mansfield*
228	Silver/dark green/black	*Local Armed Forces Heroes*
229	Silver/dark green/black	*Viv Anderson MBE*
230	Silver/dark green/black	*George Green*
231	Silver/dark green/black	*Rebecca Adlington OBE*
232	Silver/dark green/black	*William Ivory*
233	Silver/dark green/black	*Ava Lovelace*
234	Silver/dark green/black	*George Africanus*
235	Silver/dark green/black	*David Clarke*
236	Silver/dark green/black	*Sat Bains*
237	Silver/dark green/black	*Stuart Broad*

Nottingham has a tram system, the Nottingham Express Transit, that runs from Hucknall in the north to Clifton in the south and Toton in the west. On 24 February 2018, sets 228 and 232 pass at the city's railway station. *Rob France*

Trams, Metros and Light Rail

South Yorkshire Supertram

Routes: Meadowhall-Halfway, Middlewood to Herdings Park
Mileage: 21.5 miles
Number of stations: 50
Trains: 32 × five-section trams

Although understandably not quite of the scale of the Manchester Metrolink, Sheffield's 750V DC overhead tram system has slowly grown since the first section opened in 1994 from Meadowhall to Fitzalan Square in the city centre and then on to Gleadless Townend in the south of the city.

The year 1995 saw this branch extended to terminate at Halfway and Herdings Park, while a new line in the north to Middlewood, with a spur to Malin Bridge, also opened in October that year.

A new line from Meadowhall south to Parkgate via Rotherham Central opened in October 2018 and this section involved running over some Network Rail infrastructure as a first example of a new tram-train operation. For that reason, the seven trans that run on this section are Class 399s on TOPS.

The initial trams used were a fleet of 25 three-section Siemens units built in Düsseldorf, numbered 101-125. The Class 399s are Stadler Citylink three-section trams. They are numbered 201-207 for Supertram's purposes, but 399201-207 for Network Rail. The entire fleet is maintained at Nunnery depot in the north-east of the city.

Built: 1993-94 by Siemens Duewag, Düsseldorf, Germany

No.	Livery
101	Blue/red/orange
102	Blue/red/orange
103	Blue/red/orange
104	Blue/red/orange
105	Blue/red/orange
106	Blue/red/orange
107	Blue/red/orange
108	Blue/red/orange
109	Blue/red/orange
110	Blue/red/orange
111	Pretty Little Thing (advert)
112	Blue/red/orange
113	Blue/red/orange
114	Blue/red/orange
115	Blue/red/orange
116	XPO Logistics (advert)
117	Blue/red/orange
118	Pretty Little Thing (advert)
119	Blue/red/orange
120	Cream/blue
121	Blue/red/orange
122	Blue/red/orange
123	Blue/red/orange – with 21st Anniversary window vinyls
124	Blue/red/orange
125	Blue/red/orange

Built: 2015-6 by Vossloh/Stadler, Valencia, Spain
These trains share Network Rail tracks and hence are Class 399 on TOPS.

No.	Livery
399201	Blue/red/orange
399202	Blue/red/orange
399203	Blue/red/orange
399204	Blue/red/orange
399205	Blue/red/orange
399206	Blue/red/orange
399207	Blue/red/orange

The Sheffield Supertram has a fleet of seven tram trains that actually run on Network Rail infrastructure and are Class 399s on TOPS. On 30 December 2019, 399202 passes Woodbourn Road in Sheffield. *Rob France*

Tyne & Wear Metro

Routes: Newcastle-South Shields and South Hylton, St James to Whitley Bay, Newcastle-Airport
Mileage: 48
Number of stations: 60
Trains: 90 × two-car

The first example of heavy rail commuter lines being converted to light rail was the Tyne & Wear Metro, the first stage of which opened in August 1980 from Haymarket to Tynemouth.

In the north the system was gradually extended and by November 1982 it ran over all of the former loop line from Newcastle to Whitley Bay via Wallsend. South of the River Tyne estuary, from March 1984 it also ran on a new line adjacent to the BR route from Felling to South Shields, with the BR line stations closing and the track becoming freight only. Some of these freight lines have since closed.

The system was extended to run from Bank Foot to Newcastle Airport in November 1991 and from March 2002 from Pelaw to South Hylton, sharing tracks with Network Rail as far as Sunderland. The former BR freight-only line from Percy Main to Northumberland Park is mooted for being added to the system.

After two prototypes were built in 1975, the system started with these and a fleet of an additional 88 two-car units built in 1978-81. These are naturally quite dated now and in January 2020 Stadler won a contract for 42 brand new five-car trains, the first of which was due to be delivered in late 2023. There is an option for five additional trains should they be needed.

No.	Livery
4001	40th Anniversary Livery
4002	Black/yellow
4003	Black/yellow
4004	Black/yellow
4005	Black/yellow
4006	Black/yellow
4007	Black/yellow
4008	Black/yellow
4009	Black/yellow
4010	Black/yellow
4011	Black/yellow
4012	Black/yellow
4013	Black/yellow
4014	Black/yellow
4015	Black/yellow
4016	Black/yellow
4017	Black/yellow
4018	Black/yellow
4019	Black/yellow
4020	Black/yellow
4021	Black/yellow
4023	Black/yellow
4024	Black/yellow
4025	Black/yellow
4026	Black/yellow
4027	Black/yellow
4028	Black/yellow
4029	Black/yellow
4030	Black/yellow
4031	Black/yellow
4032	Black/yellow
4033	Black/yellow
4034	Artisanal Vodka (advert)
4035	Black/yellow
4036	Black/yellow
4037	Black/yellow
4038	Black/yellow
4039	Black/yellow
4040	Black/yellow
4041	Black/yellow
4042	Black/yellow
4043	Black/yellow
4044	Black/yellow
4045	Black/yellow
4046	Black/yellow
4047	Black/yellow
4048	Black/yellow
4049	Black/yellow
4050	Black/yellow
4051	Black/yellow
4052	Black/yellow
4053	Black/yellow
4054	Black/yellow
4055	Black/yellow
4056	Black/yellow
4057	Black/yellow
4058	Black/yellow
4059	Black/yellow
4060	Black/yellow
4061	Black/yellow
4062	Black/yellow
4063	Black/yellow
4064	Black/yellow
4065	Black/yellow
4066	Black/yellow
4067	Black/yellow
4068	Black/yellow
4069	Black/yellow
4070	Black/yellow
4071	Black/yellow
4072	Black/yellow
4073	Black/yellow
4074	Black/yellow
4075	Black/yellow
4076	Black/yellow
4077	Black/yellow
4078	Black/yellow
4079	Black/yellow
4080	Black/yellow
4081	Black/yellow
4082	Black/yellow
4083	Emirates Airline (advert)
4084	Black/yellow
4085	Black/yellow
4086	Black/yellow
4087	Black/yellow
4088	Black/yellow
4089	Black/yellow
4090	Black/yellow

4001/02/40/83 were not refurbished so are rarely used and can only work with each other and not any of refurbished vehicles.

Forty-six new trains are due to be built by Stadler for introduction by 2023. Numbering details are awaited.

The Tyne & Wear Metro is about to undergo a full fleet overhaul with brand-new trains to replace those that date from the 1970s. In the meantime, the old rolling stock is still in use and on 23 November 2019, sets 4082 and 4069 call at Bank Foot with a train for South Gosforth. *Rob France*

An artist's impression of the new Stadler metro trains being constructed for the Tyne & Wear Metro.

West Midlands Metro

Routes: Centenary Square-Wolverhampton St Georges
Mileage: 13
Number of stations: 29
Trains: 21 × five-section trams

This 750V DC system only has one route, from Birmingham City centre to Wolverhampton St Georges, but goes through the heart of the Black Country, serving the likes of West Bromwich, Dudley and Wednesbury.

A short extension to meet with Wolverhampton railway station is under construction, while an extension from the city centre to Edgbaston is also under construction and is open as far as Library.

It has one depot at Wednesbury to maintain the fleet. The initial 16 three-section Ansaldo trams, 01-16, have now been retired and are in store. They have been replaced by 21 newer CAF Urbos 3 five-section units, numbered 17-37.

Built: 2013-15 by CAF, Zaragoza, Spain

No.	Livery	Name
17	Blue	
18	Blue with Just Eat (advert)	
19	Silver/Blue – Marks Electrical (advert)	
20	Blue	
21	Blue with Just Eat (advert)	
22	Blue	
23	Blue	
24	Blue with Resorts World (advert)	
25	Blue with OLA (advert)	
26	Blue	
27	Blue with OLA (advert)	
28	Blue	*Jasper Carrott*
29	Blue	
30	Blue	
31	Blue	*Cyrille Regis MBE 1958-2018*
32	Blue with Brinley Place (advert)	
33	Blue	
34	Blue	
35	Blue	
36	Blue	
37	Blue	*Ozzy Osbourne*

West Midlands Metro CAF Urbos 3 trams 28 and 29 *Jasper Carrott* meet at Library in July 2021. This stop is currently the end of a new branch from Bull Street that opened in 2019 but will eventually be extended to run through to Edgbaston. *Tony Miles*

Dublin trams

The LUAS tram system in Dublin operates over two lines, with the Red line stretching from The Point in the east of the city near Docklands, westward through the city centre to Dublin Heuston, before heading south-west out of the city to a junction at Belgard. The line then splits, with one route carrying on to Saggart and the other to Tallaght.

The Green line runs north to south from Broombridge to Brides Glen, with a loop line between Dominik and O'Connell Upper to Dawson via Marlborough.

The two lines interchange via a walkway between Abbey Street on the Red line to O'Connell GPO and Marlborough on the Green Line. The trams also interchange with the national rail system at Dublin Connolly and Dublin Heuston, as well with bus connections. An extension of the Green line from Broombridge to Finglas, about 2½ miles, has been approved.

The system is operated using 750V DC overhead lines and uses standard 4ft 8½in gauge track.

The tram fleet is entirely made up of Alstom Citadis vehicles, built in La Rochelle in France. There are now a total 81 sets in service, split into three classes, which run as articulated formations.

The oldest are the 2002/03-built 3000 series, which were delivered as 30-metre, four-section sets, but extended to 40 metres in 2007-08 with an extra section added. These were for the Red line and numbered 3001-26.

At the same time, 14 Citadis 4000 series trams were delivered for the Green line, delivered as 40-metre sets, and numbered 4001-14. Both types can run on either line.

A fleet of 16 new 43-metre 5000 series sets were delivered between 2008-10 and numbered 5001-26. These are used only on the Green line, but their arrival did allow the 4000 series to move to the Red line. 5001-26 have since been extended to 55-metre sets and supplemented by a further seven new 55-metre trams, 5027-33. Eight more of these were delivered from 2020, numbered 5034-41.

No.	Livery
3001	Silver with yellow stripe
3002	Silver with yellow stripe
3003	Silver with yellow stripe
3004	Silver with yellow stripe
3005	Silver with yellow stripe
3006	Silver with yellow stripe
3007	Silver with yellow stripe
3008	Silver with yellow stripe
3009	Silver with yellow stripe
3010	Silver with yellow stripe
3011	Silver with yellow stripe
3012	Silver with yellow stripe
3013	Silver with yellow stripe
3014	Silver with yellow stripe
3015	Silver with yellow stripe
3016	Silver with yellow stripe
3017	Silver with yellow stripe
3018	Silver with yellow stripe
3019	Silver with yellow stripe
3020	Silver with yellow stripe
3021	Silver with yellow stripe
3022	Silver with yellow stripe
3023	Silver with yellow stripe
3024	Silver with yellow stripe
3025	Silver with yellow stripe
3026	Silver with yellow stripe

No.	Livery
4001	Silver with yellow stripe
4002	Silver with yellow stripe
4003	Silver with yellow stripe
4004	Silver with yellow stripe
4005	Silver with yellow stripe
4006	Silver with yellow stripe
4007	Silver with yellow stripe
4008	Silver with yellow stripe
4009	Silver with yellow stripe
4010	Silver with yellow stripe
4011	Silver with yellow stripe
4012	Silver with yellow stripe
4013	Silver with yellow stripe
4014	Silver with yellow stripe

Citadis 401 series LUAS Tram 4008 calls at Heuston on its way to The Point on the Red Line on 1 May 2019. *Tony Miles*

Trams, Metros and Light Rail

No.	Livery
5001	Silver with yellow stripe
5002	Silver with yellow stripe
5003	Silver with yellow stripe
5004	Silver with yellow stripe
5005	Silver with yellow stripe
5006	Silver with yellow stripe
5007	Silver with yellow stripe
5008	Silver with yellow stripe
5009	Silver with yellow stripe
5010	Silver with yellow stripe
5011	Silver with yellow stripe
5012	Silver with yellow stripe
5013	Silver with yellow stripe
5014	Silver with yellow stripe
5015	Silver with yellow stripe
5016	Silver with yellow stripe
5017	Silver with yellow stripe
5018	Silver with yellow stripe
5019	Silver with yellow stripe
5020	Silver with yellow stripe
5021	I've Grown capacity enhancement (advert)
5022	Silver with yellow stripe
5023	Silver with yellow stripe
5024	Silver with yellow stripe
5025	Silver with yellow stripe
5026	Silver with yellow stripe
5027	Silver with yellow stripe
5028	Silver with yellow stripe
5029	Silver with yellow stripe
5030	Silver with yellow stripe
5031	Silver with yellow stripe
5032	Silver with yellow stripe
5033	Silver with yellow stripe
5034	Silver with yellow stripe
5035	Silver with yellow stripe
5036	Silver with yellow stripe
5037	Silver with yellow stripe
5038	Silver with yellow stripe
5039	Silver with yellow stripe
5040	Silver with yellow stripe
5041	Silver with yellow stripe

Belmond

Contact details

Website: www.belmond.com/uk
Twitter: @Belmond

Overview

Belmond is a global company but in the UK has two rail operations. The British Pullman operates mostly from London on both repeat itinerary and ad hoc day trips. It is also available for hiring for corporate events.

It uses DB Cargo to haul this train, with two Class 67s, 67021/024, repainted to match the umber and cream coaches.

The Royal Scotsman operates each summer across Scotland, although it usually does an annual tour of the UK that brings it to England and Wales. This train is operated by GB Railfreight with two Class 66s, 66743/746, in matching livery.

British Pullman

These vehicles are based and maintained by Belmond at Stewarts Lane in London.

No.	Other No.	Livery	Type	Name
6313	92167	PUL	Mk 1 BG	
9502		PUL	Mk 2 BSO	
99530		PUL	Parlour First	*PERSEUS*
99531		PUL	Parlour First	*PHOENIX*
99532		PUL	Parlour First	*CYGNUS*
99534		PUL	Kitchen First	*IBIS*
99535		PUL	Parlour First	*MINERVA*
99536		PUL	Parlour First	*ZENA*
99537		PUL	Parlour First	*AUDREY*
99539		PUL	Kitchen First	*IONE*
99541		PUL	Parlour First	*LUCILLE*
99543		PUL	Parlour First	*VERA*
99545	35466	PUL	Mk 1 BSK	*Baggage Car No. 11*
99546		PUL	Kitchen First	*GWEN*

Royal Scotsman

These coaches are maintained by Assenta and housed at its Hamilton depot when not in use.

Present No.	Other No.	Livery	Type
99337		ROY	PSK
99961		ROY	State Car No. 1
99962		ROY	State Car No. 2
99963		ROY	State Car No. 3
99964		ROY	State Car No. 4
99965		ROY	Observation Car
99967		ROY	Dining Car
99968	10541	ROY	Mk 3 SLEP State Car No. 5
99969	10556	ROY	Mk 3 SLEP Service Car

Pullman Kitchen First 280 *Audrey*, part of the Belmond British Pullman rake, was at Peterborough on 17 April 2019. *Stuart West*

Eastern Rail Services

Contact details

E-mail: enquiries@easternrailservices.co.uk

Key personnel

Managing Director: James Steward

Overview

Eastern Rail Services is a new rolling stock provider that has grown rapidly over the last two years and is currently supplying vehicles to customers, including Network Rail for its infrastructure-monitoring trains.

It has acquired several redundant Mk 2/3 coaches displaced by new stock from the likes of Caledonian Sleeper and Great Anglia.

It has vehicles stored at the Mid Norfolk Railway and has recently taken on the previously redundant carriage sidings at Great Yarmouth, which it has brought back into use for storing vehicles between contracts. It has based a Class 08 shunter there. In 2020 the company also bought RMS Locotec and acquired several Class 08 shunters as a result (see page 307).

It also owns one Class 08 as an ERS asset – 08870 – and 31452, which it uses for train supply testing.

In March 2019 it obtained former Mk 3 Nightstar generator coach 96371, one of five high-output generator coaches for use on the proposed cross-Channel 'sleeper' trains that never happened. The coach contains two engine and generator sets that combined have a train heat index of 140. This is currently at UKRL's Leicester depot being modified. It has more recently acquired the other four ex-Nightstar generator coaches.

In October 2019 it bought the entire Caledonian Sleeper Mk 2 fleet consisting of two Mk 2f RFO lounges, six Mk 2f RLO lounge cars and nine Mk 2e BUO brakes. Five are in use with NR while others are stored at the Weardale Railway.

In December 2019 ERS obtained two Mk 3 kitchen cars while January 2020 saw some of the BUOs also enter service with Network Rail. More recently, ERS has acquired further sleeper coaches and Greater Anglia Mk 3s.

Class 08

08870		MBDL	ICE	ERS	GY

Class 31

31452	31228	HTLX	DCG	ERS	GY (U)

Eastern Rail Services has taken over the old carriage sidings at Great Yarmouth for stabling coaches. It has based two shunters there, including 08870, seen on 10 August 2021. *Stuart West*

Coaches

No.	Livery	Type
1210	SCP	Mk 2 RFB
1213	SCP	Mk 2 RFB
1220	SCP	Mk 2 RFB
1254	BRG	Mk 2 RFB
1692	CHC	Mk 1 RBR
3133	BRM	Mk 1 FO
3181	OXB	Mk 2 FO
3374	BRG	Mk 2 FO
3385	ICS	Mk 2 FO
3411	BCM	Mk 2 FO
5482	BRG	Mk 2 TSO
5647	BRG	Mk 2 TSO
5787	DRS	Mk 2 TSO
5866	ICS	Mk 2 TSO
5906	ICS	Mk 2 TSO
5937	DRS	Mk 2 TSO
5960	VIR	Mk 2 TSO
5989	ICS	Mk 2 TSO
6059	ICS	Mk 2 TSO
6064	DRS	Mk 2 TSO
6168	ICS	Mk 2 TSO
6173	DRS	Mk 2 TSO
6411	LUL	Mk 2 TSO
6415	BRG	Mk 2 FO
6700	CAL	Mk 2 RFO
6703	CAL	Mk 2 RFO
9448	WCR	Mk 2 BSO
9497	CAL	Mk 2 BSO
9500	ICS	Mk 2 BSO
9513	ICS	Mk 2 BSO
9709	DRS	Mk 2 DBSO
9710	DRS	Mk 2 DBSO
9800	CAL	Mk 2 BUO
9801	SCP	Mk 2 BUO
9802	CAL	Mk 2 BUO
9803	SCP	Mk 2 BUO
9805	SCP	Mk 2 BUO
9806	SCP	Mk 2 BUO
9807	SCP	Mk 2 BUO
9808	SCP	Mk 2 BUO
9810	SCP	Mk 2 BUO
10212	VIS	Mk 3 RFM
10229	GAW	Mk 3 RFM
10413	GAW	Mk 3 TSOB
10501	SCP	Mk 3 SLEP
10502	SCP	Mk 3 SLEP
10600	SCP	Mk 3 SLEP
10699	SCP	Mk 3 SLED
11078	GAW	Mk 3 FOD
11095	GAW	Mk 3 FOD
12017	CRS	Mk 3 TSO
12021	GAW	Mk 3 TSO
12036	CRS	Mk 3 TSO
12043	CRS	Mk 3 TSO
12094	CRS	Mk 3 TSO
12098	GAW	Mk 3 TSO
17164	BRG	Mk 2 BFK
86443	BRM	Mk 1 GUV
95151	BRM	Mk 1 GUV
95199	RES	Mk 1 GUV
92901	ICS	Mk 1 BG
96371	Grey	Mk 3 GV
96372	Grey	Mk 3 GV
96373	Grey	Mk 3 GV
96374	Grey	Mk 3 GV
96375	Grey	Mk 3 GV

This former Caledonian Sleeper Mk 2s 9801 and 9806 are two of several acquired by Eastern Rail Services and hired to Network Rail to provide staff accommodation on its IM trains. They were part of the 0209 King's Cross-Derby RTC test train seen at Derby station on 20 November 2021. *Pip Dunn*

Riviera Trains

Contact details

Website: www.riviera-trains.co.uk

Overview

Riviera Trains is a rolling stock provider to the charter train market and TOCs. It has a large fleet of mostly ex-BR coaches, and supplies Mk 1 and Mk 2 trains to the likes of UK Railtours, Pathfinder Tours (which it owns), Nenta Traintours and others.

It has previously supplied Mk 2 coaches to various TOCs including ScotRail, Transport for Wales and DB Schenker.

Class 08

08507	OXB	RIV	CWR	
08605	DBC	RIV	KY	*G R Walker*
08704	OXB	RIV	KY	

Coaches

Present No.	Other No.	Livery	Type
1200		BRG	Mk 2 RFB
1212	3427	BRG	Mk 2 RFB
1651		CAC	Mk 1 RBR
1657		BRG	Mk 2 TSO
1671		CHC	Mk 1 RBR
1683		BLE	Mk 1 RBR
1691		BRG	Mk 1 RBR
1813		CHC	Mk 1 RMB
1832		CAC	Mk 1 FO
3097		CHC	Mk 1 FO
3098		CHC	Mk 1 FO
3110		CHC	Mk 1 FO
3112		CHC	Mk 1 FO
3119		CHC	Mk 1 FO
3120		CHC	Mk 1 FO
3121		CHC	Mk 1 FO
3123		CAC	Mk 1 FO
3141		CHC	Mk 1 FO
3146		CHC	Mk 1 FO
3147		CHC	Mk 1 FO
3149		CHC	Mk 1 FO
3278		BRG	Mk 2 FO
3304		BRG	Mk 2 FO
3314		BRG	Mk 2 FO
3333		BRG	Mk 2 FO
3340		BRG	Mk 2 FO
3345		BRG	Mk 2 FO
3356		BRG	Mk 2 FO
3364		BRG	Mk 2 FO
3386		BRG	Mk 2 FO
3390		BRG	Mk 2 FO
3397		BRG	Mk 2 FO
4927		CAC	Mk 1 TSO

Former British Rail Mk 2f first open W3397, repainted back into its original BR blue and grey livery, is one of several owned by Riviera Trains. It was at Leeds on 12 February 2022 as part of a charter train heading to Euston. *Pip Dunn*

4946	CHC	Mk 1 TSO
4949	CHC	Mk 1 TSO
4959	CHC	Mk 1 TSO
4991	CHC	Mk 1 TSO
4998	CHC	Mk 1 TSO
5009	CHC	Mk 1 TSO
5292	CAC	Mk 2 TSO
5910	BRG	Mk 2 TSO
5921	ANG	Mk 2 TSO
5929	BRG	Mk 2 TSO
5945	SCR	Mk 2 TSO
5950	ANG	Mk 2 TSO
5952	BRG	Mk 2 TSO
5955	SCR	Mk 2 TSO
5961	BRG	Mk 2 TSO
5965	SCR	Mk 2 TSO
5976	SCR	Mk 2 TSO
5985	ANG	Mk 2 TSO
5987	SCR	Mk 2 TSO
5998	BRG	Mk 2 TSO
6006	ANG	Mk 2 TSO
6024	BRG	Mk 2 TSO
6027	SCR	Mk 2 TSO
6042	ANG	Mk 2 TSO

6051		BRG	Mk 2 TSO
6054		BRG	Mk 2 TSO
6067		BRG	Mk 2 TSO
6137		SCR	Mk 2 TSO
6141		VIR	Mk 2 TSO
6158		BRG	Mk 2 TSO
6176		BRG	Mk 2 TSO
6177		SCR	Mk 2 TSO
6183		SCR	Mk 2 TSO
6310	81448	CHC	Mk 1 BGGV
9504		BRG	Mk 2 BSO
9507		BRG	Mk 2 BSO
9509		ANG	Mk 2 BSO
9520		ANG	Mk 2 BSO
9521		DRS	Mk 2 BSO
9526		BRG	Mk 2 TSO
9527		SCR	Mk 2 TSO
9537		VIR	Mk 2 TSO
9539		SCR	Mk 2 TSO
17105		BRG	Mk 2 BFK/GV
21269		CAC	Mk 1 BCK
35469		CHC	Mk 1 BSK/GV
80041	1690	BRM	Mk 1 KC
80042	1646	BRG	Mk 1

This Mk 1 BG 6310 was built in 1958 and was one of four converted into generator coaches to provide electric train supply on Scottish 'sleeper' trains in the early 1990s. It is now owned by Riviera Trains and was at Leeds on 12 February 2022. *Pip Dunn*

SRPS

Contact details

Website: www.srps.org.uk
Twitter: @srpsrailtours

Overview

Based at the Bo'ness & Kinneil Railway, the volunteer-led SRPS has long had its own rolling stock for its charter trains, operated first by BR and then more recently by private TOCs including EWS, WCR and GBRf.

It generally has a rake of between nine and 11 Mk 1s available for its own use and to hire to other promoters. It also owns 37403 *Isle of Mull*, which is fully main-line compliant (see page 339) and is being used on some of its charters in 2022.

Coaches

Present No.	Livery	Type
1730	BRM	Mk 1 RBR
1859	BRM	Mk 1 RMB
3096	BRM	Mk 1 FO
3115	BRM	Mk 1 FO
3150	BRM	Mk 1 FO
4831	BRM	Mk 1 TSO
4832	BRM	Mk 1 TSO
4836	BRM	Mk 1 TSO
4856	BRM	Mk 1 TSO
5028	BRM	Mk 1 TSO
13229	BRM	Mk 1 FK
13230	BRM	Mk 1 FK
21241	BRM	Mk 1 BCK
35185	BRM	Mk 1 BSK

Mainline-registered support coaches

A number of coaches are used as support vehicles, mostly for steam locos. They are main-line registered and not used for carrying fare-paying passengers.

Present No.	Other No.	Livery	Type	User
14007	17007	BRM	Mk 1 BFK	NYMR
14060		BRM	Mk 2 BFK	Bahamas Loco Society/45596
14099	17099	BRM	Mk 2 BFK	GCR (ex 45305/70013)
17096	14096	PUL	Mk 2 BFK	Stewarts Lane/35028
21096		BRM	Mk 1 BCK	A4 Loco Society/60007
21249		CAC	Mk 1 BCK	A1 Steam loco Trust /60163
21268		BRM	Mk 1 BCK	Peak Rail
35453		CHC	Mk 1 BSK	61306 Mayflower
35457		BRM	Mk 1 BSK	NNR (ex 76084)
35479		BRM	Mk 1 BSK	D Buck/61306
35486		BRM	Mk 1 BSK	J. Cameron/60009
35508	17128	BRM	Mk 1 BFK	I. Riley/44871/45212/45407
35517	1708	BRM	Mk 1 BFK	I. Riley/44871/45212/45407
99041	35476	LMS	Mk 1 BSK	PRCLT MRC/6233
99953	35468	BRM	Mk 1 BSK	NRM York/60103

Heritage railways' stock

In recent years, two heritage railways have extended their services to run on the national network on otherwise sparsely used lines. This started with the North Yorkshire Moors Railway gaining access rights for the Battersby to Whitby section of the Esk Valley line. It typically runs just between Grosmont and Whitby using steam and heritage diesels but sometimes runs westwards to Battersby as well.

It has Class 25 D7628 (25278) – see page 338 – for this role, but has also hired Class 20s and 31s from other operators plus used preserved main-line diesels such as 37403, 40145, 50049, D1015 and 55022.

The North Norfolk Railway runs a few dining trains each year to Cromer. These are top-and-tailed by steam and diesel, and for this role 20227 has previously been hired as it is main-line registered. Only approved coaches can be used for these two operations.

North Yorkshire Moors Railway

The following vehicles are main-line certified for use between Battersby and Whitby only.

Present No.	Livery	Type
232	PUL	Mk 1 RK
318	PUL	Mk 1 RK
324	PUL	Mk 1 RK
328	PUL	Mk 1 RK
1823	PUL	Mk 1 RMB
3798	BRM	Mk 1 TSO
3801	BRM	Mk 1 TSO
3860	BRM	Mk 1 TSO
3872	BRM	Mk 1 TSO
3948	BRM	Mk 1 TSO
4198	BRM	Mk 1 TSO
4252	CAC	Mk 1 TSO
4286	BRM	Mk 1 TSO
4290	BRM	Mk 1 TSO
4455	BRM	Mk 1 TSO
4597	CAC	Mk 1 TSOD
4817	BRM	Mk 1 TSO
4999	BRM	Mk 1 TSO
5000	BRM	Mk 1 TSO
5001	BRM	Mk 1 TSOD
5029	BRM	Mk 1 TSO
9225	BRM	Mk 1 TSO
9267	BRM	Mk 1 TSO
9274	BRM	Mk 1 TSO
15745	BRM	Mk 1 CK
16156	CAC	Mk 1 CK
21100	CAC	Mk 1 BCK
35089	CAC	Mk 1 BSK

North Norfolk Railway

The following vehicles are main-line certified for use between Sheringham and Cromer only.

Present No.	Livery	Type
1969	CRC	Mk 1 RBR
3116	CRC	Mk 1 FO
4372	CRC	Mk 1 SO
81033	CRC	Mk 1 BG

Royal Train

Not a rolling stock provider as such, other than to the Royal family, this bespoke train is based at Wolverton and operated by DB Cargo using Class 67s, usually dedicated locos 67005/006.

No.	Other No.	Livery	Converted	Intended use
2903	11001	RTO	1977	The Queen's lounge, bedroom and bathroom
2904	12001	RTO	1977	The Duke of Edinburgh's lounge, bedroom and bathroom
2915	10735	RTO	1985	Royal Household sleeping car
2916	40512	RTO	1986	Royal Family dining car with kitchen
2917	40514	RTO	1986	Royal Household dining car with kitchen
2918	40515	RTO	1986	Royal Household car
2919	40518	RTO	1986	Royal Household car
2920	14109, 17109	RTO	1986	Royal Household couchette, diesel generator & brake van
2921	14107, 17107	RTO	1986	Royal Household couchette, kitchen & brake van
2922		RTO	New 1987	The Prince of Wales's sleeping car
2923		RTO	New 1987	The Prince of Wales's saloon

The Royal Train passes along the sea wall at Teignmouth with Her Majesty, The Queen on board, heading to Par for the G7 Summit on 11 June 2021. Dedicated locos 67006 *Royal Sovereign* and 67005 *The Queen's Messenger* top and tail the train. *Jack Boskett*

Railfreight Operators Overview

The railfreight market is mainly catered for by four major players, but a number of smaller operators are also now starting to make their presence felt. Some niche markers – such as stock movement – have been exploited by some of the newer entrants.

The main companies are DB Cargo, Freightliner, Direct Rail Services and GB Railfreight. The smaller players are Colas Rail Freight, DC Rail, SLC Operations, Rail Operations Group and Loram. West Coast Railways (page 194) has a railfreight licence and recently operated some trial timber flows from the far north of Scotland for Victa Railfreight, but as a rule, it does not undertake too much freight work.

The major player DB Cargo – which as EWS, bought five of the six British Rail freight businesses back in 1996; Transrail, Loadhaul, Mainline Freight, Railfreight Distribution and Rail Express Systems – has in the last few years has lost much business and had to trim its fleet. It still remains the largest railfreight operator in the UK.

Freightliner, now owned by American company Genesee & Wyoming, was that sixth business sold off in 1995, but has moved away from just being an intermodal operator moving containers and is now a big player in the bulk trainload market. It operates all the stone trains from the Mendips for Aggregates Industries, which saw it take ownership of 14 Class 59s. It is also heavily involved in infrastructure support trains.

Direct Rail Services started operations in 1995 with a small fleet of ex-BR Class 20s, which it had refurbished. It was created by the British Nuclear Fuels organisation with the sole purpose of hauling nuclear trains, but has also entered into the intermodal, infrastructure support and charter markets and other occasional trainload flows. It remains state owned. It later acquired ex-BR Class 37s, then 47s, 33s and 87s, before ordering Class 66s and taking on Class 57s from other operators. More recently it has acquired Stadler Class 68 and 88s and has removed all its older locos apart from a handful of 37s.

Possibly the biggest success story in the railfreight market has been the birth and growth of GB Railfreight. A new company started in 1999 with just seven Class 66s and a single contract with Railtrack, it now has well over 180 locos in its fleet, 99 of them Class 66s (with another 11 likely to join them soon) and its first Class 69s – rebuilt Class 56s with GM engines. It also has three Class 47s, a 59, ten 60s, two 67s, nine 73/1s, 11 Class 73/9s and a dozen 92s. It also has several stored locos in strategic reserve.

Other recent new entrants are Colas Railfreight, DC Rail and Rail Operations Group, and all three are growing their portfolios. Colas has Class 37s, 43s, 56s, 66s and 70s in its fleet, while DC Rail has increased its Class 60 fleet with another six or seven likely to be overhauled in the forthcoming months. Rail Operations Group, which uses Class 37s and 57s in the main, has ordered new Class 93s as well as developing freight multiple units for logistics traffic.

Having relied for so long on refurbished ex-BR locos, the days are numbered for such locos with just a handful of Class 37s still in use in the DRS fleet. On 3 July 2021, 37423 *Spirit of the Lakes* has just arrived at York with a charter train from Stanton Gate. *Pip Dunn*

Colas Rail Freight

Contact details

Website: www.colasrail.co.uk
Twitter: @ColasRailUK
Chief Executive Officer: Jean-Pierre Bertrand

Overview

Colas has grown steadily over the last decade and after starting UK operations with an initial fleet of just three Class 47s (all now sold to GB Railfreight), the company has a varied fleet of new Class 66s and 70s aided by several ex-BR Class 37 and 56s. It briefly had a fleet of ten Class 60s but these too are now part of the GBRf fleet, as are its two ex-DBC Class 67s, which were taken on by GBRf after Colas ended their lease from Beacon Rail.

The 37s and 67s are used mostly for infrastructure-monitoring trains for Network Rail, while the other locos are used for a mix of freight. Colas operates aggregates, fuel and metals traffic.

Class 08

Colas owns a Class 08 as a result of its purchase of Pullman Rail. It has since sold the business to TfW but retains ownership of the 08, which is hired to TfW and used mostly for shunting vehicles on to the wheel lathe.

08499		COL	BLU	CF	*Redlight*

Class 37

These locos are used mostly on IM trains but will support the business as required. Two locos, 37025/418, are on long-term hire and set to be joined by 37606. Colas also hires Class 37s from HNRC as required. All are in the COTS pool.

37025/418 are dual braked.

37025		BLL	STG	NM	*Inverness TMD*
37057		COL	COL	NM	*Barbera Arbon*
37099	37324	COL	COL	NM	*MERL EVANS 1947-2016*
37116		COL	COL	NM	
37175		COL	COL	NM	
37219		COL	COL	NM	*Jonty Jarvis 8-12-1998 to 18-3-2005*
37254		COL	COL	NM	*Cardiff Canton*
37418	37271	BLL	BEN	NM	*An Comunn Gaidhealach*
37421	37267	COL	COL	NM	
37606	37090, 37508	DRC	BEN	BU (U)	

Early on 20 November 2021, 37116 has arrived with the 0209 King's Cross-Derby test train and will now propel into the RTC for stabling. *Pip Dunn*

Colas has leased five HST power cars from Porterbrook to replace Class 37s on some infrastructure-monitoring trains. 43274 and 43272 stand at their old regular haunt of King's Cross, having arrived with the 0929 from Darlington on 4 October 2021. *Stuart West*

Class 43

Colas initially took two ex-EMT VP185-engined power cars to evaluate their suitability for IM train operation and while these have not been taken on long term, in their place have come five MTU-engined ex-LNER/EMR Class 43s. More locos could join this fleet, with the long-term aim to replace Class 37s on IM trains. Porterbrook has awarded a contract to the South Devon Railway to overhaul these power cars, and 43277 is the first to move to Buckfastleigh for this work.

All are in the COTS pool.

43251	43051	VEC	POR	ZA	
43257	43057	VEC	POR	ZA	
43272	43072	VEC	POR	ZA	
43274	43074	EMP	POR	ZA	
43277	43077	VEC	POR	SDR (U)	

Note: 43251/257/272/274 have Colas logos applied on their previous liveries.

Class 56

These locos are used for freight and infrastructure support work. All are in the COFS pool.

56049		COL	COL	NM	*Robin of Templecombe 1938-2013*
56051		COL	COL	NM	*Survivor*
56078		COL	COL	NM	
56087		COL	COL	NM	
56090		COL	COL	NM	
56094		COL	COL	NM	
56096		COL	COL	NM	
56105		COL	COL	NM	
56113		COL	COL	NM	
56302	56124	COL	COL	NM	*PECO The Railway Modeller 2016 70 years*

Ex-BR Class 56s are used by Colas for many of its freight flows; 56090/105 head the 1005 Preston-Lindsey through Rishton, near Blackburn, on 30 March 2021. *Tom McAtee*

There are five ex-Freightliner Class 66/8s in the Colas fleet; 66849 *Wylam Dilly* passes Brewham Summit, near Castle Cary, with the 1637 Westbury-Lostwithiel infrastructure working on 16 July 2021. 66850 is on the rear. *Glen Batten*

Class 66

These locos, formerly in use with Freightliner, are used for heavy trainload work. All are in the COLO pool.

66846	66573	COL	BEA	HJ	
66847	66574	COL	BEA	HJ	*Terry Baker*
66848	66575	COL	BEA	HJ	
66849	66576	COL	BEA	HJ	*Wylam Dilly*
66850	66577	COL	BEA	HJ	*David Maidment OBE*

Class 70

Colas ordered two batches of Class 70s, the first comprised former demonstrator loco 70099 and the nine locos from a Freightliner option, which was not taken up, while 70811-817 were ordered for business growth. 70801-804 were offered for sale but there were no suitable offers and they remain with Colas.

All are in the COLO pool.

70801	70099	COL	LOM	CF
70802		COL	LOM	CF
70803		COL	LOM	CF
70804		COL	LOM	CF
70805		COL	LOM	CF
70806		COL	LOM	CF
70807		COL	LOM	CF
70808		COL	LOM	CF
70809		COL	LOM	CF
70810		COL	LOM	CF
70811		COL	BEA	CF
70812		COL	BEA	CF
70813		COL	BEA	CF
70814		COL	BEA	CF
70815		COL	BEA	CF
70816		COL	BEA	CF
70817		COL	BEA	CF

Colas has 17 Class 70/8s in its fleet, which are used on a wide variety of freight trains across the network. 70817 passes Ais Gill on the Settle to Carlisle line on 18 May 2021 with a timber train. *Graham Roose*

DC Rail

Contact details

Website: www.dcrail.com
Twitter: @DCRail2018

Key personnel

Managing Director: Garcia Hanson
Head of Engineering: Alan Lee

Overview

DC Rail has been operating for several years now, but has always been a relatively small player in the railfreight market. However, it now has the backing of the Cappagh Group, which has seen it invest in an initial four Class 60s. These were overhauled by DB Cargo at Toton and returned to traffic. The initial deal for those locos saw 60060 acquired for spare parts.

More recently it took another 14 Class 60s from DB Cargo and the initial aspiration is that at least six or seven will be overhauled. This work is likely to be undertaken by UKRL.

The firm also has two Class 56s in its fleet, while it has also hired 50008 on occasions from the firm's MD, Garcia Hanson, and Class 20s from both HNRC and 20189 Ltd.

It operates bulk trainload flows traffic flows and has invested in new aggregates wagons for these; flows include from Scunthorpe to London. DC Rail has also undertaken stock moves.

Class 56

Ex-EWS locos, both are in the DCRO pool.

56091	DCRO	DCN	DCR	LR	*Driver Wayne Gaskell The Godfather*
56103	DCRO	DCN	DCR	LR	

DCR's two remaining active Class 56s, 56103 and 56091 pass Sherrington, between Warminster and Salisbury in the Wylye Valley, with the 1524 Westbury-Southampton train of reprocessed ballast on 26 June 2019. *Glen Batten*

Class 60

DC Rail's parent company Cappagh has bought 19 Class 60s from DB Cargo, who then overhauled an initial four locos for the company and these are now in traffic. A fifth loco was obtained for spares in 2019, while in late 2021 it acquired another 14 of which initially six or seven are expected to be overhauled.

60008	WQDA	EWS	CAP	TO (U)	
60009	WQDA	EWS	CAP	TC (U)	
60013	WQDA	TEW	CAP	TO (U)	*Robert Boyle*
60022	WQDA	EWS	CAP	TO (U)	
60028	DCRS	CAP	CAP	TO	
60029	DCRS	DCR	CAP	TO	*Ben Nevis*
60038	WQDA	EWS	CAP	TO (U)	
60046	DCRS	DCR	CAP	TO	*William Wilberforce*
60055	DCRS	DCR	CAP	TO	*Thomas Barnardo*
60057	WQDA	TEW	CAP	TO (U)	*Adam Smith*
60060	WQDA	TEW	CAP	LR (U)	
60061	WQDA	TRN	CAP	TO (U)	
60064	WQDA	TEW	CAP	TO (U)	*Back Tor*
60070	WQDA	TLH	CAP	TO (U)	
60075	WQDA	EWS	CAP	TC (U)	
60080	WQDA	EWS	CAP	TO (U)	
60090	WQDA	TEW	CAP	TC (U)	
60098	WQDA	EWS	CAP	TO (U)	
60099	WQDA	TAS	CAP	TO (U)	

DC Rail has invested in a number of ex-EWS Class 60s to boost its fleet and so far four have been overhauled. 60029 *Ben Nevis* emerges from Dundas Aqueduct, near Bath, with the 1425 Bristol FLT-Willesden stone train on 21 July 2021. *Glen Batten*

DB Cargo

Contact details

Website: https://uk.dbcargo.com
Twitter: @DBCargoUK

Key personnel

Chief Executive: Andrea Rossi
Chief Operating Officer Dr Dirk Nolte

Overview

DB Cargo is the country's largest railfreight operator in the UK, but from being the dominant player 25 years ago (as EWS) it has slowly seen its traffic levels reduced as other companies have taken work from it. Even the 280 new locos it ordered in 1996 have been trimmed to about 170.

However, with a fleet of overhauled 20-odd Class 60s, 170 Class 66s and a handful of active Class 67/90/92s, DB Cargo is still a big player on the UK railfreight scene.

It has recently returned some Class 67s to traffic for Transport for Wales, but it still has several withdrawn Class 60s, 67s, 90s and 92s on its books. The company also offers maintenance and repainting facilities to other operators at its main Toton base.

Class 60

EWS inherited the full ex-BR fleet of 100 Class 60s in 1996 and started withdrawing them from 2004. In recent times it has sold several locos to Colas, GBRf and Cappagh's DC Rail operation, while a handful have been sold for preservation and 60006 has been scrapped. That has left it with just 64 locos, of which typically only 20 are in traffic at any one time. Any loco in the WQBA/WQCA/WQDA is unlikely to run again without major expenditure. Those locos in traffic are used on any remaining heavy trainload flows in the DBC portfolio.

Of the 100 Class 60s that EWS inherited, several have been withdrawn and others sold. On 3 July 2021, active 60007 *The Spirit of Tom Kendall* was stabled at Toton. *Pip Dunn*

60001	WCAT	DBC	DBC	TO	
60003	WQDA	EWS	DBC	TY (U)	*Freight Transport Association*
60005	WQDA	EWS	DBC	TY (U)	
60007	WCBT	DBC	DBC	TO	*The Spirit of Tom Kendall*
60010	WCBT	DBC	DBC	TO	
60011	WCAT	DBC	DBC	TO	
60012	WQBA	EWS	DBC	TC (U)	
60015	WCBT	DBU	DBC	TO	
60017	WCBT	DBC	DBC	TO	
60019	WCAT	DBC	DBC	TO	*Port of Grimsby & Immingham*
60020	WCBT	DBC	DBC	TO	*The Willows*
60023	WQDA	EWS	DBC	TY (U)	
60024	WCAT	DBC	DBC	TO	*Clitheroe Castle*
60025	WQDA	EWS	DBC	TY (U)	
60027	WQDA	EWS	DBC	TY (U)	
60030	WQDA	EWS	DBC	TO (U)	
60031	WQDA	EWS	DBC	TY (U)	
60032	WQDA	TRN	DBC	TY (U)	
60033	WQCA	COR	DBC	TC (U)	*Tees Steel Express*
60034	WQBA	TEW	DBC	TO (U)	*Carnedd Llewelyn*
60035	WQCA	EWS	DBC	TO (U)	
60036	WQBA	EWS	DBC	TO (U)	*GEFCO*
60037	WQDA	EWS	DBC	TY (U)	
60039	WCAT	DBC	DBC	TO	*Dove Holes*
60040	WCAT	DBC	DBC	TO	*The Territorial Army Centenary*
60041	WQCA	EWS	DBC	TC (U)	
60042	WQDA	EWS	DBC	TY (U)	
60043	WQBA	EWS	DBC	TO (U)	
60044	WCAT	DBC	DBC	TO	*Dowlow*
60045	WQCA	EWS	DBC	TC (U)	*The Permanent Way Institution*
60048	WQCA	EWS	DBC	TO (U)	
60049	WQCA	EWS	DBC	TO (U)	
60051	WQDA	EWS	DBC	TO (U)	
60052	WQDA	EWS	DBC	TO (U)	*Glofa Tŵr – The last deep mine in Wales Tower Colliery*
60053	WQBA	EWS	DBC	TY (U)	
60054	WQAA	DBC	DBC	TO (S)	
60058	WQBA	EWS	DBC	TO (U)	
60059	WQAA	DBC	DBC	TO (U)	*Swinden Dalesman*
60062	WCAT	DBC	DBC	TO	*Stainless Pioneer*
60063	WQAA	DBC	DBC	TO	
60065	WCAT	EWS	DBC	TO	*Spirit Of Jaguar*
60066	WCAT	DRA	DBC	TO	
60067	WQBA	TEW	DBC	TY (U)	
60068	WQBA	TEW	DBC	TO (U)	
60069	WQBA	EWS	DBC	TC (U)	*Slioch*
60071	WQCA	EWS	DBC	TO (U)	*Ribblehead Viaduct*
60072	WQBA	TEW	DBC	TC (U)	
60073	WQBA	TEW	DBC	TO (U)	*Cairn Gorm*
60074	WCAT	PUM	DBC	TO	*Luke*
60077	WQBA	TEW	DBC	TC (U)	
60078	WQBA	MEW	DBC	TY (U)	
60079	WQBA	DBC	DBC	TO (U)	
60082	WQBA	TEW	DBC	CE (U)	
60083	WQBA	EWS	DBC	TY (U)	
60084	WQBA	TEW	DBC	TC (U)	
60088	WQBA	TEW	DBC	TY (U)	
60089	WQBA	EWS	DBC	TY (U)	
60091	WQAA	DBC	DBC	TO (S)	*Barry Needham*

66103	WBBE	EWS	DBC	TO	
66104	WBBT	DBC	DBC	TO	
66105	WBAR	DBC	DBC	TO	
66106	WBBE	EWS	DBC	TO	
66107	WBBT	DBC	DBC	TO	
66109	WBAR	PDP	DBC	TO	*Teesport Express*
66110	WBBE	EWS	DBC	TO	
66111	WBBE	EWS	DBC	TO	
66112	WBBE	EWS	DBC	TO	
66113	WBBE	DBC	DBC	TO	
66114	WBBT	DBS	DBC	TO	
66115	WBAE	EWS	DBC	TO	
66116	WBAE	EWS	DBC	TO	
66117	WBAE	DBC	DBC	TO	
66118	WBAE	DBS	DBC	TO	
66119	WBAE	EWS	DBC	TO	
66120	WBAE	EWS	DBC	TO	
66121	WBAE	EWS	DBC	TO	
66124	WBAR	DBC	DBC	TO	
66125	WBAE	EWS	DBC	TO	
66127	WBAT	EWS	DBC	TO	
66128	WBAE	DBC	DBC	TO	
66129	WBAR	EWS	DBC	TO	
66130	WBAR	DBC	DBC	TO	
66131	WBAE	DBC	DBC	TO	
66133	WBAE	EWS	DBC	TO	
66134	WBAE	DBC	DBC	TO	
66135	WBAE	DBC	DBC	TO	
66136	WBAE	DBC	DBC	TO	
66137	WBRT	EWS	DBC	TO	
66138	WQBA	EWS	DBC	TO (U)	
66139	WBAE	EWS	DBC	TO	
66140	WBAE	EWS	DBC	TO	
66142	WBAR	MRD	DBC	TO	*Maritime Intermodal Three*
66143	WBRT	EWS	DBC	TO	
66144	WBAR	EWS	DBC	TO	
66145	WQBA	EWS	DBC	TO (U)	
66147	WBAE	EWS	DBC	TO	
66148	WBAE	MRD	DBC	TO	*Maritime Intermodal Seven*
66149	WBAE	DBC	DBC	TO	
66150	WBAE	DBC	DBC	TO	

A handful of DBC locos have been painted in customer colours; In Maritime blue, 66148 *Maritime Intermodal*

66151	WBAE	EWS	DBC	TO	
66152	WBAE	DBS	DBC	TO	*Derek Holmes Railway Operator*
66154	WBAE	EWS	DBC	TO	
66155	WBAE	EWS	DBC	TO	
66156	WBAE	EWS	DBC	TO	
66158	WBAE	EWS	DBC	TO	
66160	WBAE	EWS	DBC	TO	
66161	WBAE	EWS	DBC	TO	
66162	WBAR	MRD	DBC	TO	*Maritime Intermodal Five*
66164	WBAE	EWS	DBC	TO	
66165	WBAR	DBC	DBC	TO	
66167	WBAE	DBC	DBC	TO	
66168	WBAR	EWS	DBC	TO	
66169	WBAR	EWS	DBC	TO	
66170	WBAE	EWS	DBC	TO	
66171	WBAR	EWS	DBC	TO	
66172	WBAE	EWS	DBC	TO	*Paul Mellany*
66174	WBAE	EWS	DBC	TO	
66175	WBAE	DBC	DBC	TO	*Rail Riders Express*
66176	WBAR	EWS	DBC	TO	
66177	WBAT	EWS	DBC	TO	
66179	WBAI	EWS	DBC	TO (S)	
66181	WBAR	EWS	DBC	TO	
66182	WQAB	DBC	DBC	TO (U)	
66183	WBAE	EWS	DBC	TO	
66185	WBRT	DBS	DBC	TO	*DP World London Gateway*
66186	WBAE	EWS	DBC	TO	
66187	WBAE	EWS	DBC	TO	
66188	WBAR	EWS	DBC	TO	
66190	WBAT	DBC	DBC	TO	
66192	WBAR	DBC	DBC	TO	
66194	WBAR	EWS	DBC	TO	
66197	WBAE	EWS	DBC	TO	
66198	WBAR	EWS	DBC	TO	
66199	WBAE	EWS	DBC	TO	
66200	WBAE	EWS	DBC	TO	
66205	WBAI	DBC	DBC	TO	
66206	WBAR	DBC	DBC	TO	
66207	WBAE	EWS	DBC	TO	
66221	WBAR	EWS	DBC	TO	
66224	WBAI	EWS	DBC	TO (S)	
66230	WQAB	DBC	DBC	TO (U)	
66244	WBAI	EWS	DBC	TO (S)	

Note: 66084/150 has been modified to run on Hydro-treated vegetable oil

One of the pair of Class 67s painted in the livery to match the Royal Train, 67005 *Queen's Messenger* stands at Peterborough on the rear of the 1655 Beverley-King's Cross charter on 18 December 2021. *Pip Dunn*

67001	WAAC	ARV	DBC	CE	
67002	WAAC	ARV	DBC	CE	
67003	WQBA	ARV	DBC	TO (S)	
67004	WQAB	DBC	DBC	TO (U)	
67005	WAAC	RTO	DBC	CE	*Queen's Messenger*
67006	WAAC	RTO	DBC	CE	*Royal Sovereign*
67007	WABC	EWS	DBC	CE	
67008	WAWC	TFW	DBC	CE	
67009	WQBA	EWS	DBC	CE (U)	
67010	WAWC	DBC	DBC	CE	
67011	WQBA	EWS	DBC	CE (U)	
67012	WAWC	CRS	DBC	CE	
67013	WAWC	DBC	DBC	CE	
67014	WAWC	TFW	DBC	CE	
67015	WAWC	CRS	DBC	CE	
67016	WAAC	EWS	DBC	CE	
67017	WAWC	TFW	DBC	CE	
67018	WQBA	DBS	DBC	CE (S)	*Keith Heller*
67019	WQBA	EWS	DBC	TO (U)	
67020	WAAC	EWS	DBC	CE	
67021	WAAC	PUL	DBC	CE	
67022	WQAA	TFW	DBC	TO (S)	
67024	WAAC	PUL	DBC	CE	
67025	WAWC	TFW	DBC	CE	
67026	WQBA	JUB	DBC	CE (U)	*Diamond Jubilee*
67028	WAAC	DBC	DBC	CE	
67029	WQAB	DMS	DBC	CE (S)	*Royal Diamond*
67030	WQBA	EWS	DBC	TO (S)	

Class 90

EWS inherited 25 Class 90s at privatisation but there is very limited work for them. Although pairs of them do work freight on the WCML intermodal trains, the company has stated publicly it wants to look at using more electric traction, which could see some more locos reinstated. Some, however, have been withdrawn for more than 15 years now and could only return following major expenditure.

Those locos in the WQAA pool, however, can be reactivated relatively quickly if required and 90021 was overhauled in February 2022.

90017		WQCA	EWS	DBC	CE (U)	
90018		WQAB	DBS	DBC	CE (U)	*Pride of Bellshill*
90019		WEDC	DBC	DBC	CE	*Multimodal*
90020		WEDC	GCR	DBC	CE	
90021		WQAA	MAB	DBC	CE	
90022	90222	WQCA	REW	DBC	CE (U)	*Freightconnection*
90023	90223	WQCA	RFE	DBC	CE (U)	
90024	90224	WEAC	MAA	DBC	CE	
90025	90125, 90225	WQCA	RFD	DBC	CE (U)	
90026	90126	WEDC	GCR	DBC	CE	
90027	90127, 90227	WQCA	RFD	DBC	CE (U)	*Allerton T&RS Depot*
90028	90128	WEDC	DBC	DBC	CE	*Sir William McAlpine*
90029	90129	WEDC	GCR	DBC	CE	
90030	90130	WQCA	EWS	DBC	CE (U)	
90031	90131	WQCA	EWS	DBC	CE (U)	*The Railway Children Partnership: Working for Street Children Worldwide*
90032	90132	WQCA	EWS	DBC	CE (U)	
90033	90133, 90233	WQCA	RFE	DBC	CE (U)	
90034	90134	WEDC	DRU	DBC	CE	
90035	90135	WEAC	DBC	DBC	CE	
90036	90136	WEDC	DBC	DBC	CE	*Driver Jack Mills*
90037	90137	WEAC	DBC	DBC	CE	*Christine*
90038	90138, 90238	WQCA	RFE	DBC	CE (U)	
90039	90139, 90239	WEDC	DBB	DBC	CE	*The Chartered Institute of Logistics and Transport*
90040	90140	WQAB	DBC	DBC	CE	
90050	90150	DHLT	TTG	ARV	CB (U)	

90037 has aircraft graphics on its DBC red.
90039 has 'I am the Backbone of the economy' branding.

DBC owns 25 Class 90s, many of which are withdrawn, but typically there are about 10-12 in the active pool at any one time. 90037 *Christine* leads 90024 – in Malcolm Group colours – on the 0707 Mossend-Daventry through Wigan North Western on 9 September 2021. *Tom McAtee*

Class 92

A fleet that is now over 25 years old, the Class 92s have been woefully underused since day one EWS inherited 30 locos and while it has redeployed some in Eastern Europe – some of which have since been sold – it is still left with 18 locos for its Channel Tunnel operations, of which it is lucky to ever need more than six at any one time.

Locos in the WQAB and WQBA are unlikely to run for DB Cargo again in the UK any time soon.

92004	WQCA	EUE	DBC	CE (U)	*Jane Austen*
92007	WQBA	EUE	DBC	CE (U)	*Schubert*
92008	WQCA	EUE	DBC	CE (U)	*Jules Verne*
92009	WQCA	DBC	DBC	CE (U)	*Marco Polo*
92011	WFBC	EUE	DBC	CE	*Handel*
92013	WQBA	EUE	DBC	CE (U)	*Puccini*
92015	WFBC	DBC	DBC	CE	
92016	WQCA	DBC	DBC	CE (U)	
92017	WQCA	STO	DBC	CE (U)	*Bart the Engine*
92019	WFBC	EUE	DBC	CE	*Wagner*
92029	WQAB	EUE	DBC	CE (U)	*Dante*
92031	WQBA	DBC	DBC	CE (U)	*The Institute of Logistics & Transport*
92035	WQCA	EUE	DBC	CE (U)	*Mendelssohn*
92036	WFBC	EUE	DBC	CE	*Bertolt Brecht*
92037	WQCA	EUE	DBC	CE (U)	*Sullivan*
92041	WFBC	EUE	DBC	CE	*Vaughan Williams*
92042	WFBC	DBC	DBC	CE	

Just six of the 30 Class 92s that DBC inherited remain in active use in the UK, working almost exclusively on HS1. 92019 *Wagner* works the 1952 Dollands Moor-Ripple Lane through the Nashenden Valley near Rochester on 22 June 2020. *David Staines*

Class 325

Four-car electric multiple units based on the Class 365 Networker units were acquired by the Royal Mail – which still owns them – and they are used for some WCML mail trains, running usually as 12-car sets. They are maintained and operated by DB Cargo.

All are in the PPMB pool.

DBC operates the remaining Class 325 freight EMUs owned by the Royal Mail. 325005/006/015 are near Rugeley working the 1621 Willesden-Shieldmuir on 21 July 2021. *Robin Ralston*

				DTPMV	MPMV	TPMV	DTPMV	
325001	ROM	ROM	CE	68300	68340	68360	68301	
325002	ROM	ROM	CE	68302	68341	68361	68303	
325003	ROM	ROM	CE	68304	68342	68362	68305	
325004	ROM	ROM	CE	68306	68343	68363	68307	
325005	ROM	ROM	CE	68308	68344	68364	68309	
325006	ROM	ROM	CE	68310	68345	68365	68311	
325007	ROM	ROM	CE	68312	68346	68366	68313	
325008	ROM	ROM	CE	68314	68347	68367	68315	*Peter Howarth CBE*
325009	ROM	ROM	CE	68316	68348	68368	68317	
325011	ROM	ROM	CE	68320	68350	68370	68321	
325012	ROM	ROM	CE	68322	68351	68371	68323	
325013	ROM	ROM	CE	68324	68352	68372	68325	
325014	ROM	ROM	CE	68326	68353	68373	68327	
325015	ROM	ROM	CE	68328	68354	68374	68329	
325016	ROM	ROM	CE	68330	68355	68375	68331	

DB Cargo Managers' Train

This short train features a driving van trailer and three Mk 3 coaches offering conference facilities and catering, and is used to take staff, customers and stakeholders to events and rail installations.

It runs with a Class 67 in push-pull mode. 67029 was the nominated loco and it was painted in the same silver to match the DVT – the coaches are maroon. However, it is presently withdrawn so any active push-pull fitted Class 67 can be used, although the train is rarely used these days.

10211	DBM	DBC	TO	Mk 3 RFM
10546	DBM	DBC	TO	Mk 3 SLEP
11039	DBM	DBC	TO	Mk 3 FO
82146	DMS	DBC	TO	Driving van trailer

Direct Rail Services

Contact details

Website: www.directrailservices.com
Twitter: @DRSgovuk

Key personnel

Managing Director: Chris Connelly
Engineering Director: Alistair Brown

Overview

DRS was set up in 1995 to haul nuclear flask traffic, which it still carries today. It has diversified into freight with Intermodal, Network Services, infrastructure, passenger hire and charter operations added to its portfolio.

The majority of the firm's flows are operated using Class 66, 68 and 88s but the firm also retain small fleets of Class 37s and 57s, both with and without train supply.

It supplies Class 68s to both Chiltern Railways (68008-015) and Transpennine Express (68019-034) for passenger operations. It also supplies four Class 57/3s (57304/307-309) to Avanti West Coast for Thunderbird duties on the WCML.

It recently took five Class 66s from DB Cargo on long-term hire. The last of its Class 20s were retired from front-line service in January 2020 and will be disposed of shortly.

The firm issued a tender for ten new locomotives some time ago, but no orders have been placed.

Class 37

DRS bought its first Class 37s in 1997 and has slowly been building up its fleet. That said, withdrawals are now happening, although some will be retained for charters, IM trains and other duties.

Class 37/0

Original locos with main generators. All have modified cabs.

37038		XWSS	DRN	DRS	KM (S)
37059		XWSS	DRN	DRS	CB (S)
37069		XWSS	DRN	DRS	CB (S)
37218		XHSO	DRN	DRS	KM
37259	37380	XWSS	DRC	DRS	KM (S)

Class 37/4

Refurbished locos fitted with electric train supply. These locos were used for saloon, charter, RHTT, snowplough and IM work. 37423 has a modified cab.

37401	37268	XHSO	BLL	DRS	KM	*Mary Queen of Scots*
37402	37274	XWSS	BLL	DRS	KM (S)	*Stephen Middlemore 23.12.1954 – 8.6.2013*
37407	37305	XWSS	BLL	DRS	KM (S)	*Blackpool Tower*
37409	37270	XWSS	BLL	DRS	KM (S)	*Lord Hinton*
37419	37291	XHSO	ICM	DRS	KM	*Carl Haviland 1954-2012*
37422	37266	XHSO	DRU	DRS	KM	*Victorious*
37423	37296	XWSS	DRX	DRS	KM (S)	*Spirit of the Lakes*
37424	37279	XHSO	BLL	DRS	KM	*Avro Vulcan XH558*
37425	37292	XHSO	REG	DRS	KM	*Sir Robert McAlpine/Concrete Bob*

Note: 37424 carries the number 37558 on its bodysides.

Class 37/6

A dozen Class 37/5s were converted for use by Eurostar but all were later sold to DRS, although most have since been sold to other operators. They have no train heating and were originally fitted with through ETS wiring, but this has been removed. They have modified cabs.

37602	37082, 37502	XSDP	DRC	DRS	ZG (U)
37605	37036, 37507	XSDP	DRC	DRS	ZA (U)

Class 37/7

Refurbished loco fitted with ballast weights and no train heating.

37716	37094	XHSO	DRN	DRS	KM

Class 57

Re-engineered Class 47s fitted with General Motors engines.

Class 57/0

Original ex-Freightliner locos without train heating. Just 57002 remains in traffic.

57002	47322	XHSO	DRN	DRS	KM	*Rail Express*
57003	47317	XWSS	DRN	DRS	KM (S)	
57007	47332	XWSS	DRN	DRS	KM (S)	*John Scott 12.5.45-22.5.12*

Class 57/3

ETH locos originally built for Virgin West Coast to act as Thunderbird rescue locos, and all modified to have retractable Dellner couplers fitted. Those locos in the XHVT pool remain as WCML standby locos. 57306 is usually on hire to FGW, while DRS still owns 57301/303/305/310/312, which were sub-leased to Rail Operations Group (see page 292).

57304	47055, 47652, 47807	XHVT	DRN	DRS	KM	*Pride of Cheshire*
57306	47242, 47659, 47814	XHAC	DRN	POR	KM	*Her Majesty's Railway Inspectorate 175*
57307	47225	XHVT	DRN	DRS	KM	*Lady Penelope*
57308	47091, 47647, 47846	XHVT	DRN	DRS	KM	*James Ferguson*
57309	47254, 47651, 47806	XHVT	DRN	DRS	KM	*Pride of Crewe*

DRS's fleet of 57s is now down to six locos, four of which are contracted as rescue locos for Avanti West Coast. 57307 *Lady Penelope* was stabled at Carlisle on 26 July 2018. *Pip Dunn*

Class 66

DRS ordered three batches of ten Class 66s before returning its first ten locos and ordering four more to give it a fleet of 24. Since then 66411-420 have been returned to their ROSCOs but the five ex-Fastline Freight Class 66/3s were taken on by the company to give it a fleet of 19 locos. This is now supplemented by five 66/0s on long-term hire from DB Cargo that have been repainted into DRS colours, but they are not being renumbered.

Class 66/0

Locos on long-term hire from DB Cargo.

66031	XWSS	DRX	DBC	KM (S)
66091	XHIM	DRX	DBC	KM
66108	XHIM	DRX	DBC	KM
66122	XHIM	DRX	DBC	KM
66126	XHIM	DRX	DBC	KM

Class 66/3

Former Fastline Freight locos. Fitted with RETB.

66301	XHIM	DRX	BEA	KM	*Kingmoor TMD*
66302	XHIM	DRX	BEA	KM	*Endeavour*
66303	XHIM	DRX	BEA	KM	*Rail Riders 2020*
66304	XHIM	DRX	BEA	KM	
66305	XHIM	DRX	BEA	KM	

DRS leases 19 Class 66s, plus has five on long term-hire from DBC. Ex-Fastline 66301 heads the 1246 Carlisle-Crewe at Wilpshire on 18 May 2021. *Tom McAtee*

Class 66/4

The original third and fourth batches of Class 66s ordered by DRS in 2007 and 2008 respectively.

66421	XHIM	DRX	AIK	KM	*Gresty Bridge TMD*
66422	XHIM	DRX	AIK	KM	
66423	XHIM	DRX	AIK	KM	
66424	XHIM	DRX	AIK	KM	
66425	XHIM	DRX	AIK	KM	
66426	XHIM	DRX	AIK	KM	
66427	XHIM	DRX	AIK	KM	
66428	XHIM	DRX	AIK	KM	*Carlisle Eden Mind*
66429	XHIM	DRX	AIK	KM	
66430	XHIM	DRX	AIK	KM	
66431	XHIM	DRX	AIK	KM	
66432	XHIM	DRX	AIK	KM	
66433	XHIM	DRX	AIK	KM	
66434	XHIM	DRX	AIK	KM	

Class 68

The first 15 locos were new Caterpillar-engined Vossloh-designed locos built in Spain. The remaining 19 were built by Stadler (which acquired Vossloh). They are fitted with electric train supply. 68010-015 are dedicated for use by Chiltern Railways (with 68008/009 as spare locos) and 68019-032 likewise for Transpennine Express (with 68033/034 as spare locos). All have push-pull capability.

The latter have caused several complains because of their noise, and hence their use on the Scarborough line could be ended soon and the locos redeployed on other TPE routes. There is also a likelihood they could ultimately be replaced by new locos and First Group has issued an invitation to tender for between 15 and 35 new bi-bode locos, 15 to 30 of which would be used for the TPE routes. When not in use with these TOCs, they may appear on general DRS duties.

68001	XHVE	DRN	BEA	CB	*Evolution*
68002	XHVE	DRN	BEA	CB	*Intrepid*
68003	XHVE	DRN	BEA	CB	*Astute*
68004	XHVE	DRN	BEA	CB	*Rapid*
68005	XHVE	DRN	BEA	CB	*Defiant*
68006	XHVE	DRG	BEA	CB	*Pride of the North*
68007	XHVE	DRB	BEA	CB	*Valiant*
68008	XHCS	DRN	BEA	CB	*Avenger*
68009	XHCS	DRN	BEA	CB	*Titan*
68010	XHCE	CRS	BEA	CB	*Oxford Flyer*
68011	XHCE	CRS	BEA	CB	
68012	XHCE	CRS	BEA	CB	
68013	XHCE	CRS	BEA	CB	
68014	XHCE	CRS	BEA	CB	
68015	XHCE	CRS	BEA	CB	
68016	XHVE	DRN	BEA	CB	*Fearless*
68017	XHVE	DRN	BEA	CB	*Hornet*
68018	XHVE	DRN	BEA	CB	*Vigilant*
68019	TPEX	TPE	BEA	CB	*Brutus*
68020	TPEX	TPE	BEA	CB	*Reliance*
68021	TPEX	TPE	BEA	CB	*Tireless*
68022	TPEX	TPE	BEA	CB	*Resolution*
68023	TPEX	TPE	BEA	CB	*Achilles*
68024	TPEX	TPE	BEA	CB	*Centaur*
68025	TPEX	TPE	BEA	CB	*Superb*
68026	TPEX	TPE	BEA	CB	*Enterprise*
68027	TPEX	TPE	BEA	CB	*Splendid*
68028	TPEX	TPE	BEA	CB	*Lord President*
68029	TPEX	TPE	BEA	CB	*Courageous*
68030	TPEX	TPE	BEA	CB	*Black Douglas*
68031	TPEX	TPE	BEA	CB	*Felix*
68032	TPEX	TPE	BEA	CB	*Destroyer*
68033	XHTP	DRN	DRS	CB	*The Poppy*
68034	XHTP	DRN	DRS	CB	

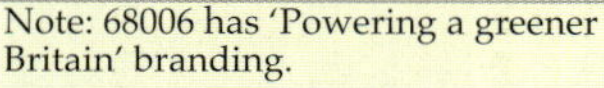
Note: 68006 has 'Powering a greener Britain' branding.

DRS was created in 1995 to haul nuclear trains but later branched out into other sectors. Two of its brand new Class 68s, with 68005 *Defiant* leading, provide overkill to move just one nuclear flask as they approach Heysham on 29 July 2020. 68006, then named *Daring*, is on the rear. *Rob France*

Class 88

Electro diesel versions of the Stadler Class 68, fitted with electric train supply used mostly for intermodal traffic on the WCML. They run off 25kV AC with a 950hp Caterpillar engine for 'last mile' operation.

88001	XHVE	DRE	BEA	KM	*Revolution*
88002	XHVE	DRE	BEA	KM	*Prometheus*
88003	XHVE	DRE	BEA	KM	*Genesis*
88004	XHVE	DRE	BEA	KM	*Pandora*
88005	XHVE	DRE	BEA	KM	*Minerva*
88006	XHVE	DRE	BEA	KM	*Juno*
88007	XHVE	DRE	BEA	KM	*Electra*
88008	XHVE	DRE	BEA	KM	*Ariadne*
88009	XHVE	DRE	BEA	KM	*Diana*
88010	XHVE	DRE	BEA	KM	*Aurora*

Coaches

DRS has several Mk 2 coaches but following the end of their passenger contracts with ScotRail, Greater Anglia and Northern, these are all currently in warm store and some are for sale.

It also owns four Mk 2 coaches used as escort vehicles for some nuclear trains. Four coaches were briefly on hire to Loram and reliveried in its colours.

No	Livery	Type
5937	DRS	Mk 2 TSO
6064	DRS	Mk 2 TSO
6173	DRS	Mk 2 TSO
9419	DRS	Mk 2 Escort coach
9428	DRS	Mk 2 Escort coach
9488	SRR	Mk 2 BSO
9506	DRS	Mk 2 Escort coach
9508	DRS	Mk 2 Escort coach
9525	LOM	Mk 2 BSO
9704	DRS	Mk 2 DBSO
9705	DRS	Mk 2 DBSO
9707	DRS	Mk 2 DBSO

9704/05/07 have been offered for sale.

DRS ordered ten Class 88s in the hope it would win the Caledonian Sleeper haulage contract, but it lost out to GBRf. The locos are now used on intermodal traffic on the WCML. 88007 *Electra* leads the 0624 Daventry-Mossend at Red Bank on 25 June 2020. *Tom McAtee*

Freightliner

Contact details

Website: www.freightliner.co.uk
Twitter: @RailFreight

Key personnel

Managing Director: Tim Shakerley

Overview

One of the original six former British Rail freight operations sold in the mid-1990s, Freightliner was a management buyout and the only FOC not bought by WCTC. Initially just an intermodal operator, it has diversified and expanded and now operates bulk trainloads in many markets.

It inherited Class 47s from BR, some of which it had rebuilt as Class 57/0s, but then it acquired new Class 66s. At one time these reached 130 locos, although several have since been redeployed in Poland or returned to their ROSCOs. It acquired 20 Class 70s, of which only 19 were successfully delivered, and has recently acquired 14 Class 59s, having won the contract for haulage of the Mendip stone trains. It owns one Class 47 used for route learning.

Until recently, its electric fleet was a mix of elderly Class 86s dating from 1965 and ten newer Class 90s dating from 1990. In 2020 it acquired 13 former Greater Anglia Class 90s, which replaced Class 86s. The latter have been stood down and kept in reserve to boost traffic growth if required, although they have now been in open storage for more than two years.

Class 08

Several Class 08s are retained for depot and terminal shunting.

08530	DDIN	FLR	FLR	SM		
08531	DDIN	FPH	FLR	FX		
08575	DHLT	FLR	FLR	LH (U)		
08585	DDIN	FLR	FLR	SM	*Vicky*	
08624	DDIN	FPH	FLR	FX	*Rambo Paul Ramsey*	
08691	DDIN	FLR	FLR	LH	*Terri*	
08785	DDIN	GWD	FLR	IP		
08891	DDIN	GWD	FLR	CX		

Freightliner is one of the few FOCs to have Class 08 shunters in its fleet and two have recently been repainted into its new livery. 08785 stands at Ipswich Freightliner depot on 10 August 2021. *Stuart West*

Class 47

Freightliner's once-sizeable Class 47 fleet is now reduced to a single loco used mostly for route learning.

Freightliner used Class 47s from its inception in 1995, but they were soon replaced by newer Class 66 and 70s. However one loco, 47830 *Beeching's Legacy*, is retained for route learning and ad hoc moves. On 3 August 2021, it passes Hawarden Bridge running light from Dee Marsh to Crewe Basford Hall as a route-learning trip. *Paul Shannon*

47830	47061, 47649	DFLH	GYP	FLR	CX	*Beeching's Legacy*

Class 59

The Class 59/0s were originally ordered by Foster Yeoman and operated by BR and then EWS/DB Cargo. 59003 was moved to Germany in 1997 but was bought by GBRf in 2014 and returned to the UK.

The 59/1s were originally Amey Roadstone locos later absorbed into the Aggregates Industries fleet along with the other 59s. The 59/2s were ordered by National Power and delivered in 1995 but sold to EWS in 1998. All 14 locos are now in the DFHG pool.

Class 59/0

Former Foster Yeoman locos.

59001	AGI	FLR	MD	*Yeoman Endeavour*
59002	AGI	FLR	MD	*Alan J Day*
59004	AGI	FLR	MD	*Paul A Hammond*
59005	AGI	FLR	LD (U)	*Kenneth J Painter*

When Freightliner won the contract for the Mendip aggregates haulage, it acquired 14 of the 15 Class 59s. One of the ex-Foster Yeoman locos, 59004 *Paul A Hammond*, passes Ealing Broadway with the 1439 Acton-Whatley Quarry empties in 7 September 2021. *Stuart West*

Class 59/1

Former ARC locos.

59101	HAN	FLR	MD	*Village of Whatley*
59102	HAN	FLR	MD	*Village of Chantry*
59103	HAN	FLR	MD	*Village of Mells*
59104	HAN	FLR	MD	*Village of Great Elm*

Class 59/2

Former National Power/EWS locos.

59201	DBU	FLR	MD	
59202	GWD	FLR	MD	
59203	GWD	FLR	MD	
59204	GWD	FLR	MD	
59205	DBU	FLR	MD	
59206	GWD	FLR	MD	*John F Yeoman Rail Pioneer*

The six Class 59/2s Freightliner took on were ex-National Power, then DBC, locos and were all in DBC red, so are being repainted into Freightliner orange livery. 59206 *John F Yeoman Rail Pioneer* approaches Swindon with a stone train on 10 May 2021. *Jack Boskett*

Class 66

The Class 66/5 and 66/6s were the original Freightliner orders, while the 66/4s are former DRS locos taken on when they became available. The 66/6s are regeared for heavy freight

Class 66/4

Former DRS locos.

66413	DFIN	GWO	AIK	LD	*Lest We Forget*
66414	DFIN	FPH	AIK	LD	
66415	DFIN	GWD	AIK	LD	*You Are Never Alone*
66416	DFIN	FPH	AIK	LD	
66418	DFIN	FPH	AIK	LD	*Patriot – In Memory Of Fallen Railway Employees*
66419	DFIN	GWD	AIK	LD	
66420	DFIN	FPH	AIK	LD	

Class 66/5

Freightliner's standard locos, ordered in multiple batches from 1999 to 2008. Several have since been redeployed in Poland.

66501	DFIM	FLR	POR	LD	*Japan 2001*
66502	DFIM	FLR	POR	LD	*Basford Hall Centenary 2001*
66503	DFIM	GWD	POR	LD	*The Railway Magazine*
66504	DFIM	FPH	POR	LD	
66505	DFIM	FLR	POR	LD	
66506	DFIM	FLR	EVS	LD	*Crewe Regeneration*
66507	DFIM	FLR	EVS	LD	
66508	DFIM	FLR	EVS	LD	
66509	DFIM	FLR	EVS	LD	
66510	DFIM	FLR	EVS	LD	
66511	DFIM	FLR	EVS	LD	
66512	DFIM	FLR	EVS	LD	
66513	DFIM	FLR	EVS	LD	
66514	DFIM	FLR	EVS	LD	
66515	DFIM	FLR	EVS	LD	
66516	DFIM	FLR	EVS	LD	
66517	DFIM	FLR	EVS	LD	
66518	DFIM	FLR	EVS	LD	
66519	DFIM	FLR	EVS	LD	
66520	DFIM	FLR	EVS	LD	
66522	DFIM	FLR	EVS	LD	
66523	DFIM	FLR	EVS	LD	
66524	DFIM	FLR	EVS	LD	
66525	DFIM	FLR	EVS	LD	
66526	DFIM	FLR	EVS	LD	*Driver Steve Dunn (George)*
66528	DFIM	FPH	POR	LD	*Madge Elliot MBE Borders Railway Opening 2015*
66529	DFIM	FLR	POR	LD	
66531	DFIM	FLR	POR	LD	
66532	DFIM	FLR	POR	LD	*P&O Nedlloyd Atlas*
66533	DFIM	FLR	POR	LD	*Hanjin Express/Senator Express*
66534	DFIM	FLR	POR	LD	*OOCL Express*
66536	DFIM	FLR	POR	LD	
66537	DFIM	FLR	POR	LD	
66538	DFIM	FLR	EVS	LD	
66539	DFIM	FLR	EVS	LD	
66540	DFIM	FLR	EVS	LD	*Ruby*
66541	DFIM	FLR	EVS	LD	
66542	DFIM	FLR	EVS	LD	
66543	DFIM	FLR	EVS	LD	
66544	DFIM	FLR	EVS	LD	
66545	DFIM	FLR	POR	LD	
66546	DFIM	FLR	POR	LD	
66547	DFIM	FLR	POR	LD	
66548	DFIM	FLR	POR	LD	
66549	DFIM	FLR	POR	LD	
66550	DFIM	FLR	POR	LD	
66551	DFIM	FLR	POR	LD	
66552	DFIM	FLR	POR	LD	*Maltby Raider*
66553	DFIM	FLR	POR	LD	
66554	DFIM	FLR	EVS	LD	
66555	DFIM	FLR	EVS	LD	
66556	DFIM	FLR	EVS	LD	
66557	DFIM	FLR	EVS	LD	
66558	DFIM	FLR	EVS	LD	

Class 66s remain the core of the Freightliner fleet; 66599 passes Ipswich on 28 July 2018 with the 1014 Felixstowe North-Crewe. *John Day/ Rail Photoprints*

66559	DFIM	FLR	EVS	LD	
66560	DFIM	FLR	EVS	LD	
66561	DFIM	FLR	EVS	LD	
66562	DFIM	FLR	EVS	LD	
66563	DFIM	FLR	EVS	LD	
66564	DFIM	FLR	EVS	LD	
66565	DFIM	FLR	EVS	LD	
66566	DFIM	FLR	EVS	LD	
66567	DFIM	FLR	EVS	LD	
66568	DFIM	FLR	EVS	LD	
66569	DFIM	FLR	EVS	LD	
66570	DFIM	FLR	EVS	LD	
66571	DFIM	FLR	EVS	LD	
66572	DFIM	FLR	EVS	LD	
66585	DFIN	FLR	AIK	LD	
66587	DFIN	ONE	HFX	LD	*As One, We Can*
66588	DFIN	FLR	HFX	LD	
66589	DFIM	FLR	HFX	LD	
66590	DFIN	FLR	HFX	LD	
66591	DFIN	FLR	HFX	LD	
66592	DFIN	FLR	HFX	LD	*Johnson Stevens Agencies*
66593	DFIN	FLR	HFX	LD	*3MG Mersey Multimodal Gateway*
66594	DFIN	FLR	HFX	LD	*NYK Spirit of Kyoto*
66596	DFIN	FLR	BEA	LD	
66597	DFIN	FLR	BEA	LD	*Viridor*
66598	DFIN	FLR	BEA	LD	
66599	DFIN	FLR	BEA	LD	

Class 66/6

Regeared locos for heavy trainload haulage. Some have been redeployed in Poland. 66623 is a low-emission loco.

66601	DFHH	FLR	POR	LD	*The Hope Valley*
66602	DFHH	FLR	POR	LD	
66603	DFHH	FLR	POR	LD	
66604	DFHH	FLR	POR	LD	
66605	DFHH	FLR	POR	LD	
66606	DFHH	FLR	POR	LD	

66607	DFHH	FLR	POR	LD	
66610	DFHH	FLR	POR	LD	
66613	DFHH	FLR	EVS	LD	
66614	DFHH	FLR	EVS	LD	*1916 Poppy 2016*
66615	DFHH	FLR	EVS	LD	
66616	DFHH	FLR	EVS	LD	
66617	DFHH	FLR	EVS	LD	
66618	DFHH	FLR	EVS	LD	*Railways Illustrated Annual Photographic Awards Alan Barnes*
66619	DFHH	FLR	EVS	LD	*Derek W. Johnson MBE*
66620	DFHH	FLR	EVS	LD	
66621	DFHH	FLR	EVS	LD	
66622	DFHH	FLR	EVS	LD	
66623	DFHH	GWD	AIK	LD	

Class 66/9

66951/952 were initial low-emission locos, then followed by 66585-599 and 66953-957.

66951	DFHH	FLR	EVS	LD	
66952	DFHH	FLR	EVS	LD	
66953	DFHH	FLR	BEA	LD	
66955	DFHH	FLR	BEA	LD	
66956	DFHH	FLR	BEA	LD	
66957	DFHH	FLR	BEA	LD	*Stephenson Locomotive Society 1909-2009*

Class 70

General Electric heavy freight locos built in Pennsylvania. Originally an option for ten more was included in the original deal, but not exercised after poor initial reliability. 70012 never made it to the Freightliner fleet after the loco was dropped during unloading at Newport Docks, severely damaging it. It was returned to the United States and is now used as a test bed loco. Many are stored.

70001	DFGI	FPH	AIK	LD	*PowerHaul*
70002	DFGI	FPH	AIK	LD	
70003	DFGI	FPH	AIK	LD	
70004	DFGI	FPH	AIK	LD	*The Coal Industry Society*
70005	DFGI	FPH	AIK	LD	
70006	DFGI	FPH	AIK	LD	
70007	DFGI	FPH	AIK	LD	
70008	DFGI	FPH	AIK	LD	

Freightliner was the first FOC to try the new General Electric Class 70s, yet they have not been as successful as the 66s, and several locos have spent long periods in store. 70008 works the 1112 Runcorn Folly Lane-Brindle Heath at Lostock Gralam on 26 April 2021. *Tom McAtee*

70009	DFGI	FPH	AIK	LD (U)
70010	DFGI	FPH	AIK	LD
70011	DFGI	FPH	AIK	LD
70013	DHLT	FPH	AIK	LD (U)
70014	DFGI	FPH	AIK	LD
70015	DFGI	FPH	AIK	LD
70016	DFGI	FPH	AIK	LD
70017	DFGI	FPH	AIK	LD
70018	DHLT	FPH	AIK	LD (U)
70019	DHLT	FPH	AIK	LD (U)
70020	DFGI	FPH	AIK	LD

Class 86

Ex-BR 25kV AC electric locos, which have been the mainstay of the Freightliner intermodal operations under the wires for 25 years. The last examples were replaced by Class 90s and retired from traffic in 2021, although most have been retained in warm store pending possible overhauls and a return to traffic. However, most have now spent two years out in the open and would need serious expenditure before they could return to traffic, so it is likely they will never run again for the company.

86251 is only a source of spares and has never run for the company.

86251		VIR	EPEX	FLR	CX (U)
86604	86004, 86404	FLR	DHLT	FLR	CX (U)
86605	86005, 86405	FLR	DHLT	FLR	CX (U)
86607	86007, 86407	FLR	DHLT	FLR	CX (U)
86608	86008, 86408, 86501	FLR	DHLT	FLR	CX (U)
86609	86009, 86409	FLR	DHLT	FLR	CX (U)
86610	86010, 86410	FLR	DHLT	FLR	CX (U)
86612	86012, 86312, 86412	FLR	DHLT	FLR	CX (U)
86613	86013, 86313, 86413	FLR	DHLT	FLR	CX (U)
86614	86014, 86314, 86414	FLR	DHLT	FLR	CX (U)
86622	86022, 86322, 86422	FPH	DHLT	FLR	RAC (U)
86627	86027, 86327, 86427	FLR	DHLT	FLR	CX (U)
86628	86028, 86328, 86428	FLR	DHLT	FLR	CX (U)
86632	86032, 86432	FLR	DHLT	FLR	CX (U)
86637	86037, 86437	FPH	DHLT	FLR	CX (U)
86638	86038, 86438	FLR	DHLT	FLR	CX (U)
86639	86039, 86439	FLR	DHLT	FLR	CX (U)

Class 90

Given there were not enough 90s to go round at privatisation, Freightliner has had to wait for 25 years for a fleet of ex-Virgin/Anglia locos to become available to supplement its original ten-strong fleet. 90016 was a direct replacement for 90050 in 2004.

90003	DFLC	GWD	FLR	CX	
90004	DFLC	GWD	FLR	CX	
90005	DFLC	GWD	FLR	CX	
90006	DFLC	GWD	FLR	CX	*Roger Ford* *Modern Railways Magazine*
90007	DFLC	GWD	FLR	CX	
90008	DFLC	GWD	FLR	CX	
90009	DFLC	GWD	FLR	CX	
90010	DFLC	GWD	FLR	CX	
90011	DFLC	GWD	FLR	CX	
90012	DFLC	GWD	FLR	CX	
90013	DFLC	GWD	FLR	CX	
90014	DFLC	GWD	FLR	CX	*Over The Rainbow*
90015	DFLC	GWD	FLR	CX	

90016		DFLC	FLR	FLR	CX
90041	90141	DFLC	FLR	FLR	CX
90042	90142	DFLC	FPH	FLR	CX
90043	90143	DFLC	FPH	FLR	CX
90044	90144	DFLC	GWD	FLR	CX
90045	90145	DFLC	FPH	FLR	CX
90046	90146	DFLC	FLR	FLR	CX
90047	90147	DFLC	GWD	FLR	CX
90048	90148	DFLC	GWD	FLR	CX
90049	90149	DFLC	FPH	FLR	CX

By taking extra Class 90s, Freightliner has been able to eliminate its elderly Class 86s. On 8 September 2021, 90044 and 90014 *Over the Rainbow* pass Wandel between Carstairs and Lockerbie with the 1633 Coatbridge-Crewe. *Robin Ralston*

GB Railfreight

Contact details

Website: www.gbrailfreight.com
Twitter: @GBRailfreight

Key personnel

Managing Director: John Smith
Engineering Director: Bob Tiller

Overview

GBRf has been operating for over 20 years now and has grown from a fleet of just seven new Class 66s to a fleet of over 180 locos of various types. Staff levels are now over 1,000.

The company operates freight trains in many markets across the country, especially in the intermodal, infrastructure, aggregates and coal sectors. As well as hauling freight, the company also has contracts to operate the Caledonian Sleeper passenger operation, supplying its own locos (Class 66s, 67s, 73/9s and 92s) to haul CS's CAF Mk 5 coaches. It also undertakes stock moves for ROSCOs and TOCs, either using its own locos to haul the trains or to provide drivers to move the units under their own power. It also has a nationwide charter licence and operates the Royal Scotsman luxury train each summer.

GBRf relies mostly on Class 66s for the majority of its operations, with Class 73s – both original and rebuilt 73/9s – for infrastructure and test train work.

It has a small fleet of thee Class 47s for moving multiple units and coaching stock for ROSCOs and TOCs, and a fleet of ten Class 60s for some of its heavier trains. A single 59/0 was also acquired for heavy trainload work.

The most exciting development within the GBRf fleet is the Class 69 project, where ex-BR Class 56s are being rebuilt with the same GM engines as used in Class 66s. Sixteen locos are authorised for conversion and for the project, GBRf acquired 18 redundant Class 56s from UK Rail Leasing and DC Rail.

The firm also continues to acquire Class 66s from other operators, with 16 so far sourced from mainland Europe (66734/747-751/790-799) and ten from DB Cargo (66780-789). Another 11 Class 66s could be added to the fleet this year and will be numbered 66350-361. A fleet of 30 new tri-model Co-Co Class 99s are also expected to be ordered in 2022.

Class 08/09

GBRf owns four ex-BR standard shunters. It supplements these locos with Class 08s hired from HNRC or RSS. They are in the GBWM pool.

08925	GWS	GBR	WK (S)
08934	GWS	GBR	WH
09002	GWS	GBR	BH (S)
09009	GWS	GBR	TP

Class 18

Clayton Equipment Bo-Bo heavy duty shunting locomotive being tested by GBRf. This is the first of an order for 15 locos by Beacon Rail. The locos are on TOPS but not passed to run on the main line.

18001	BEA	BEA	WH

Class 47

Three Class 47s were bought from Colas Rail in 2017 and are used to move rolling stock, for which they have Dellner couplers fitted. They are also occasionally used for charters. They are in the GBDF pool.

47727	47047, 47569	CAL	GBR	LR	*Edinburgh Castle/Caisteal Dhùn Èideann*
47739	47035, 47594	GBB	GBR	LR	
47749	47076, 47625	BRB	GBR	LR	*City Of Truro*

GBRf is testing Clayton Equipment's brand new Class 18 heavy duty shunter at its Whitemoor yard. The loco is not passed to run on the national network, but is on TOPS as 18001. *GBRf/Clayton.*

There are three ex-Colas Class 47/s in the GB Railfreight fleet. 47739 passes Edale Signalbox on 2 October 2019, travelling via the Hope Valley Line with 56081 in tow. *Rail Photoprints/Keith Langston*

Class 56

GBRf has acquired 18 Class 56s, of which 16 are being converted to Class 69s (see page 288). Most were acquired in an unserviceable condition, but 56081/098/104/312 have seen occasional main-line use.

56009		BLE	GBR	ZW (U)
56032	GBGS	FER	GBR	ZW (U)
56077	UKRS	LHO	GBR	LR (U)
56081	GBGD	UKR	GBR	LR

56098		GBGD	RFO	GBR	LR
56104		UKRL	UKR	GBR	LR (U)
56106		UKRS	UKR	GBR	LR (U)
56312	56003	GBGD	DCR	GBR	LR (S)

GBRf bought several Class 56s to rebuild them as Class 69s, but a handful of the better locos were returned to traffic to support the Class 47s on stock moves. On 13 May 2021, 56098 shunts a bogie at Leicester shed. *Jack Boskett*

Class 59

One of the original five Class 59/0s acquired by Foster Yeoman from 1986, 59003 was redeployed in Germany in 1997. GBRf bought the loco in 2014 and returned it to UK operation. It is in the GBYH pool.

59003	GBF	GBR	RR	*Yeoman Highlander*

Class 60

Ten Class 60s bought by Colas Rail Freight from DB Cargo in 2013 and then overhauled. In 2018 they were sold to GBRf, who then sold them to Beacon Rail and leased them back. Three locos have also been acquired directly from DBC for spares and will not return to traffic.

The ten Class 60s that Colas bought from DB Cargo were then sold to GBRf for heavy trainload work. On 14 April 2021, 60002 *Graham Farish 50th Anniversary* sets off with the 1115 Liverpool Docks-Drax. *Paul Shannon*

60002	GBTG	GBF	BEA	RR	*Graham Farish 50th Anniversary 1970-2020*
60004	WQDA	EWS	GBR	TY (U)	
60014	WQDA	TEW	GBR	TY (U)	
60018	WQDA	EWS	GBR	TY (U)	
60021	GBTG	GBF	BEA	RR	*PENYGHENT*
60026	GBTG	BEA	BEA	RR	*HELVELLYN*
60047	GBTG	COL	BEA	RR	
60056	GBTG	COL	BEA	RR	
60076	GBTG	COL	BEA	RR	*Dunbar*
60085	GBTG	COL	BEA	RR	
60087	GBTG	GBF	BEA	RR	
60095	GBTG	GBF	BEA	RR	
60096	GBTG	COL	BEA	RR	

Class 66

The mainstay of the GBRf traction fleet, the first seven locos were acquired in 2001 and since then the fleet has grown, initially to 32 then a further 14 acquired from various ROSCOs and used by DRS, Advenza Freight, Freightliner and Colas (66733-746).

A further 28 were ordered as new-build machines from Progress Rail in Muncie, Indiana (66752-779), and ten locos were acquired from DB Cargo (66780-789). By late 2021, 16 locos had been sourced from mainland Europe (66734/747-751/790-799), with more expected. 66734 is the second loco to carry this number after the original 66734 (formerly 66402) was written off in 2012 after a derailment.

66751 has a Scharfenberg coupler fitted. The GBFM locos and 66738 have RETB fitted.

GBRf is looking to acquire another 11 Class 66s from mainland Europe and these should be numbered from 66350 upwards. Contracts for the acquisitions are expected to be signed very soon.

66350	66351	66352	66353	66354	66355	66356	66357	66358	66359	66360	66361

Class 66/7

These locos are all in traffic now and used nationwide on all GBRf traffic flows. 66793-795 have different gearing and are essentially the same as 66/6s and restricted to 65mph.

66701	GBBT	GBO	EVS	RR	
66702	GBBT	GBR	EVS	RR	*Blue Lightning*
66703	GBBT	GBR	EVS	RR	*Doncaster PSB 1981-2002*
66704	GBBT	GBR	EVS	RR	*Colchester Power Signalbox*
66705	GBBT	GBR	EVS	RR	*Golden Jubilee*
66706	GBBT	GBR	EVS	RR	*Nene Valley*

66707		GBBT	GBR	EVS	RR	*Sir Sam Fay*
66708		GBBT	GBR	EVS	RR	*Jayne*
66709		GBBT	MSC	EVS	RR	*Sorrento*
66710		GBBT	GBR	EVS	RR	*Phil Packer BRIT*
66711		GBBT	AGI	EVS	RR	*Sence*
66712		GBBT	GBR	EVS	RR	*Peterborough Power Signalbox*
66713		GBBT	GBR	EVS	RR	*Forest City*
66714		GBBT	GBR	EVS	RR	*Cromer Lifeboats*

Like so many FOCs, Class 66s are the core loco in the GBRf fleet, and it is still acquiring any available locos from Europe. 66714 *Cromer Lifeboats* works the 0718 Liverpool Biomass-Drax at Plumley on 31 March 2021. *Tom McAtee*

66715		GBBT	GBR	EVS	RR	*Valour*
66716		GBBT	GBR	EVS	RR	*Locomotive & Carriage Institution Centenary 1911-2011*
66717		GBBT	GBR	EVS	RR	*Good Old Boy*
66718		GBLT	LUB	EVS	RR	*Sir Peter Hendy CBE*
66719		GBLT	GBR	EVS	RR	*Metro-Land*
66720		GBLT	EMY	EVS	RR	
66721		GBLT	LUW	EVS	RR	*Harry Beck*
66722		GBLT	GBR	EVS	RR	*Sir Edward Watkin*
66723		GBLT	GBZ	EVS	RR	*Chinook*
66724		GBLT	GBF	EVS	RR	*Drax Power Station*
66725		GBLT	GBZ	EVS	RR	*Sunderland*
66726		GBLT	GBF	EVS	RR	*Sheffield Wednesday*
66727		GBLT	MRT	EVS	RR	*Maritime One*
66728		GBLT	GBF	EVS	RR	*Institution of Railway Operators*
66729		GBLT	GBF	EVS	RR	*Derby County*
66730		GBLT	GBF	EVS	RR	*Whitemoor*
66731		GBLT	GBF	EVS	RR	*Capt. Tom Moore A True British Inspiration*
66732		GBLT	GBF	EVS	RR	*GBRf The First Decade 1999-2009 John Smith MD*
66733	66401	GBFM	GBR	POR	RR	*Cambridge PSB*
66734	PB04	GBEB	UND	BEA	ZW (Z)	
66735	66403	GBBT	GBR	POR	RR	*Peterborough United*
66736	66404	GBFM	GBR	POR	RR	*Wolverhampton Wanderers*
66737	66405	GBFM	GBR	POR	RR	*Lesia*
66738	66578	GBBT	GBR	BEA	RR	*Huddersfield Town*
66739	66579	GBFM	GBR	BEA	RR	*Bluebell Railway*
66740	66580	GBFM	GBR	BEA	RR	*Sarah*
66741	66581	GBBT	GBR	BEA	RR	*Swanage Railway*
66742	66406, 66841	GBBT	GBR	BEA	RR	*ABP Port of Immingham Centenary 1912-2012*
66743	66407, 66842	GBFM	ROY	BEA	RR	
66744	66408, 66843	GBBT	GBR	BEA	RR	*Crossrail*

On 27 November 2020, GBRf's 66738 *Huddersfield Town* approaches Spalding with a southbound intermodal train, lightly loaded due to the Covid pandemic. *Pip Dunn*

66745	66409, 66844	GBBT	GBR	BEA	RR	*Modern Railways The First 50 Years*
66746	66410, 66845	GBFM	ROY	BEA	RR	
66747		GBEB	GBF	GBR	RR	*Made In Sheffield*
66748		GBEB	GBF	GBR	RR	*West Burton 50*
66749		GBEB	GBF	GBR	RR	*Christopher Hopcroft MBE 60 Years Railway Service*
66750		GBEB	GBF	BEA	RR	*Bristol Panel Signal Box*
66751		GBEB	GBF	BEA	RR	*Inspirational Delivered Hitachi Rail Europe*
66752		GBEL	GBF	GBR	RR	*The Hoosier State*
66753		GBEL	GBF	GBR	RR	*EMD Roberts Road*
66754		GBEL	GBF	GBR	RR (U)	*Northampton Saints*
66755		GBEL	GBF	GBR	RR	*Tony Berkeley OBE RFG Chairman 1997-2018*
66756		GBEL	GBF	GBR	RR	*Royal Corps Of Signals*
66757		GBEL	GBF	GBR	RR	*West Somerset Railway*
66758		GBEL	GBF	GBR	RR	*The Pavior*
66759		GBEL	GBF	GBR	RR	*Chippy*
66760		GBEL	GBF	GBR	RR	*David Gordon Harris*
66761		GBEL	GBF	GBR	RR	*Wensleydale Railway Association 25 Years 1990-2015*
66762		GBEL	GBF	GBR	RR	
66763		GBEL	GBF	GBR	RR	*Severn Valley Railway*
66764		GBEL	GBF	GBR	RR	*Major John Poyntz Engineer & Railwayman*
66765		GBEL	GBF	GBR	RR	
66766		GBEL	GBF	GBR	RR	
66767		GBEL	GBF	GBR	RR	*King's Cross PSB 1971-2021*
66768		GBEL	GBF	GBR	RR	
66769		GBEL	PCC	GBR	RR	*League Managers Association*
66770		GBEL	GBF	GBR	RR	
66771		GBEL	GBF	GBR	RR	*Amanda*
66772		GBEL	GBF	GBR	RR	*Marie*
66773		GBNB	GBP	GBR	RR	*Pride of GB Railfreight*
66774		GBNB	GBF	GBR	RR	
66775		GBNB	GBZ	GBR	RR	*HMS Argyll*
66776		GBNB	GBF	GBR	RR	*Joanne*
66777		GBNB	GBF	GBR	RR	*Annette*
66778		GBNB	GBF	GBR	RR	*Cambois Depot 25 Years*
66779		GBEL	GYP	GBR	RR	*Evening Star*
66780	66008	GBOB	CMX	GBR	RR	*The Cemex Express*
66781	66016	GBOB	GBF	GBR	RR	
66782	66046	GBOB	GBZ	GBR	RR	
66783	66058	GBOB	BIF	GBR	RR	*The Flying Dustman*
66784	66081	GBOB	GBF	GBR	RR	*Keighley & Worth Valley Railway 50th Anniversary 1968-2018*
66785	66132	GBOB	GBF	GBR	RR	
66786	66141	GBOB	GBF	GBR	RR	
66787	66184	GBOB	GBF	GBR	RR	
66788	66238	GBOB	GBF	GBR	RR	*Locomotion 15*

66789	66250	GBOB	BLL	GBR	RR	*British Rail 1948-1997*
66790	CD66403	GBBT	GBF	BEA	RR	
66791	CD66404	GBBT	BEA	BEA	RR	
66792	CD66405	GBBT	GBF	BEA	RR	
66793	29004	GBHH	TLA	BEA	RR	
66794	29005	GBHH	TLP	BEA	RR	*Steve Hannam*
66795	29006	GBHH	GBF	BEA	RR	*Bescot LDC*
66796	29007	GBHH	GBZ	BEA	RR	*The Green Progressor*
66797	513-09	GBEB	BEA	BEA	RR	
66798	561-03	GBEB	GBF	BEA	RR	
66799	6602	GBEB	GBF	BEA	RR	

66769 carries *Paul Taylor Our Inspiration* cabside stick-on nameplates.

66773 has its GBRf logos in pride rainbow colours.

66782 has Charity Railtours branding.

Class 67

These ex-DB Cargo locos were acquired by Colas, and duly sold to Beacon, for use on IM trains that are required to operate at 100mph or faster. However, Colas has since relinquished the lease on them and they have been taken on by GB Railfreight initially for use on Caledonian Sleeper duties. Due to the lack of Dellner couplers, they will work in multiple with 73/9s on the Fort William and Inverness sleeper portions.

Long term they will be added to the GBRf fleet for general use and were expected to be repainted into GBRf livery in spring 2022.

Both are in the GBKP pool.

67023	COU	BEA	RR	*Stella*
67027	COU	BEA	RR	*Charlotte*

Class 69

This new class of loco uses Class 56 donor bodies and bogies, but the same GM engines as used in the Class 66s. The locos are undergoing construction at Longport and the first locos entered traffic in the summer of 2021. By March 2022 five locos had been built and tested. The donor locos' 'running order' is subject to change. There was an initial order for ten with an option for six more locos and the latter is now being taken up.

All will be in the GBRG pool, with those undergoing tests initially in the GBGS pool.

69001	56031	GBZ	BEA	TN	*Mayflower*
69002	56057, 56311	BLL	BEA	TN	*Bob Tiller CM&EE*
69003	56018	GBR	BEA	TN	
69004	56069	RTC	BEA	TN	
69005	56007		BEA	ZW (Z)	
69006	56128		BEA	ZW (Z)	
69007	56037		BEA	ZW (Z)	

The fourth of GBRf's Class 69s, 69004, was outshopped in a version of the old BR Railway Technical Centre livery. It sparkles in the sunshine at Eastleigh Works after being unveiled on 17 February 2022. *Bob Tiller*

69008	56038	BEA	ZW (Z)	
69009	56060	BEA	ZW (Z)	
69010	56065	BEA	ZW (Z)	
69011		BEA	ZW (Z)	
69012		BEA	ZW (Z)	
69013		BEA	ZW (Z)	
69014		BEA	ZW (Z)	
69015		BEA	ZW (Z)	
69016		BEA	ZW (Z)	

Class 73

Class 73s have been part of the GBRf fleet since 2004 and the number of locos has increased steadily. Most were put back into traffic in their original condition, but in 2013 a contract to have five rebuilt with MTU engines by Brush was agreed. This was then increased by six more locos when GBRf won the contract to haul the Caledonian Sleeper trains.

GBRf retain several 'original' 73s and has four donor locos that are held in reserve should more 73/9s overhauls be sanctioned.

Class 73/1

These are original locos with EE diesel engines. 73201/212/213 are ex-Gatwick Express locos. 73107/119/136 remain dual braked. GBRf retains 73101/110/134/139 in strategic reserve as donor bodies but no more conversions are expected.

73101	73801	GBZZ	PUL	GBR	ZG (U)	
73107		GBED	GBF	GBR	SE	*Tracy*
73109		GBED	GBZ	GBR	SE	*Battle of Britain 80th Anniversary*
73110			EBY	GBR	ZG (U)	
73119		GBED	GBR	GBR	SE	*Borough of Eastleigh*
73128		GBED	GBR	GBR	SE	*OVS Bulleid CBE*
73134		GBZZ	ICO	GBR	WK (U)	
73136		GBED	GBF	GBR	SE	*Mhairi*
73139		GBZZ	UND	GBR	ZG (U)	
73141		GBED	GBF	GBR	SE	*Charlotte*
73201	73142	GBED	BRB	GBR	SE	*Broadlands*
73212	73102	GBED	GBF	GBR	SE	*Fiona*
73213	73112	GBED	GBF	GBR	SE	*Rhodalyn*

There are nine original Class 73 in the GBRf fleet, used mostly on the southern region for infrastructure support work. Nos 73128/213 stand at the unlikely surroundings of King's Cross, having arrived worked the last few miles from Hornsey of a charter train from Swanage. *Pip Dunn*

Class 73/9

The 73/9s are new locomotives fitted with MTU engines with just the bodies and the bogies retained from the donor locos they were rebuilt from.

The first five locos are generally used for Network Rail work, either infrastructure monitoring or infrastructure support duties, and retain their third rail capabilities.

The six Caledonian Sleeper locos are used solely on sleeper trains north of Edinburgh to Inverness, Aberdeen and Fort William. For this they have Dellner couplers and snowploughs and their third rail capability is isolated. They work in pairs or in multiple with Class 66s and Class 67s on the Aberdeen, Fort William and Inverness portions with Mk 5 coaches.

	Donor loco					
73961	73120, 73209	GBNR	GBF	GBR	SE	*Alison*
73962	73125, 73204	GBNR	GBF	GBR	SE	*Dick Mabbutt*
73963	73123, 73206	GBNR	GBF	GBR	SE	*Janice*
73964	73124, 73205	GBNR	GBF	GBR	SE	*Jeanette*
73965	73121, 73208	GBNR	GBF	GBR	SE	*Des O'Brien*
73966	73005	GBCS	CAL	GBR	EC	
73967	73006, 73906	GBCS	CAL	GBR	EC	
73968	73117	GBCS	CAL	GBR	EC	
73969	73105	GBCS	CAL	GBR	EC	
73970	73103	GBCS	CAL	GBR	EC	
73971	73122, 73207	GBCS	CAL	GBR	EC	

On 17 September 2021, 73961 *Alison* and 73964 *Jeanette* work the 0830 York-Tonbridge past Spalding. *Pip Dunn*

Class 92

GBRf has 16 Class 92s, those locos originally owned by Eurostar and SNCF. Of these, 12 have been refurbished and returned to use and 92021/040/045/046 remain as spare/donor locos and are unlikely to be returned to traffic. They have recently been moved to Worksop for further storage. Of these, 92046 is expected to move to Margate for display.

The locos are used on Caledonian Sleeper work between Euston-Glasgow Central/Edinburgh, which calls on at least five locos a day. Others are used on freight work both in the UK and on Channel Tunnel duties. Several are painted in Caledonian Sleeper or GBRf livery, with just 92044 still operational in its original Eurostar livery.

92006	GBSL	CAL	GBR	WB	
92010	GBST	CAL	GBR	WB	
92014	GBSL	CAL	GBR	WB	
92018	GBST	CAL	GBR	WB	
92020	GBET	GBF	GBR	WB	*Billy Stirling*
92021	GBSD	EUK	GBR	WK (U)	*Purcell*

GBRf's Class 92s are used for freight and 'sleeper' train work; 92038 passes Peterborough on a driver training run on 13 February 2021. *Stuart West*

92023	GBSL	CAL	GBR	WB	
92028	GBST	GBF	GBR	WB	
92032	GBST	GBF	GBR	WB	*IMechE Railway Division*
92033	GBSL	CAL	GBR	WB	
92038	GBST	CAL	GBR	WB	
92040	GBSD	EUK	GBR	WK (U)	*Goethe*
92043	GBST	GBF	GBR	WB	
92044	GBST	EUK	GBR	WB	*Couperin*
92045	GBSD	EUK	GBR	WK (U)	*Chaucer*
92046	GBSD	EUK	GBR	WK (U)	*Sweelinck*

Class 99

New Stadler Tri-mode EuroDual Co-Co locos based on the Class 88/93 platform with 25kV AC overhead, diesel and battery capability. An initial order for 20, with an option for 30 more locos, was placed in March 2022.

99001	GBR
99002	GBR
99003	GBR
99004	GBR
99005	GBR
99006	GBR
99007	GBR
99008	GBR
99009	GBR
99010	GBR
99011	GBR
99012	GBR
99013	GBR
99014	GBR
99015	GBR
99016	GBR
99017	GBR
99018	GBR
99019	GBR
99020	GBR

Class 508 translator vehicles

64664	AFS	ANG	ZG	*Liwet*
64707	AFS	ANG	ZG	*Labezerin*

Rail Operations Group

Contact details
Website: www.railopsgroup.co.uk
Twitter: @railoperationsgroup

Key personnel

Chief Executive Officer: Karl Watts

Overview

The growing specialist TOC is investing in a fleet of tri-mode Class 93s – Stadler Class 88s with batteries, diesel engines and 25kV AC capability. A fleet of 30 locos is planned.

The investment will allow ROG to expand its Orion intermodal service, which started using modified Class 319 EMUs and Class 769 bi-modes.

ROG has growing expertise in moving passenger rolling stock. It regularly moves new stock after construction or delivery to the UK, or between storage sites. It also moved multiple units transferring to or from maintenance facilities before or after overhaul.

Some moves are done by using ROG staff to drive the units but most are carried out using locomotives, on long-term lease from Europhoenix or DRS.

ROG has fitted Dellner couples to some of its Class 37s to allow it to haul units without the need for barrier wagons. In 2021 it sold its six Class 47s, four to West Coast Railways and two, unserviceable, locos to HNRC.

It also offers initial shakedown testing for new trains, such as the Class 397 CAF units built for Transpennine Express.

A new venture by the company has been its Orion service, which uses ex-Thameslink Class 319 dual-voltage EMUs converted to freight use. These are being prepared as Class 326s as dual-voltage electric units or tri-mode Class 768s with diesel power as well.

Class 37

These locos are owned by Europhoenix but are currently exclusively used by ROG. They have Dellner couplers fitted and are used mostly for EMU moves. All are in the GROG pool.

37510	37112	EPX	EPX	LR	*Orion*
37601	37005, 37501	EPX	EPX	LR	*Perseus*
37608	37022, 37512	EPX	EPX	LR	*Andromeda*
37611	37171, 37690	EPX	EPX	LR	*Pegasus*
37800	37143	EPX	EPX	LR	*Cassiopeia*
37884	37183	EPX	EPX	LR	*Cepheus*

Rail Operations Group does not own any Class 37s, but has a hire agreement with Europhoenix that provides its 37/5s and 37/7s for stock moves and other work. 37800 *Cassiopeia* was at Norwich on 18 May 2019. *Pip Dunn*

Class 43

Four Class 43s have been bought by DATS but are being operated by ROG. They remain in the MBDL pool.

43052	EMB	DATS	LR (S)
43054	EMB	DATS	HQ
43066	EMB	DATS	HQ
43076	EMB	DATS	LB (S)

Rail Operations Group uses five Class 57/3s, which it has hired from DRS. On 5 October 2021, 57312 hauls 365505/537 through Northway, near Gloucester, on their way to Newport for scrapping. *Jack Boskett*

Class 57/3

These locos are sub-leased from DRS and are used mostly for stock moves. They are expected to be returned to DRS very soon. All are in the GROG pool.

57301	47069, 47638, 47845	DRC	POR	LR	*Goliath*
57303	47554, 47705	DRN	POR	LR	*Pride of Carlisle*
57305	47164, 47571, 47822	ROG	POR	LR	
57310	47037, 47563, 47831	DRU	POR	LR	*Pride of Cumbria*
57312	47330	ROG	POR	LR	

Class 91

Two Class 91s were leased from Eversholt for use on the Data Acquisition and Testing Services (DATS) train but are now stored at Crewe pending developments.

91122	91022	EROG	VEC	EVS	CX (S)
91128	91028	EROG	VEC	EVS	CX (S)

Class 93

New-build bi-mode locos based on the Class 88 platform, being built by Stadler. The first locos are now under construction.

93001	EROG	93006	EROG	93011	EROG	93016	EROG	93021	EROG	93026	EROG
93002	EROG	93007	EROG	93012	EROG	93017	EROG	93022	EROG	93027	EROG
93003	EROG	93008	EROG	93013	EROG	93018	EROG	93023	EROG	93028	EROG
93004	EROG	93009	EROG	93014	EROG	93019	EROG	93024	EROG	93029	EROG
93005	EROG	93010	EROG	93015	EROG	93020	EROG	93025	EROG	93030	EROG

Class 326

Ex-Thameslink Class 319s are being modified to haul freight in roll-cages and on pallets, and are being renumbered as Class 326 units. The first unit to be prepared is 319373, which will remain a standard Class 319 specification (apart from its internal arrangement). It has been in traffic for ROG still as a 319.

	Old No.				DMC	MS	TS	DMS
326001	319373	ORO	POR	ZG	77483	63055	71941	77482
326002			POR					
326003			POR					
326004			POR					
326005			POR					
326006			POR					
326007			POR					
326008			POR					
326009			POR					

Class 360/2

The first four of these Siemens Desiro units were delivered as four-cars for use on the Heathrow Connect service. Additional fifth cars, plus a new five-car set, were delivered in 2007 and 2005 respectively.

When they were replaced by Class 387/1s, they were essentially off lease, but have been acquired by ROG. They are currently in store pending developments. They could be converted to freight EMUs for its Orion service.

All are in the EXHW pool.

				DMS	PTS	TS	TS	DMS
360201	HEC	ROG	BC (S)	78431	63421	72431	72421	78441
360202	HEC	ROG	BC (S)	78432	63422	72432	72422	78442
360203	HEC	ROG	BC (S)	78433	63423	72433	72423	78443
360204	HEC	ROG	BC (S)	78434	63424	72434	72424	78444
360205	HEC	ROG	BC (S)	78435	63425	72435	72425	78445

Class 768

More ex-Thameslink Class 319s, which are being modified to haul freight in roll-cages and on pallets and are being renumbered as Class 768 bi-mode units. 319009/010 are the first two to be converted.

				DMC	MS	TS	DMS	
768001	319009	ORO	POR	WN	77307	62899	71780	77306
768002	319010	ORO	POR	ZG	77309	62900	71781	77308
768003			POR					
768004			POR					
768005			POR					
768006			POR					
768007			POR					
768008			POR					
768009			POR					
768010			POR					

Mk 3 Coaches

ROG has secured many of the former Arriva Trains Wales Mk 3 coaches, previously used on the Holyhead-Cardiff route, but replaced by Mk 4s.

They were bought speculatively for business growth in either the passenger or non-passenger sectors but most have since been scrapped.

10249	ARV	Mk 3 RBF	LR (S)
12176	ARV	SO	LR (S)
12180	ARV	SO	LR (S)
12182	ARV	SO	ZA (S)

Maintenance and Spot Hire Providers

The companies that supply maintenance and spot hire to the railways overlap, and hence are included together in this one section. Some of these firms provide just one aspect, be it spot hire or maintenance, while others provide both. Some have fleets of locos while others do not.

Small companies that support these two markets are not included.

Alstom West Coast Traincare

Contact details

Website: www.alstom.com/alstom-uk-and-ireland

Key personnel

Managing Director: Peter Broadley

Overview

Alstom offered maintenance as part of the package for many new train orders and duly created its West Coast Traincare operation, which took over the former InterCity depots at Polmadie (Glasgow), Longsight (Manchester), Edge Hill (Liverpool), Oxley (Wolverhampton) and Wembley (London). It also has a site at Widnes Technology Centre.

It uses these sites to maintain the Pendolino fleet, initially with Virgin and now with Avanti. It also undertakes running repairs and minor exams on other AWC stock plus is available for third-party contract work; for example, the Caledonian Sleeper coaching stock is maintained at Polmadie. AWC has ten Class 08s based at its various depots.

Alstom also runs the depot at Chester for Transport for Wales Class 175 units and also has a site at Crofton in West Yorkshire.

Class 08

Depot shunters at AWCT sites, but they are frequently moved between sites and for maintenance.

08451	ATZZ	BRW	ALS	ZG	*MA SMITH*
08454	ATLO	BRW	ALS	WX	
08611	ATLO	BRW	ALS	WB	
08617	ATLO	BRW	ALS	OX	*Steve Purser*
08696	ATLO	BRW	ALS	WB	
08721	ATLO	BRW	ALS	WD	
08764	ATLO	BRW	ALS	PO	
08790	ATLO	BRW	ALS	MA	*Longsight TMD*
08887	ATZZ	BRW	ALS	PO	
08954	ATZZ	BRW	ALS	PO	

Arlington Fleet Services

Contact details

Website: www.arlington-fleet.com
Twitter: @ArlingtonFleet

Key personnel

Managing Director: Barry Stephens
Systems Director: John Campbell

Overview

AFS holds a long-term lease on the former BREL/Alstom Eastleigh Works. It can work on most locomotive types, with this ranging from 'G-exams' to simple servicing. The bulk of the heavy work has been for GBRf, DRS and Rail Operations Group with Classes 08, 37, 43, 47, 57, 59, 66 and 73 receiving heavy exams, bogie overhauls, collision repairs and exams and modifications.

Other clients for repairs are Touax, Freightliner, Network Rail, STVA, Transport for London, LSL and VTG.

The company also owns two shunters, both based at Eastleigh; one Class 07 and one Class 08 as well as an ex-DRS Class 47 used for testing train supply.

Class 07

This shunter was inherited as part of the takeover of Eastleigh Works.

07007	MBDL	BRW	AFS	ZG

Class 08

Shunter bought from EWS.

08567	MBDL	EWS	AFS	ZG

Class 47

Bought from DRS.

47818 47240, 47663 MBDL DRU ZG (U)

Class 142

Acquired from Angel Trains, these vehicles are due to be scrapped.

				DMS	DMSL
142056	NBL	AFS	ZG	55706	55752
142089	NBL	AFS	ZG	55739	55785

Class 442

These units date from 1987-88 and were built for the Waterloo-Weymouth line after the final stretch from Bournemouth was energised in 1988. They used traction equipment recovered from Class 432 4-REP units.

They were redeployed on Gatwick Express duties and were due to be withdrawn in 2017, but as part of its franchise commitment, South Western Railway took on 18 sets. These were refurbished but never returned to traffic as in April 2021, SWR announced it would no longer be pursuing its plan to reintroduce the trains to operation, so they were sent for scrap.

The following vehicles have been acquired for reuse by Arlington with a view to forming a loco-hauled charter rake.

	Livery	owner	depot	DTS	TS	TS	DTS
442412	SWT	AFS	ZN (U)	77393	71829	71853	77417
Spare					71846	71841	77382
Spare					71822		

Class 489

EMU translator vans.

68501
68504

Boden Rail Engineering

Contact details

e-mail: Neil.boden@bodenrailengineering.co.uk

Key personnel

Managing Director: Neil Boden

Overview

BRE is based alongside the East Midlands Trains depot at Nottingham Eastcroft and undertakes most of the day-to-day maintenance and running of the Colas Rail Class 37 and 56 fleets.

BRE also owns 37240 and 50050 *Fearless*, which are both passed for main-line running and are occasionally used for spot hire work, mostly collecting Colas locomotives.

Both locos are in the COFS pool.

Class 37

37240	TRN	BOD	NM	

Class 50

50050	BRB	BOD	NM	*Fearless*

Brodie Engineering

Contact details

Website: www.brodie-engineering.co.uk

Key personnel

Managing Director: Gerry Hilferty

Overview

Founded in 1996, Brodie Engineering offers refurbishment, maintenance, overhaul and repair services to TOCs and FOCs.

It operates from a depot in Kilmarnock that has indoor capacity for up to 14 vehicles, stabling capacity for a further 11 vehicles, pits for underframe maintenance including an underframe wash area, a 28-metre paint booth, jacks for vehicle lifts, fabrication facilities and a water test facility.

The company also owns 156478. This two-car unit was written off by Angel Trains after suffering severe flood damage and sold to Brodie Engineering. It subsequently rebuilt the unit and it has returned to ScotRail, leased directly.

Carnforth Railway Restoration and Engineering Services

Contact details

Website: https://westcoastrailways.co.uk/crres

Overview

In 1990, businessman David Smith bought enough shares to take control of the operation at Carnforth from Bill McAlpine. This deal included the Carnforth Railway Restoration and Engineering Services division, part of Steamtown since 1984.

CRESS is the specialist locomotive and coach restoration arm that operates alongside West Coast Railways (also owned by Smith), and offers full or partial overhaul of steam and diesel locomotives and rolling stock as well as vehicle restoration, maintenance and repainting services.

Chrysalis Rail

Contact details

Website: chrysalisrail.com

Key personnel

Managing Director: Chris Steele

Overview

Chrysalis Rail is a UK-based rolling stock service supplier specialising in refurbishment, vehicle enhancement, modification and heavy maintenance projects designed to enhance passenger rolling stock.

Chrysalis has a site at the former GWR depot at Landore in Swansea, South Wales. It closed in 2018 but reopened in August 2019 for its new user.

Chrysalis Rail has undertaken a full interior and exterior refurbishment of the Northern Class 333s at Holbeck and has also undertaken work for Transport for Wales on its Class 153s and for ScotRail on its Class 156s.

Chrysalis undertook an interior refresh and exterior surface corrosion repair project for Siemens on 127 Class 450s and 45 Class 444s, a total of 733 vehicles, for South Western Railway.

Class 20189 Ltd

Contact details

e-mail: Owen-michael@btinternet.com

Key personnel

Managing Director: Michael Owen

Overview

A small spot hire business comprising the four Class 20s owned by Michael Owen and one loco owned by the Class 20 Locomotive Society preservation group.

Four of the locos are main-line registered and have been hired in the past to GBRf, DB Schenker and WCR. They have also been used by the London Underground, and two locos are in Metropolitan Railway maroon livery.

More recently 20227 has been on hire to the North Norfolk Railway – which operates to Cromer on the national network – while the other locos have been hired by the likes of Vintage Trains, DC Rail, SLC Operations, ROG and Loram.

It recently bought 20048 from a preservation group but presently there are no plans to add it to the main-line fleet.

Class 20

These locos are used for spot hire, apart from 20048, which is not main-line registered.

20007	MOLO	GYP	MOW	SK	
20048	MOLO	BRB	MOW	SK (S)	
20142	MOLO	MRM	MOW	SK	*Sir John Betjeman*
20189	MOLO	BRB	MOW	SK	
20205	MOLO	BRB	CTL	SK	

Ed Murray & Sons

Contact details

Website: www.edmurrays.com

Overview

Ed Murray founded the Hartlepool-based road haulage company Ed Murray and Sons in 1971. The company has been operating since then offering a range of haulage and transport services to various industries including steel and rail.

It now offers shunting locos for hire having acquired the former Wabtec and Hunslet Engine Co. fleets, giving it ten Class 08s. It also has some industrial locos in its hire fleet.

Class 08

Spot hire fleet.

08401		GRE	EMS	HHDP	
08445		MAL	EMS	DIRFT	
08472	RFSH	WAB	EMS	EC	
08571	HBSH	WAB	EMS	DIRFT	
08596	HBSH	WAB	EMS	EC	
08615	RFSH	HUN	EMS	TSS	*Uncle Dai*
08669	RFSH	WAB	EMS	ZF	*Bob Machin*
08724	HBSH	WAB	EMS	ZF	
08823	KDSD	HUN	EMS	TSS	*Kelva*
08853	RFSH	WAB	EMS	ZF	

DIRFT: Daventry International Railfreight Terminal
HHDP: Hams Hall Distribution Park
TSS: Tata Steel Shotton

Europhoenix

Contact details

Website: https://europhoenix.co.uk
Managing Director: Glenn Edwards

Overview

Europhoenix has been successful in selling and redeploying locos abroad, mostly Class 56s, 87s and 87s in Hungary and 87s in Bulgaria. It currently has two Class 91s – 91117/120 – ready to export to Hungary as trial locos for freight but their move has been held up by the Covid-19 crisis.

At home it has a fleet of nine Class 37s, six of which are in traffic and midway through a five-year hire deal to Rail Operations Group. It is looking to acquire further Class 37s if they become available on the market

Between 2008 and 2012, Europhoenix exported Class 87s to Bulmarket in Bulgaria, followed by three Class 56s and eight Class 86s to Floyd in Hungary and a further 86 to Bulgaria.

Class 37

Most Europhoenix Class 37s are on long-term hire to Rail Operations Group (see page 292). 37901 remains stored but is being worked on and is close to returning to traffic.

37901	37150	EPUK	EPX	EPX	LR (U)	*Mirrlees Pioneer*

Europhoenix has done good business exporting locomotives to Eastern Europe as well as building up a hire fleet of Class 37s for use in the UK. It took two Class 91s for evaluation for possible export, although any deals have been hampered by the pandemic. 91117 was at Doncaster on 3 September 2020, but is now at Barrow Hill, along with 91120. *Stuart West*

Class 91

Following its success selling Class 86s and 87s into mainland Europe, Class 91s are the next locos Europhoenix is looking at markets to redeploy abroad. Two locos have been acquired as demonstrators, but the project has stalled due to the pandemic.

91117	91017	EPEX	EPX	EPX	BH (S)
91120	91020	EPEX	EPX	EPX	BH (S)

Gemini Rail Services

Contact details

Website: www.geminirailgroup.co.uk

Key personnel

Chief Executive Officer: Tim Jenkins

Overview

Gemini Rail offers a wide range of services from UK centres, including full overhauls for passenger rolling stock and has a dedicated wheelshop facility offering light, medium and heavy wheelset overhauls. Its main operations are undertaken at Wolverton Works.

Hanson & Hall Rail Services

Contact details

Website: www.hansonhallrail.co.uk
Twitter: @HallRail

Key personnel

Managing Director: Garcia Hanson

Overview

Hanson & Hall Rail Services Solutions was founded in 2018 by Garcia Hanson, who owns 50008 *Thunderer*, which is used for occasional stock moves and freight wagon transfers. It bought 31106 in late 2020 with a view to returning it to the main line. The company has preserved 08663 on hire via RSS.

The company also arranges the transfer of DMUs and EMUs. It worked with SLC Operations, DC Rail or DB Cargo to provide drivers.

Class 08

08663		BRB	AVR	LE

Class 31

A single Class 31 has been bought from Howard Johnston. The loco is to be overhauled and returned to main-line capability.

31106	RVLO	BRB	GAR	LR (U)

Class 43

The company has announced an agreement with Munich-based RailAdventure, which has seen the purchase of eight Class 43s, ex-LNER 43296/308 and ex-EMR 43423/465/467/480/484. Of these, 43296/308 are spares donors, while the others will be formed as back-to-back pairs to create three twin-engine traction units.

They will be converted with translator couplings to control the electro-pneumatic brakes while Habfis coupling adapter cars will be imported from the Continent. So far 43465/468/480/484 have been repainted in RailAdventure's green and grey livery. All are in the HHPC pool.

German company RailAdventure has formed an alliance with Hanson & Hall and taken eight Class 43 Power cars from Porterbrook, six of which are expected to return to traffic. Four have been painted in the company's grey livery and in September 2021, two locos, 43480/484, were taken to Germany for display but returned to the UK in November. *Jack Boskett*

43296	43096	VEM	ZG (U)
43308	43108	VEC	ZG (U)
43423	43123	EMU	ZG (U)
43465	43065	RAV	ZG (U)
43467	43067	EMU	ZG (U)
43468	43068	RAV	ZG (U)
43480	43080	RAV	ZG (U)
43484	43084	RAV	ZG (U)

Class 50

Owned by Garcia Hanson, the loco has recently been repainted into RailAdventure grey livery. It is currently available for sport hire work and is often used by SLC Operations (see page 191).

50008	HVAC	RAV	GAR	LR	*Thunderer*

Harry Needle Railroad Company

Contact details

Website: www.harry-needle.co.uk

Key personnel

Managing Director: Harry Needle

Overview

Harry Needle Railroad Company (HNRC) specialises in locomotive spot hire, maintenance/repair and stock storage. It recently took over the 15-acre former wagon repair depot at Worksop, which is now its main base for storing vehicles. It also has a maintenance facility at Barrow Hill.

The company owns several main-line registered Class 20s, some of which are available for spot hire. It also has a large fleet Class 08/09 and several industrial shunters both on contracts or available for hire, several Class 37s for main-line spot hire – often used by Colas – and six Class 47s, although these are not presently main-line registered. It also owns examples of Classes 20, 25 and 27 which are not currently available for hire.

Class 08

HNRC has a large fleet of Class 08/09 shunters, although many are held in reserve.

08389	HNRS	EWS	HNR	Cardiff Tidal	
08417		NRY	HNR	BH (U)	
08428	HNRL	EWS	HNR	BH	
08500	HNRL	EWS	HNR	WK (U)	
08502	HNRL	NBL	HNR	EKR	
08527	HNRL	TTG	HNR	ARR	
08578	HNRS	EWS	HNR	WK (U)	
08602		BLE	HNR	WK (U)	
08623		DBS	HNR	HO (U)	
08630	HNRL	CEL	HNR	CT	*Celsa Endeavour*
08653	HNRS	EWS	HNR	LM (U)	
08676	HNRL	EWS	HNR	EKR	
08682		IND	HNR	HO (U)	
08685	HNRS	EWS	HNR	EKR	
08700	HNRS	BRW	HNR	IL	
08701	HNRS	RES	HNR	BAT	
08706		EWS	HNR	BAT	
08711	HNRS	RES	HNR	BU (U)	
08714	MBDL	EWS	HNR	HO (U)	
08742	HNRL	RES	HNR	BH (U)	
08765	HNRS	HNO	HNR	BH (U)	
08782	HNRL	COR	HNR	BH (U)	
08786	HNRS	DEP	HNR	BH (U)	
08798		EWS	HNR	BH (U)	
08799		EWS	HNR	EKR	
08802	HNRS	EWS	HNR	WK	
08804	WQDA	EWS	HNR	EKR	
08818	HNRL	GBR	HNR	WK	*Molly*
08824	HNRL	WAB	HNR	BH (U)	
08834	HNRL	HNO	HNR	AN	
08865	HNRL	EWS	HNR	CZ	
08877	HNRS	DEP	HNR	CT	
08879	HNRL	EWS	HNR	HO	
08892	HNRL	DRS	HNR	WK	
08904	HNRL	EWS	HNR	WK	
08905	HNRS	EWS	HNR	HO (U)	
08918	HNRS	DEP	HNR	BU (U)	
08924	HNRS	GBR	HNR	BH (U)	*Celsa 2*
08943	HNRL	HNO	HNR	BH (U)	
08944		BLK	HNR	ELR (U)	
08956		SEC	HNR	BH	

ARR: Attero Recycling Rossington (Doncaster)

CT: Cardiff Tidal

Class 09

Like the Class 08s, many locos are held in reserve.

09006		HNRS	EWS	HNR	BU (U)
09014		HNRS	DEP	HNR	BU (U)
09106	08759	HNRL	HNO	HNR	CT
09201	08421	HNRL	DEP	HNR	HO (U)

Class 20

HNRC has 23 Class 20s, some main-line registered, others in industrial use, in store or on loan to heritage railways.

It recently acquired six Class 20/3s from DRS, some of which may be reactivated.

Harry Needle Railroad Company has a large fleet of Class 08s, 20s and 37s available for spot hire. Railfreight grey 20118 *Saltburn-by-the-Sea* and 20132 *Barrow Hill Depot* haul a rake of VTG wagons up the Lickey Incline at Vigo Bridge en route from Long Marston to Derby Chaddesden sidings on 9 March 2021. *Jack Boskett*

20056		HNRL	COY	HNR	SS (U)	
20066			TAB	HNR	HO (U)	
20069			BRB	HNR	WK (S)	
20087		MBDL	BRB	HNR	BAT	
20110			GYP	HNR	BAT	
20118		GBEE	RSR	HNR	BH	*Saltburn-by-the-Sea*
20121		HNRS	HNO	HNR	BH (U)	
20132		GBEE	RSR	HNR	WK (I)	*Barrow Hill Depot*
20166		HNRL	HNO	HNR	WR	
20168			HOP	HNR	HO (I)	*Sir George Earle*
20301	20047	HNRS	DRC	HNR	BH (U)	
20303	20127	HNRS	DRC	HNR	BU (U)	
20304	20120	HNRS	DRC	HNR	BH (U)	
20308	20187	HNRS	DRC	HNR	BH (U)	
20309	20075	HNRS	DRC	HNR	BH (U)	
20311	20102	HNRL	HNO	HNR	WK (U)	
20312	20042	HNRS	DRC	HNR	BH (U)	
20314	20117	HNRL	HNO	HNR	WK (U)	

20901	20101	GBEE	GBU	HNR	BH	
20903	20083	HNRS	DRU	HNR	BU (U)	
20904	20041	HNRS	DRU	HNR	BU (U)	
20905	20225	GBEE	GBU	HNR	WK (U)	
20906	20219		HOP	HNR	HO (I)	

Class 25

The three Class 25s in the HNRC fleet were all recently acquired from preservation groups. A return to the main line for one or two has been mooted but has been put on hold.

25057			BRB	HNR	WK (U)
25283	25904	MBDL	GYP	HNR	WK (U)
25313			BRB	HNR	WK (U)

Class 27

Like the Class 25s, an acquisition from a preservation group, the sole HNRC Class 27 is part of its heritage fleet.

27066	27103, 27212	HNRL	BRB	HNR	BH (U)

Class 37

HNRC has bought several Class 37s from DRS and has hired some, mostly for IM work. Three of the 37/6s remain main-line registered and are regularly used by Colas Rail.

37405	37282	XSDP	DRC	HNR	BH (U)
37603	37039, 37504	XSDP	DRC	HNR	LT (U)
37604	37007, 37506	XSDP	DRC	HNR	LT (U)
37607	37103, 37511	COTS	DRU	HNR	BH
37609	37115, 37514	XSDP	DRN	HNR	LT (U)
37610	37181, 37687	COTS	BLU	HNR	BH
37612	37179, 37691	COTS	DRU	HNR	BH
37703	37067	HNRS	DRU	HNR	BH (S)

Class 47

HNRC has seven Class 47s, some of which are in use as train supply locos at Wabtec Doncaster and HNRC's own Worksop site. Three are stored, one at Barrow Hill and two at Leicester.

47703	47514	HNRS	UND	HNR	ZF (I)	
47714	47511	HNRL	ANG	HNR	ZF (I)	
47715	47502	MBDL	NSD	HNR	WK (I)	*Haymarket*
47769	47491	HNRS	UND	HNR	BH (U)	*Resolve*
47843	47090, 47623	SROG	OXB	WCR	LR (U)	
47847	47179, 47577	SROG	LLB	ROG	LR (U)	

Nemesis Rail

Contact details

Website: www.nemesisrail.com

Key personnel

Managing Director: Martin Sargent

Overview

Based at the old DB Schenker depot at Burton upon Trent, Nemesis Rail offers vehicle maintenance and overhaul services to several companies and works for Riviera Trains, HNRC and West Coast Railways among others.

It also has many locos stored at its site, including some shunters and Class 08s, 09s and 20s from HNRC. The management team also owns a collection of ex-BR locos, which are at Burton, and has just 31128 still main-line registered and occasionally used by WCR.

Class 08

The company's sole 08 is used for shunting at Burton. It is vacuum-braked only.

08168		GRW	NEM	BU	

Class 25

The two Class 25s are stored at Burton.

25067			GYP	NEM	BU (U)
25265			BRB	NEM	BU (U)

Class 26

Neither Class 26 has worked for nigh on two decades now.

26004			TLC	NEM	BU (U)
26011			BRB	NEM	BU (U)

Class 31

The only main-line registered loco in the Nemesis fleet is 31128, which is occasionally used by West Coast Railways. 31461 has not worked for over 25 years.

31128		NRLO	BRB	NEM	BU	*Charybdis*
31461	31129	NRLO	CCE	NEM	BU (U)	

Class 33

This loco was once operational at the Battlefield Line, but has returned to Burton for storage.

33019		CCE	NEM	BU (U)

Class 37

Once based at the Great Central Railway, the sole Class 37 moved to Burton in 2011 has not worked since.

37255	NRLS	CCE	NEM	BU (U)

Class 45

For a period 45112 was the only main-line registered Peak but has not worked on the national network since 2007. It has occasionally been used for shunting at Burton.

45112			BRB	NEM	BU (U)	*Royal Army Ordnance Corps*

Class 47

The only operational 47 in the fleet is 47375 on long-term hire in Hungary (see page 329). The proposal to send three others to follow it did not materialise.

47488		NRLS	GYP	NEM	BU (U)	
47640	47244	NRLO	LLB	NEM	BU (U)	
47701	47493	NRLO	BRB	NEM	BU (U)	*Waverley*
47744	47250, 47600	NRLS	EWS	NEM	BU (U)	

Class 73

This loco was also once operational at the Battlefield Line, but has returned to Burton for storage.

73114		LLB	NEM	BU (U)

Rail Support Services

Contact details

Website: www.railwaysupportservices.co.uk

Key personnel

Managing Director: Andrew Goodman
Operations Manager: Paul Fuller

Overview

Railway Support Services specialises in the spot hire of shunting locomotives and has a fleet of over 20 Class 08s available for short- or long-term hire to customers.

Three of these locos are owned by Vulcan Rail, a small spot company that hires to RSS, which then places them in industry as required. It also has facilitated a deal for preserved 08663 to be hired to Hanson & Hall (see page 301).

Class 08

RSS spot hire fleet.

08405	MBDL	EWS	DBC	NL	
08411	MBDL	BRW	RSS	WI (U)	
08441	MBDL	RSS	RSS	NC	
08460	MBDL	RSS	RSS	FX	*Spirit of the Oak*
08480	MBDL	RSS	RSS	FX	
08484	MBDL	RSS	RSS	NC	*Captain Nathaniel Darell*
08511	MBDL	RSS	RSS	EH	
08580	MBDL	EWS	RSS	GAR	
08593	MBDL	RSS	RSS	WI (U)	
08629	EPUK	KBR	RSS	ZW	
08632	MBDL	LOR	RSS	WI	
08670	MBDL	RSS	RSS	BS	
08683	MBDL	EWS	RSS	EH	
08703	MBDL	GBR	RSS	WNE	*Jermaine*
08738	MBDL	RSS	RSS	WI	
08752	MBDL	GEM	RSS	CQ	
08757		RES	RSS	TSR	
08846		BLU	RSS	WH	
08921		EWS	RSS	WI (U)	
08927		GWS	RSS	BS	
08939	MBDL	RSS	RSS	SP	

GAR: Garston Freigthliner Terminal
WNE: Willesden Europort

Rail Support Services also has a large fleet of Class 08s employed at various sites across the country. On 9 August 2021, 08511 shunts in the yard at Eastleigh. *Jack Boskett*

RMS Locotec

Contact details

Website: www.rmslocotec.com

Key personnel

Managing Director: Bill Warriner
Fleet Director: Brian Lark

Overview

Rail Management Services, trading as RMS Locotec, known for its Class 08 hire business, has recently been acquired by Proviso Holdings and so avoided administration. Proviso Holdings also owns rolling stock leasing and maintenance company Eastern Rail Services.

RMS Locotec has an extensive fleet of Class 08s, many on hire to businesses across the UK.

Class 08

Spot hire fleet.

08308	RMSX	SCP	RMS	WO
08423		RMS	RMS	PDP
08523	RMSX	RMS	RMS	WO
08573	RMSX	BLK	RMS	WO
08588	RMSX	RMS	RMS	IL
08613		RMS	RMS	WO
08622		BLK	RMS	KET

08648	RMSX	RMB	RMS	IS
08700			RMS	IL
08754	RMSX	RMS	RMS	ZN
08756	RMSX	DEP	RMS	WO
08762	RMSX	BLK	RMS	GY
08788	RMSX	RMS	RMS	PDP
08809	RMSX	RMS	RMS	KET

08847	RMSX	COT	RMS	PDP
08871	RMSX	COT	RMS	WO
08874	RMSX	SIL	RMS	PDP
08885		RMS	RMS	WO
08936	RMSX	RMS	RMS	WO

KET: Ketton Cement Works
PDP: PD Ports Middlesbrough

RMS Locotec is a big player in the shunter hire business with locos on hire to TOCs and at industrial railheads. On 16 March 2019, 08523 was at work at Inverness. *Rob France*

St Leonards Rail Engineering

Contact details

Website: www.hastingsdiesels.co.uk
E-mail: 2020@hastingsdiesels.co.uk

Key personnel

Managing Director: John White
Operations Manager: Andy Armitage
Based in the original depot used to maintain the Hastings DEMUs, St Leonards Railway Engineering has expanded beyond its original remit of maintaining its preserved stock and undertakes maintenance for TOCs.

It services Class 73s used by GBRf and Govia as well as undertaking routine servicing on the Class 171s used by Govia on its Marshlink Ashford-Hastings operations.

It also looks after Hastings Diesel Class 201 main-line registered but preserved Class 201 DEMU. It also has an ex-Southern Region Class 07 diesel shunter based at the site.

Class 07

Depot shunter.

07011	GWS	SLE	SE

Swift Express Freight

Overview

Several ex-Greater Anglia Class 321 EMUs are being converted by Gemini Rail for use as express freight multiple units. They are leased from Eversholt and are being converted at Wolverton Works.

The aim is to provide a cost-effective and low-carbon solution for transporting parcels. Swift Express Freight is in partnership with Ricardo and the first unit converted was of 321334, with five other units identified. The trains are capable of 100mph and can carry 12 tonnes of freight per vehicle. The unit was launched for trials in June 2021.

Active units are in the GMBU pool.

	Livery	Owner	Location	DTC	MS	DS	DTS
321332	GAW	EVS	WK (S)	78080	63006	71911	77884
321334	SEF	EVS	ZB	78082	63008	71913	77886
321407	GAW	EVS	WK (S)	78101	63069	71955	77949
321419	GAW	EVS	WK (S)	78113	63081	71967	77961
321428	GAW	EVS	WK (S)	78122	63090	71976	77970
321429	GAW	EVS	WK (S)	78123	63091	71977	77971

Swift Express Freight is a trial partnership with Eversholt and Ricardo to test the viability of freight multiple units and the first conversion was 321334. It was at Doncaster RMT on 18 December 2021, and has recently been undertaking ECML trials. *Pip Dunn*

Transmart Trains

Overview

Transmart owns a single Class 73, which is based at Eastleigh Works.
The loco is in the MBED pool.

Class 73

73133	TMT	TMT	ZG

UK Rail Leasing

Contact details

Website: www.ukrl.co.uk

Key personnel

Chairman: Mark Winter
Managing Director: Kristian Mengel

Overview

UKRL is a rail vehicle engineering, fleet management and traction provider to the UK rail industry. It owns and works out of the former DB Cargo depot at Leicester, where it carries out maintenance for many types of traction.

It repairs locos and wagons and looks after the Europhoenix, ROG and Hanson Traction fleets. It also services Class 170s for CrossCountry Trains on a nightly basis.

It once had a sizeable fleet of Class 56s but sold them to GB Railfreight for conversion to Class 69s. It has retained one loco, 56303, but this was damaged following an altercation at Wembley Yard, and also two Class 37/9s that were previously preserved.

It has a staff headcount of 50, a turnover of nearly £6m a year and is in talks with Wabtec to take over parts of the Brush Works at Loughborough that it has vacated. This would allow it to start vehicle overhauls while keeping Leicester for route operational exams and running repairs.

Class 37/9

Both Class 37/9s were bought from preservation groups. Main line returns have been mooted but these seems unlikely to happen any time soon.

37905	UKRM	GYP	UKR	LR (S)
37906	UKRM	RFO	UKR	BAT (S)

Class 56

Having sold all its Class 56s to GBRf, one loco was not included in the deal due to collision damage it sustained in Wembley Yard. However, GBRf retains an option to buy it.

56303	56125	HTLX	DCG	UKR	WB (U)

Artemis Intelligent Power

Contact details

Website: www.artemisip.com

Overview

Former InterCity West Coast, Virgin Trains and Chiltern Railways' driving van trailer 82113, previously owned by Porterbrook, has been converted at the Bo'ness & Kinneil Railway in Scotland into a self-propelled vehicle that combines hydrostatic transmission with regenerative braking to reduce emissions.

It has been renumbered 19001 and is regarded as a Class 19, although the vehicle is not on TOPS and is not passed for main-line use.

The project is funded by the Rail Safety and Standards Board (RSSB) in conjunction with Artemis Intelligent Power, a company specialising in hydraulic machines. The vehicle is being developed so it can store braking energy, which could then be released during acceleration.

It uses JCB diesel engines that drive hydraulic pumps, and in turn they drive hydraulic motors mounted on the axles.

82113 moved to Bo'ness in July 2017 for the conversion and the 'locomotive' has undertaken some trials on the 5-mile heritage line from August 2018.

BCRRE Class 799

Contact details

Website: www.birmingham.ac.uk/research/railway/index.aspx

Overview

The two Class 799s are four-car hydrogen-powered HydroFlex test trains rebuilt from redundant Thameslink Class 319 units. They retain 25kV AC and 750V DC capabilities.

799001 was unveiled in June 2019 at the RAIL live event at Long Marston in plain white, where it was demonstrated on a short section of track. It has since been wrapped in a promotional livery and carried out some main-line trials in September 2020.

799201 is a second hydrogen unit, which has yet to be able to work under its own power but was hauled to Glasgow Central for display as part of the COP26 Climate Change event in November 2021.

The project is being financed by Porterbrook in conjunction with the University of Birmingham's Centre for Railway Research and Education (BCRRE).

					DMS	MS	TS	DMS
799001	319001	ADV	POR	LM	77291	62891	71772	77290
799201	319382	ADV	POR	LM (S)	77975	63094	71980	77976

799001 is wrapped in a HydroFlex promotional livery of predominately green graphics on a white background.
799201 is wrapped in promotional livery of predominately light green graphics with WWW.UKCOP26.ORG and 'Together for our planet' branding.

Ex-Thameslink unit 319001 has been converted to a HydroFlex unit to run off hydrogen but still can use both 25kVAC and 750V DC. Now numbered 799201, it passes Baillieston running as the 1439 Glasgow Central-Mossend empty stock on 4 November 2021. It was in Glasgow as part of the COP26 climate change summit. *Robin Ralston*

Hydrogen Emission-free Trains

Contact details

Website: www.arcolaenergy.com

Overview

There are several projects planned to create hydrogen-powered trains and the concept is seen as offering an emission-free alternative to diesel trains on remote lines where electrification is not viable.

Hydrogen and fuel cell specialist Arcola Energy is working with Brodie Rail Engineering at Kilmarnock to convert a redundant ex-ScotRail EMU, 314209, now numbered 614209, to run emission-free using hydrogen. The project is being delivered with input from the University of St Andrews.

Arcola Energy is also working with VivaRail to offer emission-free Class 230s. The concept train will consist of two cars, one housing two battery modules and one with the fuel cell and tanks, all of which will be underneath the train.

In May 2018 plans were announced to convert some Eversholt Rail Class 321 units to HMUs, hydrogen multiple units, using Alstom hydrogen cells. The conversion was called the Breeze and designated Class 600.

The £1m project to develop Breeze trains involves reengineering BREL-built Class 321s, many of which came off lease from Greater Anglia in 2021 when replaced by new-build trains.

Alstom says the investment could allow the HMUs to be ready by 2024, supporting the government's aims to eliminate diesel-only trains from the main-line network by 2040.

The Class 600 will be converted at Alstom's Widnes Transport Technology Centre and when the project is in series production, it could create more than 200 engineering jobs.

Hydrogen trains emit no emissions besides water and Eversholt and Alstom say they are most suited to regional services on routes that are not currently electrified.

Alstom will build on the development of the Coradia iLint in Germany, the first hydrogen-powered train to operate regular passenger services in the world.

Class 614

				DMS	PTS	DMS
614209	314209	SCR	BKR	64599	71458	64600

Brush/FLEX Bi-mode MUs

In among this myriad of new developments are the FLEX trains, which are former Thameslink Class 319 dual-voltage units that have been converted into bi-mode or tri-mode power.

Following their replacement by Class 700s, the fleet of 86 dual-voltage, four-car EMUs were spare. Some were taken by TOCs, such as Northern (on a short-term basis, pending new trains), and London Midland, while 319004/451/453/455 have been scrapped.

These are all four-car sets – as per their original build – and are five 769/0s and four 769/4s for TfW, eight 769/4s for Northern, 19 769/9s for GWR and two other units for freight use for Rail Operations Group.

EMU conversions rebuilt from Class 319s

Northern	769424/426/431/434/442/450/456/458
Transport for Wales	769002/003/006-008, 769445/452
GWR	769922/923/925/927/928/930/932/935-940/943/944/946/947/949/959
ROG	319009/010

Meteor Power

Contact details

Website: www.meteorpower.com

Overview

Meteor Power is working on modifying two ex-BR locos to work on alternative power. 08649 was formerly based at Wolverton Works, and was in poor condition and had a broken crank, but is now being re-engineered with a low-emission hybrid traction package based on automotive technology.

The company is also working on a similar set-up for formerly preserved 37207 to see if the technology can be applied to a main-line loco as well.

The grant-funded work is being undertaken as part of a research programme backed by the Department for Transport. In February 2020, 08649 had its EE engine and generator removed and replaced by a new generator and battery pack. It has been fitted with a Euro 3a 6.8-litre John Deere PowerTech 6068HFU82 diesel engine rated at 285hp. This will be modified to meet Euro 5 standards.

The engine drives a generator, which trickle-charges a battery pack assembled using four modules recycled from Tesla Model 3 cars, giving a combined storage capacity of 300kWh. Each of the four batteries can be recharged by the diesel engine in less than an hour. The batteries can also capture and store regenerative braking energy.

08649 retains its original traction motors, but has been fitted with a new control system for the compressors and vacuum pumps. The cab desk is unchanged, but the control system voltage has been reduced from 400V to 24V DC to improve safety.

Meteor Power acquired 37207 from Europhoenix and is assessing the viability of rebuilding a Class 37 with similar technology to offer a low-cost 'new Type 3'. After a period in Dudley, 37207 and 08649 are now at Wolverton Works.

Class 08

08649	KBR	MET	ZN (U)

Class 37

37207	BRB	MET	ZN (U)

Industrial Locos

In the 1960s hundreds of surplus BR diesel shunters were sold for industrial use for use at collieries, steelworks, quarries and other railheads. They were sold exceptionally cheaply and allowed these railheads to modernise and replace their steam locos.

Initially it was mostly Class 02/03/04/05/10/11s followed later by Class 07/14s from the 1970s and Class 08s in the 1980s. These sales also affected many specialist builders of industrial diesel locos at the time, including Hunslet, Fowler, Ruston et al.

Demand for industrial locos had diminished dramatically due to the major reduction of heavy industry in the UK. However, some sites still require a shunting loco and several former BR/EWS Class 08/09s remain in what is best described as 'industrial use'.

Even what constitutes an 'industrial' loco is a grey area. Some sites prefer to hire locos from the likes of Rail Support Services, RMS Locotec and Harry Needle Railroad Company (see pages 302, 306, 307), while others prefer to own the locos outright.

This list details those locos in use, or owned by, industrial sites. They change over at short notice, or indeed may not work very often. It is a very fluid area, and subject to change.

This section, therefore, rounds up those locos that 'don't really fit anywhere else'.

Class 08

	Livery	Owner/user	Location	Name
08296	BLE	Aggregates Industries	Barton-under-Needwood (U)	
08375	RMS	Victoria Ports	Boston Docks	
08410	GWR	AV Dawson	Middlesbrough	
08442	LNW	Arriva	Eastleigh (U)	
08447	IND	John Russell	Assenta Rail, Hamilton (U)	
08516	ARV	Arriva	Bristol Barton Hill	*Rory*
08536	DEP	Vulcan Rail	Newton Aycliffe	
08598	POT	AV Dawson	Middlesbrough	
08600	AVD	AV Dawson	Middlesbrough	
08643	GRE	Aggregates Industries	Merehead Quarry	
08650	BLE	Aggregates Industries	Whatley Quarry	
08652	BLE	Aggregates Industries	Wishaw	
08730	ABP	ABP Ports	Tyseley	
08735	ARV	Arriva	Eastleigh	
08743	BLE	ICI Sembcorp	Wilton	*Bryan Turner*
08774	AVD	AV Dawson	Middlesbrough	*Arthur Vernon Dawson*
08783	EWS	EMR	Kingsbury (U)	
08784	DEP	Vulcan Rail	Ruddington	
08810	LNW	Arriva	Eastleigh	*Richard J. Wenham Eastleigh Depot*
08868	ARV	Arriva	Crewe LNWR	
08872	EWS	EMR	Attercliffe	
08903	BLE	ICI Sembcorp	Wilton	*John W Antill*
08912	BRB	AV Dawson	Middlesbrough (U)	
08913	MAL	EMR	Kingsbury (U)	
08922	DEP	Vulcan Rail	Asfordby	
08933	BLE	Aggregates Industries	Whatley Quarry	
08947	BLE	Aggregates Industries	Barton-under-Needwood	

Arriva Traincare has four Class 08/09 shunters, two of them based at Crewe, where 09204 was seen on 11 July 2018 proudly showing off its Arriva livery. *Jack Boskett*

Class 09

09022	BLE	Victoria Ports	Boston Docks
09023	EWS	EMR	Attercliffe (U)
09204	ARV	Arriva	Crewe LNWR

Clayton Equipment

The company has registered the Class 18 with the Rolling Stock Library on TOPS for a fleet of 15 Hybrid CBD90 Bo-Bo locomotives for Beacon Rail.

Clayton Equipment is building 15 emission-free Hybrid+ CBD90 shunting locomotives for Beacon Rail Leasing. The first example has been tested at the Chasewater Railway and is now being taken on by GB Railfreight for trials and assessment at its Whitemoor Yard.

The CBD90 is a 90-tonne Bo-Bo battery-diesel hybrid locomotive with on-board batteries that can be charged through a three-phase electric supply or on-board diesel engine that meets EU emission standards.

Clayton hopes the loco will replace elderly Class 08s, which date from the 1950s, and become the 'shunter of choice' due to its zero emissions, low noise, low maintenance and ease of operation.

Battery charging can be via a depot three-phase supply for complete emission-free performance, and/or via the on-board low-emission diesel for operating in more remote locations.

Variants are available between 1.35 tonnes and 150 tonnes, but the CBD90 is the most popular choice, with Tata Steel, Sellafield and Beacon Rail purchasing 24 so far since its launch in March 2019.

The loco is rated at 560hp, with a maximum tractive effort of 303kN. Its top speed is 20kmh, about 13mph. Clayton is also eyeing up the export market for this loco design.

The mass of new trains being introduced at the moment, plus changing legislation forcing many older vehicles off the rails, means there are several locomotives, multiple units and coaches now 'off lease'.

These are vehicles owned by the rolling stock leasing companies that currently do not have a user. Some will go for scrap, others may be redeployed with other TOCs, some will be sold for reuse or preservation and some will be held in reserve pending possible reuse elsewhere.

Any vehicle or unit listed here may be sent for scrap, sold, moved or redeployed at short notice.

Akiem

Akiem is a French company that is a relatively small player in the UK rolling stock leasing market, but owns the fleet of 30 Class 379 Bombardier Electrostar Units that were ordered by Greater Anglia and delivered in 2010-11. However, all have recently been stood down as new Class 720s come into traffic.

The Class 379s were used on the routes from Liverpool Street to Stansted and peak trains to Cambridge, although Class 745s are also now used on the Stansted route. The 379s were due to replace Class 317s to allow their withdrawal to be hastened, but as Class 720s are now entering traffic, the 379s have been returned to their ROSCO, although indications are they could move to another TOC. Locations shown for storage may change. All are in the EBHQ pool.

	Livery	Owner	Depot	DMS	MS	PTS	DMC
379001	GAW	AIK	HW (S)	61201	61701	61901	62101
379002	GAW	AIK	IL (S)	61202	61702	61902	62102
379003	GAW	AIK	IL (S)	61203	61703	61903	62103
379004	GAW	AIK	IL (S)	61204	61704	61904	62104
379005	GAW	AIK	IL (S)	61205	61705	61905	62105
379006	GAW	AIK	HW (S)	61206	61706	61906	62106
379007	GAW	AIK	IL (S)	61207	61707	61907	62107
379008	GAW	AIK	IL (S)	61208	61708	61908	62108
379009	GAW	AIK	IL (S)	61209	61709	61909	62109
379010	GAW	AIK	HW (S)	61210	61710	61910	62110
379011	GAW	AIK	IL (S)	61211	61711	61911	62111
379012	GAW	AIK	IL (S)	61212	61712	61912	62112
379013	GAW	AIK	IL (S)	61213	61713	61913	62113
379014	GAW	AIK	HW (S)	61214	61714	61914	62114
379015	GAW	AIK	HW (S)	61215	61715	61915	62115
379016	GAW	AIK	IL (S)	61216	61716	61916	62116
379017	GAW	AIK	HW (S)	61217	61717	61917	62117
379018	GAW	AIK	HW (S)	61218	61718	61918	62118
379019	GAW	AIK	HW (S)	61219	61719	61919	62119
379020	GAW	AIK	HW (S)	61220	61720	61920	62120
379021	GAW	AIK	IL (S)	61221	61721	61921	62121
379022	GAW	AIK	HW (S)	61222	61722	61922	62122
379023	GAW	AIK	HW (S)	61223	61723	61923	62123
379024	GAW	AIK	HW (S)	61224	61724	61924	62124
379025	GAW	AIK	HW (S)	61225	61725	61925	62125
379026	GAW	AIK	IL (S)	61226	61726	61926	62126
379027	GAW	AIK	IL (S)	61227	61727	61927	62127
379028	GAW	AIK	HW (S)	61228	61728	61928	62128
379029	GAW	AIK	IL (S)	61229	61729	61929	62129
379030	GAW	AIK	IL (S)	61230	61730	61930	62130

Angel

Ex-Great Western Railway and LNER HST power cars.

43017	FGB	ANG	EY (U)
43020	FGB	ANG	LA (S)
43023	FGB	ANG	LM (S)
43024	FGB	ANG	EY (U)
43025	FGB	ANG	EY (U)
43165	FGB	ANG	EY (U)
43174	FGB	ANG	EY (U)
43185	ICS	ANG	ZK (U)
43190	FGB	ANG	EY (U)
43191	FGB	ANG	LA (U)
43206	HST	ANG	EY (U)
43238	LEM	ANG	EY (U)
43295	VEM	ANG	EY (U)
43305	VEM	ANG	EY (U)
43306	VEM	ANG	EY (U)
43307	VEM	ANG	EY (U)
43309	VEM	ANG	EY (U)
43310	VEM	ANG	EY (U)
43311	VEC	ANG	EY (U)
43312	HST	ANG	EY (U)
43314	VEM	ANG	EY (U)
43315	VEC	ANG	EY (U)
43316	VEM	ANG	EY (U)
43317	VEM	ANG	EY (U)
43318	VEM	ANG	EY (U)
43319	VEM	ANG	EY (U)
43320	VEM	ANG	EY (U)
43367	VEC	ANG	EY (U)

HST Trailers

40204/205/221, 40704/715/720/734/755, 41087/091, 41100/106/118/170, 41204-206/208/209, 42024/026, 42179, 42242/243, 42347/355/357/363, 42401/402/404/405/408, 42506/569/584/585, 44034/061/063/094/098

Class 142

Former Northern railbus.

				DMS	DMSL
142047	NBL	ANG	GA (U)	55588	55638

Class 153

Former Northern Rail and East Midlands Railways single-car units.

153301	NOR	ANG	EY (S)	52301
153304	NOR	ANG	EY (S)	52304
153307	NOR	ANG	EY (S)	52307
153308	EMT	ANG	EY (S)	52308
153315	NOR	ANG	EY (S)	52315
153317	NOR	ANG	EY (S)	52317
153319	EMT	ANG	EY (S)	52319
153328	NOR	ANG	EY (S)	52328
153331	NOR	ANG	EY (S)	52331
153351	NOR	ANG	EY (S)	57351
153352	NOR	ANG	EY (S)	57352
153355	EMT	ANG	EY (S)	57355
153357	EMT	ANG	EY (S)	57357
153378	NOR	ANG	EY (S)	57378

153383 leads a line of five Porterbrook Class 153s, formerly with East Midlands Railway or West Midlands Trains, stored at Long Marston on 25 October 2021. As they have non-compliant toilets there is a limited future for these units, which are now over 30 years old. *Jack Boskett*

Class 317

	Livery	Owner	Depot	DTC	MS	TC	DTS	
317341	GAW	ANG	EY (S)	77040	62701	71617	77088	
317344	GAW	ANG	EY (S)	77043	62704	71620	77091	
317348	GAW	ANG	EY (S)	77047	62708	71624	77095	*Richard A Jenner*
317509	AGA	ANG	HW (S)	77012	62671	71587	77058	
317512	GAW	ANG	EY (S)	77015	62676	71592	77063	
317513	GAW	ANG	EY (S)	77016	62677	71593	77064	
317708	LOR	ANG	EY (S)	77007	62668	71584	77055	
317709	LOR	ANG	EY (S)	77008	62668	71585	77056	
317710	LOR	ANG	EY (S)	77009	62670	71586	77057	
317714	LOR	ANG	EY (S)	77013	62674	71590	77061	
317719	LOR	ANG	EY (S)	77018	62679	71595	77066	
317722	LOR	ANG	EY (S)	77021	62682	71598	77069	
317723	LOR	ANG	EY (S)	77022	62683	71599	77070	
317729	LOR	ANG	EY (S)	77028	62689	71605	77076	
317732	LOR	ANG	EY (S)	77031	62692	71608	77079	
317882	GAW	ANG	EY (S)	77023	62684	71600	77071	
317883	GAW	ANG	EY (S)	77000	62685	71601	77073	
317885	GAW	ANG	EY (S)	77026	62687	71603	77074	
317892	LOR	ANG	EY (S)	77035	62696	71612	77083	

Service vehicles

Current no.	Old nos.	Livery	Owner	Type	Role
6330	14084, 975629	FIR	ANG	Mk 2	HST barrier
6336	81591, 92185	FIR	ANG	Mk 1	HST barrier
6338	81581, 92180	FIR	ANG	Mk 1	HST barrier
6340	21251, 975678	BLU	ANG	Mk 1	HST barrier
6344	81263, 92080	BLU	ANG	Mk 1	HST barrier
6346	9422	BLU	ANG	Mk 2	HST barrier
6348	81233, 92963	FIR	ANG	Mk 1	HST barrier
975974	1030	BLU	ANG	Mk 1	EMU translator
975978	1025	BLU	ANG	Mk 1	EMU translator

Eversholt

The following Class 91s are in the SAXL pool and stored at either Doncaster Royal Mail Terminal or Doncaster Works. 91117/120 have been taken on by Europhoenix for possible redeployment in Europe, although the pandemic has at least delayed this move. 91122/128 have been used for the DATS operation but are also now in store in Crewe (see page 293). The first locos have been disposed of.

Class 91

91112	91012	VEC	EVS	DB (S)
91115	91015	VEC	EVS	DB (S)
91116	91016	VEC	EVS	DB (S)
91118	91018	VEC	EVS	DB (S)
91121	91021	VEC	EVS	DB (S)
91125	91025	VEC	EVS	DB (S)
91131	91031	VEC	EVS	DB (S)

Mk 4 coaches

10307/315/332, 11309/316/318, 11409/414/416/413, 12202/209/213/229/231, 12305/308/312, 12405/410/415/417/419/421/423/428/433/443/453/467/480/483/486, 12513/514/518/520/526, 82204/206/210/215/218/222

Mk 3 coaches

10551/553, 11088

Service vehicles

6376	1021, 975973	POR	Mk 1	EMU translator
6377	1042, 975975	POR	Mk 1	EMU translator
6378	1054, 975971	POR	Mk 1	EMU translator
6379	1059, 975972	POR	Mk 1	EMU translator
6393	81609, 92196	POR	Mk 1	HST barrier
6394	80878, 92906	POR	Mk 1	HST barrier

In December 2019, LNER dispensed with its last HST sets, and 43206/312 were repainted into original BR livery for a farewell tour of LNER's routes. On December 21, with the locos renumbered to their original identities, 43112 waits to leave Leeds with the final run ever – to King's Cross. Both these power cars and the set of trailers have been off lease ever since and could be destined for scrap unless a new users comes forward to take them on. *Pip Dunn*

Network Rail is responsible for managing 20 of the country's largest stations, with TOCs responsible for managing the others. This is Leeds, a busy interchange with trains from Northern, LNER, Transpennine Express, CrossCountry trains and East Midlands Railway all using it. *Pip Dunn*

Network Rail

After an ill-fated public flotation of the national infrastructure as Railtrack in the mid-1990s, following a series of tragic high-profile accidents, some caused by inadequate maintenance, in 2002 the network was renationalised as Network Rail.

Network Rail owns, operates and develops Britain's railway infrastructure of 20,000 miles of track, 30,000 bridges, tunnels and viaducts and the thousands of signals, signalboxes, level crossings and stations.

It also manages 20 of the largest stations, namely Birmingham New Street, Bristol Temple Meads, Clapham Junction, Edinburgh Waverley, Glasgow Central, Guildford, Leeds, Liverpool Lime Street, Manchester Piccadilly, Reading and most major London termini including London Bridge, Cannon Street, Charing Cross, Euston, King's Cross, Liverpool Street, Paddington, St Pancras International, Victoria and Waterloo. The other stations – of which there are over 2,500 – are managed by the train operating companies.

Network Rail created five regions: Eastern, North West & Central, Scotland's Railway, Southern and Wales & Western, each led by a managing director. They were formed in June 2019 and have the budget and capability to take on more responsibility from other parts of the business.

In August and September 2019, Network Rail created 14 new routes in the five regions. The routes are responsible for operations, maintenance and minor renewals, including the day-to-day delivery of train performance and the relationship with their local train operating companies.

Phase two of the programme started in November 2019 and is designed to strengthen NR's new regions and further build Network Services and Route Services. This new structure sets NR up for more devolution and to be more responsive to the needs of train operators, passengers and freight users by bringing its people closer to those it serves.

Network Rail Regions and Routes

- **Eastern Region**: Routes: Anglia, East Midlands, North and East and East Coast
- **North West and Central Region**: Routes: North West, Central, West Coast Mainline South
- **Scotland's Railway**
- **Southern Region**: Routes: Kent, Sussex, Wessex Route, Network Rail High Speed
- **Wales & Western Region**: Routes: Wales Route, Western

Train operation

Network Rail is not allowed to operate its own trains, and accordingly all its movements – Infrastructure Monitoring (IM), infrastructure support (ballast and material trains – the old BR Departmental sector) and track machine moves – have to be undertaken by third-party TOCs and FOCs.

However, NR does own a few locomotives. It has four Class 37s (reclassified as Class 97/3s) used for ERTMS workings west of Shrewsbury, five HST power cars for the New Measurement Train and three Class 73s for working IM trains – one standard loco and two trial re-engineered Cummins-powered Class 73/9s although 73138 is also now stored. It also owns a Class 150 DMU and Class 313 EMU and several coaches for use with IM trains.

NR has seven operations bases and five delivery depots for materials. Its IM trains are based at the former Derby RTC and maintained by Loram and operated by Colas. There are 24 seasonal treatment fleet bases, from Scotland to the South East, where Multi-Purpose Vehicles (MPVs) – two-car 'freight multiple units' – are based for weed killing and other maintenance roles. The company also has six SITT – snow-clearing trains – based in the South-East.

The IM fleet

Infrastructure monitoring allows NR to monitor the condition of the track and pre-plan any maintenance in a proactive manner rather than in a reactive 'wait for it to fail, then repair it' manner.

The New Measurement Train (NMT) monitors and records track condition information at speeds up to 125mph. It helps locate and identify faults before they become a safety issue or affect performance.

The NMT is a unique, machine that was introduced 15 years ago. It uses two HST power cars and a rake of Mk 3 trailers. It is equipped with the newest, hi-tech measurement systems, track scanners and a high-resolution camera to measure the condition of the track, especially on the busiest and fast main lines.

The NMT covers 115,000 miles in a year and will capture around 10TB of image data every 440 miles. Because it can run at up to 125mph it identifies faults faster and more accurately than ever before, so engineers can then make repairs or plan maintenance to prevent serious incidents such as derailments.

A laser sensor gives information about the profile of the railhead, measuring shape and movement optically. At the same time transducers and accelerometers mechanically measure the up and down movement of the train as it travels along the rails. This data provides information on track geometry – the shape and profile of the railhead, and the twist of the track.

NR also has a plain line pattern recognition (PLPR) system that uses a series of lasers and cameras to detect faulty track components as the train passes over them. Image analysis software uses an algorithm to compare what the cameras see with an image of how the track should look. For example, it can identify missing Pandrol clips, which secure rails to the sleepers, allowing teams on the ground to replace them quickly.

Infrastructure wagons

NR also has a High Output Ballast Cleaning (HOBC) train, and it owns around 1,000 wagons to carry aggregates such as ballast and remove waste from work sites.

Among this fleet are box wagons to remove waste; autoballaster wagons with 'trap doors' at the base that release ballast directly beneath, onto the track; and tilting wagons to position switches and crossings during track renewal work.

It has a rail delivery train that can carry 108m sections of rails from the British Steel plant in Scunthorpe, to be welded into 216m lengths at NR's long welded rail depot in Eastleigh, Hampshire. Some are welded into place at the worksites.

The ballast cleaning system (BCS) and track renewal system (TRS) trains operate in a conveyor process, renewing the track as they move along it. Both the BCS and TRS are supported by tamper/dynamic track stabiliser machines that position the track accurately and consolidate ballast after renewal.

It also owns rail grinders; a grinding train maintains the track and helps to increase the lifespan of the rail by removing small layers of metal from the railhead, helping to keep the track in good condition.

Stone blowers restore the line and level of the track by correcting its vertical (height) and lateral (left or right) profile. They measure the track to work out where it needs to be lifted to bring it level, and how much it needs to be moved side to side to ensure correct alignment.

Tampers are track maintenance machines that make sure the track is aligned correctly and has a smooth level along the rail. They help to prevent the risk of trains derailing, and ensure smooth, comfortable journeys for passengers and freight trains. Tamping machines insert large tools called tines into the ballast – the stones beneath the track – and then force the tines together to move ballast under the sleepers. The tamping machine moves additional ballast under the sleepers to raise the height of the track and can move the track sideways if required.

Specialist machines help NR's engineers to keep the railway operating all year round and, bearing in mind the effects of weather, are a challenge for the railway and can severely impact the day-to-day running of services, they provide a vital service.

Rain can result in flooding and damage infrastructure such as bridges, lineside equipment and the track, or lead to landslips, while rails can buckle and overhead lines sag in extreme summer heat. Autumn leaf fall causes operational problems for the signalling system and reduces railhead condition, which can make trains slide on the track even when braking and so run past signals at danger or merely not stop in a platform as the driver intends.

Track clearance

Poor railhead condition restricts the ability of a train to start from a station, accelerate and climb hills, which means it can lose time, thereby delaying passengers and interfering with the timetable.

As a result, each autumn NR contracts the running of hundreds of RHTTs – Railhead Treatment Trains – which are operated by the FOCs and use either its 32 MPVs or 29 sets of top-and-tailed loco-hauled wagons with RHTT tanks and equipment. These cover sections of track well known for suffering poor railhead conditions.

The snow and ice treatment trains (SITTs) help to clear snow from the conductor rails and prevent ice freezing to them again. Both MPVs and SITTs scrape ice off the conductor railhead and spray the rails with hot liquid anti-icer that prevents ice sticking to the conductor railhead.

Twenty-four MPVs are used on the third-rail routes of Kent, Sussex, Wessex and Mersey Rail, while 10 locomotive-hauled SITTs are deployed in Kent and Sussex.

The third type of winter de-icing train, the winter development vehicle (WDV), is built to blow hot air around the running rails – specifically around points and crossings – to melt snow and ice that is stopping the points from working. The WDV is also equipped with steam lances to melt the more built-up, stubborn, ice deposits.

Also in NR's arsenal to clear snow are patrols using locos fitted with a three-piece miniature snowplough (MSPs), which are hired and operated by GBRf, DRS, WCR, ROG and others. MSPs work effectively in snow depths of up to 18 inches. For heavier drifts, Beilhack V or independent snowploughs propelled by locomotives are used.

Network Rail also has two now blowers in Scotland, which are fitted with propellers that cut through and blow away drifted snow. A special hydraulic turntable within the machine makes it possible for the snow blower to turn around on its own if needed.

Eight of NR's MPVs are equipped to spray herbicides to control plant growth on the tracks. This maintains safe walking routes for workers, allowing drivers to see signals, preventing damage to trains caused by vegetation and maintaining the stability of structures.

NR is responsible for the maintenance of the network, including vegetation clearance and weed killing. For the latter it uses a fleet of multi-purpose vehicles, essentially two-car 'freight' multiple units. On 21 May 2013, MPV DR98962 and DR98912 pass Silverdale. *Rob France*

NR fleet

Class 43

There are presently five MTU-powered HST power cars, leased from Porterbrook, which are used for the NMT. 43013/014 have buffers fitted and 43290/299 are on short-term hire while the three original locos are upgraded with ETCS. Their lease is dependent on ETCS testing of two NMT Power Cars at RIDC, Old Dalby. The current plan sees 43290/299 remaining on hire until the end of August 2022, however this could be extended until the end of September when the second stage of ETCS testing is complete.

43013		NRY	POR	ZA	*Mark Carne CBE*
43014		NRY	POR	ZA	*The Railway Observer*
43062		NRY	POR	ZA	*John Armitt*
43290	43090	VEC	POR	ZA	
43299	43099	VEC	POR	ZA	

Network Rail has used an HST set for its New Measurement Train for nearly 20 years now. Two of its three dedicated power cars, 43013 and 43014, work the train through Honiton on 2 June 2016. *Tony Christie*

Class 73/1

NR still owns one standard Class 73, which has been stood down and moved to Peak Rail for storage. The long-term plan is likely to see its bogies overhauled to create a float of spares for the forthcoming Class 73 Ultra overhauls – subject to budgetary approval – which will result in the body of the loco being disposed of.

The loco is in the QADD pool.

73138	NET	NRY	PR (U)

Class 73/9

Two Class 73s were rebuilt by RVEL (later taken over by Loram) at Derby and fitted with two Cummins engines in place of their EE engine; NR refers to them as Class 73 Ultras. They are used for hauling IM trains. They have recently been modified to run off bio-LPG as well as diesel

Both are in the QADD pool.

73951	73104	NRY	NET	ZA	*Malcolm Brinded*
73952	73113, 73211	NRY	NET	ZA	*Janis Kong*

NR used Class 73/9s hired from GBRf for infrastructure monitoring trains but also has two locos of its own, Cummins-powered 73951/952 for such workings. The latter stands at Doncaster while working a route-learning trip from Derby on 20 June 2021. *Stuart West*

Class 97/3

Former Class 37/0s fitted with ERTMS for working on the Cambrian lines. The system on 97301 is different to that on 97302-304. WCR's 37668/669 also have the same ERTMS system as used on 97301 and these can also work on these lines.

They work charter trains, infrastructure support and monitoring trains, saloon visits and other traffic should it be needed such as freight, Royal trains or special workings.

All are in the QETS pool.

97301	37100	NRY	NET	ZA	
97302	37170	NRY	NET	ZA	*Ffestiniog and Welsh Highland Railways/Rheilffyrdd Ffestiniog Ac Eryri*
97303	37178	NRY	NET	ZA	*Dave Berry*
97304	37217	NRY	NET	ZA	*John Tiley*

NR has four Class 37s modified with ERTMS and renumbered as Class 97/3s, two or three of which are normally based at Shrewsbury for working Cambrian lines. Making a rare getaway from its normal duties, 97302 brings up the rear of a test train, led by 37175, at Pinchbeck on 24 June 2017. This loco has since been overhauled, fitted with new marker lights and also named. *Pip Dunn*

Class 950

This purpose-built IM test train, based on a two-car Class 150/1 Sprinter DMU, was built in 1987 and initially a Class 180 before being added to the Departmental fleet.

				DMSL	DMS
950001	NRY	NET	ZA	999600	999601

When BR was building the Class 150 Sprinter DMUs in the mid-1980s, it built an extra unit for the Railway Technical Centre for use as a track survey unit. Numbered 950001, it stands at Penzance having arrived on the 0900 from Exeter Riverside Yard on 21 August 2021. *Tom McAtee*

Class 153

Three ex-East Midlands Trains' Class 153s single-car DMUs have been taken on by NR as part of the Infrastructure Monitoring Fleet. 153311 is used for driver training, while the others, 153376/385, are currently being tested with forward-facing video and track geometry. If successful it is very likely this trial will go into production and additional units will be brought into the NR fleet.

	livery	owner	Depot	DMSL
153311	EMT	POR	ZA	52311
153376	EMT	ANG	ZA	57376
153385	EMT	POR	ZA	57385

Class 313

This former Silverlink unit is used for ERTMS testing. The vehicle is currently being used to undertake various ETCS and GSMR tests. It is planned to remain in service until it is stopped for a C4 overhaul in June 2022.

				DMS	TS	BDMS
313121	NRY	BEA	ZG	62549	71233	62613

Test Coaches

Network Rail currently has an extensive fleet of test train coaches. They are usually ex-BR coaches such as Mk 1 and 2 vehicles, although some are former multiple units and the New Measurement Train uses HST Mk 3 trailer vehicles.

Those vehicles listed as Class 488s were initially Mk 2 coaches later converted to Gatwick Express stock and reclassified as multiple units.

Current no.	Old nos	Livery	Owner	Type	Role
1256	3296	NRY	NRY	Mk 2	Plain line pattern recognition coach
5981		NRY	NRY	Mk 2	Plain line pattern recognition coach
5995		NRY	NRY	Mk 2	Test train brake force runner
6001		NRY	NRY	Mk 2	Test train brake force runner
6008		NRY	NRY	Mk 2	Test train brake force runner
6117		NRY	NRY	Mk 2	Test train brake force runner
6122		NRY	NRY	Mk 2	Test train brake force runner
6260	81450, 92116	NRY	NRY	Mk 1	Generator van
6261	81284, 92988	NRY	NRY	Mk 1	Generator van
6262	81064, 92928	NRY	NRY	Mk 1	Generator van
6263	81231, 92961	NRY	NRY	Mk 1	Generator van
6264	80971, 92923	NRY	NRY	Mk 1	Generator van
9481		NRY	NRY	Mk 2	Staff coach
9516		NRY	NRY	Mk 2	Test train brake force runner
9523		NRY	NRY	Mk 2	Test train brake force runner
9701	9528	NRY	NRY	Mk 2	Driving Brake
9702	9510	NRY	NRY	Mk 2	Driving Brake
9703	9517	NRY	NRY	Mk 2	Driving Brake
9708	9530	NRY	NRY	Mk 2	Driving Brake
9714	9536	NRY	NRY	Mk 2	Driving Brake
9801	5760	FIR	ERS	Mk 2	Test train brake coach
9803	5799	FIR	ERS	Mk 2	Test train brake coach
9806	5840	FIR	ERS	Mk 2	Test train brake coach
9808	5871	FIR	ERS	Mk 2	Test train brake coach
9810	5892	FIR	ERS	Mk 2	Test train brake coach
62287		NRY	NRY	class 421	Ultrasonic test coach
62384		NRY	NRY	class 421	Ultrasonic test coach
72612	6156	NRY	NRY	class 488/3	Test train brake force runner
72616	6007	NRY	NRY	class 488/3	Test train brake force runner
72630	6094	NRY	NRY	class 488/3	Structure gauging train coach
72631	6096	NRY	NRY	class 488/3	Plain line pattern recognition coach

OLE inspection test coach 975091 *Mentor* (Mobile Electrical Network Testing Observation and Recording) was converted in 1973 from a Mk 1 BSK 34615 and is still in use. On 21 July 2018 it was captured passing Spalding on a March to Doncaster IM train, for which it was actually not needed. *Pip Dunn*

72639	6070	NRY	NRY	class 488/3	Plain line pattern recognition coach
99666	3250	NRY	NRY	Mk 2	Structure gauging train coach
975025	60755	GRE	NRY	class 202	Inspection saloon
975091 *Mentor*	34615	NRY	NRY	Mk 1	Overhead line equipment test coach
975814	11000, 41000	NRY	NRY	HST Mk 3	New Measurement Train
975984	10000, 40000	NRY	NRY	HST Mk 3	New Measurement Train
977868	5846	NRY	NRY	Mk 2	Radio survey coach
977869	5858	NRY	NRY	Mk 2	Winter Development Coach
977969	14112, 2906	NRY	NRY	Mk 2	Staff coach
977974	5854	NRY	NRY	Mk 2	Track inspection train coach
977983	3407, 72503	NRY	NRY	Mk 2	Electrification measurement coach
977984	40501	NRY	NRY	HST Mk 3	New Measurement Train
977985	6019, 72715	NRY	NRY	class 488/3	Structure gauging train coach
977986	3189, 99664	NRY	NRY	Mk 2	Structure gauging train coach
977993	44053	NRY	NRY	HST Mk 3	New Measurement Train
977994	44087	NRY	NRY	HST Mk 3	New Measurement Train
977995	40719, 40619	NRY	NRY	HST Mk 3	New Measurement Train
977997	72613, 6126	NRY	NRY	class 488/3	Radio survey coach
999550		NRY	NRY	Mk 2	Track recording coach
999602	62483	NRY	NRY	class 421	Ultrasonic test coach
999605	62482	NRY	NRY	class 421	Ultrasonic test coach
999606	62356	NRY	NRY	class 432	Ultrasonic test coach

NR reduced costs by eliminating top and tail on some operations and replacing the second loco with a driving brake. 9701 is one of five DBSOs in the NR fleet, seen arriving at Spalding on 16 August 2017. *Pip Dunn*

Several locos have been sold, reallocated or leased for use abroad, mostly since privatisation.

When DB Schenker took over EWS, it started to redeploy surplus Class 66s in France and Poland and Class 92s in Bulgaria and Romania. It has since sold some of the latter.

It also did hire deals for Class 37s in Spain, France and Italy, deals that have all long since finished, while several 56s and 58s were also hired for infrastructure projects in France, which have also been concluded.

Several Class 58s were also moved to Spain, some of which were then sold to Transfesa and subsequently scrapped.

Freightliner has also moved several spare Class 66s to Poland, while Class 56s, 86s and 87s, redundant in the UK, have been sold abroad to Hungary and Bulgaria by Europhoenix. Class 91s could follow.

A single Class 47 has been sent to Hungary by Nemesis Rail.

Pre-privatisation exports

Many shunters of Class 03, 04, 06, 07, 08 and 14 were sold to Italy, Spain, Belgium, France and Libera in the 1960s and 1970s. All seven Class 77s DC electric locos were sold to the Netherlands in 1968.

Most have been scrapped now, although a handful have been repatriated for preservation. The following three locos are *thought* to still be intact abroad, although the Class 08s in Liberia may well have been scrapped now, having not been used for decades.

Class 03

D2156	03156	ITY	BLU

Class 08

D3047	13047	LAM	LAM
D3092	13092	LAM	LAM

Post-privatisation exports

Class 47

A single Class 47, 47375, was leased by Nemesis Rail to Hungarian operator Continental Rail solutions in 2015. The deal could have seen three other locos follow it (47488/701/744) but this never happened. The loco has since been spot-hired to a couple of different operators, and most recently has been used to haul coal trains to the Romanian border.

The loco is still on TOPS in the NRLO pool.

92 70 00 47375-5	47375	CSM	CON	HUN *Falcon*

Class 56

A fleet of 30 Class 56s were hired by EWS to Fertis between 2004-07 and all returned to EWS and were then sold.

Two of those locos, along with a third that had previously been sold for preservation, were then sold by Europhoenix to Hungarian operator Floyd. One loco is used for spares only, and the other two were used for freight trains in the country. Floyd was renamed Eurogate Rail Hungary and the ex-BR locos of Class 56 and 86 were sold in 2021 to V-Hid, a civil engineering company employed in line renewals.

92 55 0659 001-5	56101	FLY	HVID	HUN
92 55 0659 002-3	56115	FLY	HVID	HUN
92 55 0659 003-1	56117	FER	HVID	HUN (U)

Floyd took three Class 56s from the UK, of which two were reactivated. This is 659 002, formerly 56115, seen at Celldömölk on 6 June 2016. *Keith Fender*

Class 58

Class 58s were hired by EWS to ACTS in the Netherlands, Seco Rail and Fertis in France and GIF and Continental Rail in Spain. Nine of the Spanish locos were later sold to Transfesa, of which eight – 58015/020/024/029-031/043/047 – were scrapped in late 2019 and early 2020.

Of 28 locos left in Europe, 24 remain stored in France and four in Spain. The French locos are owned by DB Cargo.

	58001	WQCA	DBC	ETF	AZ (U)
	58004	WQCA	DBC	TSO	AZ (U)
	58005	WQCA	DBC	ETF	AZ (U)
	58006	WQCA	DBC	ETF	AZ (U)
	58007	WQCA	DBC	TSO	AZ (U)
	58009	WQCA	DBC	TSO	AZ (U)
	58010	WQCA	DBC	TSO	AZ (U)
	58011	WQCA	DBC	TSO	AZ (U)
	58013	WQCA	DBC	ETF	AZ (U)
	58018	WQCA	DBC	TSO	AZ (U)
	58021	WQCA	DBC	ETF	AZ (U)
	58025	WQCA	DBC	CON	AB (U)
	58026	WQCA	DBC	TSO	AZ (U)
L52	58027	WQCA	DBC	CON	AB (U)
	58032	WQCA	DBC	ETF	AZ (U)
	58033	WQCA	DBC	TSO	AZ (U)
	58034	WQCA	DBC	TSO	AZ (U)
	58035	WQCA	DBC	TSO	AZ (U)
	58036	WQCA	DBC	ETF	AZ (U)
5814	58038	WCQA	TFA	ETF	AZ (U)
5811	58039	WQCA	DBC	ETF	AZ (U)
	58040	WQCA	DBC	TSO	AZ (U)
L36	58041		TFA	CON	AB (U)
	58042	WQCA	DBC	ETF	AZ (U)
5812	58044	WQCA	DBC	ETF	WP (U)
	58046	WQCA	DBC	TSO	AZ (U)
	58049	WQCA	DBC	ETF	AZ (U)
L53	58050	WQCA	DBC	CON	AB (U)

Class 66

Surplus traction in the UK saw both EWS (now DB Cargo) and then Freightliner redeploying Class 66s. Sixty-four went to France with the former, while 15 EWS locos and 18 Freightliner locos have moved to Poland.

The DBC locos occasionally returned to the UK for heavy maintenance, although such transfers were few and far between in recent times. However, more recently DBC has started to return locos to the UK, with 66010/028/032/073/179/190/205/224/244 the first to come back from France.

The Freightliner Poland locos have been renumbered in the 660xx series for Class 66/5s and 6660x series for Class 66/6s. FPL also owns seven 'Class 66s' numbered 66001-007, which were new-build locos and did not work in the UK.

DB Cargo's Euro Cargo Rail subsidiary also operates 60 'Class 66s', numbered 77001-060, which were new-build locos and did not work in the UK. All WGEP locos have their swinghead couplers removed, while WGEA locos retain them.

92 70 0 066022-9	66022	WGEA	EWS	DBC	AZ
92 70 0 066026-0	66026	WGEA	EWS	DBC	AZ
92 70 0 066029-4	66029	WGEA	EWS	DBC	AZ
92 70 0 066033-6	66033	WGEA	EWS	DBC	AZ
92 70 0 066036-9	66036	WGEA	EWS	DBC	AZ
92 70 0 066038-5	66038	WGEA	EWS	DBC	AZ
92 70 0 066042-7	66042	WGEA	EWS	DBC	AZ
92 70 0 066045-0	66045	WGEA	EWS	DBC	AZ
92 70 0 066049-2	66049	WGEA	EWS	DBC	AZ
92 70 0 066052-6	66052	WGEA	EWS	DBC	AZ
92 70 0 066062-5	66062	WGEA	EWS	DBC	AZ
92 70 0 066064-1	66064	WGEA	EWS	DBC	AZ
92 70 0 066071-6	66071	WGEA	EWS	DBC	AZ
92 70 0 066072-4	66072	WGEA	EWS	DBC	AZ
92 70 0 066123-5	66123	WGEA	EWS	DBC	AZ
92 70 0 066146-6	66146	WGEP	EWS	DBC	PN
92 70 0 066153-2	66153	WGEP	EWS	DBC	PN
92 70 0 066157-3	66157	WGEP	EWS	DBC	PN
92 70 0 066159-9	66159	WGEP	EWS	DBC	PN
92 70 0 066163-1	66163	WGEP	DBR	DBC	PN
92 70 0 066166-4	66166	WGEP	EWS	DBC	PN
92 70 0 066173-0	66173	WGEP	EWS	DBC	PN
92 70 0 066178-9	66178	WGEP	DBR	DBC	PN
92 70 0 066180-5	66180	WGEP	EWS	DBC	PN
92 70 0 066189-6	66189	WGEP	DBR	DBC	PN
92 70 0 066191-2	66191	WGEA	EWS	DBC	AZ
92 70 0 066193-8	66193	WGEA	EWS	DBC	AZ
92 70 0 066195-3	66195	WGEA	EWS	DBC	AZ
92 70 0 066196-1	66196	WGEP	EWS	DBC	PN
92 70 0 066201-9	66201	WGEA	EWS	DBC	AZ
92 70 0 066202-7	66202	WGEA	EWS	DBC	AZ
92 70 0 066203-5	66203	WGEA	EWS	DBC	AZ
92 70 0 066204-3	66204	WGEA	EWS	DBC	AZ
92 70 0 066208-4	66208	WGEA	EWS	DBC	AZ
92 70 0 066209-2	66209	WGEA	EWS	DBC	AZ
92 70 0 066210-0	66210	WGEA	EWS	DBC	AZ
92 70 0 066211-8	66211	WGEA	EWS	DBC	AZ
92 70 0 066212-6	66212	WGEA	EWS	DBC	AZ
92 70 0 066213-4	66213	WGEA	EWS	DBC	AZ
92 70 0 066214-2	66214	WGEA	EWS	DBC	AZ
92 70 0 066215-9	66215	WGEA	EWS	DBC	AZ
92 70 0 066216-7	66216	WGEA	EWS	DBC	AZ
92 70 0 066217-5	66217	WGEA	EWS	DBC	AZ
92 70 0 066218-3	66218	WGEA	EWS	DBC	AZ
92 70 0 066219-1	66219	WGEA	EWS	DBC	AZ
92 70 0 066220-9	66220	WGEP	DBR	DBC	PN
92 70 0 066222-5	66222	WGEA	EWS	DBC	AZ
92 70 0 066223-3	66223	WGEA	EWS	DBC	AZ
92 70 0 066225-8	66225	WGEA	EWS	DBC	AZ
92 70 0 066226-6	66226	WGEA	EWS	DBC	AZ
92 70 0 066227-4	66227	WGEP	DBR	DBC	PN
92 70 0 066228-2	66228	WGEA	EWS	DBC	AZ
92 70 0 066229-0	66229	WGEA	EWS	DBC	AZ
92 70 0 066231-6	66231	WGEA	EWS	DBC	AZ
92 70 0 066232-4	66232	WGEA	EWS	DBC	AZ
92 70 0 066233-2	66233	WGEA	EWS	DBC	AZ
92 70 0 066234-0	66234	WGEA	EWS	DBC	AZ
92 70 0 066235-7	66235	WGEA	EWS	DBC	AZ
92 70 0 066236-5	66236	WGEA	EWS	DBC	AZ
92 70 0 066237-3	66237	WGEP	EWS	DBC	PN
92 70 0 066239-9	66239	WGEA	EWS	DBC	AZ
92 70 0 066240-7	66240	WGEA	EWS	DBC	AZ
92 70 0 066241-5	66241	WGEA	EWS	DBC	AZ
92 70 0 066242-3	66242	WGEA	EWS	DBC	AZ
92 70 0 066243-1	66243	WGEA	EWS	DBC	AZ
92 70 0 066245-6	66245	WGEA	EWS	DBC	AZ
92 70 0 066246-4	66246	WGEA	EWS	DBC	AZ
92 70 0 066247-2	66247	WGEA	EWS	DBC	AZ
92 70 0 066248-9	66248	WGEP	DBR	DBC	PN
92 70 0 066249-8	66249	WGEA	EWS	DBC	AZ
66013	66411	DHLT	FPH	AIK	FP
66015	66412	DHLT	FPH	AIK	FP
66014	66417	DHLT	FPH	AIK	FP
66016	66527	DHLT	FLR	EVS	FP
66017	66530	DHLT	FLR	POR	FP
66018	66535	DHLT	FLR	POR	FP
66009	66582	DHLT	FLR	EVS	FP
66010	66583	DHLT	FLR	EVS	FP
66011	66584	DHLT	FLR	EVS	FP
66008	66586	DHLT	FLR	AIK	FP
66595		DHLT	FLR	BEA	FP
66603	66608	DHLT	FLR	POR	FP
66605	66609	DHLT	FLR	POR	FP
66604	66611	DHLT	FLR	POR	FP
66606	66612	DHLT	FLR	POR	FP
66602	66624	DHLT	FLR	AIK	FP
66601	66625	DHLT	FLR	AIK	FP
66954		DHLT	FLR	BEA	FP

Class 86

Several Class 86s, made redundant by Virgin, Anglia Railways and Network Rail, have been exported to Bulgaria and Hungary.

Floyd in Hungary had nine locos, all of which have been returned to traffic, including 86424, which was originally earmarked as just being a source of spares. They were sold to H-Vid in late 2021 for use on infrastructure work.

Bulmarket in Bulgaria has a mixed Class 86/87 fleet. It has six operational locos of the former and one spares donor, 86233. All locos were exported via Europhoenix.

New No.	BR nos	Livery	Operator	Country	Name
91 52 00 85003-2	86213	BMT	BMT	BUL	*Lancashire Witch*
91 55 0450 005-6	86215	FLY	HVID	HUN	
91 55 0450 006-6	86217, 86504	FLY	HVID	HUN	
91 55 0450 004-1	86218	FLY	HVID	HUN	
91 55 0450 007-4	86228	FLY	HVID	HUN	
9152 00 85005-2	86231	BMT	BMT	BUL	*Lady of the Lake*
91 55 0450 003-3	86232	FLY	HVID	HUN	
	86233, 86506	EBY	BMT	BUL (U)	
9152 00 85006-2	86234	BMT	BMT	BUL	
9152 00 85004-7	86235	BMT	BMT	BUL	*Novelty*
91 55 0450 008-2	86242	FLY	HVID	HUN	
91 55 0450 001-7	86248	FLY	HVID	HUN	
91 55 0450 002-5	86250	FLY	HVID	HUN	
91 55 0450 009-0	86424, 86324, 86024	FLY	HVID	HUN	
91 52 00 85001-6	86701, 86503, 86205	BMT	BMT	BUL	*Orion*
91 52 00 85002-4	86702, 86260, 86048	BMT	BMT	BUL	*Cassiopeia*

Once in use in the UK as 86248, this is one of nine Class 86s sold to Floyd – and renumbered as 450.001. It was at Debrecen Csapokert on 13 September 2011, and has since been sold to H-Vid. *Keith Fender*

Class 87

Redundant former Virgin West Coast Class 87 were the first locos Europhoenix sold abroad, with 17 sent to Bulgarian rail operator BZK, although not all were returned to traffic and some have since been withdrawn from use.

Bulmarket, also in Bulgaria, has taken four locos, which are in traffic.

New No.	BR no.	Livery	Operator	Country	Name
91 52 00 87003-7	87003	BZK	BZK	BUL	
91 52 00 87004-5	87004	BRZ	BZK	BUL	*Britannia*
91 52 00 87006-0	87006	DGB	BZK	BUL (U)	
91 52 00 87007-8	87007	COT	BZK	BUL	
87008-9	87008	COT	BZK	BUL (U)	
91 52 00 87009-4	87009	BMT	BMT	BUL	
91 52 00 87010-2	87010	BZK	BZK	BUL	
91 52 00 87012-8	87012	NSE	BZK	BUL	
91 52 00 87013-6	87013	BZK	BZK	BUL (U)	
87014-7	87014	BZK	BZK	BUL (U)	
91 52 00 87017-7	87017	EPX	BMT	BUL	*Iron Duke*
91 52 00 87019-3	87019	LNR	BZK	BUL	
91 52 00 87020-1	87020	BZK	BZK	BUL	
91 52 00 87022-7	87022	DGB	BZK	BUL (U)	

Several former BR Class 87s are now working with BZK in Bulgaria. 87033 and BDZ (State Railway) Skoda 44 143 stand side-by-side at the BZK Depot in Sofia on 8 October 2011. *Keith Fender*

91 52 00 87023-5	87023	EPX	BMT	BUL	*Velocity*
91 52 00 87025-0	87025	BMT	BMT	BUL	
91 52 00 87026-8	87026	BZK	BZK	BUL	
91 52 00 87028-4	87028	DRS	BZK	BUL	
91 52 00 87029-2	87029	BZK	BZK	BUL	
91 52 00 87033-4	87033	BZK	BZK	BUL	
91 52 00 87034-2	87034	BZK	BZK	BUL	

Class 92

The story of Class 92s abroad is slightly different. When DB took over the EWS operation it inherited 30 Class 92s, of which only a fraction could be employed in the UK. Accordingly, in May 2013 it started to move surplus locos to Romania and Bulgaria.

Some of these have now been sold to Russian company LocoTech, initially for use by Transagent Spedicija in Croatia, and have been repainted into Transagent black or red. In February 2022 the locos were hauled from Zagreb to the Transmashholding (part owned by LocoTech) works at Dunakeszi in Hungary pending developments.

All remain in the WGEE pool.

New no.	BR no.	Livery	Owner	Country	Name
91 53 0 472 002-1	92001	DBR	DBC	CRO	*Mircea Eliade*
91 53 0 472 003-9	92002	DBC	LocoTech	HUN (S)	
	92003	EUE	DBC	ROM (S)	*Beethoven*
91 53 0 472 005-4	92005	CRO	LocoTech	HUN (S)	
91 53 0 472 001-3	92012	TRA	LocoTech	HUN (S)	
	88002, 92022	EUE	DBC	ROM (S)	*Charles Dickens*
91 53 0 472 004-7	92024	TRA	LocoTech	HUN (S)	
91 70 00 92025-1	92025	EUE	DBC	BUL	*Oscar Wilde*
	92026	EUE	DBC	ROM (S)	*Britten*
91 70 00 92027-7	92027	EUE	DBC	HUN (S)	*George Eliot*
91 52 16 88030-1	92030	EUE	DBC	HUN (S)	*Ashford*
91 70 00 92034-3	92034	EUE	DBC	HUN (S)	*Kipling*
91 53 0 472 006-2	92039	DBR		ROM (S)	*Eugen Ionescu*

There are several established preservation groups that have returned their locomotives to main-line registration. Some are used mostly for charters, while others have enjoyed short- and longer-term hire contracts with TOCs and FOCs. Individuals who own locos that are main-line registered and on long-term hire are included in the TOC/FOC that uses them.

Class 20 Locomotive Society

Contact details

Website: https://class20locosociety.co.uk

Overview

Although in partnership with Michael Owen, the CTLS owns three Class 20s, two of which are main-line registered: 20205/227. The former is on a long-term hire agreement to Owen, while 20227 has been on hire to the North Norfolk Railway for dining trains that the NNR runs through to Cromer on the national network, top and tailing with steam. It has had spells on hire to GBRf, DB Cargo and Vintage Trains. The loco's wheelset are due for replacement, which could see it retired from the main line or only used occasionally.

20227	MOLO	MRM	CTL	SK	*Sherlock Holmes*

On Sunday, 8 August 2021, the Class 20 Locomotive Society's 20227 *Sherlock Holmes* and Michael Owen's 20189 have just arrived with the 1252 Stratford-on-Avon-Birmingham Snow Hill 'Shakespeare Express', filling in for an unavailable steam loco. *Pip Dunn*

The Class 40 Preservation Society owns three Class 40s, one of which, 40145 (D345) is main-line registered. Following a recent repaint back into original BR green, it looks truly stunning with classmate D213 *Andania* as they pass Docker with a Crewe to Edinburgh charter on 13 November 2021. *Tom McAtee*

Class 40 Preservation Society

Contact details

Website: www.cfps.co.uk

Overview

The CFPS owns three Class 40s, one of which is main-line registered: 40145. It has been used solely for charter trains and loco-positioning moves. In August 2020, the loco went on to LSL, a deal that was ended in January 2022, when it returned to its home of the East Lancashire Railway. It currently bears the number D345.

D345	40145	CFSL	GYP	CFP	BQ

Main-line Preservation Groups

Class 50 Alliance

Contact details

Website: www.fiftyfund.org.uk

Overview

The C50A has six Class 50s under its care, of which three are main-line registered. 50044 *Exeter* returned to main-line use in 2021 and joins 50007 *Hercules* and 50049 *Defiance*, which are used for the majority of the C50A's main-line activities. As well as working charters, the locos are available to hire for other companies, most notably GB Railfreight, whose livery 50007/049 currently carry. They have been used for occasional ad hoc stock and freight work for GBRf but have also been used for route learning with CrossCountry Trains. They have also been used by Vintage Trains and DBC.

The group also owns 50031 *Hood*, 50035 *Ark Royal* and has a long-term arrangement to look after 50033 *Glorious*. Of these, 50031 was once main-line registered but has not run as such since 2007. The C50A has previously had hire contacts with Valley Lines and Arriva Trains Wales.

The first known visit by Class 50s to Stranraer was on 11 September 2021 when the Class 50 Alliance's 50049 *Defiance* and 50007 *Hercules* arrived on a Pathfinder Tours charter. *Pip Dunn*

50007	CFOL	GBR	CFA	KR	*Hercules*
50044	CFOL	BRB	CFA	KR	*Exeter*
50049	CFOL	GBR	CFA	KR	*Defiance*

Class 56 Group

Contact details

Website: www.class56group.co.uk

Overview

The C56LG owns two Class 56s: preserved 56006 based at the East Lancashire Railway and main-line registered 56301, which has been used by UK Rail leasing for spot hire. After recent wheelset and traction motor repairs, the loco is once again available for hire.

56301	56045	UKRL	JFU	CFS	LR

Crewe Diesel Preservation Group

The group owns 47712 *Lady Diana Spencer*, which is on long-term hire to Locomotive Services (see page 186).

Deltic Preservation Society

Contact details

Website: http://thedps.co.uk

Overview

The DPS owns three Class 55s, one of which is main-line registered. 55009 *Alycidon* is close to returning to main-line use following repairs at Barrow Hill after a serious failure in March 2019.

Of its other locos, 55019 *Royal Highland Fusilier* was previously main-line registered but does not have OTMR, although an appeal has been launched to return it to the main line. 55015 *Tulyar* in undergoing a protracted full rebuild and is being returned to main-line standard, but whether it has the necessary equipment fitted to allow it to operate as such is yet to be decided.

55009	DBLX	BRB	DPS	BH (U)	*ALYCIDON*

The Deltic Preservation Society returned two of its locos to the main line in 1999, and since then has usually had one of either 55009 *Alycidon* or 55019 *Royal Highland Fusilier* main-line compliant. On 16 June 2018, 55009 waits to leave with the 1523 King's Cross-Edinburgh charter. *Pip Dunn*

Main-line Preservation Groups

D05 Preservation Group

Contact details

Website: www.facebook.com/D05pres

Overview

The group owns 37688 *Great Rocks* and 47828, which are both on long-term hire to Locomotive Services (see page 186).

Diesel Traction Group

Contact details

Website: www.westernchampion.co.uk

Overview

One of the pioneer diesel preservation groups set up in the late 1970s, it owns the only surviving Class 17 Clayton D8568, Class 35 Hymek D7029, Class 42 Warship D821 *Greyhound* and Class 52 Western D1015 *Western Champion*.

The Class 52 is main-line compliant and the only Class 52 passed to haul trains on Network Rail. It has been out of use for a few years now but has returned to operation at its home, the Severn Valley Railway. It was expected the loco would return to action in 2020 – the first BR blue Western on the main line since February 1977 following a recent repaint – but it failed on a main-line test run in September and has had to be stopped for an engine swap. A main-line return is highly likely for 2022.

D1015	MBDL	BYP	DTG	KR	*Western Champion*

D1015 *Western Champion* is the only Class 52 Western passed for main-line operation. After an engine overhaul it powers through Gossington, Gloucestershire, as part of a test run piloting 66719 *Metro-Land* on an Avonmouth Docks to Penyffordd cement train. Sadly the loco did not pass the test run after experiencing an engine failure. *Jack Boskett*

Les Ross

Contact details

Website: www.86259lesross.com

Overview

Former radio DJ Les Ross owns 86259 *Peter Pan*, which is often used by WCR, especially from Euston to Preston on 'Cumbrian Mountain Express' charters. Despite WCR now acquiring 86401 *Mons Meg*, it is expected 86259 will continue to find use on these trains.

86259	86045	MBEL	EBY	LES	WN	*Les Ross/Peter Pan*

National Railway Museum

Contact details

Website: www.railwaymuseum.org.uk

Overview

The NRM at York has two locos that are main-line registered: 47798 *Prince William* and 55002 *The King's Own Yorkshire Light Infantry*. However, following a change of policy both are now on display only and not expected to see any use as such. The Class 55 is also currently only able to run on one engine, so would be precluded from hauling charter trains.

Class 09

This loco is regarded as the museum's shunter rather than a restored exhibit. It is passed to run on the national network on a restricted basis such as to York station.

09017	NRM	NRM	YK

Class 47

The ex-Royal Train loco has previously been used on the main line but is now on static display at York.

47798	47072, 47609, 47834	MBDL	RTP	NRM	YK	*Prince William*

Class 55

The loco was returned to main line registration, but only worked one train before suffering an engine failure. Restricted to one engine, it was still able to work occasional loco moves but was not allowed to run charters. Following the reorganisation of the NRM, it is now regarded as a static exhibit.

55002	DBLX	GYP	NRM	YK (S)	*The King's Own Yorkshire Light Infantry*

North Yorkshire Moors Railway

Contact details

Website: www.nymr.co.uk

Overview

The NYMR operates 'main-line' trains on the Network Rail line between Battersby and Whitby. Most such trains are actually only operated between Grosmont and Whitby and use steam, but Class 25 D7628 (25278) is also approved for running on the Battersby to Middlesbrough section of the Esk Valley Line. It is currently out of traffic undergoing an overhaul.

It is possible one of the line's two Class 24s may also be upgraded for running on this line but presently both are undergoing overhaul.

25278	MBDL	NYM	GYP	GO (U)	*Sybilla*

Main-line Preservation Groups

Scottish Class 37 Group

Contact details

Website: https://37025.tumblr.com

Overview

The S37G owns 37025 *Inverness TMD*, currently on long-term hire to Colas Rail Freight. It has worked a handful of charter trains operated by GB Railfreight. It is based at the Bo'ness & Kinneil Railway when not in use with Colas.

The loco has an operational steam heat boiler, the only main-line diesel to have this feature, and has also been modified to have through ETS wiring. The S37G also owns 37261, which is being restored, and parts donor 37214.

Scottish Railway Preservation Society

Contact details

Website: www.srps.org.uk

Overview

The SRPS owns 37403 *Isle of Mull*, which spent 2016-20 on hire to DRS. It was returned to its home at Bo'ness in July 2020 for an engine overhaul. It returned to traffic in late 2021, retains full main-line registration and is booked to haul some SRPS charters trains, in 2022, operated most likely by either GBRf or WCR. It is due to go on hire to the NYMR.

37403	37307	RAJV	BLL	SRP	BKR	*Isle of Mull*

Class 25 D7628 *Sybilla* is main-line registered but restricted to just working for the North Yorkshire Moors Railway between Battersby to Whitby. On 25 June 2015 it pauses at Glaisdale with the 1214 Battersby-Grosmont. *Pip Dunn*

Stratford Class 47 Group

Contact details

Website: www.stratford47group.co.uk

Overview

The S47G owns three Class 47s: preserved 47367 / 596 at the Mid Norfolk Railway and main-line-registered 47580 *County of Essex*. The latter has enjoyed several years available for spot hire, the vast majority of its work being with WCR where its dual brake capability was incredibly useful to the TOC.

Currently the loco is out of traffic at the Mid Norfolk Railway at Dereham and while it retains main-line registration, its hire deal with WCR has ceased. It has thin tyres, which limits what work it can do.

47580 47167, 47732 MBDL BRF SFG MNR *County of Essex*

71A Locomotive Group

Contact details

Website: www.71alocogroup.co.uk

Overview

The Swanage Railway-based group owns Class 33 D6515 *Lt Jenny Smith RN*. The loco is used mostly for occasional charters operated by either GBRf or WCR. The loco bears the number D6515. It is recently undergone a cab and body overhaul at Eastleigh Works.

33012 MBDL GYP SOA SWR *Lt Jenny Lewis RN*

Class 33 D6515 *Lt Jenny Lewis RN*, previously 33012, is owned by the 71A Locomotive Group and is used occasionally on main-line charters. On 17 August 2019 it leaves Westbury with the ECS off a Waterloo-Warminster charter top and tailed with 73107 *Tracy*. *Glen Batten*

Hastings Diesels

Contact details

Website: www.hastingsdiesels.co.uk

Overview

One of the few groups to have a main-line registered heritage multiple unit, this group owns and operates a six-car 'Hastings' Class 201 DEMU.

Only five of its seven cars are Class 201 vehicles, as the TRSB and TSL 70262 are former EMU vehicles.

The unit is used regularly on charters across the country, although usually in the south of England, with GBRf the usual train operator it uses.

			DMBS	TSL	TSL	TRSB	TSL	DMBS	
201001	GRE	SE	60116	60529	70262	69337	60501	60118	*Mountfield/Tunbridge Wells*
Spare			60000						*Hastings*

One of the few main-line preserved multiple units, Class 202 Hastings DEMU, with DMBSO S60118 *Tunbridge Wells* leading, passes Little Langford with the 'Ludgershall Legionnaire' charter on 14 September 2019. *Glen Batten*

Works, Depots and Maintenance Facilities

The old locomotive works of Swindon, Doncaster, Glasgow et al have changed dramatically over recent years. Many of the famous establishments have been closed, rationalised or changed function. Some still survive, albeit much changed.

There are several new depots as train operators often acquire rolling stock based on contracts where maintenance is provided by the manufacturer.

A lot of the codes used here are the official TOPS ones, but some are purely for the storage sites currently being used to house rolling stock.

Depot	Code	Operators	Services
Aberdeen Clayhills	AC	ScotRail, LNER	Servicing, stabling
Allerton Liverpool	AN	Northern	Maintenance
Ashford	AD	Hitachi	Maintenance
Ashford Chart Leacon	AF	Balfour Beatty	Storage
Aylesbury	AL	Chiltern Railways	Maintenance
Barrow Hill	BH	Barrow Hill Engine Shed Society	Maintenance
Barton-Under-Needwood	LH	LH Group Services (Wabtec)	Maintenance
Bedford Cauldwell Walk	BF	Thameslink	Maintenance
Bescot	BS	DB Cargo	Railhead
Bicester	BC	MoD	Storage
Birkenhead North	BD	Stadler (Merseyrail)	Maintenance
Bletchley	BY	London North Western Railway	Stabling
Bo'ness	BO	Scottish Railway Preservation Society	Maintenance
Botanic Gardens Hull	BG	Northern	Stabling
Bournemouth West	BM	South Western Railway	Maintenance
Bounds Green	BN	Hitachi (LNER)	Maintenance
Brighton Lovers Walk	BI	Govia Thameslink	Maintenance
Bristol Barton Hill	BK	Arriva Traincare	Maintenance
Bristol St Philip's Marsh	PM	Great Western Railway	Maintenance
Burton Central Rivers	CZ	Bombardier (Cross Country Trains)	Maintenance
Burton-on-Trent	BU	Nemesis Rail	Maintenance
Barrow-in-Furness		Northern	Stabling
Cambridge	CA	CrossCountry Trains	Stabling
Carnforth Steamtown	CS	West Coast Railways	Maintenance
Cardiff Canton	CF	Transport for Wales	Maintenance
Carlisle Kingmoor	KM	Direct Rail Services	Maintenance
Cheriton	CT	Eurotunnel	Stabling
Chester	CH	Alstom (Transport for Wales)	Maintenance
Clacton	CC	Greater Anglia	Storage
Clapham Junction	CJ	South Western Railway	Storage
Colchester	CR	Greater Anglia	Storage
Coquelles	CO	Eurotunnel	Maintenance
Corkerhill	CK	ScotRail	Maintenance
Craigentinny Edinburgh	EC	ScotRail, LNER	Maintenance
Crewe Basford Hall	CX	Freightliner	Maintenance
Crewe Carriage Shed	CL	Arriva	Maintenance
Crewe DMD	CD	Locomotive Services Ltd	Maintenance
Crewe International	CE	DB Cargo	Maintenance
Crewe Gresty Bridge	CB	Direct Rail Services	Maintenance
Crewe LNWR	CL	LNWR Heritage Centre	Maintenance
Crewe Works	ZC	Bombardier	Works
Crofton	XW	Bombardier	Maintenance
Derby RTC	ZA	Loram	Maintenance
Derby Litchurch Lane	ZD	Bombardier	Manufacturing
Derby Etches Park	DY	East Midlands Railway	Maintenance
Doncaster Works	ZF	Wabtec	Works
Doncaster Belmont Yard	DB	DB Cargo	Maintenance
Doncaster Carr	DN	Hitachi	Maintenance
Doncaster RMT	DM		Storage

Doncaster Roberts Road	RR	Progress Rail	Maintenance
Donnington	DO		Storage
Dundee	DE	ScotRail	Storage
Eastfield Glasgow	ED	ScotRail	Servicing, stabling
East Ham	EM	c2c	Maintenance
Eastleigh	EH	DB Cargo	Maintenance
Eastleigh Works	ZG	Arlington Fleet Group	Works
Edge Hill (Liverpool)	EG	Alstom	Maintenance
Ely Papworth	EY	Potter Group	Storage
Exeter	EX	Great Western Railway	Maintenance
Felixstowe	FX	Freightliner	Railhead
Fort William	FW	West Coast Railways, GB Railfreight	Stabling
Fratton (Portsmouth)	FR	South Western Railway	Stabling
Gascoigne Wood	GA		Storage
Gillingham	GI	Southeastern	Stabling
Great Yarmouth	GY	Eastern Rail services	Stabling
Glasgow Springburn	ZH	Knorr Bremse	Works
Harwich Parkeston Quay	HW	Greater Anglia	Storage
Haymarket Edinburgh	HA	ScotRail	Maintenance
Heaton Newcastle	HT	Northern, LNER, Grand Central	Maintenance
Headquarters	HQ		
Hornsey	HE	Thameslink	Maintenance
Holyhead	HD	Transport for Wales	Stabling
Hoo Junction	HJ	Colas Rail Freight	Stabling
Hope Cement Works	HO	Lafarge	Storage
Ilford	IL	Greater Anglia	Maintenance
Ilford Level 5 Works	ZI	Bombardier	Works
Immingham	IM	DB Cargo	Stabling
Inverness	IS	ScotRail	Maintenance
Ipswich	IP	Freightliner	Railhead
Kidderminster	KR	Class 50 Alliance	Maintenance
Kilmarnock	ZK	Wabtec	Works
Kilmarnock	ZM	Brodie Rail Engineering	Works
King's Norton	ZN	SLC Operations	Stabling
Knottingley	KY	DB Cargo	Maintenance
Leeds Midland Road	LD	Freightliner	Maintenance
Leicester	LR	UK Rail Leasing, Europhoenix	Maintenance
Loughborough	LB	Brush Traction (Wabtec)	Site pending sale
Long Marston	LM		Storage
Longport (Stoke-On-Trent)	ZW	Progress Rail	Maintenance, refurbishment
Longtown	LT		Storage
Machynlleth	MH	Transport for Wales	Maintenance
Margam	MG	DB Cargo	Maintenance
Merehead	MD	Aggregates Industries	Maintenance
Motherwell	ML	Direct Rail Services	Servicing, stabling
Manchester Ardwick	AK	Siemens (Transpennine Express)	Maintenance
Manchester Longsight Electric	LG	Northern	Maintenance
Manchester International Traincare	MA	Alstom	Maintenance
Neville Hill Leeds	NL	East Midlands Railway, LNER, Northern	Maintenance
New Cross Gate	NG	London Overground	Maintenance
Newport		CAF	Manufacturing
Newton Aycliffe		Hitachi	Manufacturing
Newton Heath Manchester	NH	Northern	Maintenance
Northampton King's Heath	NN	Siemens	Maintenance
Northam Southampton	NT	South Western Railway	Maintenance
North Pole	NP	Hitachi (Great Western Railway)	Maintenance
Norwich Crown Point	NC	Greater Anglia	Maintenance
Nottingham Eastcroft	NE	East Midlands Railway/Boden Rail Engineering	Maintenance
Old Dalby	OD	Serco	Test track/storage

Old Oak Common	OH	Heathrow Express	Maintenance
Old Oak Common	OF	Crossrail	Maintenance
Oxley	OY	Alstom	Maintenance
Oxford	OX	Great Western Railway	Stabling
Peterborough	PB	DB Cargo	Stabling
Peterborough	PG	GBRf	Maintenance
Plymouth Laira	LA	Great Western Railway	Maintenance
Polmadie	PO	Alstom	Maintenance
Penzance Long Rock	PZ	Great Western Railway	Maintenance
Ramsgate	RM	Hitachi	Maintenance
Reading	RG	Great Western Railway	Maintenance
Redhill			Servicing, stabling
Rugby	RU	Colas	Stabling
Ryde	RY	South Western Railway	Maintenance
Salisbury	SA	South Western Railway	Maintenance
Scunthorpe Steelworks	SS	Southern	Maintenance
Selhurst	SU	Southern	Maintenance
Shields Road Glasgow	GW	ScotRail	Maintenance
Slade Green	SG	Southeastern	Maintenance
Slateford Edinburgh	SF		Storage
Soho	SO	West Midlands Trains	Maintenance
Southall	SH	West Coast Railways	Stabling
Southend	SN	Greater Anglia	Storage
St Blazey	BZ	DB Cargo	Stabling
St Leonards	SE	SLRE	Maintenance
Stewarts Lane	SL	Southern, VSOE	Maintenance
Stourbridge Junction	SJ	Chiltern Railways/West Midlands Railways	Stabling
Southampton Maritime	SM	Freightliner	Railhead
Swansea Landore	LE	Chrysalis Rail	Refurbishment
Swanwick	SK	20189 Ltd	Maintenance
Taff Wells	TW	Transport for Wales	Maintenance (depot under construction)
Temple Mills	TI	Eurostar	Maintenance
Three Bridges	TB	Siemens	Maintenance
Tonbridge	TN	GB Railfreight	Stabling
Toton	TO	DB Cargo	Maintenance
Toton Compound	TC	DB Cargo	Storage
Toton Yard	TY	DB Cargo	Storage
Trafford Park	TP	GBRf	Railhead
Tyne Yard	TZ	LNER	Servicing, stabling
Tyseley	TS	West Midlands Railway	Maintenance
Tyseley Locomotive Works	TM	Vintage Trains	Maintenance
Warrington Arpley	WA	DB Cargo	Maintenance
Wembley Traincare	WB	Alstom	Maintenance
Westbury	WY	DB Cargo	Maintenance
Whitemoor Yard	WH	GBRf	Railhead
Widnes	WX	Alstom	Maintenance
Wigan Springs Branch	SP	Northern	Servicing, stabling
Willesden	WN	London Overground	Maintenance
Wimbledon	WD	South Western Railway	Maintenance
Wolsingham	WO	RMS Locotec	Maintenance
Wolverton Works	ZN	Gemini	Works
Worcester	WS	West Midlands Trains	Stabling
Worksop	WK	HNRC	Storage
Yoker	YO	ScotRail	Stabling
York	YK	Transpennine Express	Stabling

The UK 's new train fleets are mostly built by companies from abroad, although Bombardier, Hitachi and CAF assemble trains in the UK. Canadian company Bombardier does so at Derby Litchurch Lane, Japan's Hitachi at Newton Aycliffe and Spain's CAF at Newport, while the Spanish company Talgo has been looking to start operations at Longannet in Fife.

Bombardier was acquired by French company Alstom in January 2021. Alstom had previously assembled trains in the UK at the now closed Washwood Heath site in Birmingham. These included the majority of the Class 390 Pendolinos, although the bodyshells for these were imported from Italy.

Alstom

Alstom had an initial flurry of orders in the mid-1990s but delays in deliveries, followed by further reliability issues, did not help the firm. This led many TOCs to look elsewhere, at Siemens and Hitachi in particular.

While the company has had no successes in winning business in the recent tranche of new train orders, its recent acquisition of Bombardier will no doubt see the firm back in the running.

Class	Ordering TOC	Type	Unit numbers	Formations	Buying ROSCO	Delivered	Type	Subsequent redeployment
175/0	First North Western	Coradia	175001-011	11 × two-car (22)	Angel	1999-2000	DMU	Transport for Wales
175/1	First North Western	Coradia	175101-116	16 × three-car (48)	Angel	1999-2001	DMU	Transport for Wales
180	First Great Western	Adelante	180101-114	14 × five-car (70)	Angel	2000-01	DMU	180101/105/107/112/114 Grand Central. 180109-111/113 East Midlands Railway
334	ScotRail	Juniper	334001-040	40 × three-car (120)	HSBC	1999-2001		ACEMU
390/0	Virgin West Coast	Pendolino	390001-053	53 × nine-car (477)	Angel	2001-05	ACEMU	390001-034 delivered as eight-cars, increased to nine-cars in 2004-05. Some sets now increased to 11-cars
390/1	Virgin West Coast	Pendolino	390154-157	4 × 11-car (44)	Angel	2010-12	ACEMU	
390/1	Virgin West Coast	Pendolino		31 × two-car (66)	Angel	2010-12	ACEMU	Additional two cars inserted into nine-car Class 390s
458	South West Trains	Juniper	458001-030	30 × four-car (120)	Porterbrook	1998-2000	DCEMU	Sets re-formed as five-car using vehicles from Class 460s, along with six additional five-car. Fleet now being withdrawn
460	Gatwick Express	Juniper	460001-008	8 × eight-car (64)	Porterbrook	1998-99	DCEMU	Now disbanded and converted to Class 458/5s for SWT

Bombardier

Bombardier UK's headquarters are at Derby Litchurch Lane works. The company has consistently assembled new trains for the UK market since the mid-1990s. Although now owned by Alstom, for the time being it its retaining the Bombardier name.

More recently it has been producing the Aventra commuter trains, and 2,660 vehicles have been ordered for the Crossrail Elizabeth line (Class 345), London Overground (Class 710), Greater Anglia (Class 720), South Western Railway (Class 701), West Midlands Trains (Class 730) and c2c (Class 720/6) franchises. These units were designed and built in Britain. The 345s and 710s have been delivered, while the 701, 720 and 730s are still in production. The Class 701s for SWR are very late into traffic.

It built the 557 vehicles for the Class 168/170-172 Turbostar two-, three- and four-car DMU ranges, and the 2,786 vehicles for the Class 357/375-379/387 Electrostar EMU ranges, which come in 25kV AC, 750V DC and dual-voltage variants and three-, four- and five-car configurations.

Bombardier is also responsible for the Class 220, 221 and 222 Voyager/Meridian family of DEMUS that were built for Virgin CrossCountry (Later Arriva's Cross Country Trains) and Midland Mainline (later East Midlands Railways). MML also took on the 222/1s originally ordered by open access operator Hull Trains. In total 495 vehicles were built and delivered, or altered, to run in four-, five-, seven- and nine-car formations.

Away from Heavy rail, Bombardier has recently built 1,751 cars for the London Underground for its Circle, District, Hammersmith & City, Victoria and Metropolitan lines and 162 Flexity light rail vehicles for Manchester Metrolink, Croydon Tramlink and Blackpool Transport.

Class	Ordering TOC	Type	Unit numbers	Formations	Buying ROSCO	Delivered	Type	Subsequent redeployment
168/0	Chiltern Railways	Clubman	168001-005	5 × four-car (20)	Porterbrook	1997/98	DMU	
168/1	Chiltern Railways	Turbostar	168106-110	2 × four-car, 3 × three-car (17)	Porterbrook	2000-02	DMU	
168/1	Chiltern Railways	Turbostar	168111-113	3 × three- car (9)	HSBC	2000-02	DMU	
168/2	Chiltern Railways	Turbostar	168214-219	3 × three-car, 3 × four-car (21)	Porterbrook	2003-06	DMU	
170/1	Midland Mainline	Turbostar	170101-117	17 × two-car (34)	Porterbrook	1998-99	DMU	CrossCountry Trains
170/1	Midland Mainline	Turbostar	centre cars for 55101-110	10 × one-car (10)	Porterbrook	2001	DMU	CrossCountry Trains
170/2	Anglia Railways	Turbostar	170201-208	8 × three-car (24)	Porterbrook	1999	DMU	Transport For Wales
170/2	Anglia Railways	Turbostar	170270-273	4 × two-car (8)	Porterbrook	2002	DMU	Transport For Wales or East Midlands Railway
170/3	South West Trains	Turbostar	170301-309	9 × two-car (18)	Porterbrook	2000-01	DMU	Chiltern as 168321-329 170309 was previously 170399 Spent 11/07-05/15 with Transpennine Express
170/3	Central Trains	Turbostar	170397-398	2 × three-car (6)	Porterbrook	2002	DMU	CrossCountry Trains
170/3	Hull Trains	Turbostar	170393-396	4 × three-car (12)	Porterbrook	2004	DMU	ScotRail
170/4	ScotRail	Turbostar	170401-415	15 × three-car (45)	Porterbrook	1999-2001	DMU	
170/4	ScotRail	Turbostar	170416-424	9 × three-car (27)	HSBC	1999-2001	DMU	170416-420 with East Midlands Railways 170421/423 with Southern as 171201/202 (two-car) with their centre cars and those six vehicles from 170422/424 now with Southern as 171401/402 (four-car)
170/4	ScotRail	Turbostar	170425-434	10 × three-car (30)	Porterbrook	2003-05	DMU	
170/4	ScotRail	Turbostar	170450-461	12 × three-car (36)	Porterbrook	2004-05	DMU	170453-461 Northern
170/4	ScotRail	Turbostar	170470-478	9 × three-car (27)	Porterbrook	2001-05	DMU	170472-478 Northern
170/5	Central Trains	Turbostar	170501-523	23 × two-car (46)	Porterbrook	1999-2000	DMU	West Midlands Trains or East Midlands Railway
170/6	Central Trains	Turbostar	170630-639	10 × three-car (30)	Porterbrook	1999-2000	DMU	170630-635 West Midlands Trains or East Midlands Railway 170636-639 CrossCountry Trains
171/7	Southern	Turbostar	171721-730	10 × two-car (20)	Porterbrook	2003-05	DMU	
171/8	Southern	Turbostar	171801-806	6 × four-car (24)	Porterbrook	2004	DMU	
172/0	LOROL	Turbostar	172001-008	8 × two-car (16)	Angel	2009-10	DMU	West Midlands Trains
172/1	Chiltern Railways	Turbostar	172101-104	4 × two-car (8)	Angel	2009-10	DMU	On loan to West Midlands Railway

172	London Midland	Turbostar	172211-222	12 × two-car (24)	Porterbrook	2010-11	DMU	West Midlands Railway
172	London Midland	Turbostar	172331-345	15 × three-car (45)	Porterbrook	2010-11	DMU	West Midlands Railway
220	Virgin CrossCountry	Voyager	220001-034	34 × four-car (136)	HBOS	2000-01	DEMU	CrossCountry Trains
221	Virgin CrossCountry	Super Voyager	221101-144	40 × five-car 4 × four-car (216)	HBOS	2001-02	DEMU	221119-141 CrossCountry Trains 221101-118/142/143 Avanti West Coast 221144 disbanded. All five-car sets now apart from 221141
222/0	Midland Mainline	Meridian	222001-023	7 × nine-car 16 × four-car (127)	HSBC	2003-05	DEMU	East Midland Railways Re-formed as 6 × seven-car and 17 × five-car
222/1	Hull Trains		222101-104	4 × four-car (16)	HSBC	2005	DEMU	East Midland Railways
357/0	LTS Rail	Electrostar	357001-046	46 × four-car (184)	Porterbrook	1999-2001	ACEMU	c2c
357/2	LTS Rail	Electrostar	357201-228	28 × four-car (112)	Angel	2001/02	ACEMU	c2c
357212-228 now numbered 357322-328								
375/3	Connex SouthEastern	Electrostar	375301-310	10 × three-car (30)	HSBC	2001-02	ACEMU	Southeastern Trains
375/6	Connex SouthEastern	Electrostar	375601-630	30 × four-car (120)	HSBC	1999-2001	ACEMU	Southeastern Trains
375/7	Connex SouthEastern	Electrostar	375701-715	15 × four-car (60)	HSBC	2001-02	ACEMU	Southeastern Trains
375/8	Connex SouthEastern	Electrostar	375801-830	30 × four-car (120)	HSBC	2004	ACEMU	Southeastern Trains
375/9	Connex SouthEastern	Electrostar	375901-927	27 × four-car (108)	HSBC	2003-04	ACEMU	Southeastern Trains
376	Southeastern Trains	Electrostar	376001-036	36 × five-car (180)	HSBC	2004-05	ACEMU	
377/1	Connex South Central	Electrostar	377101-164	64 × four-car (256)	Porterbrook	2002-03	ACEMU	Southern, now Govia Thameslink
377/2	Connex South Central	Electrostar	377201-215	15 × four-car (60)	Porterbrook	2003-04	ACEMU	Southern, now Govia Thameslink
377/3	Connex South Central	Electrostar	377301-328	28 × three-car (84)	Porterbrook	2001-02	ACEMU	Southern Originally numbered 375311-338
377/4	Connex South Central	Electrostar	377401-475	75 × four-car (300)	Porterbrook	2004-05	ACEMU	Southern, now Govia Thameslink
377/5	Southern	Electrostar	377501-523	23 × four-car (92)	Porterbrook	2008-09	ACEMU	Govia Thameslink
377/6	Southern	Electrostar	377601-626	26 × five-car (130)	Porterbrook	2012-13	ACEMU	Govia Thameslink
378/1	LOROL	Capitalstar	377501-523	57 × five-car (285)	QW rail	2009-15	ACEMU	New as three-cars, extended to four-cars and now five-cars
379	Greater Anglia	Electrostar	379001-030	30 × four-car (120)	MQ	2010-11	ACEMU	Now off lease
345	Crossrail	Bombardier Aventra	345001-070	70 × nine-car (630)	345 Rail Leasing	2016-18	ACEMU	Option for 17 × nine-car
377/7	Southern	Electrostar	377701-708	8 × five-car (40)	Porterbrook	2013-14	DVEMU	Govia Thameslink
378	LOROL	Capitalstar	38401-457	57 × one-car (57)	QW Rail	2014-15	EMUMC	Additional cars
387/1	Thameslink	Electrostar	387101-129	29 × four-car (116)	Porterbrook	2014-15	DVEMU	Govia Thameslink
387/1	GWR	Electrostar	387130-174	45 × four-car (180)	Porterbrook	2016-17	DVEMU	387130-141 sub-leased it Heathrow Express
387/2	Southern (Gatwick Express)	Electrostar	387201-227	27 × four-car (108)	Porterbrook	2015-16	DVEMU	Govia Thameslink
387/3	c2c	Electrostar	387301-306	6 × four-car (24)	Porterbrook	2016	DVEMU	
710/1	LOROL	Aventra	710101-130	30 × four-car (120)	TfL	2019	ACEMU	
710/2	LOROL	Aventra	710256-273	18 × four-car (72)	TfL	2019	DVEMU	
710/2	LOROL	Aventra	710374-379	6 × five-car (30)	TfL	2020	DVEMU	
720/1	Greater Anglia	Aventra	720101-144	44 × five-car (220)	Angel	2020	ACEMU	Being delivered
720/5	Greater Anglia	Aventra	720501-589	89 × five-car (445)	Angel	2020	ACEMU	Being delivered
701/1	MTR South West	Aventra	701001-060	60 × ten-car (600)	Rock Rail/SL	2020	DCEMU	Being delivered

Train Builders

701/5	MTR South West	Aventra	701501-530	30 × five-car (150)	Rock Rail/SL	2020	DCEMU	Being delivered
730/0	West Midlands Trains	Aventra	730001-036	36 × three-car (108)	Corelink Rail	2020-21	ACEMU	Being delivered
730/1	West Midlands Trains	Aventra	730101-129	29 × five-car (145)	Corelink Rail	2020-21	ACEMU	Being delivered
730/2	West Midlands Trains	Aventra	730201-216	16 × five-car (80)	Corelink Rail	2020-21	ACEMU	Being delivered
720/6	c2c	Aventra	720601-612	6 × ten-car (60)	Porterbrook	2021	ACEMU	Under construction

CAF

Spanish builder CAF (Construcciones y Auxiliar de Ferrocarriles, which translates as Construction and other railway services) – sold its first trains in the UK market in the late 1990s with the Heathrow Express Class 332s, a joint venture with Siemens, that were delivered in 1997-98. They were recently withdrawn and scrapped. A similar venture was the Class 333s for Arriva Trains Northern, delivered in 2001-03.

However, more recently CAF has ramped up its delivery of new trains to the UK, and has set up an assembly plant in Newport, South Wales, which is now building the Class 196s for West Midlands Trains and Class 197 for Transport for Wales, having built some of the Class 195/331s for Northern.

CAF has also built Class 397s for Transpennine Express and the Mk 5 loco-hauled coaches for both TPE and Caledonian Sleeper.

Class	Ordering TOC	Type	Unit numbers	Formations	Buying ROSCO	Delivered	Type	Subsequent redeployment
332	Heathrow Express	Siemens/CAF	332001-014	14 × four-car (56)	HEx	1997-98	ACEMU	All now scrapped
332	Heathrow Express	Siemens/CAF	72414-418	5 × one-car (5)	HEx	2002	ACEMU	Additional centre cars added to 332005-009 All now scrapped
333	Northern Spirit	Siemens/CAF	333001-016	16 × four-car (64)	Angel	2001-03	ACEMU	Northern Delivered as 16 × three-car, upgraded to four-cars in 2002-03
195/0	Northern	Civity UK	195001-025	25 × two-car (50)	Eversholt	2017-18	DMU	
195/1	Northern	Civity UK	195101-133	33 × three-car (99)	Eversholt	2017-18	DMU	
331/0	Northern	Civity UK	331001-031	31 × three-car (93)	Eversholt	2017-19	ACEMU	
331/1	Northern	Civity UK	331101-112	12 × four-car (48)	Eversholt	2017-19	ACEMU	
397	TPE	Civity UK	397001-012	12 × five-car (60)	Eversholt	2019	ACEMU	
Mk 5	Caledonian Sleeper		15001-011					
			15101-110					
			15201-214					
			15301-340	75 × one-car (75)	Lombard Finance	2018	LHCS	
Mk 5a	TPE		11501-513					
			12701-739					
			12801-814	13 × five + one-car (66)	Beacon Rail	2018	LHCS	
Mk 5a	TPE		11801-813	13 × one-car (13)	Beacon Rail	2018	LHCS (DVT)	
196/0	West Midlands Trains	Civity UK	196001-012	12 × two-car (24)	Corelink Rail	2020	DMU	Being delivered
196/1	West Midlands Trains	Civity UK	196101-114	14 × four-car (56)	Corelink Rail	2020	DMU	Being delivered
197/0	TfW(Wales & Borders)	Civity UK	197001-051	51 × two-car (102)	SMBC	2021	DMU	Being delivered
197/1	TfW (Wales & Borders)	Civity UK	197101-126	26 × three-car (78)	SMBC	2021	DMU	Being delivered

Hitachi

Hitachi delivered its first UK trains in 2006-09 for Southeastern for the HS1 operations out of St Pancras International to Kent. A fleet of 29 six-car Class 395 Javelin units, and a new depot at Ashford to maintain them, were delivered.

The company's biggest success has been its AT300 platform, which uses similar bodyshells as the 395s and forms the InterCity Express Programme – the trains to replace the HSTs and Class 91s on the East Coast and Great Western franchises.

Between LNER and GWR, 1,237 vehicles have been ordered in a mix of electric and bi-mode units of five-, six- and nine-car formations of Classes 800, 801, 802.

The success of these trains, some of which have been built in Kasado in Japan but most of which have been assembled at Newton Aycliffe in Country Durham, has led to orders from five other operators. Transpennine Express took 19 five-car Class 802/2s and Hull Trains has five five-car Class 802/3s in traffic, as does Open Access operator Lumo (Class 803). It is now building for Avanti West Coast (Class 805/807) and East Midlands Railways (Class 810).

As well as the A300s, Hitachi also offers the AT200 series of commuter and outer suburban EMUs, which has so far only yielded one order with ScotRail taking 234 vehicles in a mix of three- and four-car sets. All are now in traffic.

Hitachi was named as part of a joint venture with Alstom to build the initial 54 eight-car HS2 trains.

Class	Ordering TOC	Type	Unit numbers	Formations	Buying ROSCO	Delivered	Type	Subsequent redeployment
395	Southeastern	Javelin	395001-029	29 × six-car (174)	Angel	2006-09	ACEMU	
385/0	ScotRail	AT200	385001-046	46 × three-car (138)	Caledonian Rail Leasing	2017-19	ACEMU	Being delivered; option for 10 × three-car sets
385/1	ScotRail	AT200	385101-124	24 × four-car (96)	Caledonian Rail Leasing	2017-19	ACEMU	
800/0	GWR	AT300	800001-036	36 × six-car (216)	Eversholt	2013-18	BMMU	
800/1	VTEC	AT300	800101-113	13 × nine-car (117)	Agility Trains	2013-18	BMMU	LNER
800/2	VTEC	AT300	800201-210	10 × five-car (50)	Agility Trains	2016-18	BMMU	LNER
801/3	GWR	AT300	800301-321	32 × nine-car (288)	Eversholt	2016-18	BMMU	
801/1	VTEC	AT300	801101-112	12 × five-car (60)	Agility Trains	2016-18	BMMU	LNER
801/2	VTEC	AT300	801201-230	30 × nine-car (270)	Agility Trains	2016-18	BMMU	LNER
802/0	GWR	AT300	802001-022	22 × five-car (110)±	Eversholt	2018	BMMU	
802/1	GWR	AT300	802101-114	14 × nine-car (126)	Eversholt	2018	BMMU	
802/0	TPE	AT300	802201-219	19 × five-car (95)	Angel	2018	ACEMU	
802/3	Hull Trains	AT300	802301-205	5 × five-car (25)	Angel	2019	BMMU	
803	East Coast Trains	AT300	803001-005	5 × five-car (25)	Beacon	2021	BMMU	
805	Avanti West Coast	AT300	805001-013	13 × five-car (65)	Rock Rail	2022	BMEMU	Under construction
807	Avanti West Coast	AT300	807001-010	10 × seven-car (70)	Rock Rail	2022	ACEMU	Under construction
810	East Midlands Railway	AT300	810001-033	33 × five-car (165)	Rock Rail	2023	BMMU	Under construction
	Avanti West Cost (HS2)			54 × eight-car (432)		2026	BMMU	Order placed

Class	TOC	Builder	Numbers	Quantity	Owner	Delivered	Type	Notes
230	London Midland (trial)	VivaRail	230001	1 × three-car (3)	VivaRail	Rebuilt 2015	DMU	Trial units
230	London Midland (trial)	VivaRail	230002	1 × two-car (3)	VivaRail	Rebuilt 2015	BEMU	Trial units, now in the USA
230	London Midland	VivaRail	230003-005	3 × two-car (6)	VivaRail	Rebuilt 2015	DMU	Delivered
230	TfW (Wales & Borders)	VivaRail	230006-010	5 × three-car (15)	VivaRail	Rebuilt 2015	BDMU	Delivered
484	SWR	VivaRail	484001-005	5 × two-car (10)	Lombard	Rebuilt 2020	DCEMU	Delivered

Parry People Movers

The Parry People Mover is a novel single car railbus of which two were ordered by London Midland as the Class 139. They work exclusively on the ¾-mile Stourbridge Town to Stourbridge Junction branch for West Midlands Trains.

Another PPM single car has been stored at Highley on the Severn Valley Railway but has yet to be placed with a customer.

Class	Ordered by	numbers	formations	ROSCOs	delivered	type	Subsequent redeployment
139	London Midland	139001/002	2 × one-car (2)	Porterbrook	2007/08	DMU	West Midlands Railway

Class 769 conversions

Class	TOC	Builder/model	Numbers	Quantity	Owner	Type	Notes
769	Northern	BREL/Brush FLEX	769424/426/431/434/442/450/456/458	8 × four-car (32)	Porterbrook	BMEMU	Delivered
769	Transport for Wales	BREL/Brush FLEX	769002/003/006-008	9 × four-car (36)	Porterbrook	BMEMU	Being delivered
769	GWR	BREL/Brush FLEX	769922/923/925/927/928/930/932/935-940/943/945/947/949/952/959	19 × four-car (76)	Porterbrook	BMEMU	Conversions ongoing, delays with entry into traffic
769	ROG	BREL/Brush FLEX	319009/010	2 × four-car (8)	Porterbrook	BMEMU	
799	Demo unit	BREL/Brush FLEX	799001	1 × four-car (8)	Porterbrook	BMEMU	

Type

DMU	Diesel Multiple Unit
ACEMU	25kV AC Electric Multiple Unit
DVEMU	Dual-voltage 25kV AC Electric Multiple Unit; 750V DC third rail and 25kV AC
EMUMC	Electric Multiple Unit motored centre car
BMMU	Electro-Diesel Bi-mode Multiple unit; 25kV AC and diesel engine(s)
DCEMU	750V DC Electric Multiple Unit

New Locomotives

While multiple units are the de facto train type for the vast majority of passenger operations, several fleets of new locos have been ordered, mostly by freight companies, since the mid-1990s privatisation process began.

Class	Ordered by	Builder	Axles	numbers	Type	Redeployed	Current user(s)	
57/0	Freightliner	Brush/GM	Co-Co	57001-012	Type 4 DE	DRS, Advenza	WCR	Rebuilt Class 47s
57/3	Virgin Trains	Brush/GM	Co-Co	57301-316	Type 4 DE	DRS, Network Rail	DRS, WCR, ROG	Rebuilt Class 47s
57/6	Porterbrook	Brush/GM	Co-Co	57601	Type 4 DE	WCR	WCR	Rebuilt Class 47
57/6	First Great Western	Brush/GM	Co-Co	57602-605	Type 4 DE	GWR	GWR	Rebuilt Class 47s
66/0	English, Welsh Scottish Railway	EMD	Co-Co	66001-250	Type 5 DE	ECR, GBRf	DB Cargo, ECR	66048 disposed of
66/3	Fastline Freight	EMD	Co-Co	66301-305	Type 5 DE	DRS	DRS	
66/4	Direct Rail Services	EMD	Co-Co	66401-434	Type 5 DE		Freightliner, GBRf, Freightliner Poland	
66/5	Freightliner	EMD	Co-Co	66501-599	Type 5 DE	Colas Rail, GBRf, Freightliner Poland	Freightliner, GBRf, Freightliner Poland, Colas Rail Freight	66521 scrapped
66/6	Freightliner	EMD	Co-Co	66601-625	Type 5 DE	Freightliner Poland	Freightliner, Freightliner Poland	
66/7	GB Railfreight	EMD	Co-Co	66701-799*	Type 5 DE		GBRf	66734 scrapped*
66/9	Freightliner	EMD	Co-Co	66951-957	Type 5 DE	Freightliner Poland	Freightliner, Freightliner Poland	
67	English, Welsh Scottish Railway	EMD	Bo-Bo	67001-030	Type 5 DE	Colas Rail Freight	DB Cargo, GBRf	
68	Direct Rail Services	Stadler	Bo-Bo	68001-034	Type 5 DE		DRS, Six locos on hire to Chiltern Railways; 14 locos on hire to Transpennine Express	
69	GB Railfreight	BREL/EMD	Co-Co	69001-016	Type 5 DE		GBRf	Rebuilt Class 56s, being delivered
70/0	Freightliner	GE	Co-Co	70001-020	Type 5 DE	Freightliner	70012 returned to the	USA
70/8	Colas Rail Freight	GE	Co-Co	70801-817	Type 5 DE		Colas Rail Freight	
73/9	Network Rail	EE/Loram	Bo-Bo	73951/952	Type 3 ED		Network Rail	Rebuilt Class 73/1s
73/9	GB Railfreight	EE/Brush	Bo-Bo	73961-965	Type 3 ED		GBRf	Rebuilt Class 73/1s
73/9	GB Railfreight	EE/Brush	Bo-Bo	73966-971	Type 3 DE		Caledonian Sleeper	Rebuilt Class 73/1s
88	Direct Rail Services	Stadler	Bo-Bo	88001-010	25kV AC ED		DRS	
93	Rail Operations Group	Stadler	Bo-Bo	93001-010	25kV AC TM			Under construction, option for 20 addition locos
99	GB Railfreight	Stadler	Co-Co	99001-020	25kV AC TM			Option for 30 additional locos

* The first 66734 (delivered as 66402 to DRS) was taken on by GBRf but scrapped in 2013 after an accident. There is now a second 66734 being prepared. 66747-752 and 66790-799 are locos delivered to Europe and acquired by GBRf and expected to be joined by 66350-361. Some GBRf and all Colas locos were previously used by DRS or Freightliner.

New Locomotives

New Trains Beyond 2025

The change of the franchising structure to management contracts, plus the fallout of reduced passenger numbers post-pandemic, means new train orders may be fewer in the immediate future.

However, there will still be some old trains that date from the 1980s, such as the Class 150/155/156/318s, that will be need to be replaced soon, as well as Class 158/165/166/320/323 units from 1990s.

Bombardier is building a fleet of 90 Class 701 Aventra 750V DC units for South Western Railway, 60 of which are ten-cars and the others are five-car sets. One of the former, 701023, is passing Eastleigh with the 0612 Eastleigh-Eastleigh via Bournemouth and Waterloo mileage accumulation run on 24 June 2021. These units have still to enter traffic with SWR. *Mark Pike*

Many heritage railways have taken advantage of the number of redundant railbuses, which have been donated or bought for next to nothing. On 23 February 2020, 142060 and 142028 stand at Redmire Station on the Wensleydale Railway. *Rob France*

If there is one rail sector that was badly affected by the pandemic, then it was the many heritage railways across the UK. Essentially tourist attractions run by dedicated volunteers, when they were able to reopen it was subject to strict social distancing regimes that seriously restricted the number of passengers they could carry. That, combined with an apprehension from their clientele on whether to travel or not, seriously drained their income.

However, with the lifting of restrictions progressively throughout 2021 and into 2022, it is hoped they will return to normality this year and will be back to running as before.

Standard Gauge Heritage Railways

Railway	Phone	Website
Appleby Frodingham RPS	07889 297271	www.afrps.co.uk
Aln Valley	0300 030 3311	www.alnvalleyrailway.co.uk
Avon Valley	0117 932 7296	www.avonvalleyrailway.co.uk
Ayrshire RPS	n/a	www.scottishindustrialrailwaycentre.org.uk
Barrow Hill Roundhouse	01246 472450	www.barrowhill.org.uk
Barry Island	01446 748816	http://barrytouristrailway.co.uk
Battlefield Line	01827 880754	www.battlefield-line-railway.co.uk
Beamish Museum & Tramway	0191 370 4000	www.beamish.org.uk
Bluebell	01825 720800	www.bluebell-railway.co.uk
Bodmin & Wenford	01208 73666	www.bodminandwenfordrailway.co.uk

Bo'ness & Kinneil	01506 822298	www.srps.org.uk/railway
Border Union	n/a	https://wrha.org.uk/border-union-railway
Bowes	0191 416 1847	www.bowesrailway.co.uk
Bristol Harbour	0117 352 6600	https://bristolharbourrailway.co.uk
Caledonian	01561 377760	www.caledonianrailway.com
Chasewater	01543 452623	www.chasewaterrailway.co.uk
Chinnor & Princes Risborough	01844 354117	www.cprra.co.uk
Cambrian Railways Trust	01691 831569	www.cambrianrailways.com
Cholsey & Wallingford	01491 835067	www.cholsey-wallingford-railway.com
Churnet Valley	01538 360522	www.churnet-valley-railway.co.uk
Colne Valley	01787 461174	www.colnevalleyrailway.co.uk
Dartmoor	01837 55367	www.dartmoor-railway-sa.org
Dean Forest	01594 845840	www.deanforestrailway.co.uk
Derwent Valley	01904 489966	https://dvlr.org.uk
East Anglian Railway Museum	01206 242524	www.earm.co.uk
East Kent	01304 832042	https://eastkentrailway.co.uk
East Lancashire	0161 764 6360	www.east-lancs-rly.co.uk
East Somerset	01749 880417	www.eastsomersetrailway.com
Ecclesbourne Valley	01629 823076	www.e-v-r.com
Eden Valley	017683 42309	www.evr-cumbria.org.uk
Elsecar Heritage	01226 746746	www.elsecarheritagerailway.co.uk
Embsay & Bolton Abbey	01756 710614	www.embsayboltonabbeyrailway.org.uk
Epping and Ongar	01277 365200	www.eorailway.co.uk
Fife Heritage	n/a	www.fifeheritagerailway.co.uk
Foxfield	01782 396210	www.foxfieldrailway.co.uk
Gloucestershire Warwickshire	01242 621405	www.gwsr.com
Great Central	01509 230726	www.gcrailway.co.uk
Great Central Nottingham	0115 940 5705	www.gcrn.co.uk
Gwili	01267 230666	www.gwili-railway.co.uk
Helston	07901 977597	www.helstonrailway.co.uk
Invergarry & Fort Augustus museum	n/a	www.invergarrystation.org.uk
Isle of Wight Steam	01983 882204	www.iwsteamrailway.co.uk
Keighley & Worth Valley	01535 645214	www.kwvr.co.uk
Keith & Dufftown	01340 821181	https://keith-dufftown-railway.co.uk
Kent & East Sussex	01580 762943	www.kesr.org.uk
Lakeside & Haverthwaite	01539 531594	www.lakesiderailway.co.uk
Lathalmond Museum	07379 914801	https://www.shed47.org
Lavender Line	01825 750515	www.lavender-line.co.uk
Lincolnshire Wolds	01507 363881	www.lincolnshirewoldsrailway.co.uk
Llanelli & Mynydd Mawr	n/a	www.llanellirailway.co.uk
Llangollen	01978 860979	www.llangollen-railway.co.uk
Mangapps Railway Museum	01621 784898	www.mangapps.co.uk
Mid Hants	01962 733810	www.watercressline.co.uk
Middleton	0113 271 0320	www.middletonrailway.org.uk
Mid Norfolk	01362 690633	www.mnr.org.uk
Midland Railway Butterley	01773 570140	www.midlandrailway-butterley.co.uk
National Railway Museum	0800 047 8124	www.railwaymuseum.org.uk
Nene Valley	01780 784444	www.nvr.org.uk
Northamptonshire Ironstone	01604 702031	www.nir.org.uk
Northampton and Lamport	01604 820327	www.nlr.org.uk
North Norfolk	01263 820801	www.nnrail.co.uk
North Yorkshire Moors	01751 472508	www.nymr.co.uk
North Tyneside	0191 200 7146	www.ntsra.org.uk
Pontypool & Blaenavon	01495 792263	http://pontypool-and-blaenavon.co.uk
Paignton & Dartmouth	01803 555872	www.dartmouthrailriver.co.uk
Peak Rail	01629 580381	www.peakrail.co.uk
Poulton & Wyre	n/a	www.pwrs.org
Radstock to Frome Railway Trust	n/a	www.radstocktofromerailwaytrust.co.uk
Ribble Steam	01772 728800	https://ribblesteam.org.uk

Rocks by Rail	07873 721941	www.rocks-by-rail.org
Royal Deeside	01330 844416	www.deeside-railway.co.uk
Rushden Transport Museum	01933 213066	https://rhts.co.uk
Severn Valley	01299 403816	www.svr.co.uk
Spa Valley	01892 537715	www.spavalleyrailway.co.uk
Somerset & Dorset	01761 411221	https://sdjr.co.uk
South Devon	01364 642338	www.southdevonrailway.org
Stainmore	07584 429481	www.kirkbystepheneast.co.uk
Strathspey	01479 810725	www.strathspeyrailway.co.uk
Swanage	01929 425800	https://swanagerailway.co.uk
Swindon & Cricklade	01793 771615	www.swindon-cricklade-railway.org
Tanfield Steam	0191 388 7545	www.tanfield-railway.co.uk
Telford Steam	07765 858348	www.telfordsteamrailway.co.uk
Tyseley Locomotive Works	0121 708 4960	www.vintagetrains.co.uk
Weardale	01388 529566	www.weardale-railway.org.uk
Wensleydale	0845 450 5474	https://wensleydale-railway.co.uk
West Somerset	01643 704996	www.west-somerset-railway.co.uk
Whitwell & Reepham	01603 871694	https://whitwellstation.com

Heritage railways thrive on recreating history with their locomotives, rolling stock and stations. Visiting from the Severn Valley Railway, 50035 *Ark Royal* calls at the superbly restored Gotherington station with the 1410 Cheltenham Racecourse-Broadway on 13 October 2019. *Pip Dunn*

Preserved Locos

How a loco is classed as 'preserved' has always been something of a grey area and open to debate, but for the purpose of this book they are those vehicles that are based at heritage railways or owned by heritage groups, regardless of whether they have been restored or not.

This includes some locos that may never have run while preserved, or realistically are unlikely to ever run again. It does not include locos located at heritage sites that are owned by FOCs or spot hire companies, which are listed under their relevant owners, unless those owners do not regard them as assets for the business; for example the locomotives owned by Jeremy Hosking that are either at, or due to move to, Margate for static display.

This list includes those locos that are owned by preservation groups for the supply of spare parts, such as 37214, 47761, and unlikely to ever be restored unless sold to another group/individual.

Those locos owned by preservation groups but are main-line registered and either on long-term hire to FOCs/TOCs or used for charter work or spot hire are included in those relevant sections.

Names are listed even if the plates are not presently fitted because the loco is partway through overhaul. Locos are listed at their home railway unless on a long-term loam, but locos do move about and visit other railways or sites. The number the vehicle normally/currently carries is in bold. Locos that carry identities of other classmates are listed by their original number(s).

Class 01

D2953	11503	GWS	PKR
D2956	11506	BLK	ELR

Class 02

02003	D2853	GWS	BH
	D2854	GWS	PKR
	D2858	GWS	MRB
	D2860	GWS	NRM
	D2866	BRW	PKR
	D2867	IND	BAT
	D2868	GWS	PKR

Class 03

03018	D2018	BRW	MRM	
03020	D2020	BRW	MRM	
03022	**D2022**	BRB	SCR	
	D2023	GWS	KES	
	D2024	IND	KES	
03027	D2027	BRW	PKR	
03037	D2037	BLK	RDR	
	D2041	BLK	CVR	
	D2046	IND	PVR	
	D2051	GNY	NNR	
03059	**D2059**	GWS	IWR	
03062	**D2062**	GWS	ELR	
03063	**D2063**	BRW	NNR	
03066	D2066	BRW	BH	
03069	**D2069**	GWS	VBR	
03072	**D2072**	GWS	LHR	
03073	D2073	BRW	RAC	
03078	**D2078**	BLK	NTR	
03079	**D2079**	BRW	DVR	
03081	D2081	BRW	MRM	
03089	D2089	GWS	MRM	
03090	**D2090**	GNY	NRS	
03094	**D2094**	GWS	RDR	
03099	D2099	BRW	PKR	
03112	**D2112**	GWS	RVR	
03113	D2113	BRW	PKR	
	D2117	MAR	LHR	
03118	D2118	BRW	GCN	
03119	D2119	BRW	EOR	
03120	**D2120**	GNY	FHR	
03128	**D2128**, 03901	BLK	PKR	
	D2133	GWS	WSR	
03134	**D2134**	IND	RDR	
	D2138	GWS	MRB	
	D2139	GNY	PKR	
03141	**D2141**	IND	PBR	
03144	D2144	BRW	WR	
03145	D2145	BRW	MOL	
	D2148	GWS	RSR	
03152	**D2152**	GRE	SCR	
03158	**D2158**	GWS	TIT	*Margaret-Ann*
03162	D2162	BRW	LLR	
03170	D2170	GWS	EOR	
	D2178	GWS	GWI	
03179	D2179	UND	RHR	
03180	D2180	BRW	PKR	
	D2182	GRN	GWR	
	D2184	BLK	CVR	
03189	D2189	BRW	RSR	
D2192	BLK	PDR	*Titan*	

Class 03 shunter 03170 rests on North Weald shed at the Epping and Ongar Railway on 27 April 2014. *Pip Dunn*

03197	D2197	BRW	MNR
	D2199	GWS	PKR
03371	D2371	BRW	PDR
03399	D2399	BRW	MRM

Class 04

	D2203	GNY	EBR
	D2205	GSW	PKR
	D2207	GWS	NYM
	D2229	GRE	PR
	D2245	GWS	DVR
	D2246	GWS	SDR
	D2271	GNY	SDR
	D2272	GRE	PKR
	D2279	BLK	PKR
	D2280	BLK	NNR
	D2284	GWS	PR
	D2289	IND	PR
	D2298	GNY	BRC
	D2302	GWS	MOL
04110	D2310	BRW	BAT
	D2324	IND	BU
	D2325	GWS	MRM
	D2334	GWS	MNR
	D2337	GWS	PKR

Class 05

05001	**D2554**, 97803	GWS	IWR
	D2578	GWS	MOL
	D2587	GWS	PKR
	D2595	GWS	RSR

Class 06

06003	**D2420**, 97804	GWS	PR

Class 07

07001	D2985	BRW	PKR
07005	D2989	IND	GCR
07010	D2994	BRW	AVR
07012	D2996	BRW	BH
07013	D2997	BRW	ELR

Class 08

	D3000	GNY	PKR	
	D3002	BLK	PVR	
	D3014	GWS	PDR	*Samson*
08011	**D3018**	GWS	CPR	*Haversham*
	D3019	UND	CRT	
08015	**D3022**	GWS	SVR	
08016	D3023	BRW	PKR	
08022	D3030	IND	CWR	*Lion*
08032	D3044	BRW	MHR	*Mendip*
08046	D3059	BRW	CAL	*Brechin City*
08054	D3067	BRB	EBR	
08060	**D3074**	IND	CWR	*Unicorn*
08064	**D3079**	BLK	NRS	
	D3101	BLK	GCR	
08102	**D3167**	GWS	LWR	
08108	**D3174**	BLK	KES	*Dover Castle*
08114	**D3180**	GRE	GCN	
08123	**D3190**	GRE	CWR	

Class 20s are popular locos at heritage railways, both as useful tools and attractions for visitors. D8137 is based at the Gloucestershire Warwickshire Railway and was joined at its home line by visiting sister loco D8098 for its 2017 diesel gala. The duo work the 1200 Cheltenham Racecourse-Toddington on 29 July. *Pip Dunn*

20098	**D8098**	GNY	GCR
20137	**D8137**	GFY	GWR
20154	**D8154**	BRB	GCN
20169	**D8169**	UND	TEB

20188	**D8188**	GYP	MHR
20214	D8314	GYP	LHR
20228	D8128	BRB	GWR

Class 23

	D5910	UND	BH

Class 24

24032	**D5032**	GYP	NYM	
24054	**D5054**, TDB968008	GNY	ELR	*Phil Southern*
24061	**D5061**, RDB968007, 97201	GNY	NYM	
24081	**D5081**	BRB	GWR	

Just four Class 24s, out of a build of 151, survived scrapping. D5054 is based at the East Lancashire Railway and works the 1421 Heywood-Rawtenstall at Heap Bridge on 24 April 2021. *Tom McAtee*

Class 25s are always popular at heritage railways. D7535 is based at the South Devon Railway but is a regular guest loco at other lines, such as on 16 May 2019 when it was at the SVR and seen leaving Arley with the 1347 Kidderminster-Bridgnorth. *Glen Batten*

Class 25

25035	**D5185**	GYP	GCR	
25059	D5209	BRB	KWV	
25072	**D5222**	GRE	CAL	
25083	D5233	BRB	CAL	
25173	**D7523**	GYP	BAT	
25185	**D7535**	BRB	SDR	
25191	D7541	GYP	SDR	
25235	D7585	BRB	BKR	
25244	**D7594**	UND	KES	
25262	**D7612**, 25901	GYP	SDR	
25279	D7629	BRB	ELR	
25309	**D7659**, 25909	GYP	PKR	
25321	**D7671**	GYP	MRB	
25322	D7672, 25912	BRU	CHV	*Tamworth Castle*

Class 26

26001	**D5301**	GNY	CAL	
26002	**D5302**	GNY	STR	
26007	D5300	RSR	BH	
26010	**D5310**	GFY	LLR	
26014	**D5314**	GNY	CAL	
26024	D5324	BRB	BKR	
26025	**D5325**	GNY	STR	
26035	D5335	BRB	CAL	
26038	D5338	BRB	BKR	*Tom Clift 1954-2012*
26040	D5340	BRB	WAV	
26043	**D5343**	BRB	GWR	

In recent years there has been a growing trend of photo charters where BR-era trains are recreated at heritage lines. 26007 was the star of an EMRPS photoshoot at Keighley on 23 October 2021. *Tom McAtee*

Class 27

27001	D5347	BRB	BKR	
27005	D5351	BRB	BKR	
27007	**D5353**	UND	CAL	
27024	D5370, ADB968028	GYP	CAL	
27050	**D5394**, 27106	GNY	STR	
27056	D5401, 27112	UND	GCR	
27059	D5410, 27123, 27205	UND	LR	

Class 28

D5705	ADB968006, S15705	GYP	ELR	

Class 31

31018	D5500	BRB	NRM	
31101	D5518	BRB	AVR	
31105	D5523	NRY	MRM	
31108	D5526	RFO	MRB	
31119	D5537	BRB	EBR	
31130	D5548	RFO	AVR	
31162	D5580	BRB	MRB	
31190	D5613	GOP	PVR	
31203	**D5627**	GNY	PBR	*Steve Ogden GM*
31206	D5630	CCE	RHR	
31207	**D5631**	GNY	NNR	
31210	D5634	RFO	DFR	
31233	D5553	NRY	MRM	
31235	D5662	BRB	DFR	
31255	D5683	EWS	MNR	
31270	D5800	REG	BU	*Athena*
31271	D5801	TLA	LLR	*Stratford 1840-2001*
31285	D5817	NRY	WR	
31289	D5821	EBP	RHR	*Phoenix*
31327	**D5862**	GYP	STR	
31418	D5522	BRB	MRB	
31430	D5695, 31265, **31530**	BRB	SPA	*Sister Dora*

The owners of 31163 have recreated the Derby RTC livery on the loco and renumbered it as 97205. It approaches Arley with the 1430 from Bridgnorth on 16 May 2019 while visiting the SVR from its home at Chinnor. *Glen Batten*

31435	**D5600**, 31179	GYE	EBR	
31438	D5557, 31139, 31538	BRB	EOR	
31454	D5654, 31228, 31554	ICM	KWH	
31459	D5684, 31256	BRB	WR	
31463	**D5380**, 31297, 31563	GOP	GCN	
31465	D5637, 31213, 31565	NRY	WR	
31466	D5533, 31115	EWS	SVR	
31514	D5814, 31414	DCE	MRB	
31601	D5609, 31186	DCR	EVR	*Devon Diesel Society*
97205	D5581, 31163	RTC	CPR	

Class 33

33002	D6501	DCE	SDR	
33008	**D6508**	GYP	BAT	*Eastleigh*
33018	D6530	BRB	MRM	
33021	D6539	POR	CHV	*Eastleigh*
33035	D6553	BRB	WEN	
33046	D6564	SWT	ELR	
33048	**D6566**	GYP	WSR	
33052	**D6570**	GNY	BLU	
33053	D6571	BRB	NLR	
33057	**D6575**	GYP	WSR	
33063	D6583	TMF	SPA	*RJ Mitchell – Designer of the Spitfire*
33065	D6585	BRB	SPA	*Sealion*
33102	D6513	BRB	CHV	*Sophie*
33103	D6514	DEP	EVR	*Swordfish*
33108	D6521	BRB	SVR	
33109	D6525	DEP	ELR	*Captain Bill Smith RNR*
33110	D6527	DEP	MRM	
33111	D6528	BRB	SWR	
33116	D6535	BRB	GCR	
33117	D6536	BRB	ELR	
33201	D6586	BRB	BAT	
33202	D6587	BRB	MNR	*Dennis G. Robinson*
33208	**D6593**	GYP	BAT	

Based at the East Lancashire Railway, 33109 *Captain Bill Smith RNR* sees use all year round on account of its working electric train heating and dual-brake capability. On 16 October 2021 it was stopped for routine maintenance at Buckley Wells shed. *Pip Dunn*

Western Region Class 35 Hymek D7076 departs Kidderminster Station, passing under the GWR Gantry next to the engine sheds alongside the Worcester-Birmingham main line, while visiting the SVR on 1 October 2015. *Jack Boskett*

Class 35

D7017	GYP	WSR	**D7029**	BRB	SVR
D7018	GYP	WSR	**D7076**	BRB	ELR

Class 37

37003	D6703	BRB	LR	
37009	D6709, 37340	BRB	GCN	
37023	D6723	BLU	PBR	
37029	**D6729**	GYP	EOR	
37032	**D6732**, 37353	GYP	NNR	
37037	D**6737**, 37321	BRB	SDR	
37042	D6742	EWS	EDR	
37075	D6775	TTG	KWV	
37097	D6797	BRB	CAL	*Old Fettercairn*
37108	D6808, 37325	BLL	RAC	
37109	D6809	BRB	ELR	
37142	D6842	BRB	BWR	
37190	D6890	BRB	MAR	
37198	**D6898**	GYP	DAL	
37214	D6914	WCR	BKR	
37215	D6915	BRB	GWR	
37216	**D6916**	GYP	PBR	
37227	D6927	TLM	CPR	
37248	**D6948**	GYP	GWR	
37250	D6950	DCE	WEN	
37261	D6961	DRU	BKR	
37263	D6963	DEP	TSR	
37264	D6964	LLB	NYM	
37275	**D6975**	BRB	PDR	
37294	D6994	BRB	EBR	
37308	D6608, 37274	UND	SVR	

There are many Class 37s preserved, and one of the finest is D6948 based at the Gloucestershire Warwickshire Railway. While on a visit to the West Somerset Railway, it approaches Williton with the 1454 from Bishops Lydeard on 22 June 2019. *Glen Batten*

37310	D6852, 37152	BRB	PKR	*British Steel Ravenscraig*
37350	**D6700**, 37119	GYP	GCR	
37503	D6717, 37017	EWS	KWH	
37674	D6869, 37169	RSR	STR	
37679	D6823, 37123	TTG	WIS	
37714	D6724, 37024	TLM	GCR	*Cardiff Canton*

Restored to its early 1980s look with domino headcodes and in BR blue, 37109 has been based at the East Lancashire Railway since sale by EWS. On 16 October 2021 it was awaiting routine maintenance at Buckley Wells. *Pip Dunn*

The Class 40 Preservation Society's D345 was repainted back into BR green in 2021. It looks magnificent as it works the 1530 Heywood-Rawtenstall at Heap Bridge on 17 September. *Tom McAtee*

Class 40

40012	**D212**, 97407	BRB	MRB	*Aureol*
40106	D306	GFY	SVR	*Atlantic Conveyor*
40118	D318, 97408	BRB	BAT	
40122	**D200**	GNY	NRM	
40135	D335, 97406	BRB	ELR	

Class 41

41001	43000, ADB975812	BRP	DY

Class 42

D821	BRB	SVR	*Greyhound*	**D832**	BRB	ELR	*Onslaught*

Only two of 71 Class 42/43 Warships survive, D821 *Greyhound* and D832 *Onslaught*. The latter is based at the ELR, has been repainted into BR Blue and spent 2020-21 running as D818 *Glory*. It arrives at Heywood with the 0910 from Bury on 26 September 2020. *Tom McAtee*

Class 43

43002		HST	NRM	*Sir Kenneth Grange*
43018		HST	RAC	
43044		ICO	GCN	
43056		FIR	GWI	
43071		FIR	CVR	
43073		EMW	CVR	
43081		EMW	RAC	
43082		EMW	CVR	
43102	43302	ICS	NRS	*The Journey Shrinker*
43159		FGB	GCN	

Class 44

44004	**D4**	BRB	MRB	*Great Gable*	44008	**D8**	GYP	PKR	*Penyghent*

Class 45

45015	D14	BRB	BAT	
45041	D53	BRB	NVR	*Royal Tank Regiment*
45060	D100	BRB	BH	*Sherwood Forester*
45105	D86	BRB	BH	
45108	D120	BRB	ELR	
45125	**D123**	GYP	GCR	*Leicestershire and Derbyshire Yeomanry*
45132	D22	BRB	EOR	
45133	D40	BRB	MRB	
45135	D99	BRB	ELR	*3rd Carabinier*
45149	D135	BRB	GWR	

Note: 45015 is expected to be scrapped soon.

Another photo charter image sees 45041 *Royal Tank Regiment* hauling a short engineers' ballast train into Loughborough station on the Great Central Railway during a EMRPS Photographic charter on 29 September 2014. This loco is now based at the Nene Valley Railway. *Jack Boskett*

Class 46

46010	D147	BRB	GCN		46045	**D182**, 97404	BYP	MRB
46035	D172, 97403	BRB	PKR					

Class 47

47004	D1524	GYP	EBR	
47077	D1661, 47613, 47840	BRB	NYM	*North Star*
47105	D1693	BRB	GWR	
47117	**D1705**	BRB	GCR	*Sparrowhawk*
47192	D1842	GYP	EVR	
47205	D1855, 47395	RFD	NLR	
47292	D1994	BLL	GCN	
47306	D1787	RFE	BWR	*The Sapper*
47367	D1886	BRB	MNR	*Kenny Cockbird*
47376	D1895	FTT	GWR	*Freightliner 1995*
47401	D1500	BRB	MRB	*North Eastern*
47402	D1501	GYP	ELR	
47417	**D1516**	GYP	MRB	
47449	**D1566**	BRB	LLR	
47484	D1662	GWR	WIS	*Isambard Kingdom Brunel*
47579	D1778, 47183, 47793	BRE	MHR	*James Nightall GC*
47596	D1933, 47255	GYP	MNR	*Aldeburgh Festival*
47635	D1606, 47029	LLB	EOR	*Jimmy Milne*
47643	D1970, 47269	IOS	BKR	
47761	D1619, 47038, 47564	RES	MRB	
47765	D1643, 47059, 47631	SCR	ELR	
47771	D1946, 47503	RES	ZG	
47785	D1909, 47232, 47665, 47820	EWS	WEN	
47799	D1654, 47070, 47620, 47835	ROY	EDR	*Prince Henry*
47841	D1726, 47134, 47622	ICS	MAR	*The Institution of Mechanical Engineers*

With a rake of period coaches in tow, Class 47 D1501 works the 0916 Bury-Heywood past Heap Bridge on 7 February 2020. *Tom McAtee*

Having previously been a main-line loco for Boden Rail, 50017 *Royal Oak* was sold back for preservation and is now based at the Great Central Railway. It is seen running round at Leicester North having arrived with the 1245 from Loughborough on 13 April 2019. *Pip Dunn*

Class 50

50002	**D402**	BRB	SDR	*Superb*
50015	D415	LLB	ELR	*Valiant*
50017	D417	NSO	GCR	*Royal Oak*
50019	D419	UND	MNR	*Ramillies*
50021	D421	LLB	ZG	*Rodney*
50026	D426	NSD	ZG	*Indomitable*
50027	D427	NSR	MHR	*Lion*
50029	D429	LLB	PKR	*Renown*
50030	D430	LLB	PKR	*Repulse*
50031	D431	ICS	SVR	*Hood*
50033	D433	LLB	SVR	*Glorious*
50035	D435	BRB	SVR	*Ark Royal*
50042	D442	LLB	BWR	*Triumph*

Class 52

D1010	MYP	WSR	*Western Campaigner*	**D1041**	UND	ELR	*Western Prince*
D1013	BRB	SVR	*Western Ranger*	**D1048**	BRB	MRB	*Western Lady*
D1023	BRB	NRM	*Western Fusilier*	**D1062**	BRB	SVR	*Western Courier*

Class 55

55015	**D9015**	GYP	BH	*Tulyar*
55016	**D9016**	GYP	MAR	*Gordon Highlander*
55019	D9019	BRB	BH	*Royal Highland Fusilier*

Class 56

56006	BRB	ELR
56097	TLC	GCN

The National Railway Museum has several ex-BR diesels in its collection, including Deltic D9002 *The King's Own Yorkshire Light Infantry*, which was on display in the Great Hall in April 2016. *Pip Dunn*

Class 58

58012	TMF	BAT	**58022**	TMF	PKR	**58048**	EWS	BAT
58016	FER	LR	**58023**	MLB	LR			

Class 60

60050	EWS	KWH	**60081**	GWR	TO	**60086**	TEW	KWH

Class 71

71001	**E5001**	BRB	NRS

Class 73

73002	E6002	LLB	ZG	
73003	**E6003**	GYP	SCR	*Sir Herbert Walker*
73118	E6024	EUS	BIR	
73129	**E6036**	EBP	GWR	
73130	E6037	EUS	BC	
73140	E6047	NSR	SPA	
73210	E6022, 73116	IGX	EVR	*Selhurst*

The only surviving LNER-deigned EM1 1,500V DC Class 76 is 26020 in the national collection at York, which has been restored into original British Railways black as it would have carried when built in the 1950s. *Pip Dunn*

Class 76

76020	**E26020**	BLK	NRM	

Class 77

1502	**E27000**	BLK	MRB	*Electra*
1505	**E27001**	DNS	MSIM	*Ariadne*
1501	E27003	DNS	TIL	*Diana*

Class 81

81002	E3003	BRB	BH	

Class 82

82008	E3054	ICO	BH	

Class 83

83012	**E3035**	EBY	BH	

Class 84

84001	E3036	BRB	BKR	

Class 85

85006	E3061, 85101	BRB	BH	

Class 87

87001		BRB	NRM	*Stephenson*
87035		VIR	RAC	*Robert Burns*

Prototype Type 5 diesel-electric Co-Co

DP1		POW	NRS	*Deltic*

Prototype 500hp diesel-electric 0-6-0

D226	D0226	GRE	KWV	

LMS 7P 46100 *Royal Scot* rounds the curve at Teignmouth on 14 September 2021 as it heads along the sea wall towards Dawlish with a Saphos Trains' excursion from Kingswear to Cardiff Central. *Jack Boskett*

There are currently four Train Operating Companies that can accommodate main-line, steam-hauled trains on the National Network; DB Cargo (page 259), West Coast Railways (Page 194), Locomotive Services Limited (page 184) and Vintage Trains (page 192).

Of those, WCR is the current leader in the field, having won much business from initially EWS (now DBC) and then FM Rail (which went bust in 2007).

While LSL, WCR and VT all own some of their own steam locos, many locos are owned by private groups and individuals, but are main-line registered and used by some or all of the TOCs that can operate steam. Many steam locos come with their own support coaches for their crew.

There are some timetabled steam operations, most notably the Fort William to Mallaig 'Jacobite', which even appears in the GB timetable. WCR also runs from Carnforth to Scarborough on its Scarborough Spas Express and its 'Dalesman' from Hellifield to Carnforth. There are also a number of repeat itinerary charter trains like the 'Torbay Express' from Bristol TM to Kingswear and the Cumbrian Mountain Express.

Registered locos

All steam locos registered for main-line operation must meet NR's standards by having TPWS, OTMR and GSMR, although locos restricted for the Battersby to Whitby line do not. Locos do not need to be dual braked, although several steam locos are now upgraded as such for ease of operation. WCR and Vintage Trains both operate vacuum-braked trains, while LSL and DB Cargo do not presently.

Other steam locos are registered to be hauled on the national network, and may be assigned TOPS numbers, but these are not included here.

As a rule, locos have a boiler examination that, if approved, gives them a ten-year ticket to run on the network, although these may be reduced in length, or indeed extended. Once a ticket has expired, however, the loco is usually either stopped for overhaul or retired for heritage line use only.

Loco	Type	Wheels	Name	Owner	TOPS no.	Brakes	Notes/restrictions
4936	GWR 49xx	4-6-0	*Kinlet Hall*	Private	98536	V	Limited to 60mph
4953	GWR 49xx	4-6-0	*Pitchford Hall*	Private	98553	V	Limited to 60mph
4965	GWR 49xx	4-6-0	*Rood Ashton Hall*	VT	98565	V	Limited to 60mph
5029	GWR 4073	4-6-0	*Nunney Castle*	VT	98728	X	
5043	GWR 4073	4-6-0	*Earl of Mount Edgcumbe*	VT	98743	V	
6024	GWR 60xx	4-6-0	*King Edward I*	Private	98824	X	
7029	GWR 4073	4-6-0	*Clun Castle*	VT	98729	V	
7752	GWR 57xx	0-6-0T		Private	98452	V	Limited to 35mph
9466	GWR 94xx	0-6-0T		Private	98466	V	Limited to 45mph
9600	GWR 8750	0-6-0T		Private	98457	V	Limited to 45mph
30777	SR N15	4-6-0	*Sir Lamiel*	Private	98577	V	
30825	SR S15	4-6-0		Private	98625	V	Limited to 45mph
30926	SR Schools	4-4-0	*Repton*	Private	98726	V	For use between Battersby and Whitby only, limited to 45mph
31806	SR U	2-6-0		Private	98406	V	Limited to 60mph
34046	SR West Country (rebuilt)	4-6-2	*Braunton*	LSL	98746	X	
34067	SR Battle of Britain	4-6-2	*Tangmere*	WCR	98767	X	
35018	SR Merchant Navy (rebuilt)	4-6-2	*British India Line*	WCR		V	
35028	SR Merchant Navy (rebuilt)	4-6-2	*Clan Line*	Private	98828	X	
44767	LMS 5MT	4-6-0		Private	98567	V	Limited to 45mph
44871	LMS 5MT	4-6-0		Ian Riley	98571	X	Limited to 60mph
44932	LMS 5MT	4-6-0		WCR	98532	V	limited to 60mph
45212	LMS 5MT	4-6-0		KWVR	98512	V	Limited to 60mph
45231	LMS 5MT	4-6-0	*The Sherwood Forester*	Private	98531	X	Limited to 60mph
45305	LMS 5MT	4-6-0	*Alderman A E Draper*	Private	98505	V	Limited to 60mph
45407	LMS 5MT	4-6-0	*The Lancashire Fusilier*	Private	98507	X	Limited to 60mph
45428	LMS 5MT	4-6-0		Private	98528	V	For use between Battersby and Whitby only, limited to 45mph
45596	LMS 5XP	4-6-0	*Bahamas*	Private	98696	V	
45690	LMS 5XP	4-6-0	*Leander*	Private	98690	V	
45699	LMS 5XP	4-6-0	*Galatea*	WCR	98699	V	
46100	LMS 7P	4-6-0	*Royal Scot*	LSL		X	
46115	LMS 7P	4-6-0	*Scots Guardsman*	WCR	98715	V	
46201	LMS 8P	4-6-2	*Princess Elizabeth*	Private	98801	X	
46233	LMS 8P	4-6-2	*Duchess of Sutherland*	Private	98834	X	
48151	LMS 8F	2-8-0		Private	98851	V	Limited to 50mph
60007	LNER A4	4-6-2	*Sir Nigel Gresley*	Private	98898	X	
60103	LNER A3	4-6-2	*Flying Scotsman*	NRM	98872	X	
60163	LNER A1	4-6-2	*Tornado*	A1SLT	98863	X	
61264	LNER B1	4-6-0		Private	98564	V	
61306	LNER B1	4-6-0	*Mayflower*	Private	98506	V	
62005	LNER K1	2-6-0		Private	98605	V	Limited to 50mph
70000	BR 7MT	4-6-2	*Britannia*	LSL	98700	X	
70013	BR 7MT	4-6-2	*Oliver Cromwell*	Private	98713	V	
71000	BR 8P	4-6-2	*Duke of Gloucester*	Private	98802	X	
75029	BR 4MT	4-6-0		Private	98429	V	For use between Battersby and Whitby only, limited to 45mph
76079	BR 4MT	2-6-0		Private	98476	X	For use between Battersby and Whitby only, limited to 60mph
76084	BR 4MT	2-6-0		Private	98484	V	limited to 60mph

Charter Promoters

There are several established charter promoters who run trains throughout the year. As a rule they buy in train operation and coach hire and merely hire the train, sell tickets and make their profit that way. There are several types of charter trains;

Day excursions: These usually head to cities or towns of interest for sightseeing, sometimes with off-train options to local attractions.

Land cruises: Typically three or four days and often to Scotland, the emphasis is on scenery, at-seat fine dining, quality hotels and off-train options.

Multi-traction tours: Lots of different, usually freight, locos hauling short legs of the trip.

Heritage traction: Preserved or old ex-BR locos likes 20s, 37s, 40145, D1015, Class 50s and the like working high-mileage, long days, often taking locos to their former stamping grounds.

Freight line tours: Trips to routes, branch line, curves, spurs, loops and yards that do not normally have passenger trains. Sometimes may use rare traction.

Steam tours: Day excursion trains hauled in part or totally by steam traction.

Branch Line Society

07785 112044

www.branchline.org.uk

The BLS specialises in covering rare track using a variety of TOCs, including loco-hauled trains and multiple units. It also raises thousands of pounds each year for good causes.

InterCity

0800 038 5364

InterCity is one of the several in-house charter promotion operations in the LSL stable, and runs day excursions using LSL's fleet of restored heritage diesel and electric locos including 20s, 37s, 40, 47s, 86s and 87s and – when they have been returned to traffic – 45118 and 55022.

InterCity is one of LSL's 'in house' promoters and concentrates on day trips using classic ex-BR diesel and electrics. A sight not seen for over 37 years was a pair of Class 40s on the main line, but on 30 October 2021, InterCity took D213 *Andania* and D345 from Preston to Plymouth and back. The duo wait to leave Crewe on the return leg. D213 is in long-term use by LSL, while D345 was on a short-term hire. *Pip Dunn*

Midland Pullman

0800 038 5360

www.midlandpullman.com

Another of LSL's in-house promoters, it runs a busy summer programme using a refurbished HST set painted in Blue Pullman colours. It has achieved a number of notable trips for HSTs, taking a set to Mallaig and Kyle of Lochalsh for the first time.

Nenta Traintours

01692 406152

www.nentatraintours.co.uk

Nenta runs typically three or four loco-hauled trains from East Anglia to the likes of Carlisle, Plymouth or Scarborough. It tends to use Riviera Trains as its rolling stock provider and DRS as its train operator, with top-and-tail Class 68s its usual traction. It has previously operated with EWS, WCR and FMR. However, it understandably ran no trains during the pandemic and there are none planned for 2022.

Pathfinder Tours

01453 835414

www.pathfindertours.co.uk

One of the busiest promoters, Pathfinder typically runs between 30 and 35 trains a year with a mix of day excursions, enthusiast railtours, land cruises and steam-hauled trains.

It was started by Peter Watts as F&W Railtours in 1973 and has run more than 1,000 excursions in the following 47 years. It uses Riviera Trains with DRS, GBRf, WCR and DB Cargo as its main TOCs.

Railway Touring Company

01553 661500

www.railwaytouring.net

King's Lynn-based promoter RTC that runs mostly steam-hauled trains across the country. It also runs rail trips abroad. It exclusively uses West Coast Railways for train operation. Each year it runs its nine-day 'Great Britain' trip, which covers all four corners of the country behind steam.

It likes to recreate days of old, taking steam locos back to their old haunts, but also enjoys delivering 'firsts'; taking a steam loco to a route or destination they never visited in their first main-line careers.

Railway Touring Company works exclusively with West Coast Railways to deliver steam-hauled charters across the length and breadth of the UK. On 24 June 2017, 60103 *Flying Scotsman* heads south near Pinchbeck, between Sleaford and Spalding, with a Scarborough-King's Cross day trip. *Pip Dunn*

Retro Railtours

0161 3309055

www.retrorailtours.co.uk

Retro typically runs just one or two trains a year for local clientele in the Huddersfield and Manchester areas. Day trips, ideally with exotic heritage traction, to an interesting destination is the usual recipe and by running infrequently, the company has gained a loyal customer base.

Saphos Trains/Rail Charter Services

0800 038 5320

https://saphostrains.com

Running its first trains in 2018, Saphos is another of the charter promotion arms of Locomotive Services, and promotes mostly steam-hauled trains.

Also part of the same business is Rail Charter Services, which uses more modern rolling stock and modern traction for day trips. It ran the 'Staycation Express', a series of Skipton to Appleby day trips in the summer of 2020. These were some of the first socially distanced charters to run during the Covid pandemic. However, they have been cancelled for 2022.

Scottish Railway Preservation Society

01698 263814

The SRPS is one of the few promoters that owns its own coaches and has a rake of Mk 1s based at Bo'ness.

Once a big EWS customer, it moved a lot of its business to WCR in the mid-00s, although it now uses a mix of WCR, GBRf and DB Cargo to run its trains. It runs a few steam-hauled trains each summer, using mostly WCR as its TOC.

It also runs typically as many as seven or eight diesel-hauled charters a year mainly to Mallaig, Kyle of Lochalsh and Oban, but also to popular English destinations including York, Chester or Keighley. Trains start from a mix of locations as far south as Carlisle, but typically from the central belt.

SRPS also owns main-line registered 37403 *Isle of Mull* and has access to 37025 *Inverness TMD*. It aims to run trains using these 37s and its coaches.

The SRPS runs several train each summer, typically up to eight trains a year. This year it should see its own 37403 *Isle of Mull* used on some trains. On 2 June 2017, 37025 *Inverness TMD* and 37421 pause at Brora on a returning Wick-Edinburgh trip. *Pip Dunn*

Statesman Rail's 1608 Blaenau Ffestiniog-Hull waits to leave behind 47593 *Galloway Princess* on 19 May 2021, a train operated by LSL. *Tom McAtee*

Statesman Rail

0845 310 2458

www.statesmanrail.com

Statesman Rail now operates with LSL and runs a series of day excursions that are mostly diesel-hauled, although steam sometimes features.

Steam Dreams

01483 209888

www.steamdreams.com

As its names suggest, Steam Dreams is essentially a stem charter promoter. It uses WCR as its sole train operator, covers predominately the south of England and likes repeat-itinerary trains.

UK Railtours

01438 715050

www.ukrailtours.com

UKR is one of the best-known names in the UK charter market and has been in business since 1975, starting as the Lea Valley Railway Club, then Hertfordshire Railtours and now UK Railtours.

The year 2020 saw the sad passing of its founder and driving force John Farrow. However, his daughter Liz now runs the business and it returned to operating trains in 2021. It specialises in day excursions to destinations of interest, land cruises and freight line enthusiasts tours. As a rule, it uses DB Cargo or GBRF as its TOCs and Riviera Trains as its rolling stock provider.

Vintage Trains

0121 708 4960

www.vintagetrains.co.uk

Vintage Trains started as a promoter specialising in steam-hauled trains from its base in Tyseley in Birmingham. It had an emphasis on steam, but also ran a few diesel-hauled trains.

In 2018 it gained its own train operating licence and it can now operate its own trains, plus for third parties (see page 192).

It still runs several in house steam-hauled charters, mostly from the West Midlands. It is initially restricting its operations to a few preferred routes, such as the Welsh Marches, Stratford-upon-Avon and Oxford.

It has also run some third-party trains for the likes of the Branch Line Society and Polar Express.

West Coast Railways

0844 850 4685

www.westcoastrailways.co.uk

As well as being a nationwide train operator in its own right (see page 194), WCR promotes many trains itself, including repeat itinerary steam-hauled trains such as the Jacobite, Scarborough Spas Express and Dalesman trains.

It also runs its Spirit of the Lakes programme and other day excursions, hauled by both steam and diesel traction, and occasionally electric.

West Coast Railways runs many in-house charter trains, including its 'Spirit of the Lakes' programme. On 12 May 2018, 47772 pauses at Peterborough working the 0614 Skegness-Winchester day trip. *Pip Dunn*

Scrapyards

There are several scrapyards heavily involved with the disposal of old railway assets, locomotives, coaches, multiple, units and wagons.

Currently few locos are being scrapped but yards have been busy dealing with disposals as a result of the massive changes to passenger fleets that has seen several redundant DMU, EMU and coaches being sent for scrap.

Those vehicles have not had the necessary modifications to make them compliant for the railway and are deemed too old to be work investing in, leaving scrap as the only alternative.

The main breakers involved at the present time are CF Booth at Rotherham, Sims Metals at Newport and European Metal Recycling at Kingsbury.

Ron Hull Jnr in Rotherham and Raxstar in Cardiff are also used for scrapping vehicles, the latter tending to do so 'on site' at the location of the condemned vehicle.

Please be aware that visits to these sites, if allowed at all, are on a controlled basis and by appointment only.

CF Booth

www.cfbooth.com

CF Booth in Rotherham is one of the most active scrapyards when it comes to disposal of redundant railway assets. Recently it has been breaking up redundant LNER Mk 4 coaches, FGW Mk 3 HST trailers and various multiple units.

Ron Hull

www.ronhull.co.uk

Another Rotherham-based scrap dealer, in recent years it has scrapped examples of Classes 08, 31, 37, 47, 56, 73 and 86 but after a lengthy period of not disposing of much railway rolling stock, it recently acquired 57008/011/012 from DRS although it promptly sold them to WCR.

European Metal Recycling

www.emrgroup.com

EMR has several sites across the UK, but the site that undertakes the most disposal of railway stock is at Kingsbury, next to the Birmingham to Tamworth main line.

It recent years it has scrapped examples of 08, 09, 11, 20, 31, 33, 37, 47, 56, 58, 86 and 87, plus the Class 373 Eurostar sets.

It still has four Class 08/09 shunters – two at Kingsbury and two at its Attercliffe site in Sheffield – pending disposal or sale. They are 08783/913 at the former and 08872 and 09023 at the latter.

Raxstar

Raxstar – not to be confused with a Luton rapper – was a prolific breaker in the early 2000s, disposing of examples of Class 08, 33, 37, 47, 50, 56 and 58.

The company has been much reduced in its railway activities of late, especially concerning locomotives.

However, it has recently disposed of Mk 3 coaches, Class 442 units, track machines and wagons at its permanent base at Eastleigh.

Sims Metal Management

www.simsmm.co.uk

This company has been busy scrapping many redundant coaches and multiple units at its Newport site. Recently passing through for disposal have been Class 43s, Class 142 DMUs and Class 313, 314, 315, 365, 508 EMUs as well as ex LNER and FGW Mk 3 and 4 coaches.

The saga of HS2 – High Speed 2 – rumbles on, and sadly recently has been cut back in scope after a monumental Governmental U-turn.

It is the new high-speed line from London to Birmingham (phase 1) and while the aspiration to then extend to Manchester in the North-West remains as part of phase 2, the other phase 2 line to Leeds in West Yorkshire was axed in November 2021, despite being a Government manifesto pledge as part of its plan to 'level up' the county. HS1 is the 67-mile line from London St Pancras to the Channel Tunnel.

The project remains as controversial as it has been from day one, especially over its ballooning cost predictions, which have risen from £16bn to £33bn (just for phase 1) and £36bn to £89bn for Phase 1 and 2. Other estimates suggest it could cost as much as £106bn.

As well as the ballooning construction, planning and consultation costs, and contradictory figures of its overall cost – which inevitably will be more as they tend to be on projects of this size – there is also a difference of opinion over the benefits it will bring for the outlay in cost.

However, even the 'worst-case' scenarios suggest a cost benefit ratio of 1.3:1 over 60 years, so for every pound spent, £1.30 will be generated. However, other models suggest this will be more like 1.5:1 or as much as 2.3:1.

That suggests the project will ultimately be worthwhile. It will also give a huge boost to the construction industry.

Initial poor publicity by the pro-HS2 bodies did not help the public perception of the project, mainly highlighting time savings on journeys rather than what is actually HS2's biggest benefit – creating capacity for new trains and in doing so allowing fast trains to be taken off the existing Manchester-Birmingham NS-Euston West Coast Mail Line, thereby releasing that route for more local and freight trains.

So while some people believe they may not benefit from HS2 as such because it does not stop in their towns, they should benefit from a better service on the existing line.

HS2 is using a new station in Birmingham built on the site of the old Curzon Street and new platforms at London Euston. It will serve Old Oak Common in west London and Birmingham Interchange, near Solihull. It is expected it will take just 49 minutes to cover the 100 miles from Euston to Curzon Street.

Much of Phase 1 is due to be in tunnels – which have ramped up the cost – with 8 miles underground in London and another 10 miles in the Chiltern Hills to placate residents in that area.

On phase 2, if approved, a new station at Toton, halfway between Derby and Nottingham, is planned and Manchester Interchange will link with other transport networks. It will also connect with other trains at Crewe.

There will be some HS2 trains that continue off the high-speed network on to the existing infrastructure, which sadly means the trains will have to be built to the current loading gauge and not a European gauge that would allow more seats and more room for passengers.

An artist's impression of one of the new HS2 trains that have now been ordered.

How HS2 trains at London Euston might look.

However, this does offer operational flexibility, especially in offering through services and so opening up the accessibility of HS2 to other parts of the network. In light of the recent cutting back of the HS network, and the political 'hot potato' it has been for more than a decade now, it is difficult to see there being any plans to extend the high-speed network to either Leeds, Liverpool or Newcastle or north of Manchester to Scotland any time soon.

Once opened, HS2 plans to offer capacity of up to nine trains an hour from its opening in 2029 (but may be as late as 2033), but that should rise to 18 trains an hour to and from London from 2033 when phase 2 opens. Initially, services on HS2 will be part of the Avanti West Coast franchise and it will be responsible for running all aspects of the service including ticketing, trains and the maintenance of the infrastructure.

A rolling stock depot will be built in Washwood Heath, Birmingham. The phase 1 infrastructure maintenance depot will be north of Aylesbury in Buckinghamshire, between Steeple Claydon and Calvert. Staveley in Derbyshire and Stone, in Staffordshire, are identified as similar sites for phase 2.

Train contract awarded

In November 2021 it was announced that Hitachi and Alstom had won the order to build and maintain the 225mph High Speed 2 trains.

The fleet will be 100 per cent electric and should be one of the world's most energy efficient very high-speed trains due to the lower train mass per passenger, aerodynamic design, regenerative power and latest energy efficient traction technology.

The HAH-S joint venture will manufacture the 54 eight-car trains in County Durham, Derby and Crewe. They will be 'predominantly UK-designed' and built, tested and maintained in Britain

The design will use unprecedented levels of smart digital technology. The inbuilt digital system includes state-of-the-art innovative sensors that enable potential faults to be identified and ensure reliable and seamless passenger journeys.

The project features seamless, high-capacity wireless internet, digital seat reservations and travel information delivered through multiple channels, with the trains future-proofed to accommodate technological advances.

The rolling stock contract is expected to create and sustain more than 2,500 jobs directly, as well as create 9,000 jobs in the supply chain. Some 505 people will be directly employed by the two companies in the Midlands and the North in the design and manufacturing phase, including 49 apprentices and graduates working on the prestigious programme.

In addition, maintenance will create more than 100 new jobs at Washwood Heath in Birmingham.

As part of the award, Hitachi and Alstom are making new multi-million-pound investments in advanced welding and bogie manufacturing. Hitachi Rail recently invested £8.5m in bespoke welding and painting facilities at its Newton Aycliffe factory, taking its total investment up to £110m. The site opened in 2015 and employs around 700 staff. The new welding facility will carry out vehicle body assembly and fit out, before each one is transported to the East Midlands.

Once in Alstom's factory in Derby the trains will be fitted with all remaining components, including interiors, electrics and bogies. The bogies will be made by Alstom at its Crewe facility.

Abbreviations in Tables

BDMS	Battery Driving Motor Standard
BDMSO	Battery Driving Motor Standard Open
DM	Driving Motor
DMBO	Driving Motor Brake Open
DMBS	Driving Motor Brake Second
DMC	Driving Motor Composite
DMCL	Driving Motor Composite Lavatory
DMCO	Driving Motor Composite Open
DMF	Driving Motor First
DMFLO	Driving Motor First Luggage Open
DMOCL	Driving Motor Open Composite Lavatory
DMOCL(W)	Driving Motor Open Composite Lavatory (wheelchair accessible)
DMOS	Driving Motor Open Standard
DMOSL	Driving Motor Open Standard Lavatory
DMRFO	Driving Motor Restaurant First open
DMS	Driving Motor Standard
DMSL	Driving Motor Standard Lavatory
DMSO	Driving Motor Standard Open
DTC	Driving Trailer Composite
DTCO	Driving Trailer Composite Open
DTPMV	Driving Trailer Parcels Mailvan
DTS	Driving Trailer Standard
DTSO	Driving Trailer Standard Open
MBC	Motor Brake Composite
MBSO	Motor Brake Standard Open
MC	Motor Composite
MF	Motor First
MFL	Motor First Lavatory
MOS	Motor Open
MOSL	Motor Open lavatory
MPMV	Motor Parcels Mail Van
MS	Motor Second
MSL	Motor Standard Lavatory
MSLRB	Motor Standard Lavatory Restaurant Buffet
MSO	Motor Standard Open
MSRB	Motor Standard Lavatory Restaurant
MSRMB	Motor Standard Restaurant Micro Buffet
PDTF	Pantograph Driving trailer First
PDTS	Pantograph Driving trailer standard
PDTRBF	Pantograph Driving trailer Restaurant Buffet First
PTF	Pantograph trailer First
PTOSL	Pantograph Driving Open standard lavatory
PTS	Pantograph Driving standard
PTSO	Pantograph Trailer Standard Open
PTSRMB	Pantograph Trailer Standard Restaurant Micro buffet
Pres	Preserved
RB	Restaurant Buffet
ROSCO	Rolling stock leasing company
TBFO	Trailer Brake First Open
TC	Trailer Composite
TCO	Trailer Composite Open
TFO	Trailer First open
TGS	Trailer Guards standard
TOC	Train Operating Company
TPMV	Trailer Parcels Mail Van
TS	Trailer Standard
Tso	Trailer standard open
TSRB	Trailer Standard restaurant buffet
TSRMB	Trailer Standard Restaurant Micro Buffet

Depot codes

Code	Depot	Operator
AB	Albacete (Spain)	Continental Rail
AC	Alicante (Spain)	Transfesa
AH	Asfordby Technical Centre	Network Rail
AN	Allerton	Alstom
AT	Attercliffe	European Metal Recycling
AZ	Alizay Nr. Rouen – France	DB Cargo
BB	Billingham	Sembcorp Utilities
BD	Boston Docks	Victoria Group
BH	Barrow Hill Roundhouse	
BK	Bristol Barton Hill	LNWR
BL	Loughborough Falcon Works	Wabtec/Bruch Traction
BM	Bournemouth West	South Western Railway
BN	Bounds Green	Virgin Trains East Coast
BO	Bo'ness	Scottish Railway Preservation Society
BQ	Bury	East Lancs Railway
BS	Bescot	DB Cargo
BU	Burton-On-Trent	Nemesis Rail
BZ	Sofia, Bulgaria	BZK БЪлгарска ЖеиезолЪтна Компания
CA	Castle Donington	storage site
CB	Crewe Basford Hall	Network Rail
CC	Cardiff Steelworks	Celsa
CD	Crewe DMD	Locomotive Services Limited
CE	Crewe International Electric	DB Cargo
CF	Cardiff Canton	Colas Rail Freight
CL	Crewe	LNWR
CM	Cambridge	Arriva CrossCountry
CO	Coquelles (France)	Eurotunnel
CQ	Crewe	Railway Age Trust
CP	Crewe Carriage Shed	Arriva
CR	Crewe Gresty Bridge	Direct Rail Services
CS	Carnforth	West Coast Railways
CZ	Burton Central Rivers	Bombardier
DC	Craiova Romania exported locos	DB Cargo
DD	Daventry International Railfreight Terminal	Malcolm Rail
DF	Derby RTC	Loram (UK)
DG	Dagenham	
DK	Konkar Bulgaria exported locos	DB Cargo
DL	Dean Lane Manchester	
DM	Dollands Moor	Eurotunnel
DS	Deanside	Deanside Transit
DY	Derby Etches Park	East Midlands Trains
EC	Craigentinny	Virgin Trains East Coast
EG	Liverpool Edge Hill	Alstom
EK	Shepherdswell	East Kent Railway
EH	Eastleigh	DB Cargo
EY	Ely	Potter Group Logistics
FD	Mobile Maintenance, Mainline Diesels	Freightliner
FE	Mobile Maintenance, Mainline Electrics	Freightliner

FG	Garston	Ford
FP	Mobile Maintenance, Poland	Freightliner
FS	Mobile Maintenance, Diesel Shunters	Freightliner
FT	Fréthun (France)	DB Cargo
FX	Felixstowe	Freightliner
GO	Grosmont	North Yorkshire Moors Railway
HA	Haymarket	ScotRail
HH	Hams Hall	Associated British Ports
HO	Hope	Hope Construction
HQ	Headquarters	
HT	Newcastle Heaton	Northern/Grand Central
HUN	Hungary	Floyd
IS	Inverness	ScotRail
KM	Carlisle Kingmoor	Direct Rail Services
KR	Kidderminster	Severn Valley Railway
KT	Ketton Cement Works	Heidelburg Cement Group
KY	Knottingley	DB Cargo
LA	Plymouth Laira	Great Western Railway
LD	Leeds Midland Road	Freightliner
LE	Swansea Landore	Great Western Railway
LG	Manchester Longsight Electric	Alstom
LH	Barton-Under-Needwood	LH Group Services (Wabtec)
LM	Long Marston	Quiton Rail
LO	Manchester Longsight Diesel	Alstom
LR	Leicester	UK Rail Leasing
LW	Longtown (Smalmstown)	MoD
MA	Manchester International Traincare	Alstom
MB	Middlesbrough	AV Dawson
MD	Merehead	Aggregated Industries
MG	Margam	DB Cargo
MQ	Machen Quarry	Hanson Aggregates
NC	Norwich Crown Point	Abellio Greater Anglia
NL	Neville Hill	East Midlands Trains
OO	Old Oak Common HST	Great Western Railway
OY	Oxley	Alstom
PD	Teesport	PD Ports
PG	Peterborough	GB Railfreight
PM	Bristol St. Philip's Marsh	Great Western Railway
PN	Rybnik/Poznan – Poland	DB Cargo
PO	Polmadie	Alstom
PU	Immingham	Puma Energy
PZ	Penzance	Great Western Railway
RR	Doncaster Roberts Road	EMD/GB Railfreight
RU	Rugby Rail Plant	Colas Rail Freight
SB	Shrewsbury Coleham Yard	Network Rail
SC	Scunthorpe Steelworks	Tata Steel
SE	St. Leonards	St Leonards Engineering
SH	Southall Railway Centre	West Coast Railways
SI	Soho	London Midland

SK	Swanwick	Midland Railway Butterley
SL	Stewarts Lane	Southern
SM	Southampton Maritime	Freightliner
SP	Wigan Springs Branch	DB Cargo
SS	Shotton Steelworks	Tata Steel
TC	Toton Training Compound	DB Cargo
TI	Temple Mills International	Eurostar
TM	Tyseley Locomotive Works	Vintage Trains
TO	Toton	DB Cargo
TR	Trostre Steelworks	Tata Steel
TS	Tyseley	London Midland
TX	Thuxton	Mid Norfolk Railway
TY	Toton Yard	DB Cargo
WB	Wembley Traincare	Alstom
WC	Washwood Heath	Cemex
WD	Widnes	Alstom
WG	Whitemoor Yard	GB Railfreight
WH	Washwood Heath	Boden Rail Engineering/DC Rail
WI	Wishaw	Moveright International
WN	Willesden	GBRf/LOROL
WO	Wolsingham	Weardale Railway
WP	Woippy (France)	DB Cargo
WQ	Headquarters	DB Cargo
WR	Leeming Bar	Wensleydale Railway
WY	Westbury Yard	DB Cargo
WZ	Warsaw – Poland	Freightliner
YK	York	National Railway Museum
ZA	Derby RTC Business Park	Loram
ZB	Doncaster Works	Wabtec
ZC	Crewe Works,	Bombardier Transportation
ZD	Derby Litchurch Lane Works	Bombardier Transportation
ZG	Eastleigh Works	Arlington Fleet Group
ZH	Glasgow Springburn, Works,	Knorr Bremse Rail Systems
ZI	Ilford Level 5 Works	Bombardier Transportation
ZK	Kilmarnock Works	Wabtec
ZN	Wolverton Works	Knorr Bremse Rail Systems
ZO	Kingsbury	European Metal Recycling
ZR	Rotherham	CF Booth Ltd
ZS	Wakefield	RMS Locotec
ZW	Stoke On Trent Works	Axiom Rail/Turners/Marcroft

Livery codes

ADV Advertising / promotional livery
ADZ Advenza Freight blue
AFS Arlington Fleet Services green
AGI Aggregates industries turquoise and silver
ANG Anglia Railways turquoise
ARB Arriva turquoise (DMUs)
ARV Arriva Trains Wales turquoise, unbranded
AVD AV Dawson red
AWC Avanti West Coast dark green
BAF Bardon Aggregates blue with Freightliner branding
BBM Battle of Britain Memorial Flight graphics
BCM Blue and cream
BDB Boston Docks blue
BEA Beacon Rail blue
BGC British Railways green (coaches)
BIF Biffa red bodyside, orange cabs (GBRf)
BLE Unspecified plain blue
BLK British Railways black
BLL British Rail 'large logo' blue with yellow cabs
BLU British Rail blue with yellow cabsides
BMT Bulmarket red (Bulgaria)
BPU Blue Pullman Nanking blue / white
BRB British Rail blue with full yellow ends
BRE British Rail blue with large numbers and emblems
BRF British Rail blue with Union flags
BRG British Rail blue and grey
BRL British Rail blue, large logo blue with black roof
BRM British Railways maroon
BRP British Rail blue grey prototype HST
BRW British Rail blue with wasp stripes
BRY British Railways blue with Foster Yeoman branding
BRZ BZK BR blue
BYP British Rail blue with small yellow panels
BZK BZK (БЪлгарска ЖеиезолЪтна Компания) green and yellow
C2C c2c white
CAC Carmine & Cream
CAL Caledonian Sleeper dark turquoise
CAP Cappagh blue
CAS Castle Cement light grey
CCE Civil engineers' grey / yellow 'Dutch'
CCT Civil engineers' grey / yellow 'Dutch' with Transrail logos
CEL Celsa black with orange cab
CEM Cemex white and blue
CFD Chemins de fer Départméntaux orange
CHC Chocolate & Cream
CLI Avanti WC Climate change
CMX Cemex white (GBRf)
COL Colas Rail Freight orange, yellow and black
CON Continental rail blue (Spain)
COR Corus silver
COT Cotswold Rail silver
COU Colas Rail Freight orange, yellow and black unbranded
COY Corus yellow
CRC Crimson & Cream
CRO Original Chiltern Railways with blue band
CRS Chiltern Railways silver / grey
CRW Chiltern Railways with blue light blue and grey bands
CSM Continental Railway Solution maroon
DBB DB Cargo black 'backbone of Britain'
DBC DB Cargo red
DBM DB Cargo Manager's Train maroon
DBR Deutsche Bahn all-over red
DBS DB Schenker red
DBU DB red unbranded
DCG Devon & Cornwall Railways green
DCN DC Rail revised light grey
DCR DC Rail grey
DEP BR Departmental grey
DGB BZK DRS blue with orange cab
DMS DB Cargo manager's train silver
DNS Nederlandse Spoorwegen grey / yellow
DRA Drax silver
DRB DRS base blue with compass logos
DRC DRS blue with Compass logos
DRE DRS blue with new Compass logos (Class 88)
DRG DRS green

DRN	DRS blue with new Compass logos
DRS	DRS blue, original
DRU	DRS blue, unbranded
DRX	DRS blue with smaller white Compass logos
DST	Deanside Transit lilac
EBY	Electric Blue with yellow panels
ECR	Euro Cargo Rail light grey
EMB	East Midlands Trains plain blue
EMP	East Midlands Railway all over purple
EMR	East Midlands Railway purple / white
EMT	East Midlands Trains blue / red / orange
EMU	East Midlands Trains unbranded
EMW	East Midlands Trains white / red / orange
EMY	GBRf with Emily Woodman graphics
EPX	Europhoenix silver and grey
ESB	Eurostar blue / grey / yellow
ESO	Eurostar original grey
ETF	Eurovia Travaux Ferroviaires yellow
EUE	Eurostar grey with EWS logos
EUK	Eurostar grey
EUT	Eurotunnel grey
EUY	Eurotunnel grey / yellow
EWR	EWS maroon and gold with RSS logos
EWS	EWS maroon and gold
FEC	First East Coast
FER	Fertis grey
FEU	Fertis grey unbranded
FGB	First Great Western blue
FGO	Fragonset black unbranded
FGU	First Great Western blue unbranded
FGS	First Great Western 'special' graphics
FIR	First Group plain blue
FLR	Freightliner green with yellow cabs
FLY	Floyd black
FPG	Freightliner green unbranded
FPH	Freightliner 'Powerhaul' green, yellow and grey
FRG	Fragonset black
FTF	Fall the Fallen graphics
GAG	Greater Anglia grey
GAW	Greater Anglia white
GBB	GB Railfreight blue with orange numbers
GBF	GB Railfreight blue and orange
GBP	GB Railfreight with pride rainbows
GBR	GB Railfreight blue and orange Europorte style
GBO	GB Railfreight blue and orange original style
GBU	GB Railfreight umbranded
GBZ	GB Railfreight blue and orange with minor variations
GCH	DB Cargo Climate hero green
GCR	Grand Central black / orange
GEM	Gemini Rail Group dark grey (RSS)
GEX	Gatwick Express red
GFY	British Railways green full yellow ends
GLA	Glaxo chemicals blue and dark grey
GOP	Golden Ochre with yellow panels
GNY	British Railways green with no yellow ends
GRC	Green & cream
GRE	Unspecified plain green
GRW	British Railways green with wasp stripes
GWA	Great Western Railway all over green with graphics
GWD	Genesee & Wyoming darker orange and black
GWE	Great Western Railway (pre-BR) green
GWO	Genesee & Wyoming lighter orange and black
GWS	British Rail green with wasp stripes
GWR	Great Western Railway all over green
GYP	British Railways green with yellow panels
HAN	Hanson aggregates blue and silver
HAR	'Harry Patch' black graphics
HEC	Heathrow Express Connect
HEX	Heathrow Express silver
HNO	Harry Needle Railroad Company orange
HOP	Hope Construction white with purple solebar
HST	Original blue, grey, yellow HST
HUL	Hull Trains blue
HUN	Hunslet blue and orange
ICE	BR InterCity Executive
ICM	BR InterCity 'Mainline'
ICO	BR InterCity original style
ICS	BR InterCity Swallow style

IGX	BR InterCity Gatwick Express
IOS	BR InterCity original with ScotRail branding
IND	unspecified Industrial livery
ISL	Island Line red
JUB	DB Cargo Diamond Jubilee silver
JFU	Jarvis Fastline unbranded grey
KBR	Knorr Bremse green, white and blue
KER	Kernow black
LAB	'Laira' blue with grey roof
LAM	Lamco orange
LEM	LNER red with EMR branding
LHO	Loadhaul
LMR	London Midland Railway dark green / black / white
LMS	London Midland Scottish railway maroon
LNE	LNER white / red
LNR	BZK L&NWR blackberry black
LNW	LNWR dark green / grey
LOM	Loram red, white and grey
LOR	London Overground
LSW	LSWR black
LUB	GBRf London Transport Museum black with graphics
LUL	LUL brown
LUM	Lumo all over blue
LUW	GBRf London Transport Museum white with graphics
LWR	L&NWR blackberry black
LZY	LZY green, white and purple (Eurostar)
MAA	DB Cargo with WH Malcolm graphics
MAL	WH Malcolm, green, yellow and blue
MAR	Lakeside & Haverthwaite Railway lined maroon
MER	Merseyrail grey / yellow
MEW	Mainline Freight blue with EWS logos
MFY	Maroon with full yellow ends
MID	Midland Railway maroon
MLB	Mainline Freight blue
MRD	DB Cargo Maritime blue
MRM	Metropolitan Railway maroon
MRT	GBRf Maritime blue
MSC	GBRf with Medite Sorrento graphics
MWS	Maroon with wasp stripes
MYP	Maroon with yellow panels
NBL	Northern plain blue
NBU	Northern Belle umber and cream unbranded
NCB	National Coal Board blue
NEX	National Express
NOB	Northern Belle umber and cream
NOR	Northern white with dark blue cabs
NRA	National Railway Museum advertising wrap
NRB	National Railway Museum light blue
NRM	National Railway Museum maroon
NRY	Network Rail yellow
NSD	Network SouthEast revised darker blue
NSE	BZK Network SouthEast red, white and blue
NSO	Network SouthEast original with white window frames
NSR	Network SouthEast revised with blue window frames
ONE	Ocean Network Express magenta
ORO	Orion Freight blue (ROG)
OXB	Oxford blue
PCC	Prostate Cancer UK GBRf
PDP	PD Ports blue
POW	Powder blue
PRI	Pride vinyl
PUM	Puma Energy dark grey
PUL	Pullman umber and cream
RAV	RailAdventure grey
RCA	Railcare red, white and blue
RCG	Railcare grey and white
RCS	Rail Charter Services green
RED	Unspecified plain red
REG	Regional Railways blue and grey
RES	Rail Express systems red and dark grey
REW	Railfreight Distribution 'European' two tone grey with EWS logos
RFD	two tone grey with Railfreight Distribution logos
RFE	Railfreight Distribution 'European' two tone grey with RfD logos
RFO	Railfreight Original grey
RFS	RFS grey
RMB	RMS Locotec black
RMS	RMS Locotec blue
ROG	Rail Operations Group blue
ROM	Royal Mail postal red

ROU	Rail Operations Group unbranded blue
ROY	Royal Scotsman plum
RSR	Railfreight Red stripe
RSS	Rail Support Services grey
RTC	Railway Technical Centre red and blue
RTO	Royal Train plum
RTP	Royal Train Res style
SCO	British Rail ScotRail
SCP	ScotRail purple
SCR	ScotRail Saltire blue
SCT	ScotRail Seven Cities
SEB	Southeastern blue
SEC	Serco green
SEF	Swift Express Freight blue
SEW	Southeastern white
SIL	Silverlink green, purple and white
SOU	Southern green and white
SRG	Southern Railway green
STO	Stobart Rail blue and white
SWB	South West Trains (Stagecoach) blue
SWM	South West Trains Metro (Stagecoach) red
SWO	South Western Railway (initial trial livery)
SWR	South Western Railway grey/white
SWU	South West Trains blue unbranded
SWW	South West Trains (Stagecoach) white
TAB	Tata Blue
TAS	Tata Silver
TEK	Teak
TEW	Trainload grey with EWS logos
TFL	Transport for London
TFW	Transport for Wales
TLA	Trainload grey with Aggregates logos
TLC	Trainload grey with Coal logos
TLH	Trainload grey with Loadhaul logos
TLK	Thameslink white
TLM	Trainload grey with Metals logos
TLP	Trainload grey with Petroluem logos
TMF	Trainload grey with Mainline Freight logo
TMT	Transmart Trains green
TPE	Transpennine Express silver/blue/purple
TTG	Two tone unbranded Railfreight grey
TRN	Transrail grey
TSO	TSO yellow
UKR	UK Rail Leasing grey with yellow cabs
UND	Undercoat/unpainted/primer
VEC	Virgin Trains East Coast with LNER branding
VEM	Virgin Trains East Coast with EMR branding
VFS	Virgin Trains Flying Scotsman (now LNER)
VIR	Virgin Trains red
VIS	Virgin Trains silver
WAB	Wabtec Black
WCR	West Coast Railways maroon with small yellow panels
WHI	plain unbranded white
WMG	West Midlands Trains gold/purple
XCT	CrossCountry Trains
XRL	Crossrail

Pool codes

ACAC	AC Locomotive Group AC Electric Locomotives
ACXX	AC Locomotive Group AC Electric Locomotives stored
ATLO	Alstom Traincare Locomotives
ATZZ	Alstom Traincare Locomotives For Disposal
AWCA	West Coast Railway Operational Diesel Locomotives
AWCX	West Coast Railway Stored Diesel Locomotives
BREL	Boden Rail Engineering
CDJD	Central Services/Serco Railtest Ex Serco Shunters
CFOL	Class 50 Operations Ltd
CFSL	Class 40 Stored Locos
COFS	Colas Rail Freight
COLO	Colas Rail Freight Hire Locomotives
COLS	Colas Rail Freight Stored Locomotives
COTS	Colas Rail Freight Locomotives For Refurbishing
DBLX	Deltic Preservation Society
DCRO	DCR Class 50/56
DCRS	DCR Class 60
DDIN	Freightliner Shunter Fleet
DFGI	Freightliner Class 70
DFHG	Freightliner Class 59

DFHH	Freightliner Class 66/6 Heavy Haul
DFIM	Freightliner Intermodal Modified
DFIN	Freightliner Intermodal Low Emission
DFLC	Freightliner Intermodal
DFLH	Freightliner Heavy Haul
DFNC	Freightliner Awaiting Maintenance
DHLT	Freightliner Stored/Not In Main Line Use Locomotives
EFOO	First Great Western Class 57/6
EFPC	First Great Western Class 43
EFSH	First Great Western Shunters
EHPC	Arriva CrossCountry HST Power Cars
EJLO	London Midland Shunters
ELRD	East Lancashire Railway Operational Locomotives
EMSL	East Midlands Trains Shunters
EPEX	Europhoenix For Scrap/Export
EPUK	Europhoenix UK Locomotives
EROG	Eversholt/Rail Operations Group Class 91
ERSL	Eastern Rail Services Locomotives
FAMT	Private Owner Class 37
GBBR	GBRf Class 73/9 – Brush Repowered
GBBT	GBRf UK Cab – Long Range Fuel Tanks
GBCH	GBRf Caledonian Sleeper
GBCS	GBRf Re-Engineered
GBCT	GB Railfreight Class 92 UK/Channel Tunnel
GBDF	GB Railfreight Class 47
GBEB	GBRf Euro Cab – Long Range Fuel Tanks
GBED	GBRf Electro Diesel Locos For Hire
GBEL	GBRf Euro Cab -Standard Fuel Tanks
GBET	GB Railfreight Class 92
GBFM	GBRf RETB Fitted Locomotives
GBGD	GBRf Class 56 operational
GBGS	GBRf Class 56 stored
GBHH	GB Railfreight Class 66 heavy Haul (66793-796)
GBKP	GB Railfreight Class 67
GBLT	GBRf UK Cab – Standard Fuel Tanks
GBNB	GBRf New Build Locos
GBNL	GB Railfreight Class 66 ex Netherlands
GBNR	GBRf For Network Rail Use
GBOB	GB Railfreight Class 66 Big Fuel Tanks & Buckeye Couplers
GBRG	GB Railfreight Class 69
GBRT	GBRf Restricted Locos
GBSD	GBRf Stored Locos
GBSL	GBRf Caledonian Sleepers
GBST	GBRf Caledonian Sleepers/Channel Tunnel
GBTG	GB Railfreight Class 60
GBWM	GBRf Shunting Duties
GBYH	GBRf General Pool
GPSS	Eurostar UK Operate From TI (DBC Maintained)
GROG	Rail Operations Group Operational Locos
HAPC	ScotRail Class 43
HBSH	LNER Class 08
HISE	Rail Vehicle Engineering East Midlands Trains Shunters
HTLX	Hanson Traction Operational Locomotives
HVAC	Hanson & Hall Class 50
HYWD	South Western Railway Thunderbird Locos
ICHP	125 Group HST Power Cars
IECA	Virgin Trains East Coast Operational Locomotives
KDSD	Bombardier Doncaster
LSLO	Locomotive Services Limited, operational locos
LSLS	Locomotive Services Limited, stored locos
MBDL	Non TOC Private Owner – Diesel locos
MBED	Non TOC Private Owner – Class 73
MBEL	Non TOC Private Owner – Electric locos
MOLO	RT Rail Limited Hired Fleet Shunter locos
MRLO	RMS Locotec Ex-FM Rail Operational Locos
MRLS	RMS Locotec Ex-FM Rail Stored locos
MRSO	RMS Locotec Ex-FM Rail Operational Shunters
NRLO	Nemesis Rail Locomotives On Hire
NRLS	Nemesis Rail Ex-FM Rail Stored locos
QACL	Network Rail Load Bank
QADD	Network Rail Diesel Locos
QCAR	Network Rail HST Power Cars
QETS	Network Rail European Signalling

RAJV Bo'ness & Kinneil Railway Locomotives
RCZH Knorr Bremse Rail Systems Springburn Works Shunters
RCZN Knorr Bremse Rail Systems Wolverton Works Shunters
RFSH Wabtec Rail Locomotives
RMSX RMS Locotec Locomotives
RTSO Riviera Trains Operational Shunters
RVLO Railway Vehicle Engineering Derby Operational Locomotives
SAXL Eversholt Rail Off Lease Locos
SBXL Porterbrook Leasing Off Lease Locos
SCEL Angel Train Contracts Off Lease Locos
SROG Rail Operations Group Stored Locomotives
TPCF Transpennine Class 68
TPEX Transpennine Class 68
TTLS Traditional Traction/Railway Support Services
TTTC Private Owner
UKRL UK Rail Leasing On Lease
UMRM UK Rail Leasing Not Main Line
UKRS UK Rail Leasing Stored
WAAC DB Cargo UK
WABC DB Cargo UK RETB Fitted
WAWC DB Cargo UK Arriva Wales Hire
WBAE DB Cargo UK Fitted With Stop/Start Technology
WBAR DB Cargo UK Remote Condition Monitoring Equipment
WBAT DB Cargo UK General
WBBE DB Cargo UK RETB & Stop/Start Technology Fitted
WBBT DB Cargo UK RETB Fitted
WBLE DB Cargo UK Lickey Bankers With Stop/Start Technology
WBLT DB Cargo UK
WBTT DB Cargo UK RHTT – Tripcock Fitted
WCAT DB Cargo UK Standard Fuel Range
WCBT DB Cargo UK Extended Fuel Rail
WDAM DB Cargo UK
WEAC DB Cargo UK
WFBC DB Cargo UK HS1 Equipped
WGEA DB Cargo UK Euro Cargo Rail
WGEE DB Cargo UK Eastern Europe
WGEP DB Cargo UK Poland
WQAA DB Cargo UK Locomotives Stopped Serviceable – Group 1A
WQAB DB Cargo UK Stored Locomotives Group 1B
WQBA DB Cargo UK Stored Locomotives Stored Serviceable – Group 2
WQCA DB Cargo UK Stored Locos For Component Recovery – Group 3
WQDA DB Cargo UK Stored Locomotives Surplus – Group 4
XHAC Direct Rail Services Operational Locos – ETS Equipped
XHCE Direct Rail Services Hire To Chiltern Railways
XHCS DRS Class 68 Chiltern Support
XHIM Direct Rail Services Intermodal Locos
XHSO Direct Rail Services Supply Chain Operations
XHTP Direct Rail Services Locos For Transpennine Express
XHVE Direct Rail Services Vossloh Locos
XHVT Direct Rail Services West Coast Thunderbird Locos
XSDP Direct Rail Services Locomotives Stored/For Disposal
XWSS Direct Rail Services Stored Locos

Owner codes

ACL AC Locomotive Group
AGI Agility Trains
AGO Andrew Goodman
ALS Alstom
ANG Angel Trains
ARV Arriva Group
BEA Beacon Rail
BEV Beaver Sports
BEN Steve Beniston
BOD Neil Boden
BRO Brodie Rail Engineering
CAP Cappagh Group (DC Rail)
CDP Crewe Diesel Preservation Group
CFA Class 50 Alliance
CFP Class 40 Preservation Society
CFS Class 56 Group
CLT Cross London Trains
COL Colas Rail Freight
CRL Caledonian Rail Leasing

CTL	Class 20 Locomotive Society
D05	D05 Preservation Group
DBC	DB Cargo
DCR	DC Rail
DfT	Department for Transport
DPS	Deltic Preservation Society
DRS	Direct Rail Services
DTG	Diesel Traction Group
EEG	English Electric Group
EMT	East Midlands Trains
EPX	Europhoenix
EUK	Eurostar
EVS	Eversholt Leasing
FIR	First Group
FLI	Freightliner
GAR	Garcia Hanson
GBR	GB Railfreight
GWR	Great Western Railway
HAN	Hanson Aggregates
HFX	Halifax Leasing
HJE	Howard Johnston
HNR	Harry Needle Railroad Company
LES	Les Ross
LOL	London Overground
LOM	Lombard Finance
LON	London Midland
LSL	Locomotive Services Limited
AIK	Akiem
MET	MerseyTravel
MOW	Michael Owen
NEM	Nemesis Rail
NET	Network Rail
NRM	National Railway Museum
POR	Porterbrook
ROC	Rock Rail
ROG	Rail Operations Group
SFG	Stratford Class 47 Group
SHW	Shawn Wright
SOA	71A Locomotives
SRP	Scottish Railway Preservation Society
STG	Scottish Class 37 Group
SWR	South Western Railway
UKR	UK Rail Leasing
VIN	Vintage Trains
WCR	West Coast Railways
XRLL	Crossrail

Spot hire/industrial owners

AFS	Arlington Fleet Services
AVD	AV Dawson
BOM	Bombardier
BMT	Bulmarket (Bulgaria)
BZK	BZK (БЪлгарска ЖеиезолЪтна Компания – Bulgaria)
CON	Continental Rail Solutions (Hungary)
CRB	Chris Beet
EMD	Electromotive Diesels
EMR	European Metal Recycling
FLY	Floyd (Hungary)
ITY	Private owner in Italy
KBR	Knorr Bremse
LAM	Lamco Mining
LOR	Loram
NYM	North Yorkshire Moors Railway
RSS	Railway Support Services (Traditional Traction)
SEC	Serco
SLE	St Leonards Engineering
TFA	Transfesa
TMT	Transmart Trains
TLW	Tyseley Locomotive Works
VIC	Victoria Group (Boston Docks)
WAB	Wabtec

Heritage railway locations

ALL	Allely's yard, Studley
ALN	Aln Valley Railway
AVR	Avon Valley Railway
BAT	Battlefield Line
BH	Barrow Hill Roundhouse
BKR	Bo'ness & Kinneil Railway
BIR	Barry Island Railway
BLU	Bluebell Railway
BRC	Buckingham Railway Centre
BWR	Bodmin & Wenford Railway
BU	Burton upon Trent
CAL	Caledonian Railway
CHR	Chasewater Railway
CHV	Churnet Valley Railway
CPR	Chinnor & Princes Risborough Railway
CRT	Cambrian Railways Trust
CWR	Cholsey & Wallingford Railway
CVR	Colne Valley Railway

DFR	Dean Forest Railway
DAL	Darlington Railway Museum
DAR	Dartmoor Railway
DRC	Didcot Railway Centre
DVR	Derwent Valley Light Railway
EBR	Embsay Steam Railway
EDR	Eden Valley Railway
EKR	East Kent Railway
ELP	Port Elphinstone, private site
ELR	East Lancashire Railway
EOR	Epping and Ongar Railway
ESR	East Somerset Railway
EVR	Ecclesbourne Valley Railway
FHR	Fawley Hill Railway
GWR	Gloucestershire Warwickshire Railway
GCR	Great Central Railway
GCN	Great Central Railway Nottingham
GWI	Gwili Railway
IWR	Isle of Wight Steam Railway
KWH	Kinsley Warehouse storage
KWV	Keighley & Worth Valley Railway
KES	Kent & East Sussex Railway
LHR	Lakeside & Haverthwaite Railway
LMR	Llanelli & Mynydd Mawr Railway
LLR	Llangollen Railway
LWR	Lincolnshire Wolds Railway
MAL	Private site Malton
MAR	Margate
MHR	Mid Hants Railway
MNR	Mid Norfolk Railway
MOL	Moreton-on-Lugg
MRB	Midland Railway – Butterley
MRM	Mangapps Railway Museum
NRM	National Railway Museum York
NRS	National Railway Museum Shildon
NLR	Northampton and Lamport Railway
NNR	North Norfolk Railway
NVR	Nene Valley Railway
NYM	North Yorkshire Moors Railway
NTR	North Tyneside Railway
PBR	Pontypool & Blaenavon Railway
PDR	Paignton & Dartmouth Railway
PKR	Peak Rail
PVR	Plym Valley Railway
RAC	Railway Age Crewe
RDR	Royal Deeside Railway
RHR	Rushden Heritage Railway
RSR	Ribble Steam Railway
RVR	Rother Valley Railway
SCR	Swindon & Cricklade Railway
SDR	South Devon Railway
SPA	Spa Valley Railway
STM	Stainmore Railway
STR	Strathspey Railway
SWR	Swanage Railway
SVR	Severn Valley Railway
TBR	Trawsfynydd & Blaenau Railway
TEB	Tebay
TIT	Titley Junction
TSR	Telford Steam Railway
VBR	Vale of Berkeley Railway
WH	Washwood Heath
WIS	Wishaw
WEN	Weardale Railway
WEA	Wensleydale Railway
WSR	West Somerset Railway

Vehicle check list

This list is the locos and multiple units (by unit number) that remain either in the UK or abroad. Their TOPS numbers are listed regardless of what identity they carry. Eurotunnel/Eurostar trains are also listed, while preserved locos are included at the end.

Two locos included here but not in the main body of text are the remains of collision write-offs 66048 and 70012. Their damaged bodies remain intact at Longport and Erie (Pennsylvania) respectively but the vehicles are not owned by a FOC, TOC or ROSCO.

Vehicle check list codes

Code	Name
71A	71A Locomotives Ltd
ABB	ABB Ports
AFS	Arlington Fleet Services
AGA	Abellio Greater Anglia
AGI	Aggregates Industries
AIK	Akiem Off Lease
ANG	Angel Trains Off Lease
ARV	Arriva
AVD	AV Dawson
AWC	Avanti West Coast
BEN	Steve Beniston
BRE	Boden Rail Engineering
C50A	Class 50 Alliance
CAL	Caledonian Sleeper
CFPS	Class 40 Preservation Society
CHR	Chiltern Railways
COL	Colas Rail
CRS	Continental Rail Hungary
CTLS	Class 20 Preservation Society
DBC	DB Cargo
DCR	DC Rail
DPS	Deltic Preservation Society
DRS	Direct Rail Services
DTG	Diesel Traction Group
EMR	East Midlands Railway
EMS	Ed Murray & Sons
ERS	Eastern Rail Services
EUS	Eurostar
EUT	Eurotunnel
EUX	Rail Operations Group
EV	Eversholt Off Lease
EXP	Exported
FLR	Freightliner
GBR	GB Railfreight
GCR	Grand Central
GTR	Govia Thameslink Railway
GWR	Great Western Railway
HAN	Hanson & Hall Traction
HEX	Heathrow Express
HDL	Hastings Diesels
HNRC	Harry Needle Railroad Company
HUL	Hull Trains
LNER	London North Eastern Railway
LNWR	London Midland Trains
LOL	London Overground
LSL	Locomotive Services Limited
LUM	Lumo
MER	Merseyrail
MET	Meteor Power
MO	20189 Ltd
NEM	Nemesis Rail
NOR	Northern
NR	Network Rail
NRL	Northumbrian Rail Limited
NRM	National Railway Museum
PBK	Porterbrook Off Lease
Pres	Preserved
RAV	Rail Adventure
RIV	Riviera Trains
RMS	RMS Locotec
ROG	Rail Operations Group
RSS	Rail Support Services
S37G	Scottish Class 37 Group
SCR	ScotRail
SEM	Sembcorp Utilities
SER	Southeastern
SFG	Stratford Class 47 Group
SLRE	St Leonards Rail Engineering
SRPS	Scottish Railway Preservation Society
SWR	South Western Railway
(S)	Stored
TBC	To Be Converted
TFL	Transport for London
TFW	Transport for Wales
TMT	Transmart Trains
TPE	Transpennine Express
UKRL	UK Rail Leasing
VP	Victoria Ports
VT	Vintage Trains
VUL	Vulcan Rail
WCR	West Coast Railways
WCTC	Alstom West Coast Traincare
WMT	West Midlands Trains
XCT	CrossCountry Trains

Fleet list

Class 01
D2953 Pres
D2956 Pres

Class 02
02003 Pres
D2854 Pres
D2858 Pres
D2860 Pres
D2866 Pres
D2867 Pres
D2868 Pres

Class 03
03018 Pres
03020 Pres
03022 Pres
D2023 Pres
D2024 Pres
03027 Pres
03037 Pres
D2041 Pres
D2046 Pres
D2051 Pres
03059 Pres
03062 Pres
03063 Pres
03066 Pres
03069 Pres
03072 Pres
03073 Pres
03078 Pres
03079 Pres
03081 Pres
03084 WCR
03089 Pres
03090 Pres
03094 Pres
03099 Pres
03112 Pres
03113 Pres
D2117 Pres
03118 Pres
03119 Pres
03120 Pres
D2133 Pres
03134 Pres
D2138 Pres
D2139 Pres
03141 Pres
03144 Pres
03145 Pres
D2148 Pres
03152 Pres
03158 Pres
03162 Pres
03170 Pres
D2178 Pres
03179 Pres
03180 Pres
D2182 Pres
D2184 Pres
03189 Pres
D2192 Pres
03196 WCR
03197 Pres
D2199 Pres
03371 Pres
D2381 WCR
03399 Pres
03901 Pres

Class 04
D2203 Pres
D2205 Pres
D2207 Pres
D2229 Pres
D2245 Pres
D2246 Pres
D2271 Pres
D2272 Pres
D2279 Pres
D2280 Pres
D2284 Pres
D2289 Pres
D2298 Pres
D2302 Pres
D2310 Pres
D2324 Pres
D2325 Pres
D2334 Pres
D2337 Pres

Class 05
05001 Pres
D2578 Pres
D2587 Pres
D2595 Pres
Class 06
06003 Pres
Class 07
07001 Pres
07005 Pres
07007 AFS
07010 Pres
07011 SLRE
07012 Pres
07013 Pres
Class 08
D3000 Pres
D3002 Pres
D3014 Pres
08011 Pres
D3019 Pres
08015 Pres
08016 Pres
08021 VT
08022 Pres
08032 Pres
08046 Pres
08054 Pres
08060 Pres
08064 Pres
D3101 Pres
08102 Pres
08108 Pres
08114 Pres
08123 Pres
08133 Pres
08164 Pres
08168 NEM
D3255 Pres
D3261 Pres
08195 Pres
08202 Pres
08220 Pres
08238 Pres
08266 Pres
08288 Pres
08296 AGI
08308 RMS
08331 Pres
08359 Pres
08375 VP
08377 Pres
08389 HNRC
08401 EMS
08405 RSS
08410 AVD
08411 RSS
08417 HNRC
08418 WCR
08423 RMS
08428 HNRC
08436 Pres
08441 RSS
08442 ARV
08443 Pres
08444 Pres
08445 EMS
08447 John Russell
08451 WCTC
08454 WCTC
08460 RSS
08471 Pres
08472 EMS
08473 Pres
08476 Pres
08479 Pres
08480 RSS
08483 LSL
08484 RSS
08485 WCR
08490 Pres
08495 Pres
08499 COL
08500 HNRC
08502 HNRC
08503 Pres
08507 RIV
08511 RSS
08516 ARV
08523 RMS
08525 EMR
08527 HNRC
08528 Pres
08530 FLR
08531 FLR
08536 VUL
08556 Pres
08567 AFS
08568 Pres
08571 EMS
08573 RMS
08575 FLR
08578 HNRC
08580 RSS
08585 FLR
08588 RMS
08590 Pres
08593 RSS
08596 EMS
08598 AVD
08600 AVD
08602 HNRC
08604 Pres
08605 RIV
08611 WCTC
08613 RMS
08615 EMS
08616 WMT
08617 WCTC
08622 RMS
08623 HNRC
08624 FLR
08629 RSS
08630 HNRC
08631 LSL
08632 RSS
08633 Pres
08635 Pres
08641 GWR
08643 AGI
08644 GWR
08645 GWR
08648 RMS
08649 MET
08650 AGI
08652 AGI
08653 HNRC
08663 Pres
08669 EMS
08670 RSS
08676 HNRC
08678 WCR
08682 HNRC
08683 RSS
08685 HNRC
08690 EMR
08691 FLR
08694 Pres
08696 WCTC
08700 RMS
08701 HNRC
08703 RSS
08704 RIV
08706 HNRC
08709 Pres
08711 HNRC
08714 HNRC
08721 WCTC
08724 EMS
08730 ABB
08735 ARV
08737 LSL
08738 RSS
08742 HNRC
08743 SEM
08752 RSS
08754 RMS
08756 RMS
08757 RSS
08762 RMS
08764 WCTC
08765 HNRC
08767 Pres
08769 Pres
08772 Pres
08773 Pres
08774 AVD
08780 LSL
08782 HNRC
08783 EMR
08784 VUL
08785 FLR
08786 HNRC
08788 RMS
08790 WCTC
08795 Pres
08798 HNRC
08799 HNRC
08802 HNRC
08804 HNRC
08805 WMT
08809 RMS
08810 ARV
08818 HNRC
08822 GWR
08823 EMS
08824 HNRC
08834 HNRC
08825 Pres
08830 Pres
08836 GWR
08846 RSS
08847 RMS
08850 Pres
08853 EMS
08865 HNRC
08868 ARV
08870 ERS
08871 RMS
08872 EMR
08874 RMS
08877 HNRC
08879 HNRC
08881 Pres
08885 RMS
08887 WCTC
08888 Pres
08891 FLR
08892 HNRC
08896 Pres
08899 EMR
08903 SEM
08904 HNRC
08908 EMR
08905 HNRC
08907 Pres
08911 Pres
08912 AVD
08913 EMR
08915 Pres
08918 HNRC
08921 RSS
08922 VUL
08924 HNRC
08925 GBR
08927 RSS
08933 AGI
08934 GBR
08936 RMS
08937 Pres
08939 RSS
08943 HNRC
08944 HNRC
08947 AGI
08948 EUS
08950 EMR
08954 WCTC
08956 HNRC
08993 Pres
08994 Pres
08995 Pres
Class 09
09001 Pres
09002 GBR
09004 Pres
09006 HNRC
09007 LOL
09009 GBR
09010 Pres
09012 Pres
09014 HNRC
09015 Pres
09017 NRM
09018 Pres
09019 Pres
09022 VP
09023 EMR
09024 Pres
09025 Pres
09026 Pres
09106 HNRC
09107 Pres
09201 HNRC
09204 ARV
Class 10
D3452 Pres
D3489 Pres
D4067 Pres
D4092 Pres
Class 11
12052 Pres
12077 Pres
12082 Pres
12083 Pres
12088 Pres
12093 Pres
12099 Pres
12131 Pres
Class 12
15224 Pres

Unclassified locos

D2511 Pres
D2767 Pres
D2774 Pres
D226 Pres
18000 Pres
DP1 Pres
PWM/Class 97
97650 Pres
97651 Pres
97654 Pres
Class 14
D9500 Pres
D9502 Pres
D9504 Pres
D9513 Pres
D9516 Pres
D9518 Pres
D9520 Pres
D9521 Pres
D9523 Pres
D9524 Pres
D9525 Pres
D9526 Pres
D9529 Pres
D9531 Pres
D9537 Pres
D9539 Pres
D9551 Pres
D9553 Pres
D9555 Pres
Class 15
D8233 Pres
Class 17
D8568 Pres
Class 20
20001 Pres
20007 MO
20016 Pres
20020 Pres
20031 Pres
20048 MO
20050 Pres
20056 HNRC
20057 Pres
20059 Pres
20063 Pres
20066 HNRC
20069 HNRC
20081 Pres
20087 HNRC
20088 Pres
20096 LSL
20098 Pres
20107 LSL

20110 HNRC
20118 HNRC
20121 HNRC
20132 HNRC
20137 Pres
20142 MO
20154 Pres
20166 HNRC
20168 HNRC
20169 Pres
20188 Pres
20189 MO
20205 MO
20214 Pres
20227 CTLS
20228 Pres
20301 HNRC
20302 LSL
20303 HNRC
20304 HNRC
20305 LSL
20308 HNRC
20309 HNRC
20311 HNRC
20312 HNRC
20314 HNRC
20901 HNRC
20903 HNRC
20904 HNRC
20905 HNRC
20906 HNRC

Class 23

D5910 Pres

Class 24

24032 Pres
24054 Pres
24061 Pres
24081 Pres

Class 25

25035 Pres
25057 HNRC
25059 Pres
25067 NEM
25072 Pres
25083 Pres
25173 Pres
25185 Pres
25191 Pres
25235 Pres
25244 Pres
25262 Pres
25265 NEM
25278 Pres
25279 Pres
25283 HNRC
25309 Pres
25313 HNRC
25321 Pres
25322 Pres

Class 26

26001 Pres
26002 Pres
26004 NEM
26007 Pres
26010 Pres
26011 NEM
26014 Pres
26024 Pres
26025 Pres
26035 Pres
26038 Pres
26040 Pres
26043 Pres

Class 27

27001 Pres
27005 Pres
27007 Pres
27024 Pres
27050 Pres
27056 Pres
27059 Pres
27066 HNRC

Class 28

D5705 Pres

Class 31

31018 Pres
31101 Pres
31105 Pres
31106 HAN
31108 Pres
31119 Pres
31128 NEM
31130 Pres
31162 Pres
31163 Pres
31190 Pres
31203 Pres
31206 Pres
31207 Pres
31210 Pres
31233 Pres
31235 Pres
31255 Pres
31270 Pres
31271 Pres
31285 Pres
31289 Pres
31327 Pres
31418 Pres
31430 Pres
31435 Pres
31438 Pres
31452 ERS
31454 Pres
31459 Pres
31461 NEM
31463 Pres
31465 Pres
31466 Pres
31514 Pres
31601 Pres

Class 33

33002 Pres
33008 Pres
33012 71A
33018 Pres
33019 NEM
33021 Pres
33025 WCR
33029 WCR
33030 WCR
33035 Pres
33046 Pres
33048 Pres
33052 Pres
33053 Pres
33057 Pres
33063 Pres
33065 Pres
33102 Pres
33103 Pres
33108 Pres
33109 Pres
33110 Pres
33111 Pres
33116 Pres
33117 Pres
33201 Pres
33202 Pres
33207 WCR
33208 Pres

Class 35

D7017 Pres
D7018 Pres
D7029 Pres
D7076 Pres

Class 37

37003 Pres
37009 Pres
37023 Pres
37025 S37G
37029 Pres
37032 Pres
37037 Pres
37038 DRS
37042 Pres
37057 COL
37059 DRS
37069 DRS
37075 Pres
37097 Pres
37099 COL
37108 Pres
37109 Pres
37116 COL
37142 Pres
37165 WCR
37175 COL
37190 Pres
37198 Pres
37207 MET
37214 Pres
37215 Pres
37216 Pres
37218 DRS
37219 COL
37227 Pres
37240 BRE
37248 Pres
37250 Pres
37254 COL
37255 NEM
37259 DRS
37261 Pres
37263 Pres
37264 Pres
37275 Pres
37294 Pres
37308 Pres
37310 Pres
37350 Pres
37401 DRS
37402 DRS
37403 SRPS
37405 HNRC
37407 DRS
37409 DRS
37418 BEN
37419 DRS
37421 COL
37422 DRS
37423 DRS
37424 DRS
37425 DRS
37503 BEN
37510 EUX
37516 WCR
37517 WCR
37518 WCR
37521 LSL
37601 EUX
37602 DRS
37603 HNRC
37604 HNRC
37605 DRS
37606 BEN
37607 HNRC
37608 EUX
37609 HNRC
37610 HNRC
37611 EUX
37612 HNRC
37667 LSL
37668 WCR
37669 WCR
37674 Pres
37676 WCR
37679 Pres
37685 WCR
37688 LSL
37703 HNRC
37706 WCR
37712 WCR
37714 Pres
37716 DRS
37800 EUX
37884 EUX
37901 EUX
37905 UKRL
37906 UKRL

Class 40

40012 Pres
40013 LSL
40106 Pres
40118 Pres
40122 Pres
40135 Pres
40145 CFPS

Class 41

41001 Pres

Class 42

D821 Pres
D832 Pres

Class 43

43002 Pres
43003 SCR
43004 GWR
43005 GWR
43009 GWR
43010 GWR
43012 SCR
43013 NR
43014 NR
43015 SCR
43016 GWR
43017 PBK (S)
43018 Pres
43020 PBK (S)
43021 SCR
43022 GWR
43023 PBK (S)
43024 PBK (S)
43025 PBK (S)
43026 SCR
43027 GWR
43029 GWR
43028 SCR
43030 SCR
43031 SCR
43032 SCR
43033 SCR
43034 SCR
43035 SCR
43036 SCR
43037 SCR
43040 GWR
43041 GWR
43042 GWR
43043 PBK (S)
43044 Pres
43045 NRL
43046 LSL
43047 LSL
43048 Pres
43049 LSL
43050 PBK (S)
43052 DATS
43054 DATS
43055 LSL
43056 Pres
43058 LSL
43059 LSL
43060 NRL
43062 NR
43063 PBK (S)
43064 PBK (S)
43066 DATS
43069 GWR
43071 Pres
43073 Pres
43076 DATS
43078 GWR
43081 Pres
43082 Pres
43083 LSL
43086 GWR
43087 GWR
43088 GWR
43089 Pres
43091 PBK (S)
43092 GWR
43093 GWR
43094 GWR
43097 GWR
43098 GWR
43102 Pres
43122 GWR
43124 SCR
43125 SCR
43126 SCR
43127 SCR
43128 SCR
43129 SCR
43130 SCR
43131 SCR
43132 SCR
43133 SCR
43134 SCR
43135 SCR
43136 SCR
43137 SCR
43138 SCR
43139 SCR
43141 SCR
43142 SCR
43143 SCR
43144 SCR
43145 SCR
43146 SCR
43147 SCR
43148 SCR
43149 SCR
43150 SCR
43151 SCR
43152 SCR

66534 FLR
66535 EXP
66536 FLR
66537 FLR
66538 FLR
66539 FLR
66540 FLR
66541 FLR
66542 FLR
66543 FLR
66544 FLR
66545 FLR
66546 FLR
66547 FLR
66548 FLR
66549 FLR
66550 FLR
66551 FLR
66552 FLR
66553 FLR
66554 FLR
66555 FLR
66556 FLR
66557 FLR
66558 FLR
66559 FLR
66560 FLR
66561 FLR
66562 FLR
66563 FLR
66564 FLR
66565 FLR
66566 FLR
66567 FLR
66568 FLR
66569 FLR
66570 FLR
66571 FLR
66572 FLR
66582 EXP
66583 EXP
66584 EXP
66585 FLR
66586 EXP
66587 FLR
66588 FLR
66589 FLR
66590 FLR
66591 FLR
66592 FLR
66593 FLR
66594 FLR
66595 EXP
66596 FLR
66597 FLR
66598 FLR
66599 FLR
66601 FLR
66602 FLR
66603 FLR
66604 FLR
66605 FLR
66606 FLR
66607 FLR
66608 EXP
66609 EXP
66610 FLR
66611 EXP
66612 EXP
66613 FLR
66614 FLR
66615 FLR
66616 FLR
66617 FLR
66618 FLR
66619 FLR
66620 FLR
66621 FLR
66622 FLR
66623 FLR
66624 EXP
66625 EXP
66701 GBR
66702 GBR
66703 GBR
66704 GBR
66705 GBR
66706 GBR
66707 GBR
66708 GBR
66709 GBR
66710 GBR
66711 GBR
66712 GBR
66713 GBR
66714 GBR
66715 GBR
66716 GBR
66717 GBR
66718 GBR
66719 GBR
66720 GBR
66721 GBR
66722 GBR
66723 GBR
66724 GBR
66725 GBR
66726 GBR
66727 GBR
66728 GBR
66729 GBR
66730 GBR
66731 GBR
66732 GBR
66733 GBR
66734 GBR
66735 GBR
66736 GBR
66737 GBR
66738 GBR
66739 GBR
66740 GBR
66741 GBR
66742 GBR
66743 GBR
66744 GBR
66745 GBR
66746 GBR
66747 GBR
66748 GBR
66749 GBR
66750 GBR
66751 GBR
66752 GBR
66753 GBR
66754 GBR
66755 GBR
66756 GBR
66757 GBR
66758 GBR
66759 GBR
66760 GBR
66761 GBR
66762 GBR
66763 GBR
66764 GBR
66765 GBR
66766 GBR
66767 GBR
66768 GBR
66769 GBR
66770 GBR
66771 GBR
66772 GBR
66773 GBR
66774 GBR
66775 GBR
66776 GBR
66777 GBR
66778 GBR
66779 GBR
66780 GBR
66781 GBR
66782 GBR
66783 GBR
66784 GBR
66785 GBR
66786 GBR
66787 GBR
66788 GBR
66789 GBR
66790 GBR
66791 GBR
66792 GBR
66793 GBR
66794 GBR
66795 GBR
66796 GBR
66797 GBR
66798 GBR
66799 GBR
66846 COL
66847 COL
66848 COL
66849 COL
66850 COL
66951 FLR
66952 FLR
66953 FLR
66954 EXP
66955 FLR
66956 FLR
66957 FLR

Class 67

67001 DBC
67002 DBC
67003 DBC
67004 DBC
67005 DBC
67006 DBC
67007 DBC
67008 DBC
67009 DBC
67010 DBC
67011 DBC
67012 DBC
67013 DBC
67014 DBC
67015 DBC
67016 DBC
67017 DBC
67018 DBC
67019 DBC
67020 DBC
67021 DBC
67022 DBC
67023 GBR
67024 DBC
67025 DBC
67026 DBC
67027 GBR
67028 DBC
67029 DBC
67030 DBC

Class 68

68001 DRS
68002 DRS
68003 DRS
68004 DRS
68005 DRS
68006 DRS
68007 DRS
68008 DRS
68009 DRS
68010 DRS/CHR
68011 DRS/CHR
68012 DRS/CHR
68013 DRS/CHR
68014 DRS/CHR
68015 DRS/CHR
68016 DRS
68017 DRS
68018 DRS
68019 DRS/TPE
68020 DRS/TPE
68021 DRS/TPE
68022 DRS/TPE
68023 DRS/TPE
68024 DRS/TPE
68025 DRS/TPE
68026 DRS/TPE
68027 DRS/TPE
68028 DRS/TPE
68029 DRS/TPE
68030 DRS/TPE
68031 DRS/TPE
68032 DRS/TPE
68033 DRS
68034 DRS

Class 69

69001 GBR
69002 GBR
69003 GBR
69004 GBR
69005 GBR
69006 GBR
69007 GBR
69008 GBR
69009 GBR
69010 GBR
69011 GBR
69012 GBR
69013 GBR
69014 GBR
69015 GBR
69016 GBR

Class 70

70001 FLR
70002 FLR
70003 FLR
70004 FLR
70005 FLR
70006 FLR
70007 FLR
70008 FLR
70009 FLR
70010 FLR
70011 FLR
70012 EXP
70013 FLR
70014 FLR
70015 FLR
70016 FLR
70017 FLR
70018 FLR
70019 FLR
70020 FLR
70801 COL
70802 COL
70803 COL
70804 COL
70805 COL
70806 COL
70807 COL
70808 COL
70809 COL
70810 COL
70811 COL
70812 COL
70813 COL
70814 COL
70815 COL
70816 COL
70817 COL

Class 71

71001 Pres

Class 73

73001 LSL
73002 Pres
73003 Pres
73101 GBR
73107 GBR
73109 GBR
73110 GBR
73114 NEM
73118 Pres
73119 GBR
73128 GBR
73129 Pres
73130 Pres
73133 TMT
73134 GBR
73136 GBR
73138 NR
73139 GBR
73140 Pres
73141 GBR
73201 GBR
73202 GTR
73210 Pres
73212 GBR
73213 GBR
73235 SWR
73951 NR
73952 NR
73961 GBR
73962 GBR
73963 GBR
73964 GBR
73965 GBR
73966 GBR/CAL
73967 GBR/CAL
73968 GBR/CAL
73969 GBR/CAL
73970 GBR/CAL
73971 GBR/CAL

Class 76

76020 Pres

Class 77

1502 Pres
1505 Pres
1501 Pres

Class 81

81002 Pres

Class 82

82008 Pres

Class 83
83012 Pres
Class 84
84001 Pres
Class 85
85006 Pres
Class 86
86101 LSL
86213 EXP
86215 EXP
86217 EXP
86218 EXP
86228 EXP
86231 EXP
86232 EXP
86233 EXP
86234 EXP
86235 EXP
86242 EXP
86248 EXP
86250 EXP
86251 FLR
86401 WCR
86424 EXP
86604 FLR
86605 FLR
86607 FLR
86608 FLR
86609 FLR
86610 FLR
86612 FLR
86613 FLR
86614 FLR
86622 FLR
86627 FLR
86628 FLR
86632 FLR
86637 FLR
86638 FLR
86639 FLR
86701 EXP
86702 EXP
Class 87
87001 Pres
87002 LSL
87003 EXP
87004 EXP
87006 EXP
87007 EXP
87008 EXP
87009 EXP
87010 EXP
87012 EXP
87013 EXP
87014 EXP
87017 EXP
87019 EXP
87020 EXP
87022 EXP
87023 EXP
87025 EXP
87026 EXP
87028 EXP
87029 EXP
87033 EXP
87034 EXP
87035 Pres
Class 88
88001 DRS
88002 DRS
88003 DRS
88004 DRS
88005 DRS
88006 DRS
88007 DRS
88008 DRS
88009 DRS
88010 DRS
Class 89
89001 LSL
Class 90
90001 LSL
90002 LSL
90003 FLR
90004 FLR
90005 FLR
90006 FLR
90007 FLR
90008 FLR
90009 FLR
90010 FLR
90011 FLR
90012 FLR
90013 FLR
90014 FLR
90015 FLR
90016 FLR
90017 DBC
90018 DBC
90019 DBC
90020 DBC
90021 DBC
90022 DBC
90023 DBC
90024 DBC
90025 DBC
90026 DBC
90027 DBC
90028 DBC
90029 DBC
90030 DBC
90031 DBC
90032 DBC
90033 DBC
90034 DBC
90035 DBC
90036 DBC
90037 DBC
90038 DBC
90039 DBC
90040 DBC
90041 FLR
90042 FLR
90043 FLR
90044 FLR
90045 FLR
90046 FLR
90047 FLR
90048 FLR
90049 FLR
90050 DBC
Class 91
91101 LNER
91105 LNER
91106 LNER
91107 LNER
91109 LNER
91110 LNER
91111 LNER
91112 EVS (S)
91114 LNER
91115 EVS (S)
91116 EVS (S)
91117 EUX
91118 EVS (S)
91119 LNER
91120 EUX
91121 EVS (S)
91124 LNER
91125 EVS (S)
91127 LNER
91130 LNER
91131 EVS (S)
Class 92
92001 EXP
92002 EXP
92003 EXP
92004 DBC (S)
92005 EXP
92006 GBR
92007 DBC (S)
92008 DBC (S)
92009 DBC (S)
92010 GBR
92011 DBC
92012 EXP
92013 DBC (S)
92014 GBR
92015 DBC
92016 DBC (S)
92017 DBC (S)
92018 GBR
92019 DBC
92020 GBR
92021 GBR (S)
92022 EXP
92023 GBR
92024 EXP
92025 EXP
92026 EXP
92027 EXP
92028 GBR
92029 DBC (S)
92030 EXP
92031 DBC (S)
92032 GBR
92033 GBR
92034 EXP
92035 DBC (S)
92036 DBC
92037 DBC (S)
92038 GBR
92039 EXP
92040 GBR (S)
92041 DBC
92042 DBC
92043 GBR
92044 GBR
92045 GBR (S)
92046 GBR (S)
Class 93
93001 ROG
93002 ROG
93003 ROG
93004 ROG
93005 ROG
93006 ROG
93007 ROG
93008 ROG
93009 ROG
93010 ROG
93011 ROG
93012 ROG
93013 ROG
93014 ROG
93015 ROG
93016 ROG
93017 ROG
93018 ROG
93019 ROG
93020 ROG
93021 ROG
93022 ROG
93023 ROG
93024 ROG
93025 ROG
93026 ROG
93027 ROG
93028 ROG
93029 ROG
93030 ROG
Class 99
99001 GBR
99002 GBR
99003 GBR
99004 GBR
99005 GBR
99006 GBR
99007 GBR
99008 GBR
99009 GBR
99010 GBR
99011 GBR
99012 GBR
99013 GBR
99014 GBR
99015 GBR
99016 GBR
99017 GBR
99018 GBR
99019 GBR
99020 GBR
Class 97
97301 NR
97302 NR
97303 NR
97304 NR
Class 139
139001 WMT
139002 WMT
Class 150
150001 NOR
150002 NOR
150003 NOR
150004 NOR
150005 NOR
150006 NOR
150101 NOR
150102 NOR
150103 NOR
150104 NOR
150105 NOR
150106 NOR
150107 NOR
150108 NOR
150109 NOR
150110 NOR
150111 NOR
150113 NOR
150114 NOR
150115 NOR
150118 NOR
150119 NOR
150120 NOR
150121 NOR
150122 NOR
150123 NOR
150124 NOR
150125 NOR
150126 NOR
150127 NOR
150128 NOR
150129 NOR
150130 NOR
150131 NOR
150132 NOR
150133 NOR
150134 NOR
150135 NOR
150136 NOR
150137 NOR
150138 NOR
150139 NOR
150140 NOR
150141 NOR
150142 NOR
150143 NOR
150144 NOR
150145 NOR
150146 NOR
150148 NOR
150149 NOR
150150 NOR
150201 NOR
150202 GWR
150203 NOR
150204 NOR
150205 NOR
150206 NOR
150207 GWR
150208 TFW
150210 NOR
150211 NOR
150213 TFW
150214 NOR
150215 NOR
150216 GWR
150217 TFW
150218 NOR
150219 GWR
150220 NOR
150221 GWR
150222 NOR
150224 NOR
150225 NOR
150226 NOR
150227 TFW
150228 NOR
150229 TFW
150230 TFW
150231 TFW
150232 GWR
150233 GWR
150234 GWR
150235 TFW
150236 TFW
150237 TFW
150238 GWR
150239 GWR
150240 TFW
150241 TFW
150242 TFW
150243 GWR
150244 GWR
150245 TFW
150246 GWR
150247 GWR
150248 GWR
150249 GWR
150250 TFW
150251 TFW
150252 TFW
150253 TFW
150254 TFW
150255 TFW
150256 TFW

150257 TFW
150258 TFW
150259 TFW
150260 TFW
150261 GWR
150262 TFW
150263 GWR
150264 TFW
150265 GWR
150266 GWR
150267 TFW
150268 NOR
150269 NOR
150270 NOR
150271 NOR
150272 NOR
150273 NOR
150274 NOR
150275 NOR
150276 NOR
150277 NOR
150278 TFW
150279 TFW
150280 TFW
150281 TFW
150282 TFW
150283 TFW
150284 TFW
150285 TFW

Class 153

153301 ANG (S)
153303 TFW
153304 ANG (S)
153305 SCR
153307 ANG (S)
153308 ANG (S)
153311 NR
153312 TFW
153315 ANG (S)
153316 PBK (S)
153317 ANG (S)
153319 ANG (S)
153320 TFW
153323 TFW
153325 TFW
153324 PBK (S)
153327 TFW
153328 ANG (S)
153329 TFW
153330 PBK (S)
153331 ANG (S)
153332 PBK (S)
153333 TFW
153334 PBK (S)
153351 ANG (S)
153352 ANG (S)
153353 TFW
153354 PBK (S)
153355 ANG (S)
153356 PBK (S)
153357 ANG (S)
153358 PBK (S)
153359 PBK (S)
153360 PBK (S)
153361 TFW
153362 TFW
153363 PBK (S)
153365 PBK (S)
153367 TFW
153369 TFW
153370 SCR
153371 PBK (S)
153373 SCR
153374 TFW
153375 PBK (S)
153376 NR
153377 SCR
153378 ANG (S)
153379 PBK (S)
153380 SCR
153381 PBK (S)
153383 PBK (S)
153384 PBK (S)
153385 NR
153906 TFW
153909 TFW
153910 TFW
153913 TFW
153914 TFW
153918 TFW
153921 TFW
153922 TFW
153926 TFW
153935 TFW
153968 TFW
153972 TFW
153982 TFW

Class 155

155341 NOR
155342 NOR
155343 NOR
155344 NOR
155345 NOR
155346 NOR
155347 NOR

Class 156

156401 NOR
156402 NOR
156403 EMR
156404 EMR
156405 EMR
156406 EMR
156407 EMR
156408 EMR
156409 EMR
156410 EMR
156411 EMR
156412 EMR
156413 EMR
156414 EMR
156415 NOR
156416 EMR
156417 EMR
156418 EMR
156419 EMR
156420 NOR
156421 NOR
156422 EMR
156423 NOR
156424 NOR
156425 NOR
156426 NOR
156427 NOR
156428 NOR
156429 NOR
156430 SCR
156431 SCR
156432 SCR
156433 SCR
156434 SCR
156435 SCR
156436 SCR
156437 SCR
156438 NOR
156439 SCR
156440 NOR
156441 NOR
156442 SCR
156443 NOR
156444 NOR
156445 SCR
156446 SCR
156447 NOR
156448 NOR
156449 NOR
156450 SCR
156451 NOR
156452 NOR
156453 SCR
156454 NOR
156455 NOR
156456 SCR
156457 SCR
156458 SCR
156459 NOR
156460 NOR
156461 NOR
156462 SCR
156463 NOR
156464 NOR
156465 NOR
156466 NOR
156467 SCR
156468 NOR
156469 NOR
156470 EMR
156471 NOR
156472 NOR
156473 EMR
156474 SCR
156475 NOR
156476 SCR
156477 SCR
156478 SCR
156479 NOR
156480 NOR
156481 NOR
156482 NOR
156483 NOR
156484 NOR
156485 NOR
156486 NOR
156487 NOR
156488 NOR
156489 NOR
156490 NOR
156491 NOR
156492 SCR
156493 SCR
156494 SCR
156495 SCR
156496 NOR
156497 EMR
156498 EMR
156499 SCR
156501 SCR
156502 SCR
156503 SCR
156504 SCR
156505 SCR
156506 SCR
156507 SCR
156508 SCR
156509 SCR
156510 SCR
156511 SCR
156512 SCR
156513 SCR
156514 SCR

Class 158

158701 SCR
158702 SCR
158703 SCR
158704 SCR
158705 SCR
158706 SCR
158707 SCR
158708 SCR
158709 SCR
158710 SCR
158711 SCR
158712 SCR
158713 SCR
158714 SCR
158715 SCR
158716 SCR
158717 SCR
158718 SCR
158719 SCR
158720 SCR
158721 SCR
158722 SCR
158723 SCR
158724 SCR
158725 SCR
158726 SCR
158727 SCR
158728 SCR
158729 SCR
158730 SCR
158731 SCR
158732 SCR
158733 SCR
158734 SCR
158735 SCR
158736 SCR
158737 SCR
158738 SCR
158739 SCR
158740 SCR
158741 SCR
158745 GWR
158747 GWR
158749 GWR
158750 GWR
158752 NOR
158753 NOR
158754 NOR
158755 NOR
158756 NOR
158757 NOR
158758 NOR
158759 NOR
158760 GWR
158762 GWR
158763 GWR
158765 GWR
158766 GWR
158767 GWR
158769 GWR
158770 EMR
158773 EMR
158774 EMR
158777 EMR
158780 EMR
158782 NOR
158783 EMR
158784 NOR
158785 EMR
158786 NOR
158787 NOR
158788 EMR
158789 NOR
158790 NOR
158791 NOR
158792 NOR
158793 NOR
158794 NOR
158795 NOR
158796 NOR
158797 NOR
158798 GWR
158799 EMR
158806 EMR
158810 EMR
158812 EMR
158813 EMR
158815 NOR
158816 NOR
158817 NOR
158818 TFW
158819 TFW
158820 TFW
158821 TFW
158822 TFW
158823 TFW
158824 TFW
158825 TFW
158826 TFW
158827 TFW
158828 TFW
158829 TFW
158830 TFW
158831 TFW
158832 TFW
158833 TFW
158834 TFW
158835 TFW
158836 TFW
158837 TFW
158838 TFW
158839 TFW
158840 TFW
158841 TFW
158842 NOR
158843 NOR
158844 NOR
158845 NOR
158846 EMR
158847 EMR
158848 NOR
158849 NOR
158850 NOR
158851 NOR
158852 EMR
158853 NOR
158854 EMR
158855 NOR
158856 EMR
158857 EMR
158858 EMR
158859 NOR
158860 NOR
158861 NOR
158862 EMR
158863 EMR
158864 EMR
158865 EMR
158866 EMR
158867 NOR
158868 NOR
158869 NOR
158870 NOR
158871 NOR
158872 NOR
158880 SWR

158881 SWR
158882 SWR
158883 SWR
158884 SWR
158885 SWR
158886 SWR
158887 SWR
158888 SWR
158889 SWR
158890 SWR
158901 NOR
158902 NOR
158903 NOR
158904 NOR
158905 NOR
158906 NOR
158907 NOR
158908 NOR
158909 NOR
158910 NOR
158950 GWR
158951 GWR
158956 GWR
158957 GWR
158958 GWR
158959 GWR

Class 159

159001 SWR
159002 SWR
159003 SWR
159004 SWR
159005 SWR
159006 SWR
159007 SWR
159008 SWR
159009 SWR
159010 SWR
159011 SWR
159012 SWR
159013 SWR
159014 SWR
159015 SWR
159016 SWR
159017 SWR
159018 SWR
159019 SWR
159020 SWR
159021 SWR
159022 SWR
159101 SWR
159102 SWR
159103 SWR
159104 SWR
159105 SWR
159106 SWR
159107 SWR
159108 SWR

Class 165

165001 CHR
165002 CHR
165003 CHR
165004 CHR
165005 CHR
165006 CHR
165007 CHR
165008 CHR
165009 CHR
165010 CHR
165011 CHR
165012 CHR
165013 CHR
165014 CHR
165015 CHR
165016 CHR
165017 CHR
165018 CHR
165019 CHR
165020 CHR
165021 CHR
165022 CHR
165023 CHR
165024 CHR
165025 CHR
165026 CHR
165027 CHR
165028 CHR
165029 CHR
165030 CHR
165031 CHR
165032 CHR
165033 CHR
165034 CHR
165035 CHR
165036 CHR
165037 CHR
165038 CHR
165039 CHR
165101 GWR
165102 GWR
165103 GWR
165104 GWR
165105 GWR
165106 GWR
165107 GWR
165108 GWR
165109 GWR
165110 GWR
165111 GWR
165112 GWR
165113 GWR
165114 GWR
165116 GWR
165117 GWR
165118 GWR
165119 GWR
165120 GWR
165121 GWR
165122 GWR
165123 GWR
165124 GWR
165125 GWR
165126 GWR
165127 GWR
165128 GWR
165129 GWR
165130 GWR
165131 GWR
165132 GWR
165133 GWR
165134 GWR
165135 GWR
165136 GWR
165137 GWR

Class 166

166201 GWR
166202 GWR
166203 GWR
166204 GWR
166205 GWR
166206 GWR
166207 GWR
166208 GWR
166209 GWR
166210 GWR
166211 GWR
166212 GWR
166213 GWR
166214 GWR
166215 GWR
166216 GWR
166217 GWR
166218 GWR
166219 GWR
166220 GWR
166221 GWR

Class 168

168001 CHR
168002 CHR
168003 CHR
168004 CHR
168005 CHR
168106 CHR
168107 CHR
168108 CHR
168109 CHR
168110 CHR
168111 CHR
168112 CHR
168113 CHR
168214 CHR
168215 CHR
168216 CHR
168217 CHR
168218 CHR
168219 CHR
168321 CHR
168322 CHR
168323 CHR
168324 CHR
168325 CHR
168326 CHR
168327 CHR
168328 CHR
168329 CHR

Class 170

170101 XCT
170102 XCT
170103 XCT
170104 XCT
170105 XCT
170106 XCT
170107 XCT
170108 XCT
170109 XCT
170110 XCT
170111 XCT
170112 XCT
170113 XCT
170114 XCT
170115 XCT
170116 XCT
170117 XCT
170201 TFW
170202 TFW
170203 TFW
170204 TFW
170205 TFW
170206 TFW
170207 TFW
170208 TFW
170270 TFW
170271 TFW
170272 TFW
170273 EMR
170393 SCR
170394 SCR
170395 SCR
170396 SCR
170397 XCT
170398 XCT
170401 SCR
170402 SCR
170403 SCR
170404 SCR
170405 SCR
170406 SCR
170407 SCR
170408 SCR
170409 SCR
170410 SCR
170411 SCR
170412 SCR
170413 SCR
170414 SCR
170415 SCR
170416 EMR
170417 EMR
170418 EMR
170419 EMR
170420 EMR
170425 SCR
170426 SCR
170427 SCR
170428 SCR
170429 SCR
170430 SCR
170431 SCR
170432 SCR
170433 SCR
170434 SCR
170450 SCR
170451 SCR
170452 SCR
170453 NOR
170454 NOR
170455 NOR
170456 NOR
170457 NOR
170458 NOR
170459 NOR
170460 NOR
170461 NOR
170470 SCR
170471 SCR
170472 NOR
170473 NOR
170474 NOR
170475 NOR
170476 NOR
170477 NOR
170478 NOR
170501 WMT
170502 WMT
170503 EMR
170504 WMT
170505 WMT
170506 WMT
170507 WMT
170508 WMT
170509 WMT
170510 WMT
170511 EMR
170512 WMT
170513 WMT
170514 WMT
170515 EMR
170516 WMT
170517 EMR
170530 EMR
170531 EMR
170532 EMR
170533 WMT
170534 EMR
170535 WMT
170618 XCT
170619 XCT
170620 XCT
170621 XCT
170622 XCT
170623 XCT
170636 XCT
170637 XCT
170638 XCT
170639 XCT

Class 171

171201 GTR
171202 GTR
171401 GTR
171402 GTR
171721 GTR
171722 GTR
171723 GTR
171724 GTR
171725 GTR
171726 GTR
171727 GTR
171728 GTR
171729 GTR
171730 GTR
171801 GTR
171802 GTR
171803 GTR
171804 GTR
171805 GTR
171806 GTR

Class 172

172001 WMT
172002 WMT
172003 WMT
172004 WMT
172005 WMT
172006 WMT
172007 WMT
172008 WMT
172101 WMT
172102 WMT
172103 WMT
172104 WMT
172211 WMT
172212 WMT
172213 WMT
172214 WMT
172215 WMT
172216 WMT
172217 WMT
172218 WMT
172219 WMT
172220 WMT
172221 WMT
172222 WMT
172331 WMT
172332 WMT
172333 WMT
172334 WMT
172335 WMT
172336 WMT
172337 WMT
172338 WMT
172339 WMT
172340 WMT
172341 WMT
172342 WMT
172343 WMT
172344 WMT
172345 WMT

Class 175
175001 TFW
175002 TFW
175003 TFW
175004 TFW
175005 TFW
175006 TFW
175007 TFW
175008 TFW
175009 TFW
175010 TFW
175011 TFW
175101 TFW
175102 TFW
175103 TFW
175104 TFW
175105 TFW
175106 TFW
175107 TFW
175108 TFW
175109 TFW
175110 TFW
175111 TFW
175112 TFW
175113 TFW
175114 TFW
175115 TFW
175116 TFW
Class 180
180101 GCR
180102 GCR
180103 GCR
180104 GCR
180105 GCR
180106 GCR
180107 GCR
180108 GCR
180109 EMR
180110 EMR
180111 EMR
180112 GCR
180113 EMR
180114 GCR
Class 185
185101 TPE
185102 TPE
185103 TPE
185104 TPE
185105 TPE
185106 TPE
185107 TPE
185108 TPE
185109 TPE
185110 TPE
185111 TPE
185112 TPE
185113 TPE
185114 TPE
185115 TPE
185116 TPE
185117 TPE
185118 TPE
185119 TPE
185120 TPE
185121 TPE
185122 TPE
185123 TPE
185124 TPE
185125 TPE
185126 TPE
185127 TPE
185128 TPE
185129 TPE
185130 TPE
185131 TPE
185132 TPE
185133 TPE
185134 TPE
185135 TPE
185136 TPE
185137 TPE
185138 TPE
185139 TPE
185140 TPE
185141 TPE
185142 TPE
185143 TPE
185144 TPE
185145 TPE
185146 TPE
185147 TPE
185148 TPE
185149 TPE
185150 TPE
185151 TPE
Class 195
195001 NOR
195002 NOR
195003 NOR
195004 NOR
195005 NOR
195006 NOR
195007 NOR
195008 NOR
195009 NOR
195010 NOR
195011 NOR
195012 NOR
195013 NOR
195014 NOR
195015 NOR
195016 NOR
195017 NOR
195018 NOR
195019 NOR
195020 NOR
195021 NOR
195022 NOR
195023 NOR
195024 NOR
195025 NOR
195101 NOR
195102 NOR
195103 NOR
195104 NOR
195105 NOR
195106 NOR
195107 NOR
195108 NOR
195109 NOR
195110 NOR
195111 NOR
195112 NOR
195113 NOR
195114 NOR
195115 NOR
195116 NOR
195117 NOR
195118 NOR
195119 NOR
195120 NOR
195121 NOR
195122 NOR
195123 NOR
195124 NOR
195125 NOR
195126 NOR
195127 NOR
195128 NOR
195129 NOR
195130 NOR
195131 NOR
195132 NOR
195133 NOR
Class 196
196001 WMT
196002 WMT
196003 WMT
196004 WMT
196005 WMT
196006 WMT
196007 WMT
196008 WMT
196009 WMT
196010 WMT
196011 WMT
196012 WMT
196101 WMT
196102 WMT
196103 WMT
196104 WMT
196105 WMT
196106 WMT
196107 WMT
196108 WMT
196109 WMT
196110 WMT
196111 WMT
196112 WMT
196113 WMT
196114 WMT
Class 197
197001 TFW
197002 TFW
197003 TFW
197004 TFW
197005 TFW
197006 TFW
197007 TFW
197008 TFW
197009 TFW
197010 TFW
197011 TFW
197012 TFW
197013 TFW
197014 TFW
197015 TFW
197016 TFW
197017 TFW
197018 TFW
197019 TFW
197020 TFW
197021 TFW
197022 TFW
197023 TFW
197024 TFW
197025 TFW
197026 TFW
197027 TFW
197028 TFW
197029 TFW
197030 TFW
197031 TFW
197032 TFW
197033 TFW
197034 TFW
197035 TFW
197036 TFW
197037 TFW
197038 TFW
197039 TFW
197040 TFW
197041 TFW
197042 TFW
197043 TFW
197044 TFW
197045 TFW
197046 TFW
197047 TFW
197048 TFW
197049 TFW
197050 TFW
197051 TFW
197101 TFW
197102 TFW
197103 TFW
197104 TFW
197105 TFW
197106 TFW
197107 TFW
197108 TFW
197109 TFW
197110 TFW
197111 TFW
197112 TFW
197113 TFW
197114 TFW
197115 TFW
197116 TFW
197117 TFW
197118 TFW
197119 TFW
197120 TFW
197121 TFW
197122 TFW
197123 TFW
197124 TFW
197125 TFW
197126 TFW
Class 201
201001 HDL
Class 220
220001 XCT
220002 XCT
220003 XCT
220004 XCT
220005 XCT
220006 XCT
220007 XCT
220008 XCT
220009 XCT
220010 XCT
220011 XCT
220012 XCT
220013 XCT
220014 XCT
220015 XCT
220016 XCT
220017 XCT
220018 XCT
220019 XCT
220020 XCT
220021 XCT
220022 XCT
220023 XCT
220024 XCT
220025 XCT
220026 XCT
220027 XCT
220028 XCT
220029 XCT
220030 XCT
220031 XCT
220032 XCT
220033 XCT
220034 XCT
Class 221
221101 AWC
221102 AWC
221103 AWC
221104 AWC
221105 AWC
221106 AWC
221107 AWC
221108 AWC
221109 AWC
221110 AWC
221111 AWC
221112 AWC
221113 AWC
221114 AWC
221115 AWC
221116 AWC
221117 AWC
221118 AWC
221119 XCT
221120 XCT
221121 XCT
221122 XCT
221123 XCT
221124 XCT
221125 XCT
221126 XCT
221127 XCT
221128 XCT
221129 XCT
221130 XCT
221131 XCT
221132 XCT
221133 XCT
221134 XCT
221135 XCT
221136 XCT
221137 XCT
221138 XCT
221139 XCT
221140 XCT
221141 XCT
221144 XCT
221142 AWC
221143 AWC
Class 222
222001 EMR
222002 EMR
222003 EMR
222004 EMR
222005 EMR
222006 EMR
222007 EMR
222008 EMR
222009 EMR
222010 EMR
222011 EMR
222012 EMR
222013 EMR
222014 EMR
222015 EMR
222016 EMR
222017 EMR
222018 EMR
222019 EMR
222020 EMR
222021 EMR
222022 EMR
222023 EMR
222101 EMR

222102 EMR
222103 EMR
222104 EMR

Class 230

230001 VivaRail
230002 EXP
230003 LNWR
230004 LNWR
230005 LNWR
230006 TFW
230007 TFW
230008 TFW
230009 TFW
230010 TFW

Class 231

231001 TFW
231002 TFW
231003 TFW
231004 TFW
231005 TFW
231006 TFW
231007 TFW
231008 TFW
231009 TFW
231010 TFW
231011 TFW

Class 313

313121 NR
313201 GTR
313202 GTR
313203 GTR
313204 GTR
313205 GTR
313206 GTR
313207 GTR
313208 GTR
313209 GTR
313210 GTR
313211 GTR
313212 GTR
313213 GTR
313214 GTR
313215 GTR
313216 GTR
313217 GTR
313219 GTR
313220 GTR

Class 315

315801 EVS (S)
315802 EVS (S)
315807 EVS (S)
315809 EVS (S)
315815 EVS (S)
315837 TFL
315838 TFL
315839 TFL
315847 TFL
315848 TFL
315853 TFL
315856 TFL
315857 TFL

Class 317

317337 AGA
317338 AGA
317339 ANG (S)
317340 ANG (S)
317341 ANG (S)
317342 AGA
317343 AGA
317344 ANG (S)
317345 AGA
317346 AGA
317347 AGA
317348 ANG (S)
317501 AGA
317502 AGA
317503 AGA
317504 AGA
317506 AGA
317507 AGA
317508 AGA
317510 ANG (S)
317511 AGA
317512 ANG (S)
317513 ANG (S)
317514 AGA
317515 AGA
317651 AGA
317652 AGA
317653 AGA
317654 AGA
317659 ANG (S)
317660 ANG (S)
317662 ANG (S)
317663 ANG (S)
317668 ANG (S)
317671 ANG (S)
317708 ANG (S)
317709 ANG (S)
317710 ANG (S)
317714 AGA
317719 ANG (S)
317722 ANG (S)
317723 AGA
317729 AGA
317732 AGA
317881 AGA
317882 ANG (S)
317883 ANG (S)
317884 ANG (S)
317885 ANG (S)
317886 ANG (S)
317887 ANG (S)
317888 ANG (S)
317889 ANG (S)
317890 ANG (S)
317891 ANG (S)

Class 318

318250 SCR
318251 SCR
318252 SCR
318253 SCR
318254 SCR
318255 SCR
318256 SCR
318257 SCR
318258 SCR
318259 SCR
318260 SCR
318261 SCR
318262 SCR
318263 SCR
318264 SCR
318265 SCR
318266 SCR
318267 SCR
318268 SCR
318269 SCR
318270 SCR

Class 319

319005 LNWR
319011 PBK (S)
319012 LNWR
319013 LNWR
319214 LNWR
319215 LNWR
319216 LNWR
319217 LNWR
319218 LNWR
319219 LNWR
319220 LNWR
319361 NOR
319362 PBK (S)
319363 PBK (S)
319364 PBK (S)
319365 PBK (S)
319366 NOR
319367 NOR
319368 NOR
319369 NOR
319370 NOR
319371 PBK (S)
319372 NOR
319374 PBK (S)
319375 NOR
319376 PBK (S)
319377 PBK (S)
319378 NOR
319379 NOR
319380 PBK (S)
319381 NOR
319383 NOR
319384 NOR
319385 NOR
319386 NOR
319429 LNWR
319433 LNWR
319441 PBK (S)
319454 PBK (S)
319457 LNWR
319460 LNWR

Class 320

320301 SCR
320302 SCR
320303 SCR
320304 SCR
320305 SCR
320306 SCR
320307 SCR
320308 SCR
320309 SCR
320310 SCR
320311 SCR
320312 SCR
320313 SCR
320314 SCR
320315 SCR
320316 SCR
320317 SCR
320318 SCR
320319 SCR
320320 SCR
320321 SCR
320322 SCR
320401 SCR
320403 SCR
320404 SCR
320411 SCR
320412 SCR
320413 SCR
320414 SCR
320415 SCR
320416 SCR
320417 SCR
320418 SCR
320420 SCR

Class 321

321301 AGA
321302 AGA
321303 AGA
321304 AGA
321305 AGA
321306 AGA
321307 AGA
321308 AGA
321309 AGA
321310 AGA
321311 AGA
321312 AGA
321313 AGA
321314 AGA
321315 AGA
321316 AGA
321317 AGA
321318 AGA
321319 AGA
321320 AGA
321321 AGA
321322 AGA
321323 AGA
321324 AGA
321325 AGA
321326 AGA
321327 AGA
321328 AGA
321329 AGA
321330 AGA
321331 EVS (S)
321333 EVS (S)
321335 EVS (S)
321336 EVS (S)
321337 EVS (S)
321338 AGA
321339 EVS (S)
321340 EVS (S)
321341 EVS (S)
321342 AGA
321343 EVS (S)
321402 EVS (S)
321405 EVS (S)
321406 EVS (S)
321407 EVS (S)
321408 EVS (S)
321409 EVS (S)
321410 EVS (S)
321419 EVS (S)
321421 EVS (S)
321423 EVS (S)
321424 EVS (S)
321426 EVS (S)
321427 EVS (S)
321428 EVS (S)
321429 EVS (S)
321430 EVS (S)
321431 EVS (S)
321432 EVS (S)
321433 EVS (S)
321434 EVS (S)
321436 EVS (S)
321439 AGA
321440 EVS (S)
321441 EVS (S)
321443 AGA
321444 AGA
321445 AGA
321447 EVS (S)
321901 AGA
321902 AGA
321903 EVS (S)

Class 322

322481 AGA
322482 AGA
322483 AGA
322484 EVS (S)
322485 AGA

Class 323

323201 WMT
323202 WMT
323203 WMT
323204 WMT
323205 WMT
323206 WMT
323207 WMT
323208 WMT
323209 WMT
323210 WMT
323211 WMT
323212 WMT
323213 WMT
323214 WMT
323215 WMT
323216 WMT
323217 WMT
323218 WMT
323219 WMT
323220 WMT
323221 WMT
323222 WMT
323223 NOR
323224 NOR
323225 NOR
323226 NOR
323227 NOR
323228 NOR
323229 NOR
323230 NOR
323231 NOR
323232 NOR
323233 NOR
323234 NOR
323235 NOR
323236 NOR
323237 NOR
323238 NOR
323239 NOR
323240 WMT
323241 WMT
323242 WMT
323243 WMT

Class 325

325001 DBC
325002 DBC
325003 DBC
325004 DBC
325005 DBC
325006 DBC
325007 DBC
325008 DBC
325009 DBC
325011 DBC
325012 DBC
325013 DBC
325014 DBC
325015 DBC
325016 DBC

Class 326

326001 ROG
326002 ROG
326003 TBC
326004 TBC
326005 TBC
326006 TBC
326007 TBC
326008 TBC
326009 TBC

377604 GTR
377605 GTR
377606 GTR
377607 GTR
377608 GTR
377609 GTR
377610 GTR
377611 GTR
377612 GTR
377613 GTR
377614 GTR
377615 GTR
377616 GTR
377617 GTR
377618 GTR
377619 GTR
377620 GTR
377621 GTR
377622 GTR
377623 GTR
377624 GTR
377625 GTR
377626 GTR
377701 GTR
377702 GTR
377703 GTR
377704 GTR
377705 GTR
377706 GTR
377707 GTR
377708 GTR

Class 378

378135 LOL
378136 LOL
378137 LOL
378138 LOL
378139 LOL
378140 LOL
378141 LOL
378142 LOL
378143 LOL
378144 LOL
378145 LOL
378146 LOL
378147 LOL
378148 LOL
378149 LOL
378150 LOL
378151 LOL
378152 LOL
378153 LOL
378154 LOL
378201 LOL
378202 LOL
378203 LOL
378204 LOL
378205 LOL
378206 LOL
378207 LOL
378208 LOL
378209 LOL
378210 LOL
378211 LOL
378212 LOL
378213 LOL
378214 LOL
378215 LOL
378216 LOL
378217 LOL
378218 LOL
378219 LOL
378220 LOL
378221 LOL
378222 LOL
378223 LOL
378224 LOL
378225 LOL
378226 LOL
378227 LOL
378228 LOL
378229 LOL
378230 LOL
378231 LOL
378232 LOL
378233 LOL
378234 LOL
378255 LOL
378256 LOL
378257 LOL

Class 379

379001 AIK (S)
379002 AIK (S)
379003 AIK (S)
379004 AIK (S)
379005 AIK (S)
379006 AIK (S)
379007 AIK (S)
379008 AIK (S)
379009 AIK (S)
379010 AIK (S)
379011 AIK (S)
379012 AIK (S)
379013 AIK (S)
379014 AIK (S)
379015 AIK (S)
379016 AIK (S)
379017 AIK (S)
379018 AIK (S)
379019 AIK (S)
379020 AIK (S)
379021 AIK (S)
379022 AIK (S)
379023 AIK (S)
379024 AIK (S)
379025 AIK (S)
379026 AIK (S)
379027 AIK (S)
379028 AIK (S)
379029 AIK (S)
379030 AIK (S)

Class 380

380001 SCR
380002 SCR
380003 SCR
380004 SCR
380005 SCR
380006 SCR
380007 SCR
380008 SCR
380009 SCR
380010 SCR
380011 SCR
380012 SCR
380013 SCR
380014 SCR
380015 SCR
380016 SCR
380017 SCR
380018 SCR
380019 SCR
380020 SCR
380021 SCR
380022 SCR
380101 SCR
380102 SCR
380103 SCR
380104 SCR
380105 SCR
380106 SCR
380107 SCR
380108 SCR
380109 SCR
380110 SCR
380111 SCR
380112 SCR
380113 SCR
380114 SCR
380115 SCR
380116 SCR

Class 385

385001 SCR
385002 SCR
385003 SCR
385004 SCR
385005 SCR
385006 SCR
385007 SCR
385008 SCR
385009 SCR
385010 SCR
385011 SCR
385012 SCR
385013 SCR
385014 SCR
385015 SCR
385016 SCR
385017 SCR
385018 SCR
385019 SCR
385020 SCR
385021 SCR
385022 SCR
385023 SCR
385024 SCR
385025 SCR
385026 SCR
385027 SCR
385028 SCR
385029 SCR
385030 SCR
385031 SCR
385032 SCR
385033 SCR
385034 SCR
385035 SCR
385036 SCR
385037 SCR
385038 SCR
385039 SCR
385040 SCR
385041 SCR
385042 SCR
385043 SCR
385044 SCR
385045 SCR
385046 SCR
385101 SCR
385101 SCR
385102 SCR
385103 SCR
385104 SCR
385105 SCR
385106 SCR
385107 SCR
385108 SCR
385109 SCR
385110 SCR
385111 SCR
385112 SCR
385113 SCR
385114 SCR
385115 SCR
385116 SCR
385117 SCR
385118 SCR
385119 SCR
385120 SCR
385121 SCR
385122 SCR
385123 SCR
385124 SCR

Class 387

387101 GTR
387102 GTR
387103 GTR
387104 GTR
387105 GTR
387106 GTR
387107 GTR
387108 GTR
387109 GTR
387110 GTR
387111 GTR
387112 GTR
387113 GTR
387114 GTR
387115 GTR
387116 GTR
387117 GTR
387118 GTR
387119 GTR
387120 GTR
387121 GTR
387122 GTR
387123 GTR
387124 GTR
387125 GTR
387126 GTR
387127 GTR
387128 GTR
387129 GTR
387130 HEX
387131 HEX
387132 HEX
387133 HEX
387134 HEX
387135 HEX
387136 HEX
387137 HEX
387138 HEX
387139 HEX
387140 HEX
387141 HEX
387142 GWR
387143 GWR
387144 GWR
387145 GWR
387146 GWR
387147 GWR
387148 GWR
387149 GWR
387150 GWR
387151 GWR
387152 GWR
387153 GWR
387154 GWR
387155 GWR
387156 GWR
387157 GWR
387158 GWR
387159 GWR
387160 GWR
387161 GWR
387162 GWR
387163 GWR
387164 GWR
387165 GWR
387166 GWR
387167 GWR
387168 GWR
387169 GWR
387170 GWR
387171 GWR
387172 GWR
387173 GWR
387174 GWR
387201 GTR
387202 GTR
387203 GTR
387204 GTR
387205 GTR
387206 GTR
387207 GTR
387208 GTR
387209 GTR
387210 GTR
387211 GTR
387212 GTR
387213 GTR
387214 GTR
387215 GTR
387216 GTR
387217 GTR
387218 GTR
387219 GTR
387220 GTR
387221 GTR
387222 GTR
387223 GTR
387224 GTR
387225 GTR
387226 GTR
387227 GTR
387301 C2C
387302 C2C
387303 C2C
387304 C2C
387305 C2C
387306 C2C

Class 390

390001 AWC
390002 AWC
390005 AWC
390006 AWC
390008 AWC
390009 AWC
390010 AWC
390011 AWC
390013 AWC
390016 AWC
390020 AWC
390039 AWC
390040 AWC
390042 AWC
390043 AWC
390044 AWC
390045 AWC
390046 AWC
390047 AWC
390049 AWC
390050 AWC
390103 AWC
390104 AWC
390107 AWC
390112 AWC
390114 AWC

390115 AWC
390117 AWC
390118 AWC
390119 AWC
390121 AWC
390122 AWC
390123 AWC
390124 AWC
390125 AWC
390126 AWC
390127 AWC
390128 AWC
390129 AWC
390130 AWC
390131 AWC
390132 AWC
390134 AWC
390135 AWC
390136 AWC
390137 AWC
390138 AWC
390141 AWC
390148 AWC
390151 AWC
390152 AWC
390153 AWC
390154 AWC
390155 AWC
390156 AWC
390157 AWC

Class 395

395001 SER
395002 SER
395003 SER
395004 SER
395005 SER
395006 SER
395007 SER
395008 SER
395009 SER
395010 SER
395011 SER
395012 SER
395013 SER
395014 SER
395015 SER
395016 SER
395017 SER
395018 SER
395019 SER
395020 SER
395021 SER
395022 SER
395023 SER
395024 SER
395025 SER
395026 SER
395027 SER
395028 SER
395029 SER

Class 397

397001 TPE
397002 TPE
397003 TPE
397004 TPE
397005 TPE
397006 TPE
397007 TPE
397008 TPE
397009 TPE
397010 TPE
397011 TPE
397012 TPE

Class 398

398001 TFW*
398002 TFW*
398003 TFW*
398004 TFW*
398005 TFW*
398006 TFW*
398007 TFW*
398008 TFW*
398009 TFW*
398010 TFW*
398011 TFW*
398012 TFW*
398013 TFW*
398014 TFW*
398015 TFW*
398016 TFW*
398017 TFW*
398018 TFW*
398019 TFW*
398020 TFW*
398021 TFW*
398022 TFW*
398023 TFW*
398024 TFW*
398025 TFW*
398026 TFW*
398027 TFW*
398028 TFW*
398029 TFW*
398030 TFW*
398031 TFW*
398032 TFW*
398033 TFW*
398034 TFW*
398035 TFW*
398036 TFW*

Class 444

444001 SWR
444002 SWR
444003 SWR
444004 SWR
444005 SWR
444006 SWR
444007 SWR
444008 SWR
444009 SWR
444010 SWR
444011 SWR
444012 SWR
444013 SWR
444014 SWR
444015 SWR
444016 SWR
444017 SWR
444018 SWR
444019 SWR
444020 SWR
444021 SWR
444022 SWR
444023 SWR
444024 SWR
444025 SWR
444026 SWR
444027 SWR
444028 SWR
444029 SWR
444030 SWR
444031 SWR
444032 SWR
444033 SWR
444034 SWR
444035 SWR
444036 SWR
444037 SWR
444038 SWR
444039 SWR
444040 SWR
444041 SWR
444042 SWR
444043 SWR
444044 SWR
444045 SWR

Class 450

450001 SWR
450002 SWR
450003 SWR
450004 SWR
450005 SWR
450006 SWR
450007 SWR
450008 SWR
450009 SWR
450010 SWR
450011 SWR
450012 SWR
450013 SWR
450014 SWR
450015 SWR
450016 SWR
450017 SWR
450018 SWR
450019 SWR
450020 SWR
450021 SWR
450022 SWR
450023 SWR
450024 SWR
450025 SWR
450026 SWR
450027 SWR
450028 SWR
450029 SWR
450030 SWR
450031 SWR
450032 SWR
450033 SWR
450034 SWR
450035 SWR
450036 SWR
450037 SWR
450038 SWR
450039 SWR
450040 SWR
450041 SWR
450042 SWR
450043 SWR
450044 SWR
450045 SWR
450046 SWR
450047 SWR
450048 SWR
450049 SWR
450050 SWR
450051 SWR
450052 SWR
450053 SWR
450054 SWR
450055 SWR
450056 SWR
450057 SWR
450058 SWR
450059 SWR
450060 SWR
450061 SWR
450062 SWR
450063 SWR
450064 SWR
450065 SWR
450066 SWR
450067 SWR
450068 SWR
450069 SWR
450070 SWR
450071 SWR
450072 SWR
450073 SWR
450074 SWR
450075 SWR
450076 SWR
450077 SWR
450078 SWR
450079 SWR
450080 SWR
450081 SWR
450082 SWR
450083 SWR
450084 SWR
450085 SWR
450086 SWR
450087 SWR
450088 SWR
450089 SWR
450090 SWR
450091 SWR
450092 SWR
450093 SWR
450094 SWR
450095 SWR
450096 SWR
450097 SWR
450098 SWR
450099 SWR
450100 SWR
450101 SWR
450102 SWR
450103 SWR
450104 SWR
450105 SWR
450106 SWR
450107 SWR
450108 SWR
450109 SWR
450110 SWR
450111 SWR
450112 SWR
450113 SWR
450114 SWR
450115 SWR
450116 SWR
450117 SWR
450118 SWR
450119 SWR
450120 SWR
450121 SWR
450122 SWR
450123 SWR
450124 SWR
450125 SWR
450126 SWR
450127 SWR

Class 455

5701 SWR
5702 SWR
5703 SWR
5704 SWR
5705 SWR
5706 SWR
5707 SWR
5708 SWR
5709 SWR
5710 SWR
5711 SWR
5712 SWR
5713 SWR
5714 SWR
5715 SWR
5716 SWR
5717 SWR
5718 SWR
5719 SWR
5720 SWR
5721 SWR
5722 SWR
5723 SWR
5724 SWR
5725 SWR
5726 SWR
5727 SWR
5728 SWR
5729 SWR
5730 SWR
5731 SWR
5732 SWR
5733 SWR
5734 SWR
5735 SWR
5736 SWR
5737 SWR
5738 SWR
5739 SWR
5740 SWR
5741 SWR
5742 SWR
5750 SWR
455801 GTR
455802 GTR
455803 GTR
455804 GTR
455805 GTR
455806 GTR
455807 GTR
455808 GTR
455809 GTR
455810 GTR
455811 GTR
455812 GTR
455813 GTR
455814 GTR
455815 GTR
455816 GTR
455817 GTR
455818 GTR
455819 GTR
455820 GTR
455821 GTR
455822 GTR
455823 GTR
455824 GTR
455825 GTR
455826 GTR
455827 GTR
455828 GTR
455829 GTR
455830 GTR
455831 GTR
455832 GTR
455833 GTR
455834 GTR
455835 GTR
455836 GTR
455837 GTR
455838 GTR

455839 GTR
455840 GTR
455841 GTR
455842 GTR
455843 GTR
455844 GTR
455845 GTR
455846 GTR
5847 SWR
5848 SWR
5849 SWR
5850 SWR
5851 SWR
5852 SWR
5853 SWR
5854 SWR
5855 SWR
5856 SWR
5857 SWR
5858 SWR
5859 SWR
5860 SWR
5861 SWR
5862 SWR
5863 SWR
5864 SWR
5865 SWR
5866 SWR
5867 SWR
5868 SWR
5869 SWR
5870 SWR
5871 SWR
5872 SWR
5873 SWR
5874 SWR
5901 SWR
5902 SWR
5903 SWR
5904 SWR
5905 SWR
5906 SWR
5907 SWR
5908 SWR
5909 SWR
5910 SWR
5911 SWR
5912 SWR
5913 SWR
5914 SWR
5915 SWR
5916 SWR
5917 SWR
5918 SWR
5919 SWR
5920 SWR

Class 456

456002 PBK (S)
456003 PBK (S)
456004 PBK (S)
456005 PBK (S)
456006 PBK (S)
456007 PBK (S)
456008 PBK (S)
456009 PBK (S)
456010 PBK (S)
456011 PBK (S)
456012 PBK (S)
456013 PBK (S)
456014 PBK (S)
456015 PBK (S)
456016 PBK (S)
456017 PBK (S)
456018 PBK (S)
456019 PBK (S)
456020 PBK (S)
456021 PBK (S)
456022 PBK (S)
456023 PBK (S)
456024 PBK (S)

Class 458

458501 SWR
458502 SWR
458503 SWR
458504 SWR
458505 SWR
458506 SWR
458507 SWR
458508 SWR
458509 SWR
458510 SWR
458511 SWR
458512 SWR
458513 SWR
458514 SWR
458515 SWR
458516 SWR
458517 SWR
458518 SWR
458519 SWR
458520 SWR
458521 SWR
458522 SWR
458523 SWR
458524 SWR
458525 SWR
458526 SWR
458527 SWR
458528 SWR
458529 SWR
458530 SWR
458531 SWR
458532 SWR
458533 SWR
458534 SWR
458535 SWR
458536 SWR

Class 465

465001 SER
465002 SER
465003 SER
465004 SER
465005 SER
465006 SER
465007 SER
465008 SER
465009 SER
465010 SER
465011 EVS (S)
465012 SER
465013 SER
465014 SER
465015 SER
465016 SER
465017 SER
465018 SER
465019 EVS (S)
465020 SER
465021 SER
465022 SER
465023 SER
465024 SER
465025 SER
465026 SER
465027 SER
465028 SER
465029 SER
465030 SER
465031 SER
465032 SER
465033 SER
465034 SER
465035 SER
465036 SER
465037 SER
465038 SER
465039 SER
465040 SER
465041 SER
465042 SER
465043 SER
465044 SER
465045 SER
465046 SER
465047 SER
465048 SER
465049 SER
465050 SER
465151 SER
465152 SER
465153 SER
465154 SER
465155 SER
465156 SER
465157 SER
465158 SER
465159 SER
465160 SER
465161 SER
465162 SER
465163 SER
465164 SER
465165 SER
465166 SER
465167 SER
465168 SER
465169 SER
465170 SER
465171 SER
465172 SER
465173 SER
465174 SER
465175 SER
465176 SER
465177 SER
465178 SER
465179 SER
465180 SER
465181 SER
465182 SER
465183 SER
465184 SER
465185 SER
465186 SER
465187 SER
465188 SER
465189 SER
465190 SER
465191 SER
465192 SER
465193 SER
465194 SER
465195 SER
465196 SER
465197 SER
465235 ANG (S)
465236 ANG (S)
465237 ANG (S)
465238 ANG (S)
465239 ANG (S)
465240 ANG (S)
465241 ANG (S)
465242 ANG (S)
465243 ANG (S)
465244 ANG (S)
465245 ANG (S)
465246 ANG (S)
465247 ANG (S)
465248 ANG (S)
465249 ANG (S)
465250 ANG (S)
465901 SER
465902 SER
465903 SER
465904 SER
465905 SER
465906 SER
465907 SER
465908 SER
465909 SER
465910 SER
465911 SER
465912 SER
465913 SER
465914 SER
465915 SER
465916 SER
465917 SER
465918 SER
465919 SER
465920 SER
465921 SER
465922 SER
465923 SER
465924 SER
465925 SER
465926 SER
465927 SER
465928 SER
465929 SER
465930 SER
465931 SER
465932 SER
465933 SER
465934 SER

Class 466

466001 SER
466002 SER
466003 SER
466004 ANG (S)
466005 SER
466006 SER
466007 SER
466008 SER
466009 SER
466010 ANG (S)
466011 SER
466012 SER
466013 SER
466014 SER
466015 SER
466016 ANG (S)
466017 SER
466018 SER
466019 SER
466020 SER
466021 SER
466022 SER
466023 SER
466024 ANG (S)
466025 SER
466026 SER
466027 SER
466028 SER
466029 SER
466030 SER
466031 SER
466032 SER
466033 SER
466034 SER
466035 SER
466036 SER
466037 SER
466038 SER
466039 SER
466040 SER
466041 SER
466042 SER
466043 ANG (S)

Class 484

484001 SWR
484002 SWR
484003 SWR
484004 SWR
484005 SWR

Class 507

507001 MER
507002 MER
507003 MER
507004 MER
507005 MER
507007 MER
507008 MER
507009 MER
507010 MER
507011 MER
507012 MER
507013 MER
507014 MER
507015 MER
507016 MER
507017 MER
507018 MER
507019 MER
507020 MER
507021 MER
507023 MER
507024 MER
507025 MER
507026 MER
507027 MER
507028 MER
507029 MER
507030 MER
507031 MER
507032 MER
507033 MER

Class 508

508103 MER
508104 MER
508108 MER
508111 MER
508112 MER
508114 MER
508115 MER
508117 MER
508120 MER
508122 MER
508123 MER
508124 MER
508125 MER
508126 MER
508127 MER
508128 MER
508130 MER
508131 MER

508136 MER
508137 MER
508138 MER
508139 MER
508140 MER
508141 MER
508143 MER

Class 700

700001 GTR
700002 GTR
700003 GTR
700004 GTR
700005 GTR
700006 GTR
700007 GTR
700008 GTR
700009 GTR
700010 GTR
700011 GTR
700012 GTR
700013 GTR
700014 GTR
700015 GTR
700016 GTR
700017 GTR
700018 GTR
700019 GTR
700020 GTR
700021 GTR
700022 GTR
700023 GTR
700024 GTR
700025 GTR
700026 GTR
700027 GTR
700028 GTR
700029 GTR
700030 GTR
700031 GTR
700032 GTR
700033 GTR
700034 GTR
700035 GTR
700036 GTR
700037 GTR
700038 GTR
700039 GTR
700040 GTR
700041 GTR
700042 GTR
700043 GTR
700044 GTR
700045 GTR
700046 GTR
700047 GTR
700048 GTR
700049 GTR
700050 GTR
700051 GTR
700052 GTR
700053 GTR
700054 GTR
700055 GTR
700056 GTR
700057 GTR
700058 GTR
700059 GTR
700060 GTR
700101 GTR
700102 GTR
700103 GTR
700104 GTR
700105 GTR
700106 GTR
700107 GTR
700108 GTR
700109 GTR
700110 GTR
700111 GTR
700112 GTR
700113 GTR
700114 GTR
700115 GTR
700116 GTR
700117 GTR
700118 GTR
700119 GTR
700120 GTR
700121 GTR
700122 GTR
700123 GTR
700124 GTR
700125 GTR
700126 GTR
700127 GTR
700128 GTR
700129 GTR
700130 GTR
700131 GTR
700132 GTR
700133 GTR
700134 GTR
700135 GTR
700136 GTR
700137 GTR
700138 GTR
700139 GTR
700140 GTR
700141 GTR
700142 GTR
700143 GTR
700144 GTR
700145 GTR
700146 GTR
700147 GTR
700148 GTR
700149 GTR
700150 GTR
700151 GTR
700152 GTR
700153 GTR
700154 GTR
700155 GTR

Class 701

701001 SWR
701002 SWR
701003 SWR
701004 SWR
701005 SWR
701006 SWR
701007 SWR
701008 SWR
701009 SWR
701010 SWR
701011 SWR
701012 SWR
701013 SWR
701014 SWR
701015 SWR
701016 SWR
701017 SWR
701018 SWR
701019 SWR
701020 SWR
701021 SWR
701022 SWR
701023 SWR
701024 SWR
701025 SWR
701026 SWR
701027 SWR
701028 SWR
701029 SWR
701030 SWR
701031 SWR
701032 SWR
701033 SWR
701034 SWR
701035 SWR
701036 SWR
701037 SWR
701038 SWR
701039 SWR
701040 SWR
701041 SWR
701042 SWR
701043 SWR
701044 SWR
701045 SWR
701046 SWR
701047 SWR
701048 SWR
701049 SWR
701050 SWR
701051 SWR
701052 SWR
701053 SWR
701054 SWR
701055 SWR
701056 SWR
701057 SWR
701058 SWR
701059 SWR
701060 SWR
701501 SWR
701502 SWR
701503 SWR
701504 SWR
701505 SWR
701506 SWR
701507 SWR
701508 SWR
701509 SWR
701510 SWR
701511 SWR
701512 SWR
701513 SWR
701514 SWR
701515 SWR
701516 SWR
701517 SWR
701518 SWR
701519 SWR
701520 SWR
701521 SWR
701522 SWR
701523 SWR
701524 SWR
701525 SWR
701526 SWR
701527 SWR
701528 SWR
701529 SWR
701530 SWR

Class 707

707001 SER
707002 SER
707003 SER
707004 SER
707005 SER
707006 SER
707007 SER
707008 SER
707009 SER
707010 SER
707011 SER
707012 SER
707013 SER
707014 SWR
707015 SWR
707016 SWR
707017 SWR
707018 SWR
707019 SWR
707020 SWR
707021 SWR
707022 SWR
707023 SWR
707024 SWR
707025 SER
707026 SER
707027 SER
707028 SER
707029 SER
707030 SWR

Class 710

710101 LOL
710102 LOL
710103 LOL
710104 LOL
710105 LOL
710106 LOL
710107 LOL
710108 LOL
710109 LOL
710110 LOL
710111 LOL
710112 LOL
710113 LOL
710114 LOL
710115 LOL
710116 LOL
710117 LOL
710118 LOL
710119 LOL
710120 LOL
710121 LOL
710122 LOL
710123 LOL
710124 LOL
710125 LOL
710126 LOL
710127 LOL
710128 LOL
710129 LOL
710130 LOL
710256 LOL
710257 LOL
710258 LOL
710259 LOL
710260 LOL
710261 LOL
710262 LOL
710263 LOL
710264 LOL
710265 LOL
710266 LOL
710267 LOL
710268 LOL
710269 LOL
710270 LOL
710271 LOL
710272 LOL
710273 LOL
710374 LOL
710375 LOL
710376 LOL
710377 LOL
710378 LOL
710379 LOL

Class 717

717001 GTR
717002 GTR
717003 GTR
717004 GTR
717005 GTR
717006 GTR
717007 GTR
717008 GTR
717009 GTR
717010 GTR
717011 GTR
717012 GTR
717013 GTR
717014 GTR
717015 GTR
717016 GTR
717017 GTR
717018 GTR
717019 GTR
717020 GTR
717021 GTR
717022 GTR
717023 GTR
717024 GTR
717025 GTR

Class 720

720101 AGA
720102 AGA
720103 AGA
720104 AGA
720105 AGA
720106 AGA
720107 AGA
720108 AGA
720109 AGA
720110 AGA
720111 AGA
720112 AGA
720113 AGA
720114 AGA
720115 AGA
720116 AGA
720117 AGA
720118 AGA
720119 AGA
720120 AGA
720121 AGA
720122 AGA
720123 AGA
720124 AGA
720125 AGA
720126 AGA
720127 AGA
720128 AGA
720129 AGA
720130 AGA
720131 AGA
720132 AGA
720133 AGA
720134 AGA
720135 AGA
720136 AGA
720137 AGA
720138 AGA
720139 AGA

EUT locos

9005 EUT
9007 EUT
9011 EUT
9013 EUT
9015 EUT
9018 EUT
9022 EUT
9024 EUT
9026 EUT
9029 EUT
9033 EUT
9036 EUT
9037 EUT
9701 EUT
9702 EUT
9703 EUT
9704 EUT
9705 EUT
9706 EUT
9707 EUT
9708 EUT
9709 EUT
9710 EUT
9711 EUT
9712 EUT
9713 EUT
9714 EUT
9715 EUT
9716 EUT
9717 EUT
9718 EUT
9719 EUT
9720 EUT
9721 EUT
9722 EUT
9723 EUT
9801 EUT
9802 EUT
9803 EUT
9804 EUT
9806 EUT
9808 EUT
9809 EUT
9810 EUT
9812 EUT
9814 EUT
9816 EUT
9819 EUT
9820 EUT
9821 EUT
9823 EUT
9825 EUT
9827 EUT
9828 EUT
9831 EUT
9832 EUT
9834 EUT
9835 EUT
9838 EUT
9840 EUT

EUT Locos

0001 EUT
0002 EUT
0003 EUT
0004 EUT
0005 EUT
0006 EUT
0007 EUT
0008 EUT
0009 EUT
0010 EUT
0031 EUT
0032 EUT
0033 EUT
0034 EUT
0035 EUT
0036 EUT
0037 EUT
0038 EUT
0039 EUT
0040 EUT
0041 EUT
0042 EUT